Compendium of Plant Genomes

Series Editor

Chittaranjan Kole, Raja Ramanna Fellow, Government of India, ICAR-National Research Center on Plant Biotechnology, Pusa, New Delhi, India

Whole-genome sequencing is at the cutting edge of life sciences in the new millennium. Since the first genome sequencing of the model plant Arabidopsis thaliana in 2000, whole genomes of about 100 plant species have been sequenced and genome sequences of several other plants are in the pipeline. Research publications on these genome initiatives are scattered on dedicated web sites and in journals with all too brief descriptions. The individual volumes elucidate the background history of the national and international genome initiatives; public and private partners involved; strategies and genomic resources and tools utilized; enumeration on the sequences and their assembly; repetitive sequences; gene annotation and genome duplication. In addition, synteny with other sequences, comparison of gene families and most importantly potential of the genome sequence information for gene pool characterization and genetic improvement of crop plants are described.

Interested in editing a volume on a crop or model plant?

Please contact Prof. C. Kole, Series Editor, at ckoleorg@gmail.com

More information about this series at http://www.springer.com/series/11805

Long-Xi Yu • Chittaranjan Kole
Editors

The Alfalfa Genome

Springer

Editors
Long-Xi Yu
USDA-ARS, Plant Germplasm
Introduction and Testing Research
Prosser, WA, USA

Chittaranjan Kole
ICAR-National Institute
for Plant Biotechnology
Raja Ramanna Fellow
New Delhi, India

ISSN 2199-4781 ISSN 2199-479X (electronic)
Compendium of Plant Genomes
ISBN 978-3-030-74465-6 ISBN 978-3-030-74466-3 (eBook)
https://doi.org/10.1007/978-3-030-74466-3

This Springer imprint is published by the registered company Springer Nature Switzerland AG
The registered company address is: Gewerbestrasse 11, 6330 Cham, Switzerland

This book series is dedicated to my wife Xin Shen,
our daughter Jing and grandsons Merrick and Morgan

Long-Xi Yu

This book series is dedicated to my wife Phullara and our children Sourav and Devleena

Chittaranjan Kole

Preface to the Series

Genome sequencing has emerged as the leading discipline in the plant sciences coinciding with the start of the new century. For much of the twentieth century, plant geneticists were only successful in delineating putative chromosomal location, function, and changes in genes indirectly through the use of a number of "markers" physically linked to them. These included visible or morphological, cytological, protein, and molecular or DNA markers. Among them, the first DNA marker, the RFLPs, introduced a revolutionary change in plant genetics and breeding in the mid-1980s, mainly because of their infinite number and thus potential to cover maximum chromosomal regions, phenotypic neutrality, absence of epistasis, and codominant nature. An array of other hybridization-based markers, PCR-based markers, and markers based on both facilitated construction of genetic linkage maps, mapping of genes controlling simply inherited traits, and even gene clusters (QTLs) controlling polygenic traits in a large number of model and crop plants. During this period, a number of new mapping populations beyond F2 were utilized and a number of computer programs were developed for map construction, mapping of genes, and for mapping of polygenic clusters or QTLs. Molecular markers were also used in the studies of evolution and phylogenetic relationship, genetic diversity, DNA fingerprinting, and map-based cloning. Markers tightly linked to the genes were used in crop improvement employing the so-called marker-assisted selection. These strategies of molecular genetic mapping and molecular breeding made a spectacular impact during the last one and a half decades of the twentieth century. But still they remained "indirect" approaches for elucidation and utilization of plant genomes since much of the chromosomes remained unknown and the complete chemical depiction of them was yet to be unraveled.

Physical mapping of genomes was the obvious consequence that facilitated the development of the "genomic resources" including BAC and YAC libraries to develop physical maps in some plant genomes. Subsequently, integrated genetic–physical maps were also developed in many plants. This led to the concept of structural genomics. Later on, emphasis was laid on EST and transcriptome analysis to decipher the function of the active gene sequences leading to another concept defined as functional genomics. The advent of techniques of bacteriophage gene and DNA sequencing in the 1970s was extended to facilitate sequencing of these genomic resources in the last decade of the twentieth century.

As expected, sequencing of chromosomal regions would have led to too much data to store, characterize, and utilize with the-then available computer software could handle. But the development of information technology made the life of biologists easier by leading to a swift and sweet marriage of biology and informatics, and a new subject was born—bioinformatics.

Thus, the evolution of the concepts, strategies, and tools of sequencing and bioinformatics reinforced the subject of genomics—structural and functional. Today, genome sequencing has traveled much beyond biology and involves biophysics, biochemistry, and bioinformatics!

Thanks to the efforts of both public and private agencies, genome sequencing strategies are evolving very fast, leading to cheaper, quicker, and automated techniques right from clone-by-clone and whole-genome shotgun approaches to a succession of second-generation sequencing methods. The development of software of different generations facilitated this genome sequencing. At the same time, newer concepts and strategies were emerging to handle sequencing of the complex genomes, particularly the polyploids.

It became a reality to chemically—and so directly—define plant genomes, popularly called whole-genome sequencing or simply genome sequencing.

The history of plant genome sequencing will always cite the sequencing of the genome of the model plant Arabidopsis thaliana in 2000 that was followed by sequencing the genome of the crop and model plant rice in 2002. Since then, the number of sequenced genomes of higher plants has been increasing exponentially, mainly due to the development of cheaper and quicker genomic techniques and, most importantly, the development of collaborative platforms such as national and international consortia involving partners from public and/or private agencies.

As I write this preface for the first volume of the new series "Compendium of Plant Genomes," a net search tells me that complete or nearly complete whole-genome sequencing of 45 crop plants, eight crop and model plants, eight model plants, 15 crop progenitors and relatives, and three basal plants is accomplished, the majority of which are in the public domain. This means that we nowadays know many of our model and crop plants chemically, i.e., directly, and we may depict them and utilize them precisely better than ever. Genome sequencing has covered all groups of crop plants. Hence, information on the precise depiction of plant genomes and the scope of their utilization are growing rapidly every day. However, the information is scattered in research articles and review papers in journals and dedicated Web pages of the consortia and databases. There is no compilation of plant genomes and the opportunity of using the information in sequence-assisted breeding or further genomic studies. This is the underlying rationale for starting this book series, with each volume dedicated to a particular plant.

Plant genome science has emerged as an important subject in academia, and the present compendium of plant genomes will be highly useful to both students and teaching faculties. Most importantly, research scientists involved in genomics research will have access to systematic deliberations on the plant genomes of their interest. Elucidation of plant genomes is of interest not only for the geneticists and breeders, but also for practitioners of an array of plant science disciplines, such as taxonomy, evolution, cytology, physiology, pathology, entomology, nematology, crop production, biochemistry, and obviously bioinformatics. It must be mentioned that information regarding each plant genome is ever-growing. The contents of the volumes of this compendium are, therefore, focusing on the basic aspects of the genomes and their utility. They include information on the academic and/or economic importance of the plants, description of their genomes from a molecular genetic and cytogenetic point of view, and the genomic resources developed. Detailed deliberations focus on the background history of the national and international genome initiatives, public and private partners involved, strategies and genomic resources and tools utilized, enumeration on the sequences and their assembly, repetitive sequences, gene annotation, and genome duplication. In addition, synteny with other sequences, comparison of gene families, and, most importantly, the potential of the genome sequence information for gene pool characterization through Genotyping by Sequencing (GBS) and genetic improvement of crop plants have been described. As expected, there is a lot of variation of these topics in the volumes based on the information available on the crop, model, or reference plants.

I must confess that as the series editor, it has been a daunting task for me to work on such a huge and broad knowledge base that spans so many diverse plant species. However, pioneering scientists with lifetime experience and expertise on the particular crops did excellent jobs editing the respective volumes. I myself have been a small science worker on plant genomes since the mid-1980s and that provided me the opportunity to personally know several stalwarts of plant genomics from all over the globe. Most, if not all, of the volume editors are my longtime friends and colleagues. It has been highly comfortable and enriching for me to work with them on this book series. To be honest, while working on this series I have been and will remain a student first, a science worker second, and a series editor last. And I must express my gratitude to the volume editors and the chapter authors for providing me the opportunity to work with them on this compendium.

I also wish to mention here my thanks and gratitude to the Springer staff, particularly Dr. Christina Eckey and Dr. Jutta Lindenborn for the earlier set of volumes and presently Ing. Zuzana Bernhart for all their timely help and support.

I always had to set aside additional hours to edit books beside my professional and personal commitments—hours I could and should have given to my wife, Phullara, and our kids, Sourav and Devleena. I must mention that they not only allowed me the freedom to take away those hours from them but also offered their support in the editing job itself. I am really not sure whether my dedication of this compendium to them will suffice to do justice to their sacrifices for the interest of science and the science community.

New Delhi, India Chittaranjan Kole

Preface

Alfalfa (*Medicago sativa* L.) is an important legume forage crop grown on approximately 30 million hectares worldwide and is the third most valuable crop in the U.S. after corn and soybean. One of the most important properties of alfalfa is its high nutritional quality as animal feed, thus alfalfa is also called the "Queen of the Forages". Alfalfa hay or silage provides nutrient components for most dairy and other animal rations with high protein, valuable fiber, and excellent source of vitamins and minerals to promote animal health. Alfalfa is also an important source of biological nitrogen fixation. The average rate of nitrogen fixation of alfalfa is about 200 kg of nitrogen per acre per year, thus reducing the need to apply expensive nitrogen fertilizers. In addition to the traditional uses of alfalfa as an animal feed, alfalfa can be consumed by humans in the form of alfalfa sprouts and as health food products. Moreover, alfalfa has the potential to be used as a biofuel crop and as a factory for the production of industrial enzymes such as lignin peroxidase, alpha-amylase, cellulase, and phytase.

Changing trends in multipurpose uses increase the demand for alfalfa. However, production of alfalfa is challenged by endemic and emerging diseases and adverse environmental factors. On a national level, diseases and abiotic stresses affecting alfalfa production result in severe economic losses. It is imperative to improve yields in alfalfa to keep alfalfa production economical and sustainable. Newer genomics tools and biotechnologies, as well as new management systems with remote sensing to predict yield and other traits, provide new ways to develop high-yielding cultivars and increasing yield potential.

This volume intends to cover multiple aspects of alfalfa from production to breeding. Different chapters highlight recent progress in alfalfa improvement and how the availability of the genomics approaches has enabled new research in this important forage crop. Chapter 1 highlighted the economic importance, practical limitations to production, management, and breeding targets of alfalfa. Chapter 2 described alfalfa yield improvement and factors influencing yield and quality. Chapters 3 and 4 summarized the origin, evolution, germplasm collection, genetic diversity, and gene pools of alfalfa. Chapter 5 reviewed biotechnology advances and summarized genetically modified alfalfa and the use of constitutive, tissue-specific, and inducible promoters in developing transgenic alfalfa for desired agronomic traits. Chapter 6 highlighted the recent progress on the alfalfa genome sequencing and assembly. Chapter 7 focused on transcription factors, the regulatory components of the alfalfa genome that coordinate gene expression, and introduced a web-based database of the AlfalfaTFDB. Chapter 8 summarized the factors of forage quality of alfalfa and the application of genomics tools to improve quality traits. Chapter 9 described the physiological, morphological, biochemical, and genetic responses of alfalfa to salinity. Chapter 10 reviewed the development of single nucleotide polymorphic markers for genomics assessments in autotetraploid alfalfa. Chapter 11 Introduced the "Breeding Insight", a public-sector initiative that has been put in place to aid breeders and hasten the adoption of new technologies, including high-throughput genotyping, to accelerate breeding and pre-breeding efforts. Chapter 12 described the strategy of Genomic Selection (GS) for higher yield and quality of alfalfa and discussed various factors that may affect the prediction ability and the cost-efficient exploitation of GS in breeding programs. Chapter 13 reviewed the identification and characterization of disease resistance genes in alfalfa and *Medicago truncatula* for breeding improved cultivars and summarized the cases in which the mechanisms of disease resistance and inheritance of resistance in the two species. Chapter 14 described the strategy of genomic approaches and reviewed recent progress on genomics assessments for improving alfalfa resilience to drought. Chapter 15 reviewed self-incompatibility, inbreeding depression, and potential to develop inbred lines toward hybrid alfalfa. Chapter 16 described the advanced strategies of targeted mutagenesis and gene editing in alfalfa. While each chapter is prepared to be an independent read, we hope that the diverse topics addressed in this book will be of interest to researchers within the specific field as well as in other areas of plant genetics and genomics.

We are thankful to all the contributors for their critical contributions to this book volume.

Prosser, WA, USA — Long-Xi Yu
New Delhi, India — Chittaranjan Kole

Contents

Contributors

Biswa R. Acharya US Salinity Laboratory (USDA-ARS), Riverside, CA, USA;
College of Natural and Agricultural Sciences, University of California Riverside, Riverside, CA, USA

P. Annicchiarico Centre for Animal Production and Aquaculture, Council for Agricultural Research and Economics, Lodi, Italy

David Combs University of Wisconsin, Madison, WI, USA

Shaun J. Curtin United States Department of Agriculture, Plant Science Research Unit, St. Paul, MN, USA;
Department of Plant Pathology, University of Minnesota, St. Paul, MN, USA

Melinda R. Dornbusch United States Department of Agriculture, Plant Science Research Unit, St. Paul, MN, USA

Andrew D. Farmer Department of Bioinformatics, National Center for Genome Resources, Santa Fe, NM, USA

Jorge F. S. Ferreira US Salinity Laboratory (USDA-ARS), Riverside, CA, USA

Stephanie L. Greene USDA ARS National Laboratory for Genetic Resources Preservation, Fort Collins, CO, USA

Samuel Grinstead United States Department of Agriculture, Beltsville Agricultural Research Center, National Germplasm Resources Laboratory, Beltsville, MD, USA

Juan Gutierrez-Gonzalez Departamento de Biología Molecular, Universidad de León, León, Spain

Brian M. Irish USDA ARS Plant Germplasm Introduction and Research Testing Unit, Prosser, WA, USA

Sen Lin USDA-ARS Plant Germplasm Introduction Testing and Research, Prosser, WA, USA

Cesar Augusto Medina Irrigated Agriculture Research and Extension Center, Washington State University, Prosser, WA, USA

Maria Katherine Mejia-Guerra Breeding Insight, Cornell University, Ithaca, NY, USA

Susan S. Miller United States Department of Agriculture, Plant Science Research Unit, St. Paul, MN, USA

Ali M. Missaoui Department of Crop and Soil Sciences, Institute of Plant Breeding Genetics and Genomics, The University of Georgia, Athens, GA, USA

Joann Mudge Department of Bioinformatics, National Center for Genome Resources, Santa Fe, NM, USA

N. Nazzicari Centre for Animal Production and Aquaculture, Council for Agricultural Research and Economics, Lodi, Italy

Lev G. Nemchinov United States Department of Agriculture, Beltsville Agricultural Research Center, Molecular Plant Pathology Laboratory, Beltsville, MD, USA

Steve Norberg Regional Forage Specialist and Irrigated Cropping Systems, Franklin County Extension Office, Washington State University, Pasco, WA, USA

Atit Parajuli Department of Crop and Soil Sciences, Washington State University, Pullman, WA, USA

L. Pecetti Centre for Animal Production and Aquaculture, Council for Agricultural Research and Economics, Lodi, Italy

Michael Peel United States Department of Agriculture-Agricultural Research Service, Forage & Range Research Lab, Logan, UT, USA

Olga A. Postnikova National Institute of Health, Bethesda, MD, USA

Daniel H. Putnam Department of Plant Sciences, University of California Davis, CA, Davis, USA

Muhammet Şakiroğlu Department of Bioengineering, AlparslanTürkeş Science and Technology University, Adana, Turkey

Deborah A. Samac USDA-ARS-Plant Science Research Unit, St. Paul, MN, USA

Devinder Sandhu US Salinity Laboratory (USDA-ARS), Riverside, CA, USA

Deven See Wheat Health, Genetics, and Quality Research Unit, USDA-ARS and Washington State University, Pullman, WA, USA

Jonathan Shao United States Department of Agriculture, Beltsville Agricultural Research Center, Molecular Plant Pathology Laboratory, Beltsville, MD, USA

Moira J. Sheehan Breeding Insight, Cornell University, Ithaca, NY, USA

Stephen J. Temple Forage Genetics International, West Salem, WI, USA

Dan Undersander Department of Agronomy, University of Wisconsin-Madison, Madison, WI, USA

Steve Wagner Alforex Seeds, West Salem, WI, USA

Zan Wang College of Grassland Science and Technology, China Agricultural University, Beijing, China

Long-Xi Yu United States Department of Agriculture-Agricultural Research Service, Plant Germplasm Introduction and Testing Research, Prosser, WA, USA

Zhiwu Zhang Department of Crop and Soil Sciences, Washington State University, Pullman, WA, USA

Dongyan Zhao Breeding Insight, Cornell University, Ithaca, NY, USA

Abbreviations

5AmMC6	5' Amino Modifier C6
A3	Alternative 3′ splice-site
A5	Alternative 5′ splice-site
ABA	Abscisic acid
ADF	Acid detergent fiber
ADL	Acid detergent lignin
AF	Alternative first exon
AFLP	Amplified fragment length polymorphism
AL	Alternative last exon
AM	Association mapping
aNDF	Neutral detergent fiber analyzed with amylase
ANOVA	Analysis of variance
AP2	APETALA2
APG	Australian Pastures Genebank
APHIS	Animal and Plant Health Inspection Service
ASE	Alternative splicing events
ATP	Adenosine triphosphate
BAC	Bacterial artificial chromosome
BAM	Binary alignment map
BLUE	Best linear unbiased estimate
BLUP	Best linear unbiased prediction
BMR	Brown mid-rib
BRs	Brassinosteroids
BSA	Bulked segregation analysis
BUP	Barcoded universal primers
BUSCO	Benchmarking universal single-copy orthologs
CADL	Cultivated alfalfa at the diploid level
CaMV	Cauliflower mosaic virus
CAPS	Cleaved amplified polymorphic sequences
CAT	Catalase
Catalog.str	Catalog of known short tandem repeats
CCOMT	Caffeoyl-CoA 3-O-methytransferase
CCS	Circular consensus sequencing
CDD	Conserved Domains Database
CGC	Crop Germplasm Committee

CGIAR	Consultative Group for International Agricultural Research
cGMP	3', 5'-cyclic guanosine monophosphate
CK	Check
CLC	QIAGEN CLC Genomics
CLP	Primer name
CLR	Continuous long read
CMS	Cytoplasmic male sterility
CmYLCV	Cestrum yellow leaf curling virus
CNVs	Copy number variants
CoA	Hydroxycinnamoyl-coenzyme A
COMT	Caffeic acid 3-O-methyltransferase
CP	Crude protein
CPP	Cystein-rich polycomb-like protein
CRC	Clover root curculio
CRISPR	Clustered regularly interspaced short palindromic repeats
CWR	Crop wild relative
DAS	Days after sowing
DBD	DNA-binding domain
DDM	Digestible dry matter
DM	Dry matter
DMI	Dry matter intake using NDF
DMI1	Dry matter intake using NDF
dNDF48	48-hour digestible NDF
DRI	Drought resistance index
DSI	Drought susceptible index
DS	Drought stress
dS m-1	DeciSiemens per meter
DSB	Double-stranded break
EA	Environmental Assessment
EC	Electrical conductivity
ECe	EC of saturation paste extract
ECiw	EC of the irrigation water
ECPGR	European Cooperative Programme for Plant Genetic Resources
EIS	Environmental Impact Statement
ENE	Estimated net energy
EPA	Environmental Protection Agency
EPSPS	5-Enolpyruvylshikimate-3-phosphate synthase
ER	Endoplasmic reticulum
ERF	Ethylene-responsive factor
EST	Expressed Sequence Tag
EURISCO	European Search Catalogue for Plant Genetic Resources
FAO	Food and Agriculture Organization

FDA	Food and Drug Administration
FGI	Forage Genetics International
FISH	Fluorescence in situ hybridization
FMV	Figwort mosaic virus
GA	Gibberellin
GAB	Genomics-assisted breeding
G-BLUP	Genomic best linear unbiased prediction
GBS	Genotyping-by-sequencing
GE	Genetically engineered
GEBV	Genome-estimated breeding value
GEI	Genotype × environment interaction
GM	Genetically modified
GOI	Gene of interest
GRF	Growth-regulating factor
GRIN	USDA Germplasm Resources Information Network
GS	Genomic selection
GUS	β-Glucuronidase
GWAS	Genome-wide association studies
H	Haplotypes
HCT	hydroxycinnamoyl transferase
HERA	Highly efficient repeat assembly
HKT	High-affinity potassium transporter
HLH	Basic helix-loop-helix
HM	HapMap
HMM	Hidden Markov model
HMT	Hydroxycinnamoyl transferase
HR	Host resistance
HRT	Hairy-related transcription factor
HSP	Heat shock protein
IAA	Auxin indole-3-acetic acid
IBPGR	International Board for Plant Genetic Resources
ICARDA	International Centre for Agricultural Research in Dry Areas
ID	Inbreeding depression
Indels	Insertion or deletions
INRA	Institut National de la Recherche Agronomique
IP	Intellectual property
IP3	Inositol trisphosphate
IPK	Leibniz-InstitutFürPflanzengenetik Und Kulturpflanzenforschung
IPTG	Isopropyl β-D-1-thiogalactopyranoside
IR	Intron retention
ISAAA	International Service for the Acquisition of Agri-biotech Applications
Iso-Seq	Isoform sequencing
ISSR	Inter-simple sequence repeat

ITPGRFA	International Treaty for Plant Genetic Resources for Food and Agriculture
IVDDM30	30-hour *in vitro* digestible dry matter
IVDDM48	48-hour *in vitro* digestible dry matter
JA	Jasmonic acid
KASP	Kompetitive allele specific PCR
K-Matrix	Kinship matrix
LAA	Long amplicon analysis
LACHESIS	Ligating adjacent chromatin enables scaffolding in situ
LB	Luria-Bertani liquid medium
LD	Linkage disequilibrium
LG	Linkage group
lncRNAs	Long non-coding RNAs
LRR	leucine-rich repeat
LRS	Long read sequencing
LTR	Long terminal repeat
MAFFT	Multiple alignments using fast Fourier transform
MAMP	Microbe-associated molecular patterns
MAS	Marker-assisted selection
maxAlnsPerStartPos	Maximum number of alignments allowed to start at the same reference site
maxBaseQS	Maximum base pair quality score
MCPD	Multi-crop Passport Descriptors
MDA	Malondialdehyde
ME	Metabolizable energy
MECAT	Mapping, Error Correction, and de novo Assembly Tools
minMAF	Minimum minor allele frequency
minMQ	Minimum mapping quality to call an alignment unique
miRNAs	MicroRNAs
MLP	Multi-layer perceptron
MOD	Module
MSI	Minnesota Supercomputer Institute
MT	Mitochondrial genome
MTA	Material transfer agreement
MTA	Material Transfer Agreement
MTB	Mycobacterium tuberculosis
MX	Mutually exclusive exons
MYB	Myeloblastosis
MySQL	Open-source relational database management system
NAAIC	North American Alfalfa Improvement Conference
NAC	No apical meristem (NAM), ATAF1-2, and CUC2 (cup-shaped cotyledon)
NAFA	National Alfalfa & Forage Alliance

NBS	Nucleotide-binding site
NBT	New breeding technology
NDF	Neutral detergent fiber
NDF30	30-hour digestible NDF
NDFD	Neutral detergent fiber digestibility
NDFD48	48-hour NDFD
NEB	New England Bioscience
NEG	Net energy for gain
NEL	Net energy for lactation
NEM	Net energy for maintenance
NF	Nuclear factor
NFC	Nonfibrous carbohydrates
NF-X1	Nuclear transcription factor, X-box binding 1
NGS	Next-generation sequencing
NHEJ	Nonhomologous DNA end-joining
NHR	Nonhost resistance
NHX	Na+/H+ exchanger
NLGRP	National Laboratory for Genetic Resources Preservation
NMD	Nonsense-mediated mRNA decay
NPGS	National Plant Germplasm System
NPTII	Neomycin phosphotransferase II
NSID	Natural selection by inbreeding depression
NTP	Nucleoside triphosphates
OD	Optical density
ONT	Oxford Nanopore Technologies
ORF	Open reading frame
PacBio	Pacific biosciences
PAM	Protospacer adjacent motif
PAMP	Pathogen-associated molecular patterns
PCA	Principal Components Analysis
PCR	Polymerase chain reaction
PEG	Polyethylene glycol
PGK	Phosphoglycerate kinase
PGR	Plant genetic resources
PGRFA	Plant Genetic Resources for Food and Agriculture
PHD	Plant homeodomain
PI	Plant introduction
PIC	Polymorphism information content
PLCSEM	Partition-ligation-combination-subdivision expectation maximization algorithm
Pn	Photosynthetic rate
POD	Peroxidase
PPO	Polyphenol oxidase
PS	Phenotypic selection
QC	Quality control

QTL	Quantitative trait loci
RAD	Restriction-site associated DNA
RADseq	Restriction site-associated DNA sequencing
RAPD	Random amplified polymorphic DNA
REP	Repeat
RF	Random forest
RFLP	Restriction fragment length polymorphism
RFQ	Relative forage quality
RFV	Relative feed value
RIL	Recombinant inbred line
RKHS	Reproducing kernel Hilbert space
RMSE	Root mean squared error
ROS	Reactive oxygen species
RRA	Roundup Ready® alfalfa
RR-BLUP	Ridge regression best linear unbiased prediction
RRM	Not needed: the sentence will be changed in proofs
RUP	Rumen undegradable protein
RWC	Relative leaf water content
SANT	SWI3, ADA2, N-CoR, and TFIIIB B domain
SAP	STERILE APETALA
SARA	Subacute ruminal acidosis
SARDI	South Australian Research and Development Institute
SBP	S-protein binding protein
SCINet	USDA Scientific Computing Initiative
SCR	S-locus cystine-rich protein
SGSV	Svalbard Global Seed Vault
SI	Self-incompatibility
S-Locus	Self-compatibility locus
SMRT	Single molecule real time
SMTA	Standard Material Transfer Agreement
SNP	Single nucleotide polymorphism
SNV	Single nucleotide variant
SOD	Superoxide dismutase
SOS	Salt overly sensitive
SPADA	Small peptide alignment discovery application
SPL	Squamosa promoter binding protein-like
SRAP	Sequence-related amplified polymorphism
SRK	S-locus receptor kinase
S-RNase	S-locus ribonuclease
SS	Salt stress
SSR	Simple sequence repeat
STI	Salt tolerance index
STRs	Short tandem repeats
SVM	Support vector machine
SVR-gau	Support vector regression with gaussian kernel function

SVR-lin	Support vector regression linear with linear kernel function
TACAS	Targeted clone & sequencing assay
TAIR	The Arabidopsis Information Resource
TAL	Transcription-activator-like
TALEN	Transcription-activator-like effector nuclease
TDN	Total digestible nutrients
TDNL	Total digestible nutrients for legume
TE	Transposable Element
TF	Transcription factor
TFDB	Transcription factor database
TIR	Toll interleukin-1 receptor
TPM	Transcripts per million
Tr	Transpiration rate
TRANS	Transformation
TREX	Transcription and export
TRNA	Transfer ribonucleic acid
UBC	Ubiquitin conjugating E2 enzyme
UBQ	Ubiquitin
UMGC	University of Minnesota Genomics Center
UNEAK	Universal network-enabled analysis kit
UPGMA	Unweighted pair-group method with arithmetic means
UV	Ultraviolet
VCF	Variant call format
VIR	N.I. Vavilov Research Institute of Plant Industry, Russia
VOZ	Vascular plant One Zinc finger protein
VW	Verticillium wilt
WRKY	WRKY domain, defined by conserved amino acid sequence WRKYGQK
WUE	Water use efficiency
YA	Nuclear Factor Y subunit NF-YA
YB	Nuclear Factor Y subunit NF-YB
YC	Nuclear Factor Y subunit NF-YC
YEB	Yeast Extract Beef (YEB) Broth
ZF-HD	Zinc finger homeodomain protein
ZFN	Zinc-finger nuclease

1 Economic Importance, Practical Limitations to Production, Management, and Breeding Targets of Alfalfa

Dan Undersander

Abstract

Alfalfa (*Medicago sativa* L.) is the most valuable forage crop grown on approximately 30 m ha (74 million acres) worldwide. It is the third most valuable crop harvested in the U.S. behind only corn and soybeans. Alfalfa is widely recognized as a significant component of most dairy and other animal rations because of the multiple benefits it provides: Alfalfa adds valuable fiber, has a faster rate of fiber digestion than grasses, is very palatable, is high in protein, provides needed vitamins, is a good buffer in the rumen, and promotes animal health. In addition, it provides legume nitrogen credits and has rotational yield boost to following crops. It provides environmental benefits in soil remediation, reduced erosion (compared to row crops), and supports varied wildlife by being a source of cover and food from insects, spiders, mites, and earthworms. A major issue is low yield, where the record alfalfa yield in the U.S. is 26 t/a DM with irrigation in Arizona while the average alfalfa yield on a farm in the U.S. is about 4.5 tons/acre DM. The yield problem is a combination of environmental and management issues. Breeding objectives should continue to be resistance to diseases and insects. However, a major long-term direction should be to increase alfalfa yield. Possibly each of the following breeding objectives could contribute: raise limits to natural photosynthesis, reduce dark respiration, develop cultivars that have less leaf loss during harvest, and minimizing the impact of wheel traffic on stands. An additional breeding objective should be to increase the bypass protein of alfalfa. While this will not affect yield, it will increase alfalfa value to users.

D. Undersander (✉)
Department of Agronomy, University of Wisconsin-Madison, 3804 Dolphin Dr, Madison, WI 53719, USA
e-mail: djunders@wisc.edu

1.1 Introduction

Alfalfa (*Medicago sativa* L.) is the most valuable forage crop worldwide because of its high quality as an animal feed. In 2009, alfalfa was grown on approximately 30 million hectares (74 million acres) worldwide; of this North America produced 41%, Europe produced 25%, South America produced 23%, and Asia produced 8% (Basigalup et al. 2018). In 2017, international trade of alfalfa hay reached 8.3 million metric tons for a total value of 2.3 billion U.S. dollars (Basigalup et al. 2018, Adiyaman and Ayhan 2015).

1.1.1 Economic Importance

It is the third most valuable crop harvested (Table 1.1) in the U.S. behind only corn (52.9

L.-X. Yu and C. Kole (eds.), *The Alfalfa Genome*, Compendium of Plant Genomes,
https://doi.org/10.1007/978-3-030-74466-3_1

Table 1.1 Alfalfa Hay and Haylage Acres Harvested, Value, Yield/a, and Total Production in U.S

Data items	2019	2018	2017	2016	2015	2014
Acres harvested	11,730,000	11,515,000	12,065,000	12,466,000	13,486,000	13,510,000
Production, total dollar value	10,881,984,000	10,144,379,000	9,677,862,000			
Yield, t/a, DM	3.61	3.64	3.71	3.92	3.72	3.79
Total production, tons DM	42,343,000	41,935,000	44,816,000	48,927,000	50,108,000	51,153,000

Data from USDA National Agricultural Statistical Service (2017)

billion $ in 2019) and soybeans (31.2 billion $ in 2019) (USDA National Agricultural Service).

Alfalfa is also important due to its high biomass production. The record U.S. annual yield of alfalfa is 12 tons/acre (Delaware State College) without irrigation and 24 tons/acre (University of Arizona) with irrigation (LuKwick 2000).

Alfalfa seed is primarily grown in the U.S. in the states of California, Idaho, Montana, Nevada, Washington, and Wyoming. The approximate production of alfalfa seed in 2015 for the U.S. is 50.5 million pounds. A fringe benefit to the production of alfalfa seed is the production of honey from bees. In the U.S., $327.1 million dollars worth of honey is produced from all flower sources each year (2017 USDA National Agricultural Statistics Service https://www.nass.usda.gov/).

In addition to the traditional uses of alfalfa as an animal feed, alfalfa is beginning to be used as a biofuel for the production of electricity, bioremediation of soils with high levels of nitrogen, and as a factory for the production of industrial enzymes such as lignin peroxidase, alpha-amylase, cellulase, and phytase (Mueller et al. 2008).

1.1.2 Value of Alfalfa in Animal Diets

Alfalfa is widely recognized as a significant component of most dairy and other animal rations because of the multiple benefits it provides.

Alfalfa adds valuable fiber: Low fiber rations result in acid accumulation in the rumen causing acute and subacute ruminal acidosis (SARA) and displaced abomasums (Zebeliet al. 2008; Heinrichs and Kmicikewycz 2016). In general, rumen starch digesting microbes tolerate low pH levels, but the fiber digesters are inhibited by low pH. This is particularly true in dairy rations where concentrate and corn silage add much nonfibrous carbohydrate, resulting in low rumen pH.

Horses do well on alfalfa since it is lower in digestible fiber than grass hay and supplies more calories per pound (Shewmaker et al. 2005). Alfalfa can be a good high-energy ratio for lactating mares, growing horses, and working horses. The biggest benefit of alfalfa for horses is that it tends to be more nutrient-dense than most grasses when harvested at the same stage of maturity. It typically contains more digestible energy, more crude protein and calcium, and fewer nonstructural carbohydrates (especially fructose and fructans) than grasses. When fructose and fructans reach the hindgut of horses, they are fermented by bacteria that produce lactic acid. Excessive amounts of lactic acid are not absorbed efficiently from the hindgut and the accumulation of lactic acid is one cause of colic and laminitis in some pasture-kept horses. The biggest issue with alfalfa for horses is weight gain in horses that don't have adequate exercise.

Alfalfa has a faster rate of fiber digestion than grasses (including corn silage) which allows the energy from fiber fractions to become available rapidly over 12–24 h but not so rapidly to acidify the rumen (as from grain) (Stensig and Robinson 1997).

Alfalfa has a higher rate of passage through to digestive tract allowing similar dry matter intake to other forages in spite of the larger indigestible NDF fraction and, consequently, higher physical fill in the rumen (Krizsan et al. 2010).

Alfalfa is very palatable: Numerous trials have shown that alfalfa harvested before flower is very palatable to cattle and many other animals (Jennings et al. 2005). It is more palatable than small grain silage and mixes well with corn silage. Alfalfa that is flowering, diseased, or rained on in the windrow will be less palatable. Owners know that horses typically love alfalfa—a highly palatable hay, that is popular when balanced with a blend.

Alfalfa is high in protein: Protein is the most expensive component of most rations. Good quality alfalfa runs 16–20% crude protein while corn silage is 7–9% crude protein and most grasses run 12–14% crude protein (grasses may be somewhat higher if harvested when very immature) (Dairyland Laboratories Summaries 2020). A milking dairy cow needs 16–17% crude protein (National Research Council 2001) and growing animals need 12–6% crude protein (National Academies of Sciences, Engineering, and Medicine 2016). Thus, alfalfa can go further to meeting the protein needs of animals than most other forages.

Alfalfa provides needed minerals: Alfalfa contains a greater concentration of calcium, phosphorus, potassium, magnesium, sulfur, iron, zinc, and selenium than grasses (Foster et al. 2009). It is a good source of calcium, for all animals but especially for horses.

Alfalfa provides needed vitamins: Leafy, green alfalfa hay is unusually high in carotene, the precursor of Vitamin A (Foster et al. 2009; Uildiz et al. 2020). Vitamin A is the most common beef cow vitamin deficiency. Good quality alfalfa hay can furnish all the Vitamin A needs of beef animals. In addition to the many dietary functions of Vitamin A, this vitamin also may have some therapeutic value and be a contributing factor in preventing "shipping fever complex" and other disorders associated with animal stress. Alfalfa is usually a good source of Vitamin E and selenium, when grown on soils of good nutrient status. "White muscle disease" which sometimes causes serious losses of calves is caused by a deficiency of Vitamin E and selenium. Sun-cured alfalfa hay is also a source of Vitamins D and K as well as riboflavin and niacin.

Alfalfa is a good buffer in the rumen: Research suggests that rumen pH below 6.0 decreases rumen fiber digestion (Hoover 1986). Diets with alfalfa hay or haylage as the sole forage generally don't need added buffers (Erdman 1988), unless high grain feeding depresses fat test, off-feed problems occur, or finely chopped haylage is fed. Alfalfa, with a higher natural buffering capacity and higher fiber level than corn silage, also reduces the problem of ration adaptation when feed sources are changed.

Alfalfa promotes animal health: The high mineral and vitamin contents of alfalfa are so important to animal health. However, other benefits are known:

- cattle diets lacking adequate fiber can damage the rumen wall so that fluid is drawn from the systemic circulation into the rumen, resulting in dehydration and possibly hypovolemic shock. Further, damage to the rumen epithelium can result in leakage of bacteria and toxins into the circulatory system. Chronic rumen acidosis can lead to fungal rumenitis and occasionally formation of liver abscesses (Beef Cattle Research Council 2019).
- Alfalfa is also suitable for horses prone to gastric ulcers because the extra calcium acts as a buffer against stomach acid (Brown-Douglas 2012).
- Swine have been shown to need fewer antibiotics when fed some alfalfa (Undersander unpublished).

Alfalfa is often considered expensive when evaluating feedstuffs for a single component (e.g., fiber or energy) but is often the most economical feed when all valuable components are considered.

1.1.3 Crop Rotation Value

Alfalfa does not need to be fertilized with nitrogen, which saves money and energy for growing the crop. Further, when the alfalfa stand is terminated, it provides nitrogen for the succeeding crop (Table 1.2).

In addition to the nitrogen benefit researchers have described a "rotational benefit" of approximately 20% greater corn yields following alfalfa than corn following corn (Porter et al. 1997). Similar yield benefits have been reported for wheat and canola grown on fields where alfalfa had been terminated. The reason for the benefits is unknown.

1.1.4 Environmental Value

Alfalfa is beneficial in soil remediation. Alfalfa can help prevent ground water contamination by removing excess nitrogen from the soil before it can enter water table and streams (Russelle et al. 2001). Alfalfa can also reduce soil contamination of petroleum hydrocarbons and metals (Carolina 2014) and organic compounds (Chekol and Vough 2001).

Additionally, alfalfa provides habitat to a wide range of wildlife, including numerous beneficial insects, (Hartman and Kyle 2010). It is a direct food source; granivorous birds such as White crowned Sparrows (*Zonotrichialeucophrys*), Golden crowned Sparrows (*Zonotrichiaatricapilla*), and Horned Larks (*Eremophilaalpestris*) that consume newly planted seeds and seedlings (Clark 1976). It is a habitat for the invertebrate prey of many birds, including insects, spiders, mites, and earthworms. Alfalfa may be particularly good habitat for earthworms which, in turn, is an important food source for many birds. Alfalfa contributes nitrogen to the soil which promotes earthworm growth (Evans 1948) and increases their protein content (Stribling and Doerr 1985). Increased abundance of more protein-rich earthworms is one hypothesis for preferential use of alfalfa over other irrigated crops by some waterbirds (Bray and Klebenow 1988). Further, alfalfa often supports an abundant small mammal community that is exploited by various birds of prey. Because these small mammals consume the alfalfa crop, these birds of prey provide a pest-control service to the alfalfa grower.

The physical structure of the alfalfa plant provides many birds with cover from predators as well as a site for nesting. While some have considered this a bird trap when the alfalfa is harvested before the nesting is complete, longer intervals between harvests with reduced lignin cultivars and other technologies may alleviate this problem.

1.2 Practical Limitations to Production

As stated earlier, the record alfalfa yield in the U. S. is 26 t/a DM with irrigation in Arizona (Lukwick 2000) while the average alfalfa yield in the U.S. is about 4.5 tons/acreDM (USDA

Table 1.2 First-year nitrogen credits following alfalfa termination

Stand density	Medium/Fine soil		Sandy soils	
	————Regrowth after last cutting————			
	>8 inches	<8 inches	>8 inches	<8 inches
	——— lb nitrogen /acre———			
Good, > 4 plt/ft^2	190	150	140	100
Fair, 1.5 to 4 plt/ft^2	160	120	110	60
Poor, < 1.5 plt/ft^2	130	90	80	40

Data from Carrie Laboski (2019) Nitrogen credits following winter-killed alfalfa. University of Wisconsin Crop Manager Vol 26 Number 3

National Agricultural Statistical Service 2017). Several factors are responsible for the yield limitations resulting in the U.S. average yield being so far below the record—some are due to the growing environment and some due to management.

1.2.1 Environmental Yield Limitations

1.2.1.1 Amount of Sunlight

The sunniest place in the lower 48 U.S. states is Imperial County, Calif. Overall, the Southwest gets the most sun in an average day, along with select regions on the Gulf Coast and the southern tip of Florida (Fig. 1.1).

1.2.1.2 Length of Growing Season

The length of the growing season is determined by days between the last 32 °F frost of spring and the first 32 °F frost of fall. As might be expected the length of the growing season lengthens north to south across the U.S. with some southern sites having 365-day growing seasons some year (Fig. 1.2).

1.2.1.3 Temperature

While temperature can have a significant effect on plant growth, it is important to remember that different stages are optimized at different temperatures. To begin with the optimum temperature for seedling development was 65–77 °F with some variation due to the fall dormancy of the cultivar (Undersander et al. 2011). Evenson (1979) reported that maximum predicted aboveground growth occurred at a 90 °F crown temperature. Robison and Massengale (1969) reported that high night temperatures reduced growth, possibly because dark respiration remained higher; thus, some high-altitude locations (where temperatures are low at night) often have higher than expected yield.

1.2.1.4 Drought

On the average alfalfa uses 4- to 5-acre-inches of water for every ton of forage produced (Sloan 2009). Water shortages or less than optimum application timing can greatly reduce alfalfa yield. The efficiency of water use depends on temperature of the irrigation water (should be 60–70 °F), salinity (salt level) of the irrigation water and salt build-up in the soil, length of time the field is flooded, and amount of water applied, and growth stage and age of alfalfa (Sloan 2009).

1.2.1.5 Diseases

Many diseases can attack alfalfa (Samac et al. 2015). Some can kill or severely limit growth under certain conditions. Little chemical control is available and most practices for minimizing yield loss relate to management described below.

1.2.2 Mitigating Yield Losses Through Management

Operator management can have a large effect on yield, both directly and in mitigating environmental effects.

1.2.2.1 Irrigation

One of the most common limitations to alfalfa production is improper irrigation management. To optimize response water must be applied by appropriate methodology including system design, irrigation rate and length, and other factors. In one study top growth was reduced by 50% with 4 days of flooding at 70 °F, with 3 days of flooding at 80 °F, and 2 days of flooding at 90 °F (Sloan 2009).

1.2.2.2 Salinity

Salinity increases in the soil as irrigation water carrying salts is applied and then water is used in evapotranspiration leaving salts behind. Salinity is of greatest concern in soils that are irrigated with water high in salts, poorly drained, allowing for too much evaporation from the soil surface, in areas where the water table (the level or depth to free-flowable water in the soil) is shallow; or in seepage zones, which are areas where water from other locations (normally up slope) seep out. Numerous tools are available to minimize yield loss by proper irrigation management (Sloan 2009).

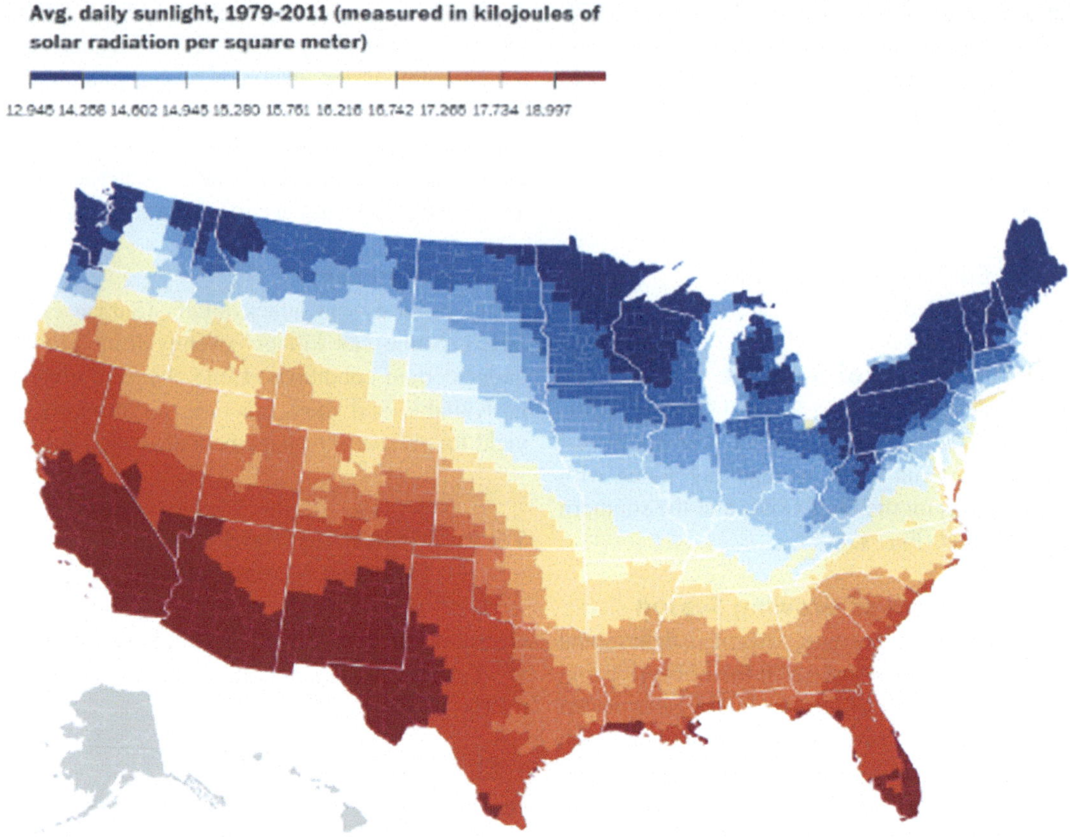

Fig. 1.1 Amount of sunlight received annually

Recently some cultivars of alfalfa have been developed which tolerate higher salinity levels than standard alfalfa.

1.2.2.3 Diseases

Significant research has gone into developing cultivars with resistance to many yield- and stand-limiting diseases (National Alfalfa and Forage Alliance 2021). Careful selection of cultivars with resistance to locally occurring diseases can greatly reduce yield loss.

1.2.2.4 Insects

Some genetic resistance exists for certain insect pests of alfalfa. These should be used when the insect is a potential problem. Certain pesticides are also labeled for insect control in alfalfa (consult state listings of available products) once the insect has been identified and determined to be at a control threshold.

1.2.2.5 Thin Stands

High yields only occur when leaf density is adequate to optimize light interception and photosynthesis. All stands naturally thin over time. Some have used plant density to estimate when the stand thinning resulted in yield reductions, however, research indicates that stem density is a better indicator of yield than plant density (Undersander and Cosgrove 2007). The reason is that large plants will produce more yield than small plants of the same number. The recommendation is to consider that yield losses occur when the stem density falls below 55 stems/ft^2. Slightly fewer stems may be recommended for water limiting situations.

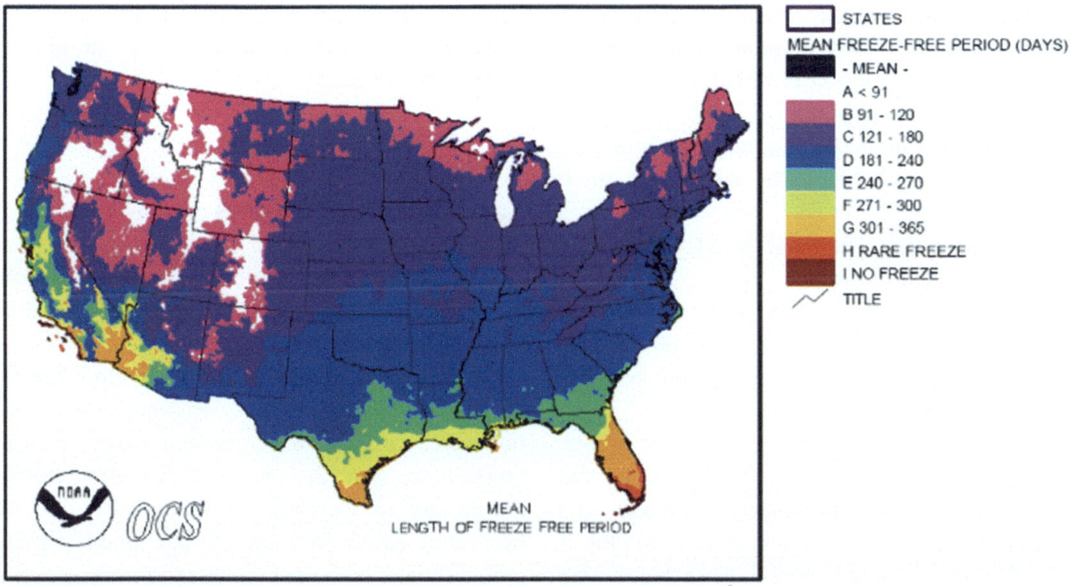

Fig. 1.2 Length of the season between last frost of spring and first frost of fall

1.2.2.6 Wheel Traffic

Wheel traffic is known to increase soil compaction which, on some soils, reduces macropore air permeability, soil water infiltration, and root development of alfalfa (Hamza and Anderson 2005; Lipiec et al. 2006). Driving over a field to harvest or apply chemicals can break stems of regrowth and (in some cases) damage crowns. The largest effect of wheel traffic is to break off regrowing alfalfa stems there by reducing the next cutting yield (Undersander 2008).

The following management recommendations are made to reduce wheel traffic damage to alfalfa:

(1) Use small tractors when possible to reduce soil compaction, i.e., don't use larger tractor than necessary or raking, or leave loader on tractor when harvesting.
(2) Avoid unnecessary trips across the field when harvesting:
   a. Mowing and conditioning in a single operation.
   b. Loaded wagons/trucks should be driven off the field in as little distance as possible.
   c. If bales are dropped, collect with least driving possible and as soon as possible.
   d. Do not drive on alfalfa field when harvesting crop of adjacent field.
(3) Consider using larger harvesting equipment to reduce the percent of field covered with wheel tracks (however, the affected area has greater weight applied to it).
(4) Avoid use of tractors with dual wheels.
(5) Harvest (drive on field) as soon after cutting as possible:
   a. Make silage from higher yielding fields, hay from lower yielding fields.
   b. Use wide swath to allow hay/haylage to dry faster.
   c. Make wrapped bales to allow harvest of wetter hay.
   d. Apply manure immediately after harvest.

1.3 Management

The management for high yield alfalfa includes the following:

1. Select field carefully as alfalfa requires a well-drained soil for optimum production.
   a. Ensure that alfalfa has been out of alfalfa for 2 years to minimize autotoxicity.
   b. Ensure that alfalfa planting restrictions are followed for herbicides applied to previous crop.
   c. Ensure that perennial weeds have been controlled in previous crop.
   d. Soil test and apply lime (if needed) and other nutrients. Lime and phosphorus should be applied 1 year before planting alfalfa.

2. Select alfalfa cultivar with appropriate fall dormancy, winterhardiness, disease, and insect resistance for region. Select cultivar as well for high yield potential. Note that in over 300 trials conducted over 30 years in Wisconsin the highest yielding variety has averaged 2.3 t/a DM more than the lowest yielding cultivar (Fig. 1.3).
3. Till field so that it is flat for ease of harvesting.
4. Inoculate seed (if not pre-inoculated). Place seed at ¼ to ½ deep and pack into firm soil.
5. Harvest a spring alfalfa seeding about 60 days after planting.
6. When harvesting established stands for dairy take first cutting when alfalfa at 27-inch height and later cuttings harvested at 28-day intervals (35 days for reduced lignin alfalfa). If harvesting for growing animals take first cutting at 30 inches and later cuttings at 25% bloom.
7. Mow (with conditioning for hay) and place into wide swath to enhance drying rate. This allows the forage to be removed quickly from the field to minimize wheel traffic damage and to begin irrigating as soon as possible.
8. If irrigating, begin season with full soil water profile. Irrigate as soon after harvest as possible
9. Fertilize annually to replace nutrients removed with hay. A split application is often recommended to avoid luxury consumption of nutrient by the alfalfa.
10. Take last cutting either early enough (6 weeks before killing frost) or at killing frost.

1.4 Breeding Targets

Alfalfa breeders have done a wonderful job breeding for resistance to diseases and this should continue.

However, the major breeding direction has to be breeding for yield increases since alfalfa has not increased in yield like its major dairy forage competitor, corn (*Zea mays* L). To some extent the failure to increase yield has been due to harvesting earlier for higher forage quality which reduces yield, both per cutting and for the season. This may be alleviated by reduced lignin alfalfa which can be harvested less frequently with consequence yield increases. Another option is to consider that the forage quality is in the leaves which the stem makes good biomass. As such, breeding for and developing a system for bushy alfalfa where leaves are separated from stems as the crop is harvested can be fruitful (Lamb et al. 2007; Shinners et al. (2007). Such may provide additional benefits of more consistent forage quality (from stripped leaves) and reduced number of harvests per growing season.

A longer term effort may be to develop germplasm with higher limits on natural photosynthesis (van Grondelle and Boeker 2017). Such alfalfa could produce more photosynthate in less than optimal conditions. Along the same lines, leaf design (characters such as vein density, leaf thickness, and cell shape) has been found to be strongly correlated with the hydraulic conductivity and maximum photosynthetic rate of foliage (Brodribb et al. 2007).

Another approach to increasing yield would be the reduction of dark respiration. Similar to photosynthesis, dark respiration is impacted by temperature. When nighttime temperatures are below 70 °F, the entire system runs smoothly and highly efficient. As temperatures increase more energy is required by the plant to stay alive and

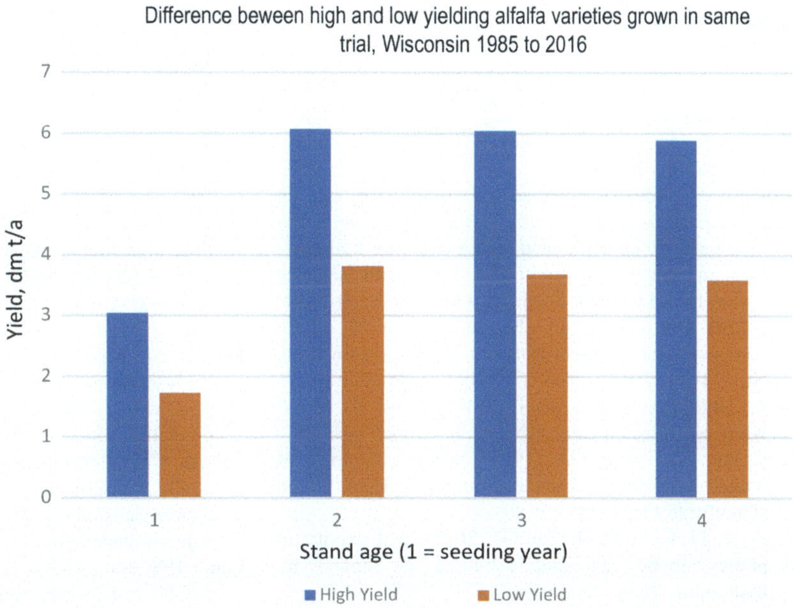

Fig. 1.3 Difference between high and low yielding alfalfa varieties grown in same trial, Wisconsin, MI 1985–2016 *Source* from Dan Undersander, University of Wisconsin

remain cool. Dark respiration begins to rapidly accelerate as the nightly temperatures begin to climb above the mid 70 s. It has been observed that high-altitude areas that cool significantly overnight tend to have higher growth rates than areas where the temperature is higher overnight.

A breeding effort already underway is to increase the bypass protein of alfalfa. Alfalfa, as a legume, is naturally high in protein (approximately 20% at the bud stage). However, this protein is largely albumins that are rapidly digested by microbial bacteria. This causes two problems: (1) excessive degradation of forage proteins in the rumen leads to ammonia-N (NH_3-N) production. When NH_3-N exceeds microbial requirement, it is absorbed from the rumen and converted to urea in the liver. Conversion NH_3-N to urea in the liver requires energy. Urea excretion in urine rather than protein used by the ruminant is both a metabolic burden to the animal and an economic loss to the farmer since nitrogen is the most expensive component of animal diets. (2) rumen microbes convert plant protein into microbial protein. This protein is lower in certain amino acids than required by the animal. If more plant protein bypasses rumen degradation to be degraded in the intestine and the plant amino acids can be absorbed directly by the animal. This is the only way to get more sulfur-containing amino acids to the cow.

The primary approaches to reduced rumen degradation of alfalfa protein are to produce tannins in alfalfa leaves (which cause bypass protein in birdsfoot trefoil and sainfoin) or to attempt to move genes from red clover to alfalfa that contain polyphenol oxidase enzymes that inhibit protein breakdown.

Breeding for a trait to minimize leaf loss during harvesting would be of great benefit to users since most energy and protein are in the leaves. Leaf loss is also a yield loss. In studies (Undersander, unpublished) the average leaf loss for harvested haylage was 9% of total dry matter with some farmers losing as much as 30%. Losses are higher if the forage is harvested as a hay.

Research and breeding efforts should continue to minimize the impact of wheel traffic on the alfalfa. While some traffic is unavoidable, current germplasm has been shown to have some variation in response and more could likely be identified and incorporated into current cultivars.

References

Adiyaman E, Ayhan V (2015) Investigation of feed value of alfalfa (*Medicago sativa* L.) harvested at different maturity stages. Legume Res 39(OF). https://doi.org/10.18805/lr.v0iof.9292

Agnello AC (2014) Potential of alfalfa for use in chemically and biologically assisted phytoremediationof soil co-contaminated with petroleum hydrocarbons and metals. Université Paris-Est, Earth Sciences

Basigalup D, del Carmen Spada M, Odorizzi A, Arolfo V (2018) Proceedings of the second world alfalfa congress global interaction for alfalfa innovation, Nov 11–14, 2018 Cordoba, Argentina. Preface p 2

Beef Cattle Research Council (2019) Acidosis http://www.beefresearch.ca/research-topic.cfm/acidosis-63

Bray MP, Klebenow DA (1988) Feeding ecology of white-faced ibises in a great basin valley, USA. Colon Waterbirds 11:24–31

Brodribb TJ, Feild TS, Jordan GJ (2007) Leaf maximum photosynthetic rate and venation are linked by hydraulics. Plant Physiol 144:1890–1898

Brown-DouglasC (2012)Use of alfalfa or lucerne and its effect on gastric ulcers. *Equinews*. Kentucky Equine Research. https://ker.com/equinews/use-alfalfa-or-lucerne-and-its-effect-gastric-ulcers/

Carolina AA (2014) Potential of alfalfa for use in chemically and biologically assisted phytoremediation of soil contaminated with petroleum hydrocarbons and metals. Ph.D Thesis,University of Paris

Chekol T, Vough LR (2001) A study of the use of alfalfa (*Medicago sativa* L.) for the phytoremediation of Organic Contaminants in Soil. Remed J 11:89–101. https://doi.org/10.1002/rem.1017

Clark DO (1976) An overview of depredating bird damage control in California. InternetCenter for wildlife damage management, Bird Control Seminars Proceedings, pp 21–27

Dairyland Laboratories Summaries https://www.dairylandlabs.com/resources/feed-and-forage/summaries

Erdman RA (1988) Dietary buffering requirements of the lactating dairy cow. J Dairy Sci 71(12):3246–3266

Evans AC (1948) Studies on the relationships between earthworms and soil fertility. Ann App Biol 35:485–493

Evenson PD (1979) Optimum crown temperatures for maximum alfalfa growth. Agron J 71(5):798–800

Foster S, McCuin G, Nelson D, Schultz B, Torell R (2009) Alfalfa for beef cows. University of Nevada Reno. https://extension.unr.edu/publication.aspx?PubID=2228

van Grondelle R, Boeker E (2017) Limits on natural photosynthesis. J Phys Chem B 121(30):7229–7234

Hamza MA, Anderson WK (2005) Soil compaction in cropping systems: a review of the nature, causes and possible solutions. Soil Till Res 82:121–145

Hartman CA, Kyle K (2010) Farming for birds: Alfalfa and forages as valuable wildlife habitat. In: California Alfalfa & Forage Symposium and Corn/Cereal Silage Mini-Symposium. https://alfalfa.ucdavis.edu/+symposium/proceedings/2010/10-61.pdf. Accessed 28 November 2019

Heinrichs J, Kmicikewycz A (2016) Sub-acute rumen acidosis and physically effective fiber. https://extension.psu.edu/sub-acute-rumen-acidosis-and-physically-effective-fiber

Hoover WH (1986) Chemical factors involved in ruminal fiber digestion. J Dairy Sci 69(10):2755–2766

Jennings J (2005) Alfalfa for dairy cattle, University of Arkansas Cooperative Extension Service FSA4000. https://www.uaex.edu/publications/pdf/FSA-4000.pdf

Krizsan SJ, Ahvenjärvi S, Huhtanen P (2010) A meta-analysis of passage rate estimated by rumen evacuation with cattle and evaluation of passage rate prediction models. J Dairy Sci 93(12):5890–5901

Laboski C (2019) Nitrogen credits following winter-killed alfalfa. Wisconsin Crop ManagerNewsletter https://ipcm.wisc.edu/blog/2019/03/nitrogen-credits-following-winter-killedalfalfa-2/

Lamb JFS, Jung H-JG, Sheaffer CC, Samac DA (2007) Alfalfa leaf protein and stem cell wall polysaccharide yields under hay and biomass management systems. Crop Sci 47(4):1407–1415

Lipiec J, Kus J, Słowinska-Jurkiewicz A, Nosalewicz A (2006) Soil porosity and water infiltration as influenced by tillage methods. Soil Till Res 89:210–220

Lukwick AE (2000) High yield alfalfa: 24 tons irrigated…12 tons non-irrigated. Better Crops 84:18–19

Mueller SC, Undersander DJ, Putnam DH (2008) Alfalfa for industrial and other uses. University of California Agriculture and Natural Resources. ANR publication 8305:1–6

National Alfalfa and Forage Alliance (2021) Alfalfa variety ratings 2021. https://www.alfalfa.org/variety Leaflet.php

National Academies of Sciences, Engineering, and Medicine (2016) Nutrient requirements of beef cattle: eighth revised edition. The National Academies Press, Washington, DC. https://doi.org/10.17226/19014

National Research Council (2001) Nutrient requirements of dairy cattle: Seventh revised edition, 2001. The National Academies Press, Washington, DC. https://doi.org/10.17226/9825

Porter PM, Crookston RK, Ford JH, Huggins DR, Lueschen WE (1997) Interrupting yield depression in monoculture corn: comparative effectiveness of grasses and dicots. Agron J 89:247–250

Robison GD, Massengale MA (1969) Effect of night temperature on growth and development of alfalfa (*Medicago sativa* L.). J Arizona Acad Sci 5(4):227–231

Russelle MP, Lamb JFS, Montgomery BR, Elsenheimer DW (2001) Alfalfa rapidly remediates excess inorganic nitrogen at a fertilizer spill site. J Environ Qual 30:30–36

Samac DA, Rhodes LH, Lamp WO (2015) Compendium of Alfalfa diseases and pests, 3rd edn. The American Phytopathological Society ISBN: 978-0-89054-448-8

Shewmaker G, Undersander DJ, Lawrence LM, Lacefield GD (2005) Alfalfa, the high quality hay for horses. National Alfalfa Alliance

Shinners KJ, Herzmann ME, Binversie BN, Digman MF (2007) Harvest fractionation of alfalfa. Trans ASABE 50(3):713–718

Sloan J (2009) How much water does alfalfa need? University of Nevada, Reno. https://extension.unr.edu/publication.aspx?PubID=2575

Stensig T, Robinson PH (1997) Digestion and passage kinetics of forage fiber in dairy cows as affected by fiber-free concentrate in the diet. J Dairy Sci 80 (7):339–1352

Stribling HL, Doerr PD (1985) Nocturnal use of fields by American woodcock. J Wildlife Manag 49:485–491

Uildiz AÖ, Sentürk ET, Olgun O (2020) Use of alfalfa meal in layer diets–a review. World's Poultry Sci J 76 (1):134–143. https://doi.org/10.1080/00439339.2020.1708839

Undersander DJ (2008) Minimizing wheel traffic damage to alfalfa. University of Wisconsin Extension Focus on Forage 10(3):1

Undersander DJ, Cosgrove D (2007) Using stem count as a decision aid in turning over alfalfa stands. Forage Grazinglands. http://www.plantmanagementnetwork.org

Undersander DJ, Hall MH, Vassalotti P, Cosgrove D (2011) Alfalfa germination and growth. University of Wisconsin A3681

USDA National Agricultural Statistical Service (2017). https://www.nass.usda.gov/Statistics_by_Subject

Zebeli Q, Dijkstra J, Tafaj M, Steingass H, Ametaj BN, Drochner W (2008) Modeling the adequacy of dietary fiber in dairy cows based on the responses of ruminal pH and milk fat production to composition of the diet. J Dairy Sci 91(5):2046–2066

Factors Influencing Yield and Quality in Alfalfa

2

Daniel H. Putnam

Abstract

Alfalfa is one of the most productive, widely adapted crops in the world, of importance to animal food production systems, to environmental health, as well as the potential for direct human food and industrial uses in the future. However, yield improvements have been painfully slow with this crop. While average farm yields have improved twofold in many environments over a 100-year period, evidence indicates a leveling off of yield improvements over the past 30 years. Disappointing yields are a frequent complaint of growers. However, much higher yields are highly attainable with this crop as evidenced by the 'yield gap'—the fact that maximum yields are two to threefold greater than the median on-farm yields in a given region. Limitations are largely due to agronomic, not genetic limits, particularly harvest scheduling, the need for quality, and poor stand persistence, but opportunities for cultivar improvement in yield are significant. There are several innate biological and agronomic tradeoffs that limit yield in alfalfa. These constraints include the need for high quality as well as yield, the need for maintenance of stand longevity as well as annual yields, and the requirement for frequent harvests involving farm machinery and wheel traffic. Agronomic strategies are likely to be most important in improving yields, including improved drainage, fertilizer management, harvest efficiency, reduced traffic, improved stand establishment, pest management, and irrigation efficiency. Key genetic characteristics for yield improvement include improved pest resistance, breaking the yield-quality-persistence tradeoff as related to harvest schedules, and improved root resistance to winterkill and wheel traffic. Improving late-season yield (summer slump) is an important goal, since late-season harvests are 40–50% of the yield of early harvests. Delayed flowering and maturity, reduced lignification of the cell wall, disease and insect resistance will be key traits. Other traits, such as the contribution of alfalfa stands to crop rotation and soil health and the environment, should be considered along with yield and quality goals. However, improved yields are the key to the future alfalfa systems that are resilient to the challenges of water supply and increased human populations.

2.1 Introduction

Alfalfa (*Medicago sativa* L.) as a crop differs significantly from most of the worlds' important crops due to several key biological and agricultural

D. H. Putnam (✉)
Department of Plant Sciences, University of California Davis, One Shields Avenue, CA Davis 957616, USA
e-mail: dhputnam@ucdavis.edu

L.-X. Yu and C. Kole (eds.), *The Alfalfa Genome*, Compendium of Plant Genomes,
https://doi.org/10.1007/978-3-030-74466-3_2

features. These distinct features impose significant constraints on the ability to improve yields in this crop, but also offer important advantages for alfalfa improvement and its contribution to overall sustainability of farming systems. These include the fact that it is a highly efficient N_2-fixing legume, a perennial, a deep-rooted crop which improves soil health and wildlife habitat (Putnam et al. 2001), and the fact that it is harvested multiple times each year. The fact that the harvest index (percentage of economically important yield to total above-ground biomass yield) of alfalfa is 100%, unlike grain crops where the harvest index ranges from 25 to 50% is an important yield advantage for alfalfa, one of the highest yielding agronomic crops. The latter point is important, since most of the historical yield improvements in grain crops have been largely through modification of harvest index, not biomass yields.

An additional constraint for alfalfa from an agricultural and economic perspective is that unlike most grain and fruiting crops, its value is realized directly through animal production systems (Nelson and Moser 1994). Thus, the need to achieve high yields must be continually balanced with the need for high forage quality (high protein, high digestibility, moderate to low fiber) which produces a desirable result in animal production systems, primarily dairy. Unfortunately, quality is highly negatively correlated with yield as well as stand longevity, both with choice of cultivar and agronomic practices. Important biological tradeoffs in alfalfa systems include 1. Tradeoff between carbon allocation to perennial structures versus above-ground production, including the need to maintain N_2 fixation, 2. the tradeoff between yield and quality characteristics due to demand.

This article reviews several of the key factors influencing constraints to yield and quality in alfalfa and suggests approaches to envisioning higher yielding systems in the future.

2.2 Historical Alfalfa Yield Trends

It's instructive to examine the history of yields to understand the potential yields in the future. Average on-farm alfalfa yields over the past 100 years in the United States have shown some improvement, approximately doubling over that period (Fig. 2.1). These broad yield estimates (USDA-NASS statistics) across 17–30 million acres (7–12 million ha, depending upon year) represent a wide range of environmental conditions, from very cold short-season non-irrigated sites in the northern regions to long-season irrigated sites in the desert Southwest, where the crop grows 365 days/year (Baron and Belanger 2020). Since silage making has been increasingly adapted in recent years, these data may underestimate actual yield levels, particularly in the upper midwestern and northeastern states. Winter grazing occurs in the milder mid-latitude and desert southwestern states, which is not reflected in hay yield statistics. Seeding-year data in addition to end-of-stand data (when densities deteriorate significantly) are typically included in such summaries, lowering the overall estimate of yield potential of alfalfa, so the maximum yield levels in different regions are likely to be higher than in Figs. 2.1 or 2.2.

However, yield improvement in alfalfa has not been particularly impressive, especially in comparison with corn (*Zea mays L.*). Corn grain yields have approximately quadrupled since 1950, advances which can be attributed to the development of high-yielding hybrid systems linked to advanced agronomic practices, pest-resistance characteristics, and genetically modified crops (Nielson 2020).

2.2.1 Examining Historical Innovations

Over this 100-year period, alfalfa has evolved from a mostly hand- and horse-harvested (and cattle-grazed) crop, to a mechanically harvested crop grown with advanced varieties with pest-resistance characteristics using modern methods of weed, insect management, and advanced harvesting methods (Fig. 2.2). In many regions, it has transitioned from an on-farm only harvested crop, fed only to one's own animals, to a highly traded cash crop bought and sold with reasonable profitability. This transition to a cash crop has

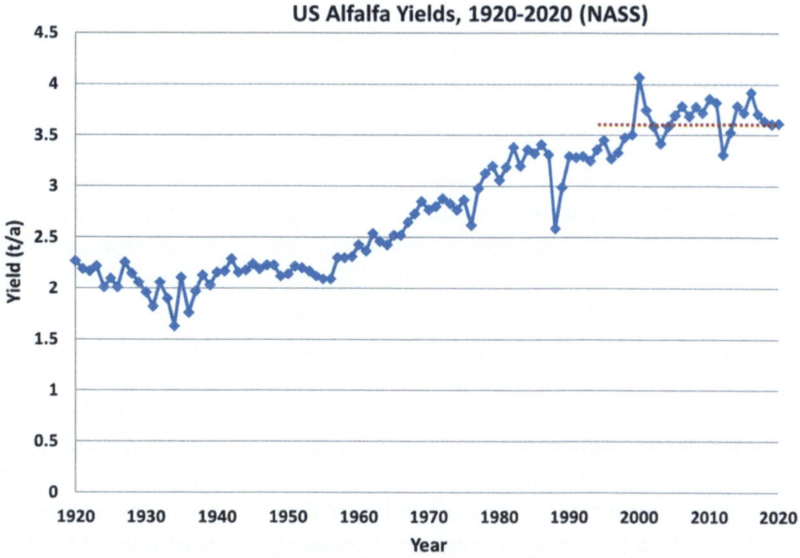

Fig. 2.1 Changes in US alfalfa hay equivalent yields over a 100-year period. Conversion of t/a to Mg ha-1, multiply × 2.242. Data from USDA-NASS

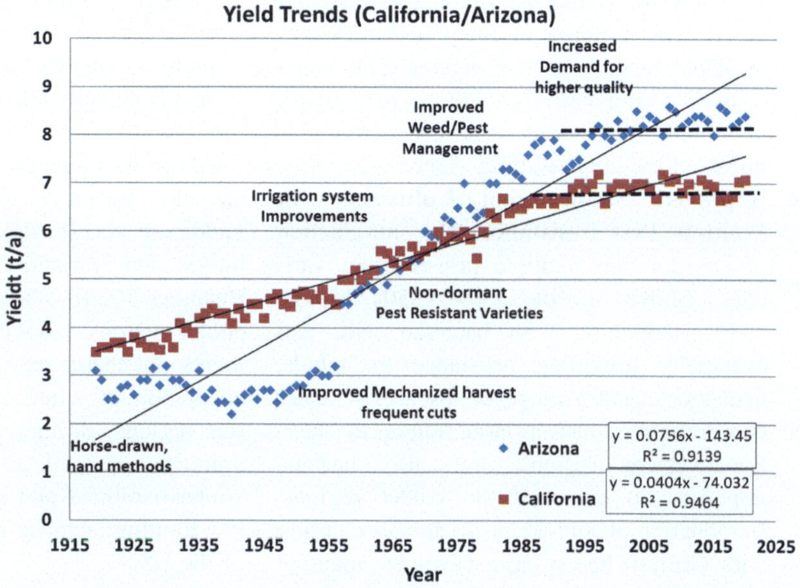

Fig. 2.2 Changes in alfalfa yields, California and Arizona over a 100-year period. Conversion of t/a to Mt/ha, multipy × 2.2417. (USDA-NASS, 2020). Important changes in technology over this period which may have impacted yields are shown

happened primarily in western areas and in regions removed from the US dairy shed concentrations of animals. Alfalfa is currently the nations' third most important economic crop in on-farm returns (USDA-NASS). Approximately 50% of the nations' alfalfa is produced under irrigation (USDA-NASS). The mechanisms for yield improvement over this period can be illustrated by examining the possible causes of yield improvements in the two highest yielding environments, Arizona and California (Fig. 2.2). Key attributes in these regions have been:

- **Mechanization and More Efficient Harvests**. Early and mid-century, the transition from horse-driven cutting, collection, and movement to mechanized harvest systems (tractor-driven swathers, rakes, and balers and automated bale pickup) has had a profound effect on yield. More rapid harvests enabled more rapid regrowth and more cuts per year, and better irrigation management. Mechanization enabled a change from a 2 or 3 cut system to a 3–4 cut system in colder regions as well as a transition from a 3-6 cut system to a 6-11 cut system in warmer regions.

- **Irrigation System Innovations**. The development of overhead pressurized irrigation systems, laser leveling for flood irrigation, and drip irrigation enabled better irrigation schedules, improving yields in Arizona and California. Development of better pumping and delivery systems was also key to more timely water applications.
- **Weed and Pest Management**. Improved stand establishment techniques utilizing herbicides, and the ability to control devastating insect pests have played an important role in yield advances since the 1950s, and currently pests have a range of integrated measures available to growers (UC-IPM 2021). Weed control during early growth is particularly important, since long-term productivity is impacted by early seedling vigor.
- **Improved Non-dormant Cultivars with Multiple Pest Resistance**. The introduction of non-dormant multiple pest-resistant varieties occurred mostly in the 1950s through 1970s, beginning with bacterial wilt, and eventually providing resistance to aphids, nematodes, and a range of diseases. Greater non-dormancy had a large effect in these Southwestern regions, but also became important in some of the colder regions. Introduction of cold-adapted varieties (starting with Grimm) had a large effect on adapting alfalfa to the colder, wetter regions of the Upper Midwest, enabling expansion of acreage and higher yields (Fig. 2.1).
- **Improved Soil Fertility, Soil Conditioning, Agronomic Practices**. Improved soil conditioning (such as deep ripping), liming and fertility management, land leveling, improved drainage, and salinity management have been important innovations, improving yields. In addition, general management considerations, such as early fall planting and improved stand establishment techniques (such as seeding depth), and optimizing harvest schedules enabled significant improvements over the latter half of the twentieth Century (Marble 1990).
- **Tradeoff with Quality**. The 1970s and 80s saw an intensification in demand for high-quality (low fiber) alfalfa hays for forage production and the introduction of testing. For example, while medium fiber (less than ADF < 31%, < NDF 41%) were considered high quality in the early 1970s (and few hay lots were tested) in California, by the late 1980s, dairies demanded very low fiber hays (ADF < 27%, NDF < 34%) and hays were increasingly tested, at least in the West. Growers responded by shortening harvest schedules. This is very likely the key explanation for the leveling of yields from the 1980 through current years in the Southwest (Fig. 2.2), as well as nationwide (Fig. 2.1). In addition, many of the innovations listed above had been fully adapted by that time.

Many of these factors described in Fig. 2.2 are also operative nationwide (Fig. 2.1)—but yields are necessarily limited by the colder climates and variation in rainfall (Baron and Belanger 2020). Winterkill is a major factor in colder regions, reducing yields. Yield improvements in these regions can be attributed to adaptation of winter-hardy germplasm in cold, wet regions, development of liming strategies, drainage, as well as mechanization. Lack of winter-hardiness and short growing season is a key limiting factor in many of the colder regions of the US.

2.3 Yield Gap

In spite of these yield improvements, there is a considerable gap between maximum potential biological yields alfalfa and average on-farm yields, even given current technology and cultivars (Brummer and Putnam 2018). In an excellent analysis of the alfalfa yield gap, Russelle (2014) concluded that there is a 2–3-fold difference between the median yield in a region and the potential yield across the US. This gap was evident in all regions examined whether one compared median yields with university yield reports or the top 10% of producers in that region (Fig. 2.3). Within a single high-yielding environment (Fresno Co. CA), on-farm county yields were 62% of those reported in university trials (Fig. 2.5). Variety differences within trials over 18 site-years averaged 29% of the mean in these trials (vertical bars, Fig. 2.5), which is typical of the yield differences observed in non-dormant alfalfa variety trials. So even in a high-yielding irrigated environment, yield gaps are commonly observed. It should be pointed out that on-farm reported yields typically include seeding year yields (which may include only a few cuttings), artificially lowering a multi-year NASS estimate of on-farm yield potential.

2.4 Key Factors Impacting Yield and Quality

Yield and quality are both economically and biologically important factors when considering overall productivity of an alfalfa-based cropping system. Additionally, important features such as the value of perenniality, value in crop rotation, contribution to soil health and conservation, N_2 fixation, and wildlife habitat (Putnam et al. 2001) should be considered when developing yield goals. They must be considered in an integrated fashion. Yields should not be considered in isolation with quality attributes or other important crop features, many of which are tradeoffs with yield.

It must be pointed out that if dry matter yield was the only criteria for choice of forage crop, we would not likely grow alfalfa. Many of our warm-season perennial C4 grasses especially (e.g., *Miscanthus spp.*, *Pennisetum* spp.) achieve much greater biomass yields (e.g., 2x) than are generally possible with alfalfa. However, these crops are considerably lower in feeding value (Horrocks and Vallentine 1999; Moore et al. 2020), are not highly acceptable as high-quality dairy feed, and furthermore are dependent upon N fertilizers (unlike alfalfa, which requires near zero N fertilizers). Similarly, quality of forage crops cannot be considered in isolation with yield, since some of the highest 'quality' forage harvests may not be economically viable due to low yields.

A range of factors impact yield and quality in alfalfa and these must be considered in line with

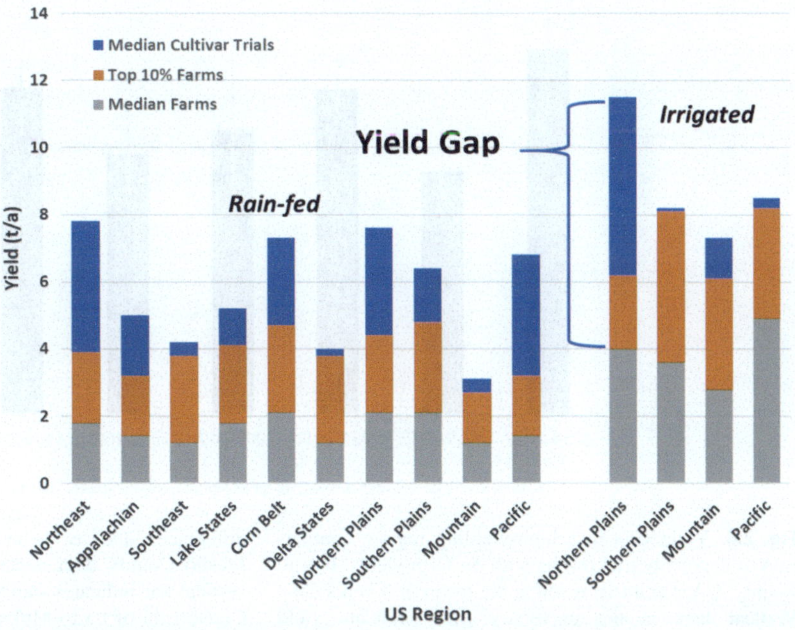

Fig. 2.3 Yield gap in various locations in the United States, based upon the differences between university cultivar trials (highest number, blue) or the top 10% (orange) of reported farm yields or the median farm yield (grey) for a region. Adapted from Russelle, 2014.Conversion of t/a to Mt/ha, multiply × 2.2417

the other attributes of importance, and the biological and agronomic tradeoffs that occur.

2.4.1 Environmental Conditions

Climate and weather set the upper limits to alfalfa production in all regions due to available solar radiation, and adequate temperatures for growth, cold and heat stress, and response to day-length (Baron and Belanger 2020). Growth rates are optimized at between 16–20 °C day and 10–15 °C night (Pearson and Hunt 1972), but others have recorded optimum growth at 27–32 °C (Evenson 1979; Leach 1971). Soil type is another important constraint, with poorly drained, acidic soils being the most problematic in addition to low phosphorus, sulfur, or potassium concentrations (Undersander et al. 2011; Brauder and Volenec 2020). For alfalfa, the most common limiting nutrients are Phosphorus, Potassium, and Sulfur, with Boron, Zinc, and Manganese occurring in some environments, but soil pH and plant drainage is a critical limiting factor in some environments (Myer et al. 2007; Orloff 2007). Adaptation to a given environment is a key aspect of quality as well as yield (Nelson and Moser 1994).

The cooler, winter-influenced high elevation intermountain regions of the West are similar in many respects to the conditions throughout the upper Midwest and Northeast US, and utilize a similar set of dormant varieties, with the exception that winterkill is a higher risk in the upper Midwest and Northeastern regions. Desert and Mediterranean zone provide the opportunity for long-season alfalfa production, but with summer heat and salinity stress. In variety trials, short-seasoned environments (Intermountain region) under optimum management produced 65% of the yield of that in a longer-seasoned Mediterranean environment (Fig. 2.4), largely a function of temperature, length of season, and solar radiation.

2.4.2 Seasonal Yield Patterns

However, annual yield patterns over the season are also of strong interest and affected by environment. In nearly all environments, late-season yields are a fraction of early season harvests, and

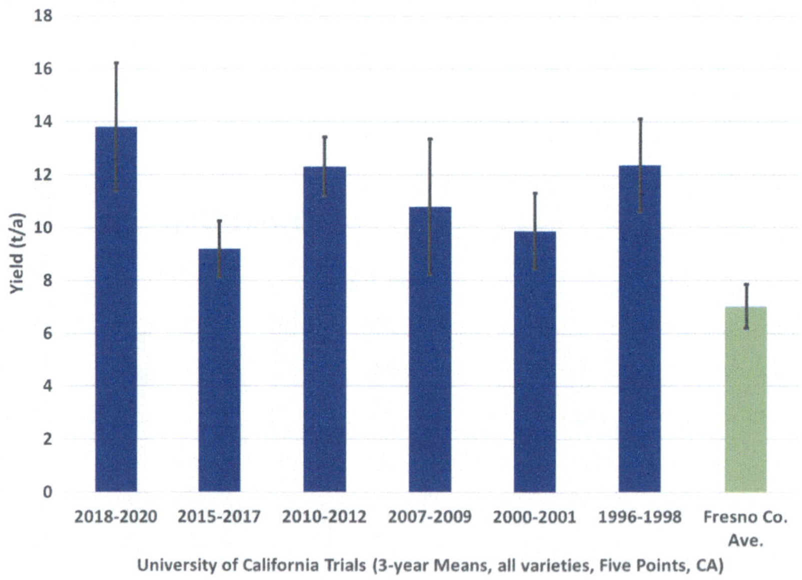

Fig. 2.4 Yield Gap in a high-yielding region: comparison of university trial results with on-farm yields, Fresno County, CA. Each trial result is the mean of 3 years data. Vertical bars in the university trials indicate yield differences due to variety. approximately 8 cuts/year. Fresno County data mean of 6 years (2013–2018) and vertical bar indicates standard deviation over that period. Conversion of t/a to Mt/ha, multiply × 2.2417

declined an average of 44% and 51% of first cutting yields, Intermountain sites, and Mediterranean (long-seasoned) sites, respectively (Fig. 2.4). In both of these trials, water was not limited, but early flowering and reduced growth rates are often observed during these late-summer periods.

Often termed 'summer slump', disappointing late-summer yields are observed in most environments, even in warm high-solar radiation environments like Arizona under full irrigation after the summer solstice (Putnam and Ottman 2013). This may be due to a combination of available solar radiation and response of varieties to shortened days (and investment of plant resources to root reserves). This suggests breeding goals for breeders and physiologists to fundamentally understand and to reduce the yield decline as affected by seasonal climatic conditions. It is likely that increased reproductive development late in the year, as well as the response of alfalfa varieties (even non-dormant varieties) to Fall Dormancy signals of temperature and day length (Teuber and Brick 1988) may be important aspects of this decline in late-season herbage yield.

2.4.3 Harvest Schedules, Crop Maturity, and the Yield-Quality-Persistence Tradeoff

It is axiomatic that in alfalfa production systems, quality (broadly defined) is nearly always negatively correlated with yield as the crop matures (Horrocks and Vallentine 1999, Fig. 2.5). Thus, those wishing to improve alfalfa yields must consider and accommodate the need for quality, since high quality is an important goal of farmers. Alfalfa as a crop really consists of a series of 'crops' harvested in 20–40 day increments over the season, with the limits to number of harvests set by the environment (the range is from 2 to 12 harvests/year). The most common number of harvests is 3–4 in short-seasoned environments, with up to 7–10 in long-seasoned environments, as illustrated in Fig. 2.5. Each of these 'crops' involves the growth from the stubble from buds originating from the crown, which emerge to develop through vegetative, bud, flower, and seed production phases (Mueller and Teuber 2007). Ten stages of plant development have been described, from stage 0 to 9, with vegetative stages (1–2), bud stages (3–4), and flowering stages (5–6); the most important for forage production (stages 7–9) describes seed pod development (Kalu and Fick1981; Teuber and Brick 1988; Tueber et al. 1988; Mitchell et al. 2020).

Normally, alfalfa for forage is harvested in vegetative through early to mid-flowering periods, which occur within a 24- to 40-day window, depending upon environment and time of year. This yield-quality tradeoff occurs during each of the harvests but is affected by temperature and other factors. In the first two harvests in Yolo Co., Ca (April and May), yields increased approximately fourfold over a 42-day period in the spring (April and May), while NDF concentrations increased from about 27 to 43% over the same period (Fig. 2.6). High fiber (ADF, NDF) and high lignin concentrations are associated with lower forage quality, and major determinants of price per unit weight for alfalfa, although digestibility of fiber is often considered equally important.

2.4.4 Mechanisms for crop maturity impacts on yield and quality

There are several powerful mechanisms that impact the increases in yield as well as the decline in quality as influenced by plant maturity: (1) the mix of leaf and stem and (2) the increased lignification of the cell wall. On a morphological level, as alfalfa plants mature, stem growth exceeds leaf growth, particularly after about 15 days. Alfalfa swards are characterized by a population of stems, ranging from newly developing stems to maturing stems which consist of a high percentage of the dry matter yield. While Alfalfa leaves do not change appreciably in quality from 15 days to harvest, stems become increasingly lignified and lower in digestibility.

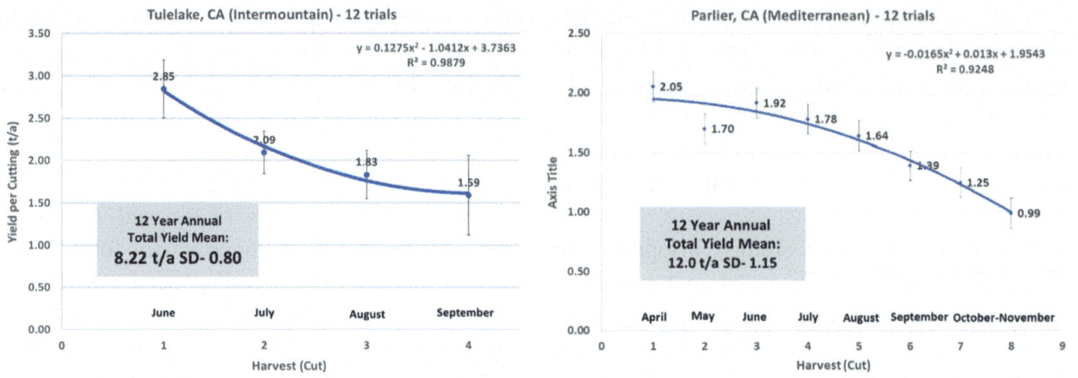

Fig. 2.5 Yield patterns in a short-season environment (Tulelake) and a long-season (Parlier) environment in California. Data compiled is average yield across multiple varieties and 12 years. Vertical bars are Standard Deviations for each cutting. Note differences in Y axes. Data from Putnam et al. (2005–2020). Conversion of t/a to Mt/ha, multiply × 2.2417 Data from https://alfalfa.ucdavis.edu/+producing/variety

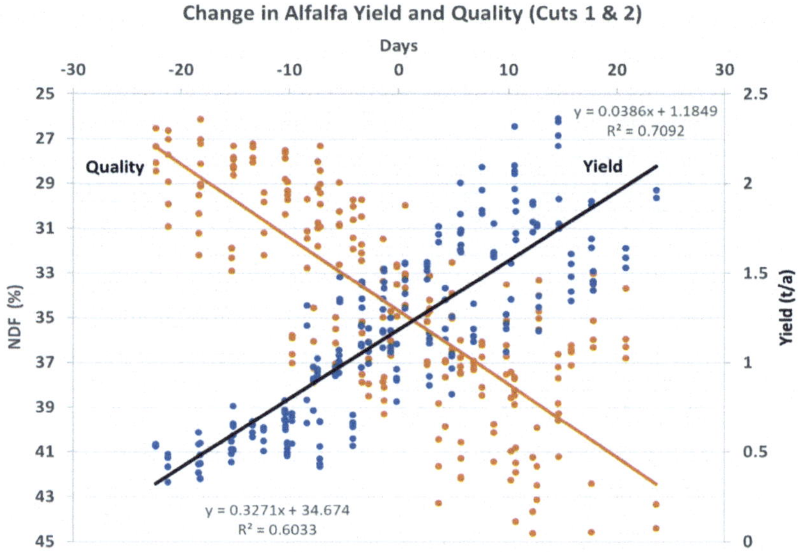

Fig. 2.6 Change in yield and quality over approximately 44 days, cuts 1 and 2, Yolo County, CA (1999 and 2000 data). Scatter represents replication, cut, and year variation. Note reverse axis on NDF. Conversion of t/a to Mt/ha, multiply × 2.2417

This twofold influence of plant maturity—increased mix of stems combined with rapidly decreasing quality of stem—is the key mechanism for the yield-quality tradeoff in alfalfa.

2.4.5 Multi-year Effects on Yield, Quality, and Stand

This tradeoff between yield and quality occurs during an individual growth period, but also over years, has been confirmed in a wide range of studies in multiple environments. In an earlier study, yields over a 3-year period, average yields improve by 55% simply by reducing the number of harvests from 9–10 (21 d interval) to 5–7 harvests (37 d interval) in a Mediterranean environment (Table 2.1). However, quality of the extended (late) harvests was significantly lower than early harvests, with increases in fiber (ADF) and declines in Crude Protein. At the same time, frequent harvests resulted in a significant stand decline over a 3-year period, with a resulting increase in weed intrusion as stands

Table 2.1 Effect of crop maturity on annual crop yield, average ADF, CP, leaf percentage, plant stand at the end of the 3-year trial, and weed percentage in year three, Fresno, CA (Marble 1990, 1974)

Maturity	Days	Number of Cuts	Yield (t/a)	ADF (%)	CP (%)	Leaf (%)	Stand (%)	Weeds (%)
Pre-bud	21	9–10	7.5	26.3	29.1	58	29	48
Mid-Bud	25	8–9	8.8	29.5	21.3	56	38	54
10% Bloom	29	7	9.9	32.2	21.3	53	45	8
50% Bloom	33	6–7	11.4	32.7	18.0	50	56	0
100% Bloom	37	5–6	11.6	35.5	16.9	47	50	0

decline (Table 2.1). Stand decline is widely attributed to lack of adequate partitioning of photosynthate to perennial structures, which are compromised by early harvests (Heichel et al. 1988).

Thus, the challenge of the yield-quality tradeoff is complicated by the influence of harvest schedule on stand persistence. While yield is typically negatively correlated with quality, higher yielding crops tend to be correlated with better stand persistence, as influenced by harvest schedule (Fig. 2.7). When crops are harvested to achieve higher quality, this frequently leads to stand decline (Table 2.1), implying a negative correlation. However, when stands do not decline (with long cutting schedules), this results in less weed intrusion, implying a positive relationship between stand persistence and yield (Table 2.1).

2.4.6 Variety Effects on the Yield-Quality Tradeoff

It is widely known that alfalfa variety can have an effect on quality as well as yield. Mechanisms may include the multi-leaf trait, the leaf percentage, or by changing the lignification of the cell wall. Fall dormancy has a fairly large effect over a wide range of fall dormancy levels, when those cultivars can be grown in a region (Fig. 2.8). More dormant cultivars normally are somewhat lower yielding, but because of their slower growth characteristics, are typically higher in quality when harvested at the same schedule as non-dormant lines. The introduction of HarvXtra varieties in 2014, as well as the HiGest varieties provided an important opportunity for breaking or at least challenging the yield-quality tradeoff in alfalfa (Grev et al. 2017). To be more precise, reduced lignin trait in alfalfa does not completely negate the yield-quality tradeoff as impacted by harvest schedules. However, it has shown to reduce the decline in digestibility normally observed with plant maturity by reducing the rate of lignification of the stem cell wall while maintaining high yields with later harvests (Grev et al. 2017). This provides the opportunity to provide higher yields while maintaining quality at levels similar to those seen at shorter harvest schedules (Sulc et al. 2016), results which have been confirmed also with non-dormant cultivars (D. H. Putnam, unpublished).

2.4.7 Economic Returns

Balancing this yield-quality-persistence challenge as related to harvest schedules is not easily resolved, and most growers choose a 'middle path' (e.g., 28-day harvests or late bud to 5% bloom), which often compromises yield and quality as well as persistence. However, this may not be the best solution, since sometimes longer or shorter schedules may be superior to a regular harvest schedule (Fig. 2.9). Economic returns are a major consideration, since both yield and

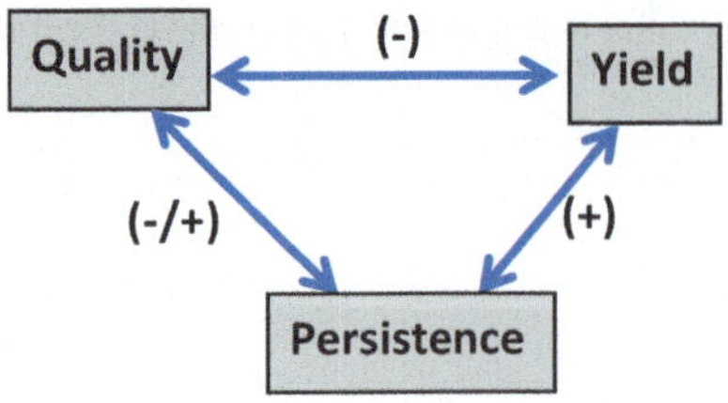

Fig. 2.7 Yield and quality as affected by harvest schedule is nearly always negatively correlated, but harvest schedules have a large influence on persistence and long-term yields

quality impact gross returns. In an analysis of economic returns over a 10-year period in comparisons of 3 versus 4 cut systems or a 6 versus 8 cut system (from 3-year research data) indicated that (1) yields nearly always were more important economically than quality, (2) Delayed harvests (for example, delaying second cutting in a 3-cut system) improved yields and economic returns (Fig. 2.9). In the delayed 2nd cutting treatment in the three-cut system, both forage quality and yields were improved (Orloff and Putnam 2010). When the improved price for high quality was considered, yields were still the most important economic factor (Orloff and Putnam 2006; Orloff and Putnam 2010). A 'staggered' approach to harvest schedule is suggested, allow individual cuts to 'go long' to improve yields and to replenish root reserves that improve stand persistence (Putnam et al. 2005; Marble 1990; Orloff and Putnam 2010). However, in terms of breeding goals, yield should take precedence over quality (Fig. 2.9).

2.4.8 Agronomic Effects on Yield and Quality

When considering the dramatic yield gap observed in essentially all alfalfa growing regions (Fig. 2.3), it is likely that much of this difference is due to a combination of environmental limitations and agronomic practices, not genetic factors, although varieties must be adapted to a region. While the yield-quality-persistence quandary is a key issue as affected by variety and harvest schedule, there are a range of practices that, in addition to this, can improve yields to allow the crop to reach its genetic potential. Individually, each of these factors can account for large portions of the yield gap observed in alfalfa, and varieties must be chosen in the context of a package of practices aimed at improving yield, quality, and persistence. Several of the most important factors are as follows:

2.4.8.1 Stand Establishment Method

Stand establishment practices to achieve a very high yielding, deep-rooted crop that allows high yields not only in the first year but in multiple years of production are a key factor, and a limitation when those methods are inadequate. Key among these are avoidance of auto-toxicity and diseases by rotating with non-alfalfa crops, deep tillage, land preparation, and proper seedling protection from diseases and weeds, seeding timing, and seeding depth (Putnam 2012; Undersander et al. 2017). Alfalfa is a relatively weak seedling and must be protected during early growth periods. Fall planting is preferred if moisture is present since fall growth encourages root development versus shoot development (Teuber et al. 1998). Time of seeding has been shown to improve yields by 20–25% in year one, and significantly in year two.

2.4.8.2 Traffic and Efficiency of Harvest

Since alfalfa is harvested multiple times per year, wheel traffic of implements is a more important factor for alfalfa than with other crops that are harvested once. Three to four operations per harvest (swathing, raking, baling, and bale pickup) are typically required, in addition to any fertilizer or pest management trips. Thus in 3–4 cut systems, 12–16 implement trips may be required, and in 8 cut regions, well over 30 trips over the field are required. The mechanism for damage is a combination of soil compaction and crushing of regrowth which slows DM accumulation. Traffic is a major issue limiting yields and is thought to affect over 60% of fields (Sheesley et al. 1974; Schmierer et al. 2004; Undersander et al. 2011). Engineering opportunities for yield improvement by confining wheel traffic, reducing compaction, as well as efficient and rapid dry-

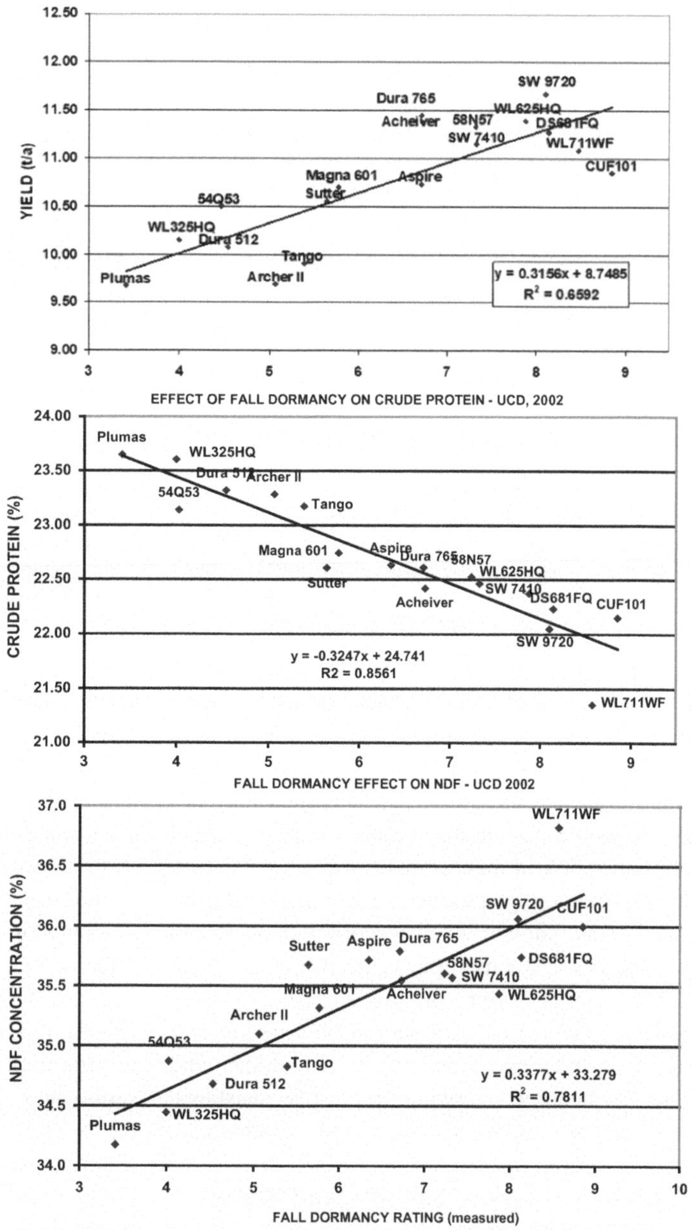

Fig. 2.8 Relationship between fall dormancy and yield, neutral detergent fiber and crude protein in a wide range of alfalfa varieties, spanning fall dormancy 3 through 9. Data are average of three harvest schedules, 3 replications, Davis, CA (2002)

down methods to hasten harvest efficiency are important yield- and quality-improving strategies for the future (Coblentz 2020).

2.4.8.3 Weed, Insect, and Disease Management

Weed competition and a wide range of diseases and insects have been shown to impact alfalfa yield and quality (Samac et al. 2016; UC-IPM 2021). Diseases and nematodes are primarily managed through cultivar resistance and farm practices such as equipment sanitation and drainage (Frate and Davis 2007), with somewhat limited chemical options, with the exception of seedling diseases, where the use of fungicide seed treatments is common. Resistance levels of existing varieties are published each year (National Alfalfa and Forage Alliance 2021).

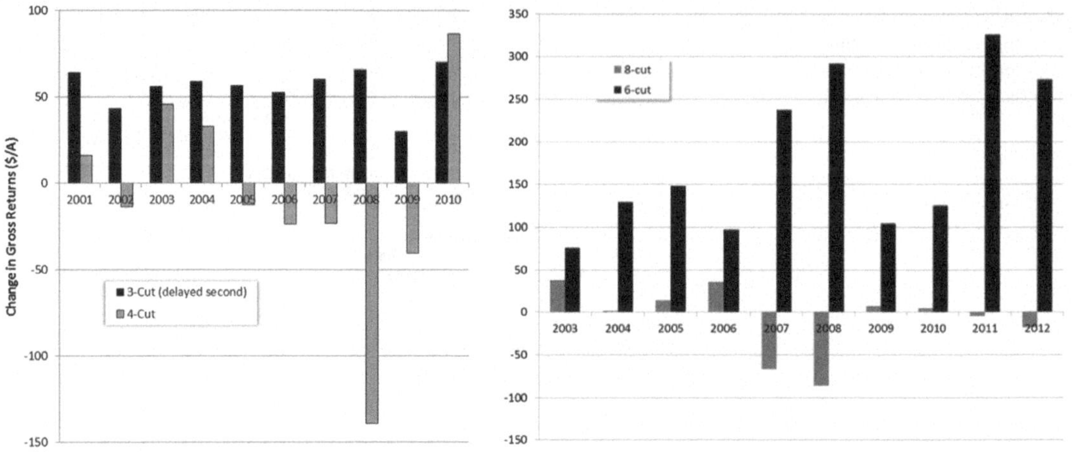

Fig. 2.9 Economic returns modelled from field studies measured yield and quality differences due to harvest schedule, Tulelake, (3–4 cut), and Davis (6–8 cut), 3 year yield and quality data. At Tulelake (left), a delayed, high yielding second cut was compared with standard 3-cut system and a 4-cut system. At Davis (right), a 35-day system (6 cuts) was compared with 24 days (8 cuts). Standard (3 cuts and 7 cuts, respectively, Tulelake and Davis) is represented by the horizontal (0) line. Quality price differences were modelled in each individual year from USDA-Market News Reports (Orloff and Putnam 2010)

Methods to document disease, nematode, and insect resistance are almost entirely based upon seedling assays (North American Alfalfa Improvement Conference 2004), which may not be adequate for more mature plants or field resistance. Weed management is a key yield- and quality-impacting practice, particularly during stand establishment when weed competition can impact yields for multiple years (Canevari et al. 2007). Development of much higher variety resistance characteristics should be a major goal of plant breeders, since when nematodes and diseases become overwhelming, 50% resistance (considered highly resistant) may be inadequate.

2.4.8.4 Irrigation, Salinity, and Water Management

Approximately 50% of US alfalfa production is irrigated (USDA-NASS 2020). It is well known that alfalfa yields are linearly related to crop evapotranspiration (Bauder et al. 1978; Hanson et al. 2007), which can be supplied through either rainfall or irrigation. In water-scarce environments, the introduction of irrigation significantly improves yields, but not all irrigation management systems adequately deliver water in a timely fashion. Improved overhead irrigation, drip irrigation, and improved surface irrigation practices offer the opportunity to improve yields by 20–30% in many long- and short-seasoned environments (Bali et al. 2017; Peters and Flury 2017; Putnam et al. 2017). Although alfalfa has sometimes been described as saline sensitive, evidence of high levels of salinity (up to 6 dS/m) tolerance have been reported (Cornacchione and Suarez 2014) and confirmed in field studies (Putnam et al. 2017; Putnam unpublished), suggesting that alfalfa is well suited to acceptance of saline and degraded irrigation water. Evidence for variety tolerance to salinity is apparent (Sandhu et al. 2017), but variety x salinity interactions are not always consistent in the field (Putnam, unpublished). Deficit irrigation strategies are important for alfalfa in given uncertainties in future water supplies (Orloff et al. 2005; Cabot et al. 2017); alfalfa has a major biological advantage during drought due to its deep-rooted characteristics and ability to regrow after extended stress (Putnam 2015, Cabot et al. 2017). Breeding goals for improved resilience to face a future of reduced water supplies and acceptance of saline and degraded waters are a critical need.

2.5 Important Goals for Plant Breeders

Key plant strategies for improvement include breaking the yield-quality tradeoff, improving late-season yields, increasing the juvenility (vegetative) phase of growth by delayed flowering, improving pest-resistance characteristics, improving rooting characteristics for resistance to traffic, stand longevity and stand persistence. In many environments, winter-hardiness is inadequate. Heat tolerance may be an important trait in many regions due to late-summer yield slump. A more fundamental understanding of the crop response to day length, temperature, and growth, yield, and quality is needed. Additionally, alfalfa has an important role in crop rotations, in mitigating climate change through stable non-tilled leguminous cropping systems, as well as providing significant soil, wildlife, and ecological benefits. Varieties are needed which push the yield ceiling while modifying harvesting and irrigation technologies to maximize yields. Breeders and geneticists should adjust breeding goals which are fully in line with innovative high-yielding agronomic practices such as improved harvesting schedules and harvesting methods, irrigation methods, and other agronomic techniques that will enable very high-yielding alfalfa systems in the future.

References

Baron VS, Belanger G (2020) Chapter 8. Climate, climate-change and forage adaptation. In: Moore K, Collins M, Nelson CJ, Redfearn DD (eds) Forages, Volume I2: The science of grassland agriculture, 7th Edition. Wiley Blackwell Publishing, UK. www.wiley.com

Bauder JW, Bauer A, Ramirez JM, Cassel DK (1978) Alfalfa water use and production on dryland and irrigated sandy loam. Agron J 72(6):1017–1020

Brauder SM, Volenec JJ (2020) Mineral nutrient acquisition and metabolism. Chapter 5. In: Moore K, Collins M, Nelson CJ, Redfearn DD (eds) Forages, the science of grassland agriculture, Volume II. Wiley Blackwell Publishing, UK. www.wiley.com

Brummer C, Putnam DH (2018) The alfalfa yield gap: What is holding us back? Hay & Forage Grower Magazine. https://hayandforage.com/article-1856-The-alfalfa-yield-gap-What%E2%80%99s-holding-us-back.html

Cabot P, Brummer J, Gautam S, Jones L, Hansen N (2017) Benefits and impacts of partial season irrigation on alfalfa production. In: Proceedings, 47th California Alfalfa & Grain Symposium, November 28-November 30 2017, Reno, Nevada, Department of Plant Sciences, University of California, Davis, CA 95616 https://alfalfa.ucdavis.edu/+symposium/proceedings/index.aspx

Coblentz WK (2020) Post-harvest physiology. In: Moore K, Collins M, Nelson CJ, Redfearn DD (eds.), Forages, the science of grassland agriculture Volume II. Wiley Blackwell Publishing, UK. www.wiley.com

Cornacchione MV, Suarez DL (2014) Emergence, forage production, and ion relations of alfalfa in response to saline waters. Crop Sci 55:444–457. https://doi.org/10.2135/cropsci2014.01.0062 Crop Science Society of America, Madison, WI

Evenson PD (1979) Optimum crown temperatures for maximum alfalfa growth. Agron J 71:798–800

Frate C, Davis RM (2007) Alfalfa diseases and management. In: Putnam DH, Summers CG (eds) Irrigated alfalfa management for mediterranean and desert zones. University of California Division of Agriculture and Natural Resources. https://alfalfa.ucdavis.edu/IrrigatedAlfalfa/

Grev AM, Wells MS, Samac DA, Martinson KL, Sheaffer CC (2017) Forage accumulation and nutritive value of reduced lignin and reference alfalfa cultivars. Agron J 109:2749–2761. https://acsess.onlinelibrary.wiley.com/doi/pdf/10.2134/agronj2017.04.0237

Hanson B, Bali K, Sanden BL (2007) Irrigating alfalfa in arid regions. In: Putnam and Summers (eds.), Irrigated alfalfa management for mediterranean and desert zones. University of California Division of Agriculture and Natural Resources. https://alfalfa.ucdavis.edu/IrrigatedAlfalfa/

Heichel GH, Delaney RH, Cralle HT(1988) Carbon assimilation, partitioning, and utilization. In: Hanson AA, Barnes DK, Hill Jr. RR (eds), Alfalfa and alfalfa improvement. Volume 29. Hanson, Publication Number 29. American Society of Agronomy, Madison, WI

Horrocks RD, Vallentine JF (1999) Chapter 13. Field harvesting hay. Harvested forages. Academic Press, San Diego, CA. https://www.elsevier.com/books/harvested-forages/horrocks/978–0-12-356255-5

Kalu BA, Fick GW (1981) Quantifying morphological stage of maturity as a predictor of alfalfa herbage quality. Crop Sci 23:267–271

Leach CJ (1971) The relation between shoot growth and temperature. Aust J Agric Res 22:49–59

Marble V (1974) How cutting schedules and varieties affect yield, quality and stand life. In: Proceedings, California Alfalfa Symposium, 4–5 December 1974, Fresno, California, Department of Agronomy and Range Science Extension, University of California, Davis, CA

MarbleV (1990) Factors to optimize alfalfa production in the 1990s. 20th California Alfalfa Symposium, 6–7 December 1990, Visalia, California, Department of Agronomy and Range Science Extension, University of California, Davis, CA

Mitchell RB, Redfearn DD, Moore KJ (2020) Chapter 7. Growth and development. In: Moore K, Collins M, Nelson CJ, Redfearn DD (eds), Forages, the science of grassland agriculture, Volume II. Wiley Blackwell Publishing, UK. www.wiley.com

Mueller S, Teuber L (2007) Alfalfa growth and development. In Putnam DH, Summers CG (eds), Irrigated alfalfa management for mediterranean and desert zones. University of California Division of Agriculture and Natural Resources. https://alfalfa.ucdavis.edu/IrrigatedAlfalfa/

Moore KJ, Curtiss CF, Fales SL (2020) Chapter 39. Factors affecting forage quality. In: Moore K, Collins M, Nelson CJ, Redfearn DD (eds), Forages, the science of grassland agriculture Volume II. Wiley Blackwell Publishing, UK. www.wiley.com

Myer RD, Marcum DB, Orloff SB, Schmierer JL (2007) Alfalfa fertilization strategies. In: Putnam DH, Summers CG (eds.), Irrigated alfalfa management for mediterranean and desert zones. University of California Division of Agriculture and Natural Resources. https://alfalfa.ucdavis.edu/IrrigatedAlfalfa/

National Alfalfa and Forage Alliance (2021) Alfalfa variety ratings: winter survival, fall dormancy and pest resistance ratings for alfalfa varieties. 2021 Edition. https://www.alfalfa.org/varietyLeaflet.php

Nelson CJ, Moser LE (1994) Chapter 3. Plant factors affecting forage quality. In: Fahey GC (ed) Forage quality, evaluation and utilization. American Society of Agronomy, Madison, WI

North American Alfalfa Improvement Conference (2004) Standard tests to characterize alfalfa cultivars Third Edition. www.alfalfa.org

Nielson RL (2020) Historical grain yields in the US. Purdue University. Corn News Network. https://www.agry.purdue.edu/ext/corn/news/timeless/YieldTrends.html

Orloff S (2007) Choosing appropriate sites for alfalfa production. In: Putnam DH, Summers CG (eds.), Irrigated alfalfa management for mediterranean and desert zones. University of California Division of Agriculture and Natural Resources. https://alfalfa.ucdavis.edu/IrrigatedAlfalfa/

Orloff S,Putnam DH, Hanson B, Carlson H (2005) Implications of deficit irrigation management of alfalfa. In: Proceedings, California alfalfa and forage symposium, 12–14 December, 2005, Visalia, CA, UC Cooperative Extension, Plant Sciences Department, University of California, Davis 95616. https://alfalfa.ucdavis.edu/+symposium/2019/index.aspx

Orloff SB, Putnam DH (2006) Cutting schedules to maximize returns. In: 36th Western Alfalfa & Forage Symposium, 11–13 December 2006, Reno, Nevada, Department of Agronomy and Range Science Extension, University of California, Davis, CA 95616

Orloff SB, Putnam DH (2010) Adjusting alfalfa cutting schedules for economic conditions. In: 40th California Alfalfa & Forage Symposium, 31 November–1–2 December 2010, Visalia, California, Department of Agronomy and Range Science, University of California, Davis, CA 95616

Pearson CJ, Hunt LA (1972) Effects of temperature on primary growth and regrowth of alfalfa. Can J Plant Sci 52:1017–1027

Putnam DH, Russelle M, Orloff S, Kuhn J, Fitzhugh L, Godfrey L, Kiess A, Long R(2001) Alfalfa, wildlife and the environment. The importance and benefits of alfalfa in the 21st century. California Alfalfa & Forage Association, Sacramento, CA. https://alfalfa.ucdavis.edu/+producing/index.aspx?cat=Environment&cat=Environment

Putnam DH, Orloff SB, Teuber LR (2005) Strategies for balancing quality and yield in alfalfa using cutting schedules and varieties. In: Proceedings, 35th California Alfalfa & Forage Symposium, Visalia, CA 12–14 December, 2005. UCCooperative Extension, Agronomy Research and Extension Center, Plant Sciences Department, University of California, Davis 95616. https://alfalfa.ucdavis.edu/+symposium/2019/index.aspx

Putnam DH, and Ottman M (2013) Your alfalfa crop got the blues? It may be 'Summer Slump'. University of California Cooperative Extension Blog. https://ucanr.edu/blogs/blogcore/postdetail.cfm?postnum=17721

Putnam DH (2015) Why alfalfa is the best crop to have in a drought. Alfalfa and forage news. University of California Cooperative Extension Blog. https://ucanr.edu/blogs/blogcore/postdetail.cfm?postnum=17721

Putnam DH, Benes S, Galdi G, Hutmacher B, Grattan S (2017) Alfalfa (*Medicago sativa* L.) is tolerant to higher levels of salinity than previous guidelines indicated: Implications of field and greenhouse studies. In: Proceedings, 19th EGU General Assembly, 23–28 April, 2017, Vienna, Austria. https://ui.adsabs.harvard.edu/abs/2017EGUGA..1918266P/abstract

Russelle MP (2014) The alfalfa yield gap: a review of the evidence. Forage and Grazinglands. https://doi.org/10.1094/fg-2013-0002-rv https://acsess.onlinelibrary.wiley.com/doi/abs/10.1094/FG-2013-0002-RV

Sandhu D, Cornacchione M, Ferreira J et al. (2017) Variable salinity responses of 12 alfalfa genotypes and comparative expression analyses of salt-response genes. Sci Rep 7, 42958. https://doi.org/10.1038/srep42958

Samac DA, Rhodes LH, Lamp WO(2016) Compendium of alfalfa diseases and pests, 3rd edn. https://doi.org/10.1094/9780890544488 https://apsjournals.apsnet.org/doi/book/10.1094/9780890544488

Schmierer J, Putnam DH, Undersander D, Liu J, Meister H (2004) Wheel traffic in alfalfa–What do we know? What can we do about it? In: Proceedings, National Alfalfa Symposium, 13–15 December, 2004, San Diego, CA, UC Cooperative Extension, University of California, Davis 95616. https://alfalfa.ucdavis.edu/+symposium/2019/index.aspx

Sheesley R, Grimes DW, McClell WD (1974) Influence of wheel on yield and stand of alfalfa. California Agriculture, October, 1974. http://hilgardia.ucanr.edu/Abstract/?a=ca.v028n10p6

Sulc M, Parker A, Albrecht K, Casida K, Hall M, Min DH, Orloff S, Undersander D, XuX (2016) Low lignin alfalfa: Wide area field test results. In: Proceedings, 2016 California Alfalfa and Forage Symposium, Reno, NV, Nov 29 Dec 1, 2016. UC Cooperative Extension, Plant Sciences Department, University of California, Davis, CA 95616. (See http://alfalfa.ucdavis.edu for this and other alfalfa conference Proceedings)

Teuber L, Brick MA (1988) Morphology and anatomy. In: Alfalfa and Alfalfa Improvement. Hanson, Barnes, and Hill, eds.) Publication Number 29. American Society of Agronomy, Madison, WI

Teuber L, Jernstedt J, Foord K (1988) Alfalfa growth and development. In: The 18th California Alfalfa Symposium, 7–8 December 1988, Modesto, California, Department of Agronomy and Range Science, University of California, Davis, CA 95616 https://alfalfa.ucdavis.edu/+symposium/2019/index.aspx

UC-IPM (2021) University of California integrated pest management guidelines for alfalfa. http://ipm.ucanr.edu/PMG/selectnewpest.alfalfa-hay.html

Undersander D, Cosgrove D, Cullen E, Grau C, Rice M, Renz M, Sheaffer C, Shewmaker G, Sulc M (2011) Alfalfa management guide. American Society of Agronomy, Madison WI. ISBN 978-0-89118-179-8

Undersander DJ (2008) Minimizing wheel traffic damage to alfalfa. Univ Wis Ext Focus Forage 10(3):1

USDA-NASS. National agriculture statistics service, US Department of Agriculture. https://www.nass.usda.gov/

The Origin, Evolution, and Genetic Diversity of Alfalfa

3

Zan Wang and Muhammet Şakiroğlu

Abstract

Alfalfa is one of the earliest domesticated forage crops that played an important economic and cultural role in the history. Cultivated alfalfa (*Medicago sativa* ssp. *sativa*) is a perennial, outcrossing, and autotetraploid ($2n = 4x = 32$) forage crop selected from *M. sativa* species complex that includes both diploid and tetraploid interfertile subspecies. There are two putative centers of diversity and origin suggested for alfalfa, Asia Minor/Caucasia (a region that contains Northwestern Iran, highlands of Armenia, Georgia, and Eastern Turkey) and Central Asia (Kazakhstan, Uzbekistan, and highland of Afghanistan). As being member of a relatively comprehensive complex, the evolution of alfalfa and the allied taxa can be traced using morphological traits as well as the cytogenetic state of ploidy and subsequent hybridization among the taxa. As the most prominent morphological traits to deduce the evolution of alfalfa, flower color and pod shape were used extensively. The purple flower color occurs in a limited number of taxa and was regarded as a rather recent evolutionary event and linked to alterations in pollinators' behaviors while the evolution of the coiled pods was reported to be the result of a selective evolution in a restricted region around Caucasia to adapt to the seed chalcid (*Bruchophagus roddi*) infestation. Both of these events possibly resulted in genetic isolation and subspeciation. Similarly, autotetraploidy within the complex has been linked to the evolution of alfalfa and allied taxa. Despite the recent efforts to unveil genetic diversity and clustering pattern in cultivated and wild alfalfa, the population genomics studies that are aiming to tackle domestication history of alfalfa and the effect of domestication and subsequent selection on the alfalfa genome lagged behind the other major crops. Recently two tetraploid alfalfa accessions and a diploid accession were sequenced, assembled, and annotated providing a great access to study population genomics of alfalfa and other members of the complex.

Z. Wang (✉)
College of Grassland Science and Technology, China Agricultural University, Beijing 100193, China
e-mail: zanwang@cau.edu.cn

M. Şakiroğlu
Department of Bioengineering, AlparslanTürkeş Science and Technology University, Adana 01250, Turkey
e-mail: msakiroglu@atu.edu.tr

L.-X. Yu and C. Kole (eds.), *The Alfalfa Genome*, Compendium of Plant Genomes,
https://doi.org/10.1007/978-3-030-74466-3_3

3.1 Introduction

As one of the earliest domesticated forage crops, alfalfa has played an important economic and cultural role in the history of human development. It is not only one of the most important forage legume crops for hay and silage production worldwide for livestock, but also an integral part of the spread of human agricultural civilization. Alfalfa (*Medicago sativa* ssp. *sativa*) is a perennial, outcrossing, and autotetraploid ($2n = 4x = 32$) forage crop. It belongs to a species complex (*M. sativa* species complex) that includes diploid and tetraploid interfertile subspecies (Lesins and Lesins 1979). Alfalfa demonstrates a wide phenotypic variation pattern that has been complicated by hybridization, polyploidy, and domestication. The origin, evolution, and genetic diversity of alfalfa are influenced by human activities, and its rich genetic resources are the result of long-term natural and artificial selection to adapt to different ecological environments. Therefore, studying the origin, evolution and genetic diversity of alfalfa is not only helpful to understand the history of its breed formation, but also helpful to reflect the history of human civilization transmission and multi-culture formation from the perspective of livestock domestication. In this paper, the origin, evolution, and genetic diversity research in alfalfa was summarized in order to provide a reference for the collection and utilization of alfalfa genetic resources and the enhancing alfalfa breeding efforts.

3.2 The Origin of Alfalfa

Alfalfa is the oldest plant grown exclusively for forage (Michaud et al. 1988) and its cultivation predates the recorded history which limits the inference to deduce the exact domestication centers (Bolton et al. 1972; Small 2010). Bolton et al. (1972) noted that brick tablets of Alacahöyük Ruins in Turkey are the oldest known record of the alfalfa cultivation (1400-1200 BC). The Babylonian texts in 700 BC were indicated to be another earlier record of alfalfa cultivation (Hendry 1923).

A number of different origins were proposed for alfalfa including Eastern Anatolia (Modern day Turkey), Iran, Armenia, Afghanistan, Central Asia, and Jammu and Kashmir (Lesins and Lesins 1979). Some of the regions were suggested based on the novel source of variations. For instance, Central Asia was suggested as the second potential origin for alfalfa for maintaining novel sources of resistance to bacterial wilt and blue alfalfa aphid (Small 2010). Therefore, the two putative centers of diversity and origin for alfalfa are broadly described as Asia Minor/Caucasia (a region that contains Northwestern Iran, highlands of Armenia, Georgia, and Eastern Turkey) and Central Asia (Kazakhstan, Uzbekistan, and highland of Afghanistan) (Lesins and Lesins 1979; Small and Brookes 1984b; Small 2010; Şakiroğlu and İlhan 2021). Having been extensively used as a horse and cattle feed during the invasions, alfalfa was introduced to Greece and wider Middle East from its proposed center(s) of origin (Asia Minor, and Central Asia) and subsequently spread to Rome (Michaud et al. 1988; Small 2010) (Fig. 3.1).

After discovery of the continent and the subsequent Spanish invasion of the South America, alfalfa was introduced to the Americas. The introduction of alfalfa to North America and Australia is considerably late and both the introduced material and the routes are traceable (Barnes 1977; Şakiroğlu and Brummer 2007). The introduction to North America was through the southwestern states from Mexico and subsequently extended to other parts of the USA (Bolton et al. 1972).

Since the alfalfa cultivation in North America is recent, both the early germplasm and the sources are recorded and analyzed extensively. A total of nine distinct alfalfa germplasm sources were introduced to the USA between 1850 and 1947. These nine germplasm sources were denoted as Falcata, Varia, Ladak, Turkistan, Flemish, Chilean, Peruvian, Indian, and African alfalfa germplasm sources (Barnes 1977). Most

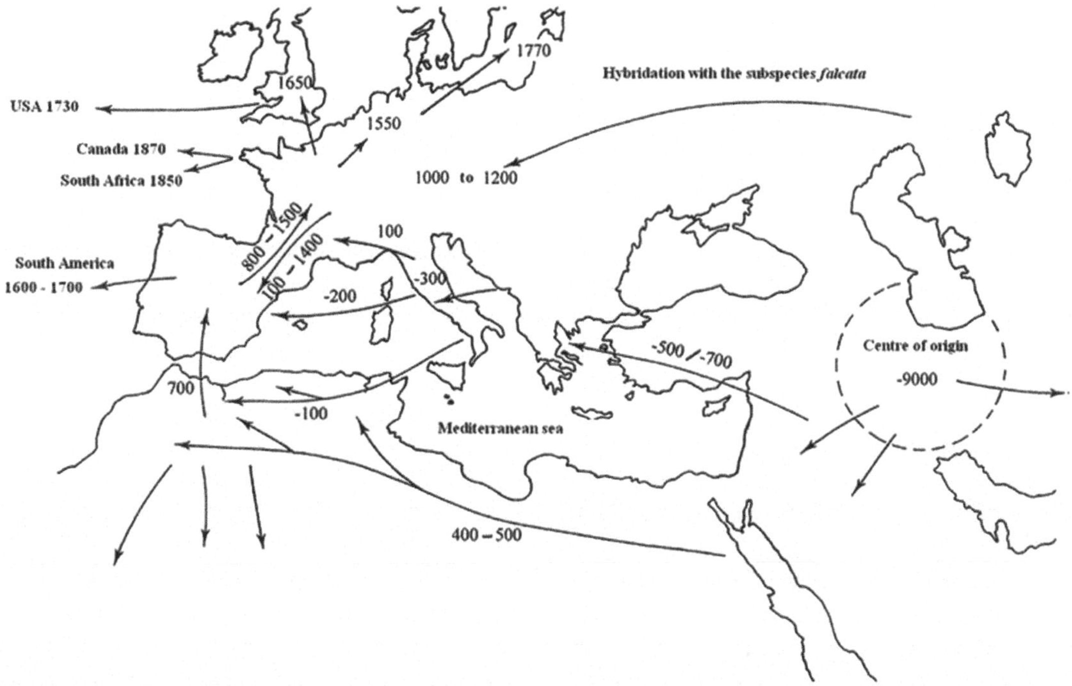

Fig. 3.1 Alfalfa historical distribution to Eurasia and subsequently to new world from its putative center of origin (Prosperi et al. 2014)

of genetic diversity currently maintained in the cultivar development programs in the USA is attributed to these nine germplasm sources. However, these nine sources were gradually intermixed during the breeding efforts. Prior to 1930, each cultivar was developed from a single germplasm source while two or three sources were combined to develop cultivars between 1941 and 1960. In the period of 1961 to 1970, three or four of the germplasm sources were combined in cultivar development process. The cultivars released after 1971 typically include genetic variation from all nine germplasm sources (Barnes 1977).

Given the contemporary advancements in genomic analyses and tools at the disposal of the researchers, domestication patterns and the consequences of domestication on the genomes of many crops have been investigated and largely unveiled (Badr and El-Shazly 2012). Unfortunately, population genomics studies that are aiming to tackle domestication history of alfalfa and the effect of domestication and subsequent selection on the alfalfa genome lagged behind the other major crops.

The earlier evaluation of the global scale natural and cultivated germplasm targeted domestication history of alfalfa using both nuclear and mtDNA variation. All these studies concluded the presence of at least two independent alfalfa domestication routes which corresponded to the putative centers of origin (Asia Minor and Central Asia). A total of 22 mitotypes were reported in 155 individual alfalfa genotypes. The strong geographical clustering within cultivated gene pool implied diverse paths of post-domestication diffusion. Moreover, a set of natural populations gathered from Spain was found to have formed an endemic wild pool that does not indicate similarities to either of the proposed centers of diversity and have introgressed partially into cultivated alfalfa. The studies also compared the effects of domestication on the diversity of alfalfa genome by comparing the DNA sequence diversity of neutral nuclear loci across the wild and domesticated pool and results indicated that the domesticated germplasm contained 31% less diversity than the wild populations. The results of the coalescent simulations revealed that the results are incongruent with the constant population size which in turn suggest a bottleneck during the alfalfa domestication (Muller et al. 2003, 2006; Prosperi et al. 2014).

The diploid *M. sativa* subsp. *caerulea* is the purple flowered taxa in the complex and regarded as the progenitor taxa for the tetraploid *M. sativa* subsp. *sativa* and remotely to the cultivated germplasm (Havananda et al. 2011; Li et al. 2012; Small 2010). Therefore, deducing the exact center of origin for subsp. *caerulea* has implications for the origin and evaluation of alfalfa. To achieve such a goal, a balanced set of 60 individual genotypes from 16 accessions were gathered from the two putative diversity centers and were subsequently genotyped using simple sequence repeat (SSR) markers. The extents of the genetic diversity present within the two putative diversity centers were compared to identify the likely center of origin. Results revealed that the Central Asian *caerulea* accessions and Caucasian/Asia Minor accessions formed separate clusters. In-depth analyses also demonstrated presence of a positive correlation between the overall genetic distance and geographical distance. The main genetic diversity parameters such as mean F_{ST} values, allele diversity, and heterozygosity were higher in Caucasian/Asia Minor accessions than the Central Asian accessions. The results comprehensively indicated that the Caucasian region is more likely the center of origin for subsp. *caerulea* (Şakiroğlu and Brummer 2013).

Although a general origin of cultivated alfalfa has been depicted and widely accepted, a clear evolutionary trajectory for the origin of the *Medicago sativa* species complex has not been set with the genomic tools. The earlier proposal provided based on the phenotype by Quiros and Bauchan (1988) needs to be investigated with the contemporary genomic tools. *M. sativa* subsp. *glomerata* was suggested the ancestral taxa and the origin of other member taxa. Nonetheless, no molecular study has tested these proposals leaving the origin of the complex inconclusive (Bena et al. 1998; Bena 2001).

Similarly, the large scale genomic evaluation of the center of diversity for all other member taxa in the *M. sativa* species complex along with the extent of genomic alterations during the domestication and subsequent selection along with the genomic regions under the selection and genome-wide patterns of the bottleneck have not been thoroughly investigated in alfalfa either.

3.3 The Evolution of Alfalfa

3.3.1 *The Role of Morphological Traits*

The evolution of alfalfa and the taxa in the *M. sativa* species complex can be traced using morphological traits as well as the cytogenetic state of ploidy and subsequent hybridization among the taxa (Havananda et al. 2011; İlhan et al. 2016; Lesins and Lesins 1979; Şakiroğlu and Brummer 2017; Şakiroğlu et al. 2010; Small 2010). The two most prominent morphological traits in the evolution of alfalfa are flower color and pod shape. The base flower color for the entire genus *Medicago* is yellow which is accompanied with the straight to falcate pods. The purple flower color occurs in a limited number of taxa which was regarded as a rather recent evolutionary event (Şakiroğlu and İlhan 2021). The deviations from the flower morphology were also linked to alterations in pollinators' behaviors that possibly resulted in genetic isolation and subspeciation (Quattrocchio et al. 1999). Being the predominant tool for the classification of the member taxa in pre-genomics era, biochemical composition and transmission genetics patterns of the flower color pigmentation was investigated and the yellow color of petals are attributed to flavonoids whereas the violet petals were found to be the products a few anthocyanin pigments (Barnes 1966; Small and Brookes 1984b; Steiner et al. 1992). Taxa with the purple flower color also have coiled fruits making the two traits interrelated. All the cultivated alfalfa varieties around the globe have purple flowers with a few yellow flowered cultivars such as

Wisfal and Anik. Yellow flowered taxa have contributed to the winter hardiness in alfalfa with a rather restricted use as direct cultivars. Thus historical and modern cultivated alfalfa have had purple flowers (Sakiroglu et al. 2011; Small 2010).

However, the inheritance of pod glandular hairs as the third morphological trait has a different pattern among taxa and could be separated from the other two traits. Evolution of the glandular hairs on the alfalfa pods was associated with the biology and geography *Rychius* species which are insect pests of alfalfa allied taxa (Small 1985). A similar pattern was also reported for the evolution of the coiled pods as a result of a selective evolution in a restricted region around Caucasia to adapt to the seed chalcid (*Bruchophagusroddi*) infestation (Small and Brookes 1982, 1984a). Presence of plant-pollinator (alfalfa leafcutter bee, *Megachilerotundata*) herbivorous coadaptation and coevolution mechanisms was also proposed that resulted in reproductive isolation and subspeciation (Small 1996).

3.3.2 *Polyploidy in Evolution of the Complex*

Another mechanism involved in the evolution of alfalfa and the members of complex is polyploidization (autotetraploidy) which has been long recognized in the evolution of the plants (Soltis et al. 2004). The role of polyploidy and subsequent hybridization in the species fitness and agronomic performance has been in the center of attention since the discovery of the polyploidy. Genome structure and nature of the polyploids have been extensively investigated owing to availability of the wide range of genetic tools (Chen and Birchler 2013).

Chromosome number variations in alfalfa were recognized as early as 1930 and the haploid chromosome numbers of $n = 8$ and $n = 16$ for different taxa were reported (Fryer 1930). Both number and morphology of chromosomes were investigated afterward and the autotetraploid segregation pattern was confirmed (Agarwal 1983; Clement and Lehman 1962; Falistocco et al. 1995; Falistocco 1987; Gillies 1968, 1970; Ledingham 1940; Stanford 1951).

Among the members of complex, the diploid subspecies ($2n = 2x = 16$) are *M. sativa* subsp. *glomerata* (yellow flowers, coiled fruits, glandular hairs), *M. sativa* subsp. 2x *falcata* (yellow flowers, straight fruits, no glandular hairs), *M. sativa* subsp. *caerulea* (purple flowers, coiled fruits, no glandular hairs), and natural hybrid between diploid subsp. *falcata* and subsp. caerulea, *M. sativa* subsp. × *hemicycla* (variegated flowers, semi-coiled fruits, no glandular hairs). The tetraploid subspecies ($2n = 4x = 32$) are *M. sativa* subsp. *glutinosa* (tetraploid analog of subsp. *glomerata* with yellow flowers, coiled fruits, glandular hairs), *M. sativa* subsp. *sativa* (tetraploid analog of subsp. *caerulea* with purple flowers, coiled fruits, no glandular hairs), *M. sativa* subsp. 4x *falcata* (yellow flowers, straight fruits, no glandular hairs), and M. sativa subsp. × varia (the tetraploid hybrid between subsp. *sativa* and subsp. *falcata* with variegated flowers, semi-coiled fruits, no glandular hairs) (Lesins and Lesins 1979; Quiros and Bauchan 1988; Small 2010; Stanford et al. 1972).

Given the fact that autopolyploidy arises from the within species via chromosome doublings, tetraploid taxa, and cytotypes in the complex are considered to have evolved from the whole genome duplications of the diploids (Gillies 1970; Gillies and Bingham 1971; Havananda et al. 2011). The mechanisms that have led to multiplication of the chromosomes were investigated in alfalfa and allies and presence of unreduced gametes largely in the form of eggs with the sporophytic chromosome number ($2n$ eggs) were reported. Genetic control of the absence of cytokinesis after restitutional meiosis in diploid alfalfa eggs was investigated and a total of five genes were proposed to control the production of unreduced gametes among the members of the complex (Barcaccia et al. 2000).

One of the major questions regarding the rise of autotetraploidy among the members of the complex is the origin of the tetraploidization event. Contemporary genomic tools have been utilized to address the question and to infer the evolutionary history of autotetraploidization events in wild alfalfa. A broad range of wild diploid and tetraploid alfalfa accessions from different subspecies were gathered and two chloroplast noncoding regions (rpl20 - rps12 and trnS - trnG spacers) used to construct phylogenies and haplotype networks (Havananda et al. 2011). The results revealed that the raise of tetraploids have different trajectories in different taxa. For instance, a single polyploidization event was detected for subsp. *sativa* from the diploid counterpart subsp. *caerulea* whereas tetraploid cytotypes of subsp. *falcata* were found to have multiple different chloroplast haplotypes that denoted multiple independent polyploidization events (Havananda et al. 2011). The origins of tetraploids were also investigated among the members of complex using comparative cytogenetic analysis across ploidy levels utilizing the fluorescence in situ hybridization (FISH) technique. The results revealed that the FISH repeats observed in diploid subsp. *caerulea* chromosomes indicated high affinity to those in tetraploid subsp. *sativa* suggesting that subsp. *sativa* was originated from the diploid subsp. *caerulea.* Very distinct FISH patterns between diploid and tetraploid subsp. *falcata* implies a complex pattern of evolution and multiple origin of autopolyploidy (Yu et al. 2017).

3.3.3 Hybridization Among Taxa as a Mean of Hybrid Subspecies Formation

After differentiation, isolation, and subspeciation in the wild alfalfa germplasm, extensive hybridization among taxa—particularly in the sympatric zones—have been reported both within and across ploidy levels (Gillies and Bingham 1971; Lesins and Gillies 1972). These hybridizations can be observed at the morphological levels as well as at the genome level. For instance, there are hybrids both at diploid and tetraploid levels where both flower color and pod shape indicate and intermediate patterns. The diploid hybrid subsp. × *hemicycla* and the tetraploid hybrid subsp. × *varia* have variegated flowers (intermediate of yellow and purple flowers) and semi-coiled fruits (intermediate of falcate pods and coiled pods) (Lesins and Lesins 1979; Small 2010). This pattern of subspecies formation has been observed in the genomics studies conducted among the members of the complex (İlhan et al. 2016; Şakiroğlu and Brummer 2013, 2017; Şakiroğlu et al. 2010).

The diploid hybrid taxon subsp. × *hemicycla* was found to have formed a distinct genomic diversity and clustering pattern from other subspecies over the course of evolution via reproductive isolation mechanisms such as flowering time, pollinator preferences, and ploidy barrier (Şakiroğlu et al. 2010). The degree of demarcation at the tetraploid hybrid subsp. × *varia* compare to other tetraploid taxa was found to be lower than that of diploid subsp. *hemicycla* (İlhan et al. 2016).

3.3.4 Self-incompatibility

Self-incompatibility or self-sterility is defined as the mechanism that imposes allogamy in plants through failure of seed production after self-pollination despite the presence of functional male and female gametes (Vogel and Lamb 2007). The nature of the mating systems in plants has been indicated to have a profound impact on genome evolution, including molecular evolutionary rates, base composition, genomic conflict, and possibly genome size (Glemin and Galtier 2012). Despite the fact that a wealth of research has been devoted to understand the genome evolution in polyploids, a relatively restricted attention has been allocated to the role of different mating systems on genomes. One of the reasons suggested for such a difference is that the effects of polyploidy on the evolution are immediate and experimentally observable while evolutionary consequences of variation in the mating systems can extend over evolutionary timescales (Wright et al. 2008). Alfalfa and allies have outbreeding mating systems that are associated with inbreeding depression when selfed, preventing the development of inbred seeds and thus inbreed lines. Alfalfa manifests a severe inbreeding depression mainly because of the loss of dominance or breakage of epistatic interactions (Li and Brummer 2009; Şakiroğlu and Brummer 2007). This has contributed to the wide adaptation of alfalfa and relatives to a broader ecogeography and the massive genetic diversity.

The evidence from the previous studies showed that single or multiple genome duplications has resulted the current autotetraploid taxa. The recurrent hybridizations in sympatric zones of Eurasia and subsequent isolation resulted the hybrid taxa. The fitness advantages of tetraploids such as elevated resistance to pathogens and pests, increased tolerance to abiotic factors have allowed a greater dispersion of tetraploids. These advantages were recognized by humans and subsequent global anthropogenic distribution have been achieved (Levin 1983; Soltis et al. 2004; Vyšniauskienė et al. 2013).

3.4 Genetic Diversity of Alfalfa

Genetic diversity is reflected in the genomic differences between different populations and individuals, which is the basis for biological evolution and adaptation to the environment. The richer the genetic diversity within a variety, the stronger the ability of the species to adapt to environmental changes and the greater the potential of cross breeding. The research on the genetic diversity of alfalfa is not only of great significance to understand the origin, evolution, migration, variety differentiation, and improvement of alfalfa, but also can provide the basis for the protection of biodiversity and sustainable utilization of biological resources. With the advancements in the various fields of life science including biochemistry, data management, and computer science, the research methods of genetic diversity assessment have shifted from traditional morphological, chromosomal, and biochemical markers to molecular genetic markers and genomics. Molecular markers, by contrast, can directly reflect the differences of genomic DNA between biological populations or individuals, and their development has gone through three main stages: the first generation of molecular markers including restriction fragment length polymorphism (RFLP) markers, etc. The second generation includes microsatellite markers (SSR), amplified fragment length polymorphism (AFLP), inter-simple sequence repeat (ISSR), sequence-related amplified polymorphism (SRAP), and so on. The third generation of including single nucleotide polymorphism (SNP) markers.

3.4.1 Genetic Diversity of Tetraploid Alfalfa

The genetic diversity of tetraploid alfalfa was well assayed by different molecular marker systems (Bhandari et al. 2011; Flajoulot et al. 2005). SSRs have been one of the most useful molecular marker systems for their co-dominant inheritance, abundance in genomes, high reproducibility, and have been extensively used in tetraploid alfalfa molecular genetic research in the past two decades. Diwan et al. (1997) have been the first to develop SSR markers in *Medicago*. They have shown how SSR can be used to describe genetic diversity and to analyze the genetic relationships among genotypes in tetraploid alfalfa. Since then, abundant SSR markers were developed by expressed sequence tag (EST) sequences (Julier et al. 2003; Wang et al. 2013, 2014) and bacterial artificial chromosome (He et al. 2009) from *M. truncatula* and *M. sativa* subsp. *sativa*, respectively.

A comparative analysis of genetic characteristics was performed at the chromosome level in cultivated tetraploid alfalfa (Qiang et al. 2015). Different chromosomes exhibited different genetic variation patterns revealing by the mean allele number per chromosome, expected heterozygosity (He), and the polymorphism information content (PIC) (Fig. 3.2). As shown in Fig. 3.2, three genetic parameters exhibited a consistent variation trend among different individual chromosomes, i.e., chromosomes 2 and 3 had the highest genetic diversity although there were no significant differences among the different chromosomes.

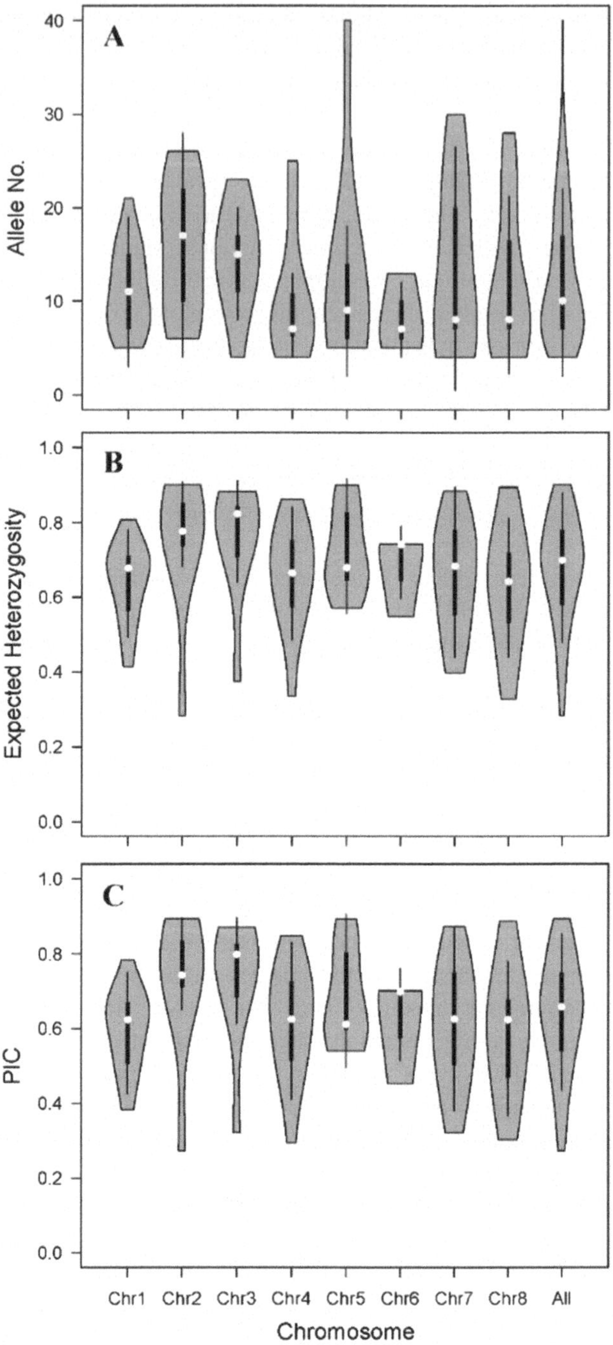

Fig. 3.2 Distribution of average allele numbers (**A**), Expected heterozygosity (**B**), and PIC values (**C**) across eight chromosomes in tetraploid alfalfa. Violin plots show density distribution of PIC values, horizontal bar indicates average value, median is shown as white circle, top and bottom of vertical bar represent the first and third quartile (Qiang et al. 2015)

Because of the mechanism of distribution, migration, and diffusion in history, alfalfa gradually formed a population adapted to local ecological environment. The genetic diversity of alfalfa also showed different regional distribution characteristics. Muller et al. (2003) reported the mtDNA diversity of the tetraploid alfalfa. The Asiatic gene pool indicated larger mtDNA diversity than Middle East, and other gene pools because Asiatic pool possesses three particular mitotypes, which were only detected in the Xinjiang province of China. Interestingly, the SSR markers-based analyses indicated that the alfalfa germplasm from China was also found to have relatively higher genetic diversity than all other places, while no significant difference was observed among different regions (Qiang et al. 2015).

Population structure analysis enables to understand genetic diversity in a given collection. Population structure of worldwide collection of tetraploid cultivated alfalfa germplasm based on 44,757 SNPs in multiple analyses (STRUCTURE, PCoA, and phylogenetic trees) showed that the genotypes from China were distinct from those collected from the other regions of the world (Fig. 3.3) (Wang et al. 2020). The patterns of putatively neutral population structure were consistent with those based on an analysis using SSR markers (Qiang et al. 2015). Alfalfa was initially introduced into China more than 2000 years ago (around 139 BC) (Hong 2009). However, Rome was reported to plant alfalfa in Europe by 100 AD, and this has been the principal kind of alfalfa cultivated in most of Europe until the sixteenth century (Russelle et al. 2001). Alfalfa has been introduced and cultivated in China for thousands of years, with little connection to the material from outside the country, and thus, Chinese alfalfa has developed a unique set of genetic characteristics (Russelle et al. 2001).

The genetic diversity of tetraploid alfalfa was also determined in different improved states, i.e., cultivars, landraces, cultivated materials, and wild germplasm. The same variation patterns were observed for different three genetic parameters tested. All the analyses yielded that the landraces had the highest genetic variation followed by wild materials, while cultivars had the lowest amount of the genetic diversity (Fig. 3.4). Additionally, a higher level structure was observed where the first division of the analyzed accessions was based on the improvement status of the germplasm. The two groups corresponded to mainly wild (including wild and landrace materials) and cultivated (including cultivars and cultivated materials) groups. The genetic diversity of wild group was significantly higher than that of cultivated group ($P < 0.05$) (Qiang et al. 2015). The reduced genetic diversity in the cultivated gene pool compared to the wild pool is attributed to the bottleneck effect of domestication (Frankel et al. 1995). The average loss of diversity is about 30%, which means that wild populations of ssp. *sativa* still contain potentially useful genetic variability, and the cultivated alfalfa accessions are the synthetic varieties that are generally formed by combining a large number of progenitors (Muller et al. 2003, 2006).

3.4.2 Genetic Diversity Comparisons Among the Subspecies

Generally, autopolyploids could theoretically maintain a greater level of diversity than their diploid progenitors, due to a doubled effective population size (Moody et al. 1993). When the genetic diversity of cultivated tetraploid alfalfa was compared to that of diploid germplasm in the *M. sativa—falcata* complex, it was slightly higher than that of its diploid progenitor *M. sativa* subsp.*caerulea,* but was slightly lower than that of the other diploid germplasm *M. sativa* subsp. *falcata* (Şakiroğlu et al. 2010; Li et al. 2012). A narrow reduction of genetic diversity was observed among cultivated tetraploid alfalfa compared to wild diploid alfalfa based on SNP markers (Li et al. 2012). It is exciting that two tetraploid alfalfa, *M. sativa* subsp. *sativa,* genomes have been assembled and annotated in 2020 (Chen et al. 2020; Shen et al. 2020). A total of 137 tetraploid alfalfa core accessions and 25 ssp. *caerulea* accessions were re-sequenced to characterize the genetic diversity, population migration history, and genetic exchange between the subspecies (Shen et al. 2020). According to average values of nucleotide polymorphism (π) and SNP differences, the genetic diversity of the tetraploid alfalfa was higher than that of the diploid ssp. *caerulea* population (Shen et al. 2020). Estimates of Tajima's D values show that selective sweep signals were rarely detected in the tetraploid alfalfa population (Shen et al. 2020). These

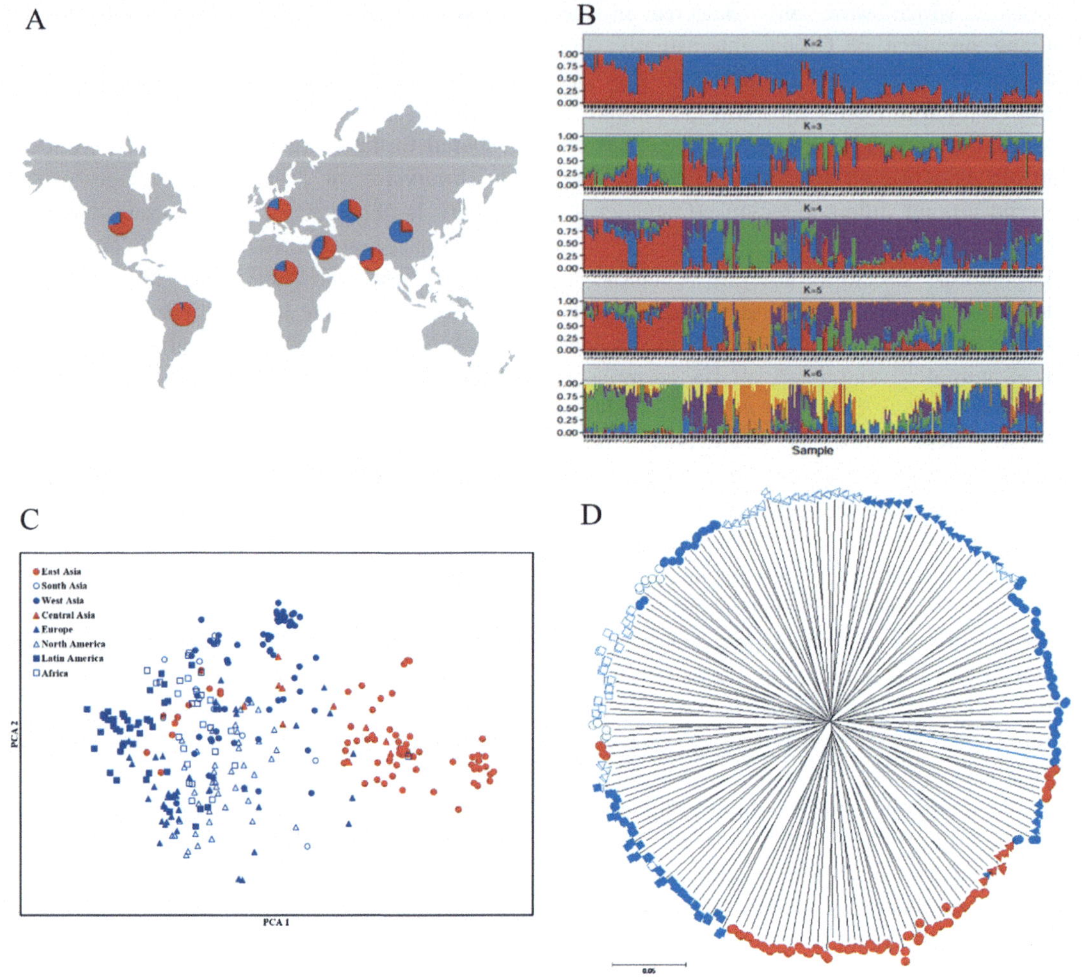

Fig. 3.3 Thepopulation structure based on Bayesian analyses (**B**) and the corresponding geographic distribution (**A**), principal coordinate analysis (**C**), and Neighbor-Joining tree (**D**) of the alfalfa association panel using the marker data generated by GBS modified from Wang et al. (2020)

results confirm that the current tetraploid alfalfa populations maintain high genetic diversity and may not experience purified selection.

3.5 Conclusion and Prospect

In recent years, a wealth of research has been devoted to different aspects of origin and evolution of domesticated and wild alfalfa. However, a number of key issues regarding the evolution of alfalfa and the *M. sativa-falcata* species complex remain undisclosed. The proposal that *M. sativa* subsp. *glomerata* is the ancestral taxa for the members of complex has not been tested with genomic tools. Due to recentness and ongoing hybridization among tetraploid taxa, the clear distinction of subsp. × *varia* from the two ancestral subspecies (İlhan et al. 2016; Şakiroğlu et al. 2010) and its relation with the diploid counterpart subsp. × *hemicycla*has not been investigated either. Furthermore, due to unavailability of adequate number of markers sufficiently covering the genome, the insufficient number of individual genotypes to represent all the members of complex, the exploration of the

Fig. 3.4 Distribution of average allele numbers (**A**), expected heterozygosity (**B**), and PIC values (**C**) germplasm groups with different improvement statue. Violin plots show density distribution of PIC values, median is shown as white circle, top and bottom of vertical bar represent the first and third quartile (Qiang et al. 2015)

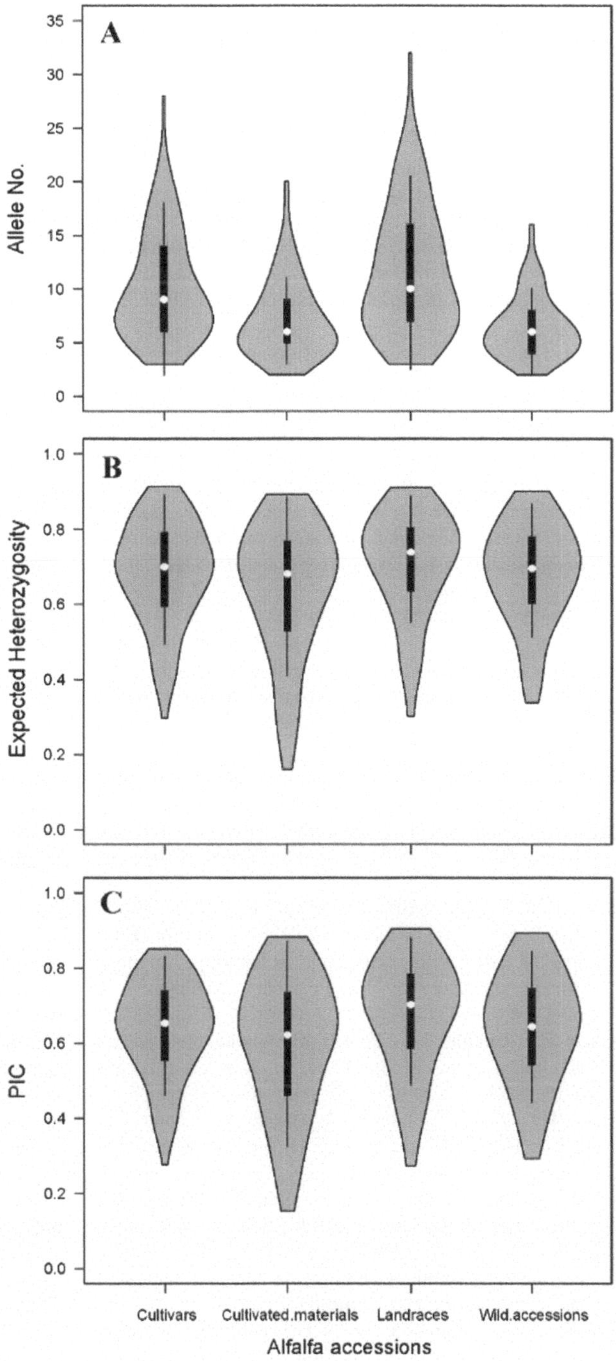

origin, evolution, and genetic diversity of alfalfa and allied taxa at the population genomics level has not been accomplished. As the assembled genomes of more subspecies are released, the patterns and structure of genomic diversity of the hybrid taxa and direct comparison within and across different ploidy levels with new genome-wide molecular diversity analyses could help to fill the knowledge gaps in the origin and evolution of alfalfa and allied taxa.

References

Agarwal K (1983) Cytological studies in the genus *Medicago* Linn. Cytologia 48(4):781–793

Badr A, El-Shazly H (2012) Molecular approaches to origin, ancestry and domestication history of crop plants: Barley and clover as examples. J Genet Eng Biotechnol 10(1):1–12

Barcaccia G, Albertini E, Rosellini D, Tavoletti S, Veronesi F (2000) Inheritance and mapping of 2n-egg production in diploid alfalfa. Genome 43(3):528–537

Barnes DK (1966) Flower color inheritance in diploid and tetraploid alfalfa: A reevaluation. US Government Printing Office

Barnes DK (1977) Alfalfa germplasm in the United States: Genetic vulnerability, use, improvement, and maintenance. Department of Agriculture, Agricultural Research Service

Bena G (2001) Molecular phylogeny supports the morphologically based taxonomic transfer of the "medicagoid" Trigonella species to the genus *Medicago* L. Plant Syst Evol 229(3–4):217–236

Bena G, Lejeune B, Prosperi JM, Olivieri I (1998) Molecular phylogenetic approach for studying life-history evolution: The ambiguous example of the genus *Medicago* L. Proc Roy Soc London B: Biol Sci 265(1401):1141–1151

Bhandari HS, Lawati AHA, Ray IM (2011) Amplified fragment length polymorphism marker diversity among high yielding perennial alfalfa (*Medicago sativa* L.) core collection accessions. Crop Sci 51:1110–1121

Bolton JL, Goplen BP, Baenziger H (1972) World distribution and historical developments. In: Hanson CH (ed) Alfalfa science and technology. American Society of Agonomy Inc Publisher, Madison, WI, pp 1–34

Chen ZJ, Birchler JA (2013) Polyploid and hybrid genomics. John Wiley & Sons, New York, New York, USA

Chen HT, Zeng Y, Yang YZ, Huang LL, Tang BL, Zhang H, Hao F, Liu W, Li YH, Liu YB, Zhang XS, Zhang R, Zhang YS, Li YX, Wang K, He H, Wang ZK, Fan GY, Yang H, Bao AK, Shang ZH, Chen JH, Wang W, Qiu Q (2020) Allele-aware chromosome-level genome assembly and efficient transgene-free genome editing for the autotetraploid cultivated alfalfa. Nat Commun 11:2494|

Clement WM, Lehman WF (1962) Fertility and cytological studies of a dihaploidplant of alfalfa. Medicago sativa L. Crop Sci 2(6):451–453

Diwan N, Bhaqwat AA, Bauchan GB, Cregan PB (1997) Simple sequence repeats DNA markers in alfalfa and perennial and annual *Medicago* species. Genome 40:887–895

Falistocco E, Falcinelli M, Veronesi F (1995) Karyotype and C-banding pattern of mitotic chromosomes in alfalfa. Medicago sativa L. Plant Breed 114(5):451–453

Falistocco E (1987) Cytogenetic investigations and karyological relationships of two *Medicago*: *M. sativa* L. (alfalfa) and *M. arborea* L. Caryologia 40(4):339–346

Flajoulot S, Ronfort J, Baudouin P, Barre P, Huguet T, Huyghe C, Julier B (2005) Genetic diversity among alfalfa (*Medicago sativa*) cultivars coming from a breeding program, using SSR markers. TheorAppl Genet 111(7):1420–1429

Frankel OH, Brown AHD, Burdon JJ (1995) The conservation of plant biodiversity. Cambridge University Press, Cambridge

Fryer JR (1930) Cytological studies in *Medicago*, *Melilotus* and *Trigonella*. Can J Plant Res 3(1):3–50

Gillies CB (1968) The pachytene chromosomes of a diploid *Medicago sativa*. Can J Genet Cytol 10 (4):788–793

Gillies CB (1970) Alfalfa chromosomes. II. pachytene karyotype of a tetraploid *Medicago sativa* L. Crop Sci 10(2):172–175

Gillies CB, Bingham ET (1971) Pachytene karyotypes of 2x haploids derived from tetraploid alfalfa (*Medicago sativa*)—Evidence for autotetraploidy. Can J Genet Cytol 13(3):397–403

Glemin S, Galtier N (2012) Genome evolution in outcrossing versus selfing versus asexual species. In: Anisimova M (ed) Evolutionary genomics. Methods in Molecular Biology (Methods and Protocols). Humana Press, Totowa, pp. 311–335

Havananda T, Brummer EC, Doyle JJ (2011) Complex patterns of autopolyploid evolution in alfalfa and allies (*Medicago sativa*; Leguminosae). Amer J Bot 98 (10):1633–1646

He C, Xia ZL, Campbell TA, Bauchan GR (2009) Development and characterization of SSR markers and their use to assess genetic relationships among alfalfa germplasms. Crop Sci 49:2176–2186

Hendry GW (1923) Alfalfa in history. Agron J 15(5):171–176

Hong FZ (2009) Alfalfa science. China Agriculture Press, Beijing, China

İlhan D, Li X, Brummer EC, Şakiroğlu M (2016) Genetic diversity and population structure of tetraploid accessions of the *Medicago sativa–falcataComplex*. Crop Sci 56(3):1146–1156

Julier B, Flajoulot S, Barre P, Cardinet G, Santoni S, Huguet T, Huyghe C (2003) construction of two genetic linkage maps in cultivated tetraploid alfalfa (*Medicago sativa*) using microsatellite and AFLP markers. BMC Plant Biol 3:9

Ledingham GF (1940) Cytological and developmental studies of hybrids between *Medicago sativa* and a diploid form of *M. falcata*. Genetics 25(1):1–15

Lesins KA, Lesins I (1979) Genus *Medicago* (Leguminosae): A TaxogeneticStudy. W. Junk bvPublishers, Boston, USA, Dr, p 228

Lesins K, Gillies CB (1972) Taxonomy and cytogenetics of *Medicago*. In: Hanson CH (ed) Alfalfa science and technology. American Society of AgonomyInc Publisher, Madison, WI, pp 53–86

Levin DA (1983) Polyploidy and novelty in flowering plants. Amer Nat 122(1):1

Li X, Brummer EC (2009) Inbreeding depression for fertility and biomass in advanced generations of inter- and intrasubspecific hybrids of tetraploid alfalfa. Crop Sci 49(1):13–19

Li X, Acharya A, Farmer AD, Crow JA, Bharti AK, Kramer RS, May GD (2012) Prevalence of single nucleotide polymorphism among 27 diverse alfalfa genotypes as assessed by transcriptome sequencing. BMC Genom 13(1):568

Michaud R, Lehman WF, Rumbaugh MD (1988) World distribution and historical development.In: Hanson AA, Barnes DK, Hill RR (eds) Alfalfa and Alfalfa Improvement. American Society of AgonomyInc Publisher, Madison, WI, pp 25–91

Moody ME, Mueller LD, Soltis DE (1993) Genetic variation and random drift in autotetraploid populations. Genetics 134:649–657

Muller MH, Poncet C, Prosperi JM, Santoni S, Ronfort J (2006) Domestication history in the Medicago sativa species complex: Inferences from nuclear sequence polymorphism. Mol Ecol 15(6):1589–1602

Muller MH, Prosperi JM, Santoni S, Ronfort J (2003) Inferences from mitochondrial DNA patterns on the domestication history of alfalfa (*Medicago sativa*). MolEcol 12(8):2187–2199

Prosperi JM, Jenczewski E, Muller MH, Fourtier S, Sampoux JP, Ronfort J (2014) Alfalfa domestication history, genetic diversity and genetic resources. Legume Perspect 4:13–14

Qiang H, Chen Z, Zhang Z, Wang X, Gao H, Wang Z (2015) Molecular diversity and population structure of a worldwide collection of cultivated tetraploid Alfalfa (*Medicago sativa* subsp. *sativa* L.) germplasm as revealed by microsatellite markers. PLoS One 10(4): e0124592

Quattrocchio F, Wing J, van der Woude K, Souer E, de Vetten N, Mol J, Koes R (1999) Molecular analysis of the anthocyanin2 gene of petunia and its role in the evolution of flower color. Plant Cell 11(8):1433–1444

Quiros CF, Bauchan GR (1988) The genus *Medicago* and the origin of the *Medicago sativa* complex. In: Hanson AA, Barnes DK, Hill RR (eds) Alfalfa and alfalfa improvement. American Society of AgonomyInc Publisher, Madison, WI, p 29

Russelle MP (2001) Alfalfa. Am Sci 89:252–261

Şakiroğlu M, Brummer EC (2007) Little heterosis between alfalfa populations derived from the midwestern and southwestern United States. Crop Sci 47 (6):2364–2371

Şakiroğlu M, Brummer EC (2013) Presence of phylogeographic structure among wild diploid alfalfa accessions (*Medicago sativa* L. subsp. *microcarpa*Urb.) with evidence of the center of origin. Genet Resour Crop Evol 60(1):23–31

Şakiroğlu M, Brummer EC (2017) Identification of loci controlling forage yield and nutritive value in diploid alfalfa using GBS-GWAS. TheorAppl Genet 130 (2):261–268

Şakiroğlu M, İlhan D (2021) Medicago sativa species complex: re-Visiting the century old problem in the light of molecular tools. Crop Sci. https://doi.org/10.1002/csc2.20316

Şakiroğlu M, Doyle JJ, Brummer EC (2010) Inferring population structure and genetic diversity of broad range of wild diploid alfalfa (*Medicago sativa* L.) accessions using SSR markers. Theor Appl Genet 121(3): 403–415

Sakiroglu M, Moore KJ, Brummer EC (2011) Variation in biomass yield, cell wall components, and agronomic traits in a broad range of diploid alfalfa accessions. Crop Sci 51(5):1956–1964

Shen C, Du HL, Chen Z, Lu HW, Zhu FG, Chen H, Meng XZ, Liu QW, Liu P, Zheng LH, Li XX, Dong JL, Liang CZ, Wang T (2020) The chromosome-level genome sequence of the autotetraploid alfalfa and resequencing of core germplasms provide genomic resources for alfalfa research. Mol Plant 13(9):1250–1261

Small E, Brookes BS (1982) Coiling of alfalfa pods in relation to resistance against seed chalcids. Can J Plant Sci 62(1):131–135

Small E, Brookes BS (1984a) Coiling of alfalfa pods in relation to resistance against seed chalcids: additional observations. Can J Plant Sci 64(3):659–665

Small E, Brookes BS (1984b) Taxonomic circumscription and identification in the *Medicago sativa-falcata* (alfalfa) continuum. Econ Bot 38(1):83–96

Small E (1985) Insect pests and the evolution of defensive glandular trichomes in alfalfa. Can J Plant Sci 65 (3):589–596

Small E (1996) Adaptations to herbivory in alfalfa (*Medicago sativa*). Can J Bot 74(6):807–822

Small E (2010) Alfalfa and relatives: evolution and classification of *Medicago*. NRC Research Press, Ottawa, Canada

Soltis DE, Soltis PS, Tate JA (2004) Advances in the study of polyploidy since plant speciation. New Phytol 161(1):173–191

Stanford EH (1951) Tetrasomic inheritance in alfalfa. Agron J 43:222–225

Stanford EH, Clement WM Jr, Bingham ET (1972) Cytology and evolution of the Medicago sativa-falcata complex. Alfalfa Sci Technol 15:87–101

Steiner JJ, Beuselinck PR, Peaden RN, Kojis WP, Bingham ET (1992) Pollinator effects on crossing and genetic shift in a three-flower-color alfalfa population. Crop Sci 32(1):73–77

Vogel KP, Lamb JFS (2007) Forage breeding. In Barnes RF, Nelsom CJ, Moore KJ, Collins M (eds) Forages, The Science of Grassland Agriculture. Iowa State University Press, Ames, IA. pp 427–438

Vyšniauskienė R, Rančelienė V, Patamsytė J, Žvingila D (2013) High genetic differentiation among wild populations of alien *Medicago sativa* in Lithuania. Open Life Sci 8(5):480–491

Wang Z, Yan HW, Fu XN, Li XH, Gao HW (2013) Development of simple sequence repeat markers and diversity analysis in alfalfa (*Medicago sativa* L.). Mol Biol Rep 40:3291–3298

Wang Z, Yu GH, Shi BB, Wang XM, Qiang HP, Gao HW (2014) Development and characterization of simple sequence repeat (SSR) markers based on RNA-sequencing of *Medicago sativa* and *in silico* mapping onto the *M. truncatula* genome. PLoS One 9(7): e103682

Wang Z,Wang XM, Zhang H, Ma L, Zhao HM, Jones CS, Chen J, Liu G (2020)A genome-wide association study approach to the identification of candidate genes underlying agronomic traits in alfalfa (*Medicago sativa* L.). Plant Biotechnol J 18:611–613

Wright SI, Ness RW, Foxe JP, Barrett SC (2008) Genomic consequences of outcrossing and selfing in plants. Intl J Plant Sci 169(1):105–118

Yu F, Wang H, Zhao Y, Liu R, Dou Q, Dong J, Wang T (2017) Karyotypic evolution of the Medicago complex: sativa-caerulea-falcata inferred from comparative cytogenetic analysis. BMC Evol Biol 17:104

Germplasm Collection, Genetic Resources, and Gene Pools in Alfalfa

4

Brian M. Irish and Stephanie L. Greene

Abstract

Popular for its feed quality and benefits to the environment, alfalfa (*Medica sativa* L.) is the most economically important forage legume crop in the world. Most agricultural crops have plant genetic resource (PGR) collections that are conserved, with their use promoted by national and international genebanks. Large alfalfa germplasm collections are held, in the form of seed, in genebanks worldwide. These same genebanks also are responsible for many other cultivated species and alfalfa crop wild relatives (CWR) in the genus *Medicago*. Once acquired, these taxonomically diverse collections are actively managed, requiring ideal storage to promote longevity, regeneration activities to resupply viable seed, and sample duplicates to safeguard against loss. Extensible database software platforms continue to expand capabilities for maintaining fidelity and public access to accession-associated information. As resources and policies allow, PGR are distributed to stakeholders (e.g., plant breeders) in public, non-profit, and private research sectors. Many of the PGR have been extensively characterized and evaluated for important phenotypic, agronomic, and genetic traits. As many of the alfalfa and other *Medicago* spp. collections are too large to evaluate at once, subsets of representative genetically distinct accessions (i.e., core collections) have been established and evaluated. Data generated in characterizations and evaluations are often linked to accessions and continue to expand, helping users identify useful germplasm. Although difficult to assess because alfalfa and many *Medicago* spp. accessions are often heterozygous outcrossing polypoid populations, comprehensive genetic diversity has been described for some of the larger collections. Diversity and accession numbers for CWR collection holdings are much lower, although the potential value in introgressing key traits from germplasm in these genepools has been shown. In the United States, early plant germplasm introductions played important roles in breeding modern alfalfa cultivars. Many of these traits (e.g., cold adaptation and biotic/abiotic resistance) have contributed to expanded growing areas and sustainable production. Historically, the production of alfalfa and other *Medicago* spp. forages has benefited from the diversity conserved in PGR

B. M. Irish (✉)
USDA ARS Plant Germplasm Introduction and Research Testing Unit, 24106 N. Bunn Rd., Prosser, WA 99352, USA
e-mail: Brian.Irish@usda.gov

S. L. Greene
USDA ARS National Laboratory for Genetic Resources Preservation, 1111 S. Mason St., Fort Collins, CO 80521, USA
e-mail: Stephanie.Greene@usda.gov

L.-X. Yu and C. Kole (eds.), *The Alfalfa Genome*, Compendium of Plant Genomes,
https://doi.org/10.1007/978-3-030-74466-3_4

collections. Continued access to these critical *Medicago* spp., PGR will only be secure if committed support from stakeholder communities endures.

4.1 Introduction

Alfalfa (*Medica sativa* L.) is known as "the queen of the forages" and is one of the most significant agricultural crops in the world. It is a high-quality perennial forage with broad environmental adaptability that brings added benefits to sustainable production such as nitrogen fixation and reduced tillage and erosion. Its long history of domestication, naturalization, and distribution as well as its economic impact have been covered in previous introductory chapters in this book. Information on the status of alfalfa and other related *Medicago* spp. plant genetic resources (PGR) exists (Bauchan and Greene 2001; Meglič et al. 2003; Boller and Greene 2010; Greene et al. 2011; Smýkal et al. 2015), although it is somewhat dated or the attention is on a broader scope of crops. Therefore, this chapter will focus on a current review of alfalfa germplasm, its crop genepools and wild relatives (e.g., other *Medicago* spp.), and the management of these important genetic resource collections. Content will include aspects of PGR management applied more broadly while adding specific context relating to alfalfa and *Medicago* spp. germplasm. An emphasis and perspectives on these PGR in the United States (U.S.) and the United States Department of Agriculture (USDA) National Plant Germplasm System (NPGS) is also highlighted.

4.2 Alfalfa Origin and Taxonomy

The precise center of origin and domestication for alfalfa is not well defined owing in part to its long evolutionary and distribution histories (Michaud and Rumbaugh 1988; Prosperi et al. 2014). There is general agreement that the genus *Medicago*, comprised of more than 80 different species, is native to the Mediterranean basin through Central and Eastern Asia (Michaud and Rumbaugh 1988; Quiros and Bauchan 1988; Small 2011). Centers of diversity include the Caucasus region, northwestern Turkey, and northeastern Iran (Quiros and Bauchan 1988). Ivanov and Brezhnev (1988) considered Central Asia to be the primary center of diversity for the *M. sativa* complex. Humphries et al. (2020) provide a thorough description of the *M. sativa* complex and discuss which members are likely to contribute to drought tolerance. The proclivity of alfalfa to naturalize where it has been cultivated has also contributed to genetic resources valuable to breeders (Lesins and Lesins 1979; Annicchiarico et al. 2015; Boe et al. 2020).

To discuss *Medicago* spp. PGR collections and the germplasm held, it is important to define taxonomic groupings. Not surprisingly, and partially owing to its long history and extensive distribution, taxonomy in alfalfa (and in the genus *Medicago*) continues to change and is multifaceted. In the Fabaceae, alfalfa (*Medicago sativa*) is in section Medicago of tribe Trifolieae. Depending on the resource, several different infraspecific taxa or subspecies comprise the *M. sativa* complex. Because of the difficulties in clearly delineating subspecies, suggestions to classify alfalfa based on phenotypic traits have been proposed (Piano et al. 1996). Several excellent books, book chapters, and publications review *Medicago* spp. and *M. sativa* taxonomy (Sinskaya 1961; Lesins and Lesins 1979; Quiros and Bauchan 1988; Steele et al. 2010; Small 2011). Other online resources (e.g., Germplasm Resources Information Network [GRIN] Taxonomy) are available that aid in assigning proper systematics and terminology. For the purposes of this chapter, the nomenclature proposed and used by GRIN Taxonomy (https://npgsweb.ars-grin.gov/gringlobal/) is applied to alfalfa. In this case, *M. sativa* is comprised of six subspecific taxa including (1) *M. s.* subsp. *caerulea* (Less. ex Ledeb.) Schmalh., (2) *M. s.* subsp. *falcata* (L.) Arcang., (3) *M. s.* subsp. *glomerata* (Balb.) Rouy, (4) *M. s.* L. subsp. *sativa*, (5) *M. s.* nothosubsp. (a hybrid between subspecies) *tunetana* Murb., and (6) *M. s.* nothosubsp. *varia* (Martyn) Arcang. Two varietal forms of *M. s.*

subsp. *falcata*, var. *falcata* (L.) Döll and var. *viscosa* (Rchb.) Posp., have been described but are considered and enumerated under subsp. *falcata* herein.

4.3 Plant Genetic Resources

Plant genetic resource collections have been assembled and established for most of the globally important agricultural crops. These large and diverse ex situ germplasm collections conserve, often in the form of seed, PGR with genes that confer traits for agricultural adaptation. The goal of these genebanks is to maximize or capture genetic diversity that is representative of a crop and make the PGR, and their associated information, readily accessible for use. Demand for PGR in germplasm collections around the world continues to increase with genebanks having difficulties meeting their mission in part owing to a lack of adequate resources (Smale and Day-Rubenstein 2002; Rubenstein et al. 2006). Another important issue of concern affecting genebanks is the need to recruit highly and specifically trained personnel and staff. In the U.S., efforts are underway to develop PGR training modules and programs to meet some of these requirements in the near future (Byrne et al. 2018; Volk et al. 2019).

Significant national collections are managed by individual countries (e.g., U.S. National Plant Germplasm System [NPGS]) and international collections are managed by the Consultative Group for International Agricultural Research (CGIAR[1]). Stakeholders of these PGR are primarily plant scientists (e.g., breeders) at public and non-profit organizations in the research and education communities, with other important users including private industry and producers (Rubenstein et al. 2006). Ultimately, it is farmers and the public who benefit from access to safe, nutritious, and high-quality goods developed by research and breeding programs that use conserved PGR.

4.4 Alfalfa/*Medicago* Spp. Genebank Collections

Genetic resource collections for alfalfa, and related species, have been assembled and are held by national and international organizations dedicated to their long-term conservation and use. Alfalfa is by far the most economically important species in *Medicago*, and as such large collections are found in genebanks, predominantly in countries where the crop has economic importance, such as the U.S., Australia, and Russia. Table 4.1 summarizes information on the number of alfalfa accessions currently (January 2021) held in ten of the largest genebanks with responsibilities for this crop. Three of these organizations (the U.S. NPGS, Plant Germplasm Introduction and Testing Research Unit [PGITRU], the South Australian Research and Development Institute [SARDI], Australian Pastures Genebank [APG], and the Nikolai I. Vavilov Research Institute of Plant Industry [VIR]) account for just over 55% of the world's more than 21,000 alfalfa germplasm accessions. In most countries, responsibilities for crop-specific PGR management are associated with a single or primary organization but collections can be held at multiple sites. Figure 4.1 offers a summary of *M. sativa* PGR held in country collections. Detailed accession-associated passport information for many of these collections can be accessed through Genesys (https://www.genesys-pgr.org/) or through publicly available online genebank-specific databases.

Although not as significant as alfalfa in terms of crop production, there are 16 other perennial and annual *Medicago* spp. that are cultivated and many wild species that are grazed for forage. According to the Genesys germplasm database, excluding *M. sativa*, there are just over 47,000 *Medicago* spp. PGR accessions held in genebanks. The three organizations with the largest non-alfalfa *Medicago* spp. collections are the

[1]The CGIAR was formerly known as the Consultative Group for International Agricultural Research and now is the Consortium of International Agricultural Research Centers.

Table 4.1 Country, genebank/organization, site, and estimated number of alfalfa (*Medicago sativa* L.) and subordinate taxa accessions held in germplasm collections

Country[a]	Genebank/organization[b]	Site	Accessions[c]
USA	National Plant Germplasm System, Plant Germplasm Introduction and Testing Research Unit	Pullman	4,083
AUS	South Australian Research and Development Institute, Australian Pastures Genebank	Adelaide	3,843
RUS	Nikolai I. Vavilov Research Institute of Plant Industry	Saint Petersburg	3,647
GBR	Genetic Resources Unit, Institute of Biological, Environmental and Rural Sciences, Aberystwyth University	Aberystwyth	1,023
HUN	Centre for Plant Diversity	Tápiószele	914
LBN	International Centre for Agricultural Research in Dry Areas	Beirut	890
DEU	External Branch North of the Department Genebank, IPK, Oil Plants and Fodder Crops	Malchow	743
ITA	Applied Biology Department, University of Perugia	Perugia	697
CZE	Research Institute of Crop Production	Prague	635
ROM	National Agricultural Research and Development Institute	Fundulea	546
OTHER	*Composite of 54 international organizations*		4,011
Total			21,032

[a]ISO 3166-1 alpha-3 three-letter country codes
[b]Primary organization with the largest number of accessions conserved within country
[c]Data derived from NPGS GRIN-Global https://npgsweb.ars-grin.gov/; APG GRIN-Global https://apg.pir.sa.gov.au/gringlobal/; Genesys https://www.genesys-pgr.org/; and the VIR Plant Genetic Resources Database http://db.vir.nw.ru/virdb/—*accessed* January, 2021

APG—(5,276 accessions); ICARDA (9,144 accessions); APG (5,76 accessions); PGITRU (4,695 accessions).

Much of the alfalfa and other *Medicago* spp. germplasm in genebanks originates from centers of genetic diversity and domestication in the Mediterranean basin through Central and Eastern Asia. The collections consist not only of wild-collected ecotypes but also include cultivars, breeding lines, landraces, and accessions with unknown improvement status. Figure 4.2 summarizes the representation of different levels of improvement status for a large proportion of the alfalfa germplasm held in the global collections.

Taxonomic characteristics and number of germplasm accessions held by the NPGS, APG, and VIR for the six *M. sativa* subspecific taxa are presented in Table 4.2. Most of the collection holdings at each of these sites correspond to *M. s.* subsp. *sativa*, which is the principally cultivated taxon with its characteristic purple flower, coiled pods, and tetraploid ($2n = 4x = 32$) genome. Two of the other taxa (subsp. *falcata* and nothosubsp. *varia*) also have a significant number of accessions represented in the collections. The smaller number of accessions for subspp. *caerulea*, *glomerata* and nothosubsp. *tunetana* might indicate a need for additional collections to fill gaps in coverage.

The diversity held in each of the *Medicago* spp. collections is extensive (Bauchan and Greene 2001; Meglič et al. 2003) (Fig. 4.3). Overlap or redundancy in accession holdings is thought to occur, but the exact extent has been difficult to quantify. An ongoing effort led by the Crop Trust (www.croptrust.org) has focused on developing a "Global strategy for the ex situ conservation of temperate forages" that includes alfalfa and other important temperate forages.

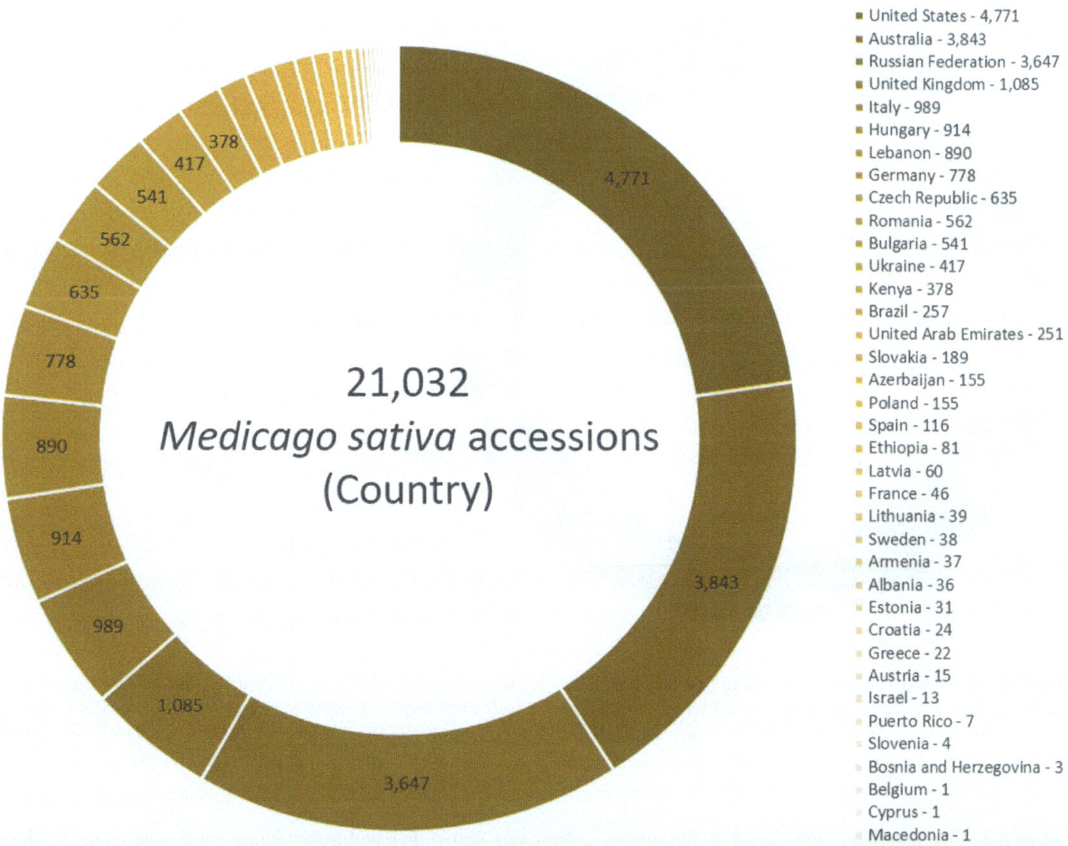

Fig. 4.1 Total number of alfalfa (*Medicago sativa* L.) germplasm accessions held in country collections worldwide. The larger number of accessions by country, when compared to those reported in Table 4.1, is likely explained by more than one institution per country conserving *M. sativa* germplasm. Data derived from NPGS GRIN-Global https://npgsweb.ars-grin.gov/; APG GRIN-Global https://apg.pir.sa.gov.au/gringlobal/; Genesys https://www.genesys-pgr.org/; and the VIR Plant Genetic Resources Database http://db.vir.nw.ru/virdb/—*accessed* January, 2021

One of the goals in the drafted proposed strategy would be to address potential redundancies and gaps in PGR coverage within and across collections.

4.5 Alfalfa/*Medicago* spp. PGR Management

Managing large plant germplasm collections is an active and technical process (Clark et al. 1997; Byrne et al. 2018), and institutional, national, and international policy complicate acquisition and distribution (Lopez-Noriega et al. 2012; Brink and van Hintum 2019). Generally, principal management activities involve the acquisition, maintenance, characterization, evaluation, and distribution of PGR and their associated information. Whenever possible, the administration of PGR should follow best practices outlined in guidelines such as those proposed by the Genebank Standards for Plant Genetic Resources for Food and Agriculture (Engels and Visser 2003; FAO 2013). Even when generalized guidelines exist for PGR, managing alfalfa germplasm is multifaceted and several crop- and species/taxon-specific approaches need to be implemented to be effective. The following sections describe general

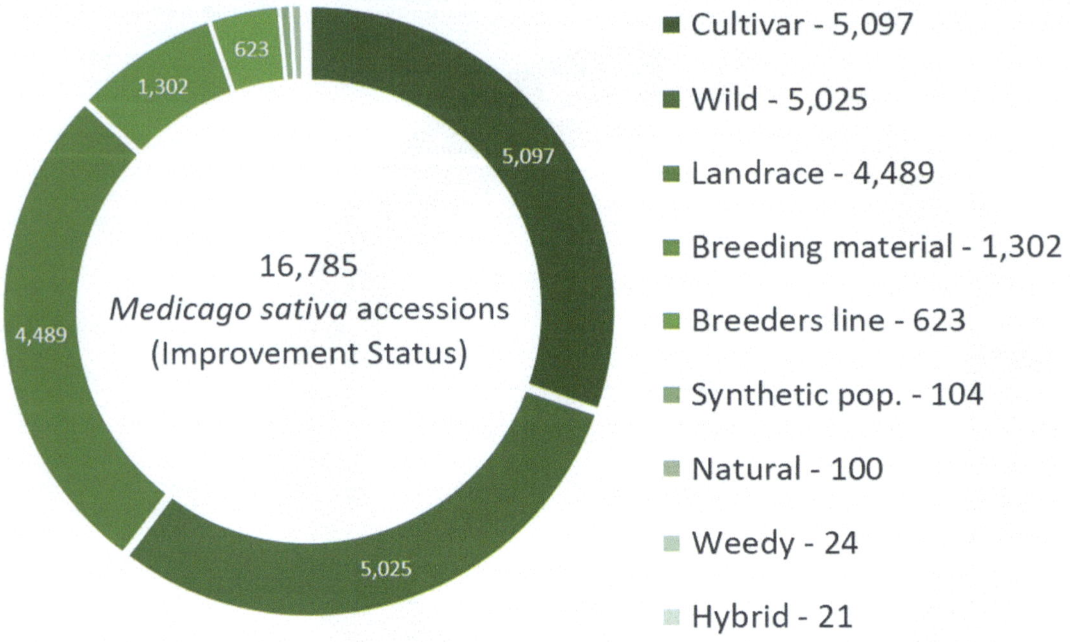

Fig. 4.2 Reported improvement status of alfalfa (*Medicago sativa* L.) conserved in worldwide collections (improvement status not reported for many accessions). Data derived from NPGS GRIN-Global https://npgsweb.ars-grin.gov/; APG GRIN-Global https://apg.pir.sa.gov.au/gringlobal/; Genesys https://www.genesys-pgr.org/; and the VIR Plant Genetic Resources Database http://db.vir.nw.ru/virdb/—*accessed* January, 2021

Table 4.2 Characteristics and number for alfalfa (*Medicago sativa* L.) and subordinate taxa accessions in three international genebanks with the largest collections

Taxon[a]	$2n$[b]	Flower	Pod	NPGS[c]	SARDI	VIR
M. s. subsp. *caerulea* (Less. ex Ledeb.) Schmalh.	16	Purple	Coiled	97	182	–
M. s. subsp. *falcata* (L.) Arcang.	16/32	Yellow	Sickle	453	302[d]	347
M. s. subsp. *glomerata* (Balb.) Rouy	16/32	Yellow	Coiled	12	37	20
M. s. L. subsp. *sativa*	32	Purple	Coiled	3,071	2,899	6
M. s. nothosubsp. *tunetana* Murb.	32	Variegated	–	7	–	21
M. s. nothosubsp. *varia* (Martyn) Arcang.	16/32	Variegated	Partly coiled	436	398	789
Other *M. sativa*[e]	–	–	–	7	25	2,464
Total				4,083	3,843	3,647

[a]Based on GRIN Taxonomy nomenclature with data derived from NPGS GRIN-Global https://npgsweb.ars-grin.gov/; APG GRIN-Global https://apg.pir.sa.gov.au/gringlobal/; Genesys https://www.genesys-pgr.org/; and the VIR Plant Genetic Resources Database http://db.vir.nw.ru/virdb/—*accessed* January, 2021
[b]Somatic chromosome number
[c]NPGS = National Plant Germplasm System; SARDI = South Australian Research and Development Institute; and VIR = Nikolai I. Vavilov Research Institute of Plant Industry
[d]Includes *M. s.* subsp. *falcata* var. *viscosa* accessions
[e]*M. sativa* accessions where subspecies not specified, most likely many *M. s.* L.subsp. *sativa*

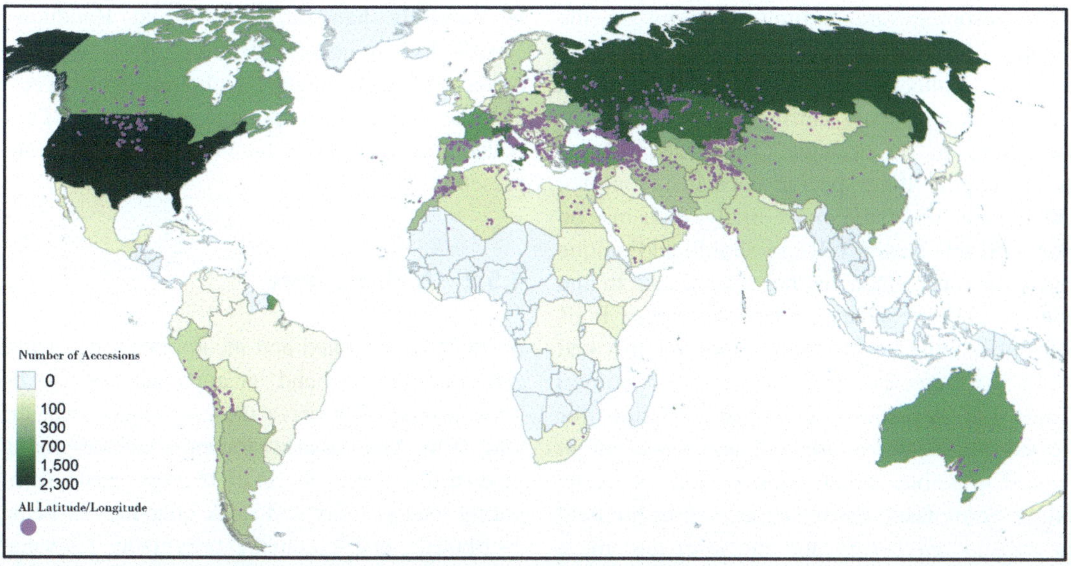

Fig. 4.3 Choropleth map assigning 16,800 alfalfa (*Medicago sativa* L.) accessions to country of origin based on source information and collection site distribution (dots) based on available coordinates for 3,449 in situ-collected wild or landrace accessions held in world genebanks. Data derived from Genesys https://www.genesys-pgr.org/ multi-crop passport descriptor data download—*accessed* January, 2021

management aspects applicable to all PGR with details that apply specifically to alfalfa and other *Medicago* spp.

4.5.1 Acquisition

Germplasm maintained in genebanks has been acquired in the past through plant exploration and field collection trips, transfers of accession between collections, and donations from research programs (e.g., breeding programs). Although germplasm is still being added to collections, major acquisition phases for alfalfa and relatives in global genebanks occurred over the past ~100 years with significant growth and expansion during the second half of the twentieth century.

Published literature on germplasm exploration and collection trips is difficult to identify and cite. The North American Alfalfa Improvement Conference (NAAIC http://naaic.org/) holds archived conference proceedings and shows alfalfa/*Medicago* collection reports from Turkey, Morocco, and Chile. Greene et al. (2005) collected alfalfa wild relatives in Kazakstan which are now part of the NPGS collections. In Spain, Prosperi et al. (2006) collected naturalized and threatened alfalfa populations including a unique genetic background called "Mielga" with rhizomes and prostrate habit. The germplasm collected in Spain was deposited in the *Medicago* genebank, of Institut National de la Recherche Agronomique (INRA) in Montpellier, France. Vaitsis (2000) discussed perennial *Medicago* collecting and breeding in Greece for subspp. *falcata* and *sativa* along with *M. arborea* L. Alfalfa germplasm was also collected and evaluated from oases in Tunisia (Loumerem et al. 2007). Unfortunately, in these last two cases, it is not clear if the germplasm collected was deposited into national or international genebanks. Bioversity International hosts a collecting mission database http://bioversity.github.io/geosite/ with information associated with *Medicago* spp. germplasm exploration and collecting trips that can be queried. In most cases, information associated with specific germplasm acquisition can be found in passport data linked to accessions in publicly accessible databases.

As resources are a limiting factor in the quantity of germplasm that can be maintained in genebank collections, and because significant diversity is conserved currently, future acquisitions need to be strategic. These acquisitions should target filling specific gaps in taxonomic and genetic diversity, most likely focusing on crop wild relatives (CWR), or should add unique traits not represented in the collections. In the primary and secondary genepools, larger PGR collections exist for the most common cultivated alfalfa subspecies (*M. s.* subsp. *sativa*) (Table 4.2). A reduced number of accessions for the subspp. *caerulea*, *falcata*, and *varia* might suggest possible underrepresentation of some unique germplasm. Far fewer accessions are held for subspp. *glomerata* and *tunetana* and more than likely efforts to acquire additional representative germplasm of these subspecies should be prioritized.

Greene et al. (2011) found gaps in ex situ collections for alfalfa CWR from mountainous areas in Central Asia and notably, Eastern Siberia, areas where cold-tolerant germplasm may occur. A project led by the Crop Trust (www.croptrust.org) has focused on "Adapting Agriculture to Climate Change" and on the collection of CWR, which includes alfalfa (Dempewolf et al. 2014). Alfalfa CWR acquisition efforts were followed by a prebreeding project that developed and evaluated lines derived from CWR. This material is available from APG (Humphries et al. 2020). Acquisition priorities include gaps in coverage for germplasm that is threatened or in locations where gene flow and genetic erosion between natural (i.e., in situ) populations and cultivated alfalfa might occur (Prosperi et al. 2006; Greene et al. 2008; Annicchiarico et al. 2015). At least in the U.S., alfalfa germplasm continues to be added to genebank holdings from a queue of registered cultivars that becomes public after intellectual property rights expire. The incorporation of registered cultivars with genetically engineered (GE) traits (e.g., glyphosate resistance and low-lignin traits) will become an important issue in long-term management as these PGR will need to be carefully handled. Tiwari and Randhawa (2010) discuss general strategies on how genebanks will need to cope with transgenic accessions as they are incorporated into existing PGR collections and how to mitigate adventitious presence.

4.5.2 Maintenance

As PGR are acquired and incorporated into active collections, a plan and an approach need to be implemented to preserve these resources over the long term. Maintenance strategies include storing seed (if this is how the PGR is propagated), monitoring seed viability and stock quantities (if being distributed), and eventually prioritizing regeneration. Of utmost importance is the management of all aspects of labeling and associated data fidelity.

4.5.2.1 Storage

Most of the world's germplasm collections are maintained in the form of orthodox seed. Orthodox seed can be dried and kept at low temperatures, even frozen, for long periods of storage time, prolonging their viability. Genebanks with active distributions often partition their collections into a working collection (4 °C), from which distributions are often made, and a base collection held under colder conditions (~ −18 °C). Many factors affect seed longevity in storage. Principal among these factors is initial seed quality, the drying process (Annicchiarico et al. 2004), and ultimately the long-term storage conditions that usually include low temperature and humidity (Solberg et al. 2020). Although conflicting results were obtained in experimental evaluations, reported half-life (P_{50})—time for a seed lot to decline in viability by half of its initial value—for alfalfa is medium to high. This means that viability does not decline substantially and seed lots have excellent survival of greater than 50 years under ideal storage conditions (Walters et al. 2007; Solberg et al. 2020). Comparable storage longevities would be expected of other hard-seeded *Medicago* spp. under ideal storage conditions.

4.5.2.2 Regeneration

In many situations, originally acquired seed lots are not of sufficient quality (i.e., low viability) or quantities for long-term storage and distribution. In other circumstances, and despite effective storage conditions, seed loses viability over time or supplies run low from germplasm that is actively being distributed. In these cases, seed enters a queue to be increased, commonly referred to as regeneration, by the genebanking community. An important concern during regeneration is to minimize genetic change during the process. Again, where appropriate, genebank standards (Engels and Visser 2003; FAO 2013) should be followed to maintain the genetic integrity of accessions. Most perennial species of *Medicago* are insect-mediated outcrossers. During regeneration, genetic integrity is maintained by using described effective population sizes (Rowe 1986) and controlled pollination using insect-proof cages (Fig. 4.4) to prevent cross contamination caused by pollinator-mediated gene flow (Brunet et al. 2019a, b).

Since the deregulation of transgenic traits and subsequent commercial-scale production of genetically engineered (GE) alfalfa in several regions of the world, additional precautions need to be put in place. These safety measures would be to prevent possible gene flow that could result in the adventitious presence of transgenic traits in original germplasm accessions. Strategies for the co-existence of organic, conventional, and transgenic commercial seed production in alfalfa have been developed. Some of these can be adopted for managing PGR seed increases in *M. sativa* and other *Medicago* spp. that could form transgenic hybrids (Van Deynze et al. 2008; Greene et al. 2015; Kesoju et al. 2020). Sentinel

Fig. 4.4 U.S. Department of Agriculture, National Plant Germplasm System field site in Prosser, Washington with insect-proof isolation cages used for increasing (regenerating) seed of cross-pollinated temperate-adapted forage legume plant genetic resources, like alfalfa (*Medicago sativa* L.)

plots for monitoring potential transgenic gene flow have been used in maize (Mezzalama et al. 2010) and could be a useful tool in alfalfa seed regeneration for monitoring the adventitious presence of GE traits and genetic integrity.

4.5.2.3 Backup

Another important aspect of PGR management is to maintain duplicate samples at a second location, termed "security backup" in genebanking communities. An optimal backup sample should contain a minimum amount of 500 seeds, for outcrossing species, so that enough is available for several regenerations (FAO 2013). More than 90% of the NPGS alfalfa and 89% of other *Medicago* spp. collections are backed up with the USDA National Laboratory for Genetic Resources Preservation (NLGRP), in Fort Collins, Colorado. The NLGRP also serves as an alternate backup site for germplasm from several of the international CGIAR genebanks. The NLGRP is the portal for backing up NPGS germplasm with the Svalbard Global Seed Vault (SGSV) in Longyearbyen, Norway. To date, 2,988 *M. sativa* and 242 other non-alfalfa *Medicago* spp. NPGS accessions have been backed with the SGSV. Close to 12.8% (1,859/14,436) of the *M. sativa* accessions in other world collections are backed up at the SGSV. For other non-alfalfa *Medicago* spp. outside of the NPGS, the proportion backed up with SGSV is considerably higher (39.5%; 18,791/47,611) because 18,564, or 98.8% of the total backed up for this cohort, are *Medicago* spp. accessions belonging to the APG. This corresponds to 73.4% of the total (25,276) non-alfalfa APG accessions. The backup status for all *Medicago* spp. in other international and national genebanks is not precisely known, but efforts to conserve quality seed under ideal conditions at alternate sites should be promoted.

4.5.2.4 Data Administration

To maximize utilization of PGR, critical and structured information linked to individual accessions must be meticulously kept and readily accessible (Weise et al. 2020). Several available online resources dedicated to storing and providing access to PGR (including *Medicago* spp.) passport and other valuable accession-associated information exist (Table 4.3). Much of the value in PGR collections comes from the information associated with accessions. For example, passport information might provide data on the collection source along with useful taxonomy. If only limited passport information exists, it still might offer insight into how genebanks might manage given accessions. As information tied to accessions grows, for instance, detailed acquisition, phenotypic, evaluation, and/or genotypic data, utilization of the conserved PGR can be more deliberate with users being able to "home in" on specific subsets of germplasm (Rubenstein et al. 2006; Weise et al. 2020). Standardization of multi-crop passport descriptors (MCPD) is being promoted in efforts to facilitate data sharing and comparisons across genebanks (Alercia et al. 2015). As *Medicago* spp. accession-associated genotype and genomic information are generated in research (discussed in detail in other book chapters, herein), PGR database platforms will need to expand capabilities for storage or be able to interoperate and network with other data sources to effectively link associated information (Mascher et al. 2019; Belzile et al. 2020; Weise et al. 2020).

To date, many different software platforms are used to administer PGR-associated data. For instance, GRIN-Global (https://www.grin-global.org/) is a database platform developed between the USDA Agricultural Research Service, Bioversity International, and the Global Crop Diversity Trust, and was implemented in late 2015 by the NPGS. It is a scalable, robust tool for managing substantial amounts of accession-associated data (Postman et al. 2010) and is being used by many of the CGIAR and national genebanks (e.g., APG https://apg.pir.sa.gov.au/gringlobal/). The only database dedicated specifically to *Medicago* spp. genetic resources is the French national European Perennial *Medicago* Database https://www.ecpgr.cgiar.org/

Table 4.3 List of some useful online resources with passport and associated information for alfalfa (*Medicago sativa* L.) and other *Medicago* spp. genetic resources

Name	Website	Function
European Cooperative Programme for Plant Genetic Resources (ECPGR)	https://www.ecpgr.cgiar.org/	Is a collaborative program among most European countries aimed at ensuring the long-term conservation and facilitating the increased utilization of plant genetic resources in Europe
European Search Catalogue for Plant Genetic Resources (EURISCO)	https://eurisco.ipk-gatersleben.de/	Provides information about more than 2 million accessions of crop plants and their wild relatives, preserved ex situ by almost 400 institutes, and represents an important effort for the preservation of world's agrobiological diversity
Genesys	https://www.genesys-pgr.org/	Is a database that allows users to explore the world's crop diversity conserved in genebanks through a single website
Germplasm Collecting Mission Database	http://bioversity.github.io/geosite/	Provides access to information on all germplasm-collecting missions in which Bioversity has been involved
Germplasm Resources Information Network (GRIN)-Global[a]	https://www.grin-global.org/	Is a database application that enables genebanks to store and manage information associated with plant genetic resources and deliver that information globally
Germplasm Resources Information Network (GRIN) Taxonomy	https://npgsweb.ars-grin.gov/gringlobal/taxon/taxonomysearch	Provides taxonomic data and underlying structure and nomenclature for accessions of the National Plant Germplasm System (NPGS)
Svalbard Global Seed Vault	https://seedvault.nordgen.org/	Provides options for searching among the stored seed samples in the Seed Vault and the depositing institutions
The Crop Wild Relatives Project	https://www.cwrdiversity.org/	Is a project to collect important species of crop wild relatives, ensure their long-term conservation, and facilitate their use in breeding new, improved crops
World Information and Early Warning System on Plant Genetic Resources for Food and Agriculture (WIEWS)	http://www.fao.org/wiews/en/	Is the information system used by FAO for the preparation of periodic, country-driven global assessments of the status of conservation and use of PGRFA

[a]GRIN-Global, a scalable version of the Germplasm Resource Information Network (GRIN) originally developed in a joint effort by the Global Crop Diversity Trust, Bioversity International, and the Agricultural Research Service of the USDA. It is suitable for use by any interested genebank in the world and is being implemented at various genebanks around the world including the NPGS and many of the CGIAR international research centers

resources/germplasm-databases. Many other online resources and database platforms are readily accessible and used to query PGR information including that of *Medicago* spp. For example, the Genesys portal was instrumental in assembling *Medicago* spp. PGR information across world genebanks.

4.5.3 Distribution

Oftentimes distribution, or making PGR available for use by stakeholders, is of lower precedence than maintenance. Frequently, resources within genebanks are limited and priorities are assigned to other "more important" tasks. If available, distribution often is of small research quantities (e.g., ~ 100–200) of seed with little to no cost to requestors. *Medicago* spp. germplasm seed is small and easily confused with seed of other potentially weedy species. Therefore, extra caution and efforts should be made to limit the inadvertent co-distribution of weed seed and potential insect pests like alfalfa seed chalcids (Peterson and Baird 1994). Additionally, seed-borne bacterial (e.g., *Clavibacter michiganensis* subsp. *insidiosus* [McCulloch] Davis et al.) and viral (e.g., *Alfalfa Mosaic Virus*) diseases are known to occur in alfalfa and might limit distribution (Samac et al. 2016). National and international regulatory procedures including quarantine and phytosanitary regulations are often in place with goals of reducing the spread of pests when moving seed propagated (and clonal) PGR (Kumar et al. 2021).

New rules and regulations, brought on by national and international policies, have been instituted to protect nations' sovereign rights over PGR and have heightened concerns for access (Lopez-Noriega et al. 2012; Brink and van Hintum 2019). For example, the International Treaty for Plant Genetic Resources for Food and Agriculture (ITPGRFA) promotes PGR availability through a multilateral system and uses a standard material transfer agreement (SMTA) as a vehicle for access and benefit-sharing. All the CGIAR genebanks and many national genebanks have adopted the use of the SMTA, with plans by the NPGS to begin its use soon.

4.6 Characterization and Evaluation

The usefulness of PGR in collections increases as characterization and evaluation information about individual accessions are accrued and become available (Rubenstein et al. 2006; Weise et al. 2020). Often, in the process of growing out accessions for regenerations (discussed above), genebank managers will collect highly heritable descriptor data (i.e., characterizations). These data are most likely collected as non-replicated single-season datapoints at several phenological stages. Descriptors have been developed for forage legumes, including alfalfa (IBPGR 1984), and for annual medics (Bioversity International 2007). Important descriptor traits in *Medicago* spp. include plant growth habit, flower color, pod characteristics, and ploidy (Brummer et al. 1999), but may cover observations on insect pest and disease reactions as well. In the U.S., historically an Alfalfa Crop Germplasm Committee (CGC)—comprised of subject matter experts and stakeholders (e.g., breeders)—has provided input on important crop descriptors to the NPGS. Digital voucher images, of different plant organs and developmental stages, can also be incorporated in the descriptor data collecting steps, with many added to NPGS accessions. Accession-associated descriptor information and digital images can be accessed regularly through online database portals. Descriptor information ultimately can be useful in assigning taxonomic relationships and estimating genetic diversity and population structure in the collections. At the same time, descriptors are useful to stakeholders looking for specific traits in narrowed-down subsets of germplasm.

Medicago spp. PGR have been extensively evaluated by genebanks and in other collaborative efforts for agronomic, biotic and abiotic stress tolerance/resistance, and forage quality traits. Some important traits have been collected in core *Medicago* spp. subsets and are discussed immediately below. Other plant characters have been gathered on unique PGR subsets as part of breeding programs with specific goals and are discussed in the corresponding section. Some alfalfa and other *Medicago* spp. evaluation data are accessible as addenda to peer-reviewed publications, while other information can be accessed through online databases (e.g., NPGS GRIN-Global descriptor site—https://npgsweb.ars-grin.gov/gringlobal/descriptors).

4.7 Core PGR Collections

When large PGR collections exist for a given crop, the sheer number and complexity of accessions often preclude detailed evaluations and effective utilization of all accessions. This is especially true when trying to include accessions in larger greenhouse- or field-based trials. To work around this problem, more manageable subsets of unique (i.e., avoiding duplication) accessions representing the genetic diversity of a crop are assembled into core collections (Frankel 1984). A number of different ways to identify and establish PGR core collections have been proposed. Odong et al. (2013) recommended that distance-based methods be used as they provide concurrent variable evaluations and innate and interpretable standards. An alternative technique for core collection development, proposed by van Treuren et al. (2009), is based on trying to optimize collection coverage by taking into account the hierarchical structuring of the genepool and assigning relative importance to its components. An additional benefit of developing core collections is that genebank managers can identify potential redundancies and/or gaps in coverage.

Core collections have been established for the U.S. NPGS alfalfa (Basigalup et al. 1995) as well as for the annual *Medicago* spp. (Diwan et al. 1994, 1995). In developing these two core collections, authors utilized combinations of passport and phenotypic evaluation data to establish subsets. Both core collections were developed with available methods and considered a reduced number of accessions in the collections at the time. Genomic resources, discussed in detail in this book, will be useful in validating original subsets or in developing more representative and updated versions of these core collections. Using a SARDI annual *Medicago* spp. core collection based on phenotypic evaluations and developed by Skinner et al. (1999), a set of six microsatellite markers was used to validated diversity in *M. truncatula* Gaertn. accessions (Ellwood et al. 2006). More recently, the van Treuren et al. (2009) approach for developing core collections was used for alfalfa, and the hierarchical structure of its genepool can be accessed (and downloaded) from Genesys (https://www.genesys-pgr.org/c/forages). By matching *Medicago* spp. accessions from within (or between) genebanks to the branches and endpoints/groups on the diversity tree, the composition of the collections could be compared.

The NPGS *Medicago* spp. core collections have been used for evaluations, with online access to descriptor data via the GRIN-Global portal. Traits the NPGS alfalfa core collection that have been evaluated include forage quality (Jung et al. 1997), acid soil tolerance (Bouton 1996), heterosis, and forage yield (Bhandari et al. 2007), to list a few. The annual *Medicago* spp. core collection has been evaluated for reaction to several diseases including anthracnose (O'Neill and Bauchan 2000), *Phoma medicaginis* Malbr. & Roum. (O'Neill et al. 2003), *Erysiphe pisi* DC. (Yaege and Stuteville 2002), and *Peronospora trifoliorum* de Barry (Yaege and Stuteville 2000).

4.8 Assessing PGR Diversity

Genetic diversity in alfalfa and other *Medicago* spp. PGR has been systematically assessed in phenotypic, phylogenetic, and population structure evaluations. This research aids in PGR management by estimating genetic relationships and population structure among and between accessions, aiding in proper identification and genetic integrity of holdings in the collections, and determining levels of redundancies (i.e., duplication) and/or gaps in collection coverage. If accessible to stakeholders, this information also aids in targeting specific germplasm for intended research.

Defined phenotypic descriptor traits for legumes and annual medics have been used to assess diversity in *Medicago* spp. PGR collections (IBPGR 1984; Bioversity International 2007). Individual descriptor data associated with alfalfa (64 descriptors) and annual medics (69 descriptors) accessions for NPGS germplasm can be accessed through GRIN-Global with options for downloading information. Descriptors include morphological (e.g., flower color and pod coiling) traits that are key in plant taxonomic definitions. Other descriptors characterize traits important to plant breeding such as phenological (e.g., flowering stages), biotic and abiotic stress resistance and forage quality traits. Outside of the NPGS, phenotypic evaluations have also been used to characterize diversity in alfalfa germplasm in Croatia, with yield, vigor, and growth habit among the most important traits in partitioning diversity (Tucak et al. 2009). Phenotyping genetic diversity was also used to effectively manage alfalfa germplasm in a Moroccan breeding program. To achieve efficiency in larger, more in-depth evaluations, accessions with similar phenotypes were pooled to reduce complexity and numbers while preventing loss in genetic diversity (Rumbaugh et al. 1988).

Molecular markers, DNA barcoding, and to a lesser extent next-generation sequencing and genomic approaches, have been used to assess genetic diversity in alfalfa and *Medicago* spp. PGR. In alfalfa, population structure is often difficult to define because it is an outcrossing, polypoid species with accessions consisting of populations with unique genotypes for each plant. For many of the diploid annual *Medicago* spp., with smaller overall collections and allogamous pollination, some of these obstacles are not as formidable.

Despite difficulties, diversity in diploid (Şakiroğlu et al. 2010) and tetraploid (Ilhan et al. 2016) alfalfa accession from the NPGS alfalfa collections have been assessed with microsatellite markers. Analyses of diploid *M. sativa* accessions showed significant variability, clear distinctions between taxa, general agreement in taxonomic classifications, and geographic and ecogeographic subpopulation structure (Şakiroğlu et al. 2010) (Fig. 4.5). In a diverse NPGS subset of wild-collected tetraploid alfalfa accessions from across the species range, Ilhan et al. (2016) found that accessions grouped into two main clusters corresponding to subsp. *falcata* and subsp. *sativa* (Fig. 4.6). Also, a significant number of admixed accessions belonged to nothosubsp. *varia* and spatial genetic structure was identified for the subsp. *falcata* accessions. In a collection of mostly NPGS and Chinese alfalfa gemplasm, genetic diversity was also assessed with microsatellite markers (Qiang et al. 2015). Here, authors found high levels of genetic diversity, but no clear relationships between clustering and geography.

Along with phenotypic characterization, DNA barcoding—using plastid or nuclear sequence regions—has been used to elucidate taxonomic, genetic, and phylogenetic relationships in alfalfa and relative PGR in *Medicago* (Steele et al. 2010). This same approach was used to clarify native and introduced *Medicago* spp. in China, with results indicating strong support for treating *M. s.* subsp. *falcata* as a valid subspecies in the *M. sativa* complex (Chen et al. 2021). Elsherif and Ibrahim (2020) found in their research that DNA barcoding approaches of *M. sativa* germplasm were useful in characterizing diversity within species with sufficient resolution to identify and separate cultivars.

The development and use of next-generation sequencing platforms and extensive genomic resources are beginning to open possibilities for more in-depth analyses of *Medicago* spp. genetic resources. For example, Wang et al. (2020) characterized alfalfa population structure using single nucleotide polymorphisms (SNP) generated from genotype-by-sequencing (GBS) in a genomewide association study (GWAS). In this work, population structure of alfalfa PGR indicated that Chinese alfalfa was distinct from other germplasm evaluated. Published genome sequences for *M. truncatula* (Tang et al. 2014), *M. s.* subsp. *caerulea* (Li et al. 2020), and for cultivated tetraploid alfalfa (Chen et al. 2020), along with other efforts developing alfalfa genomics and genotyping platforms (described in this book), will likely facilitate characterizing population structure in the *Medicago* spp. PGR germplasm collections.

4.9 Alfalfa Crop Wild Relatives

Crop wild relatives (CWR) are species genetically close enough to the domesticated crop that can be used by plant breeders for improvement. The productive alfalfa cultivars grown today

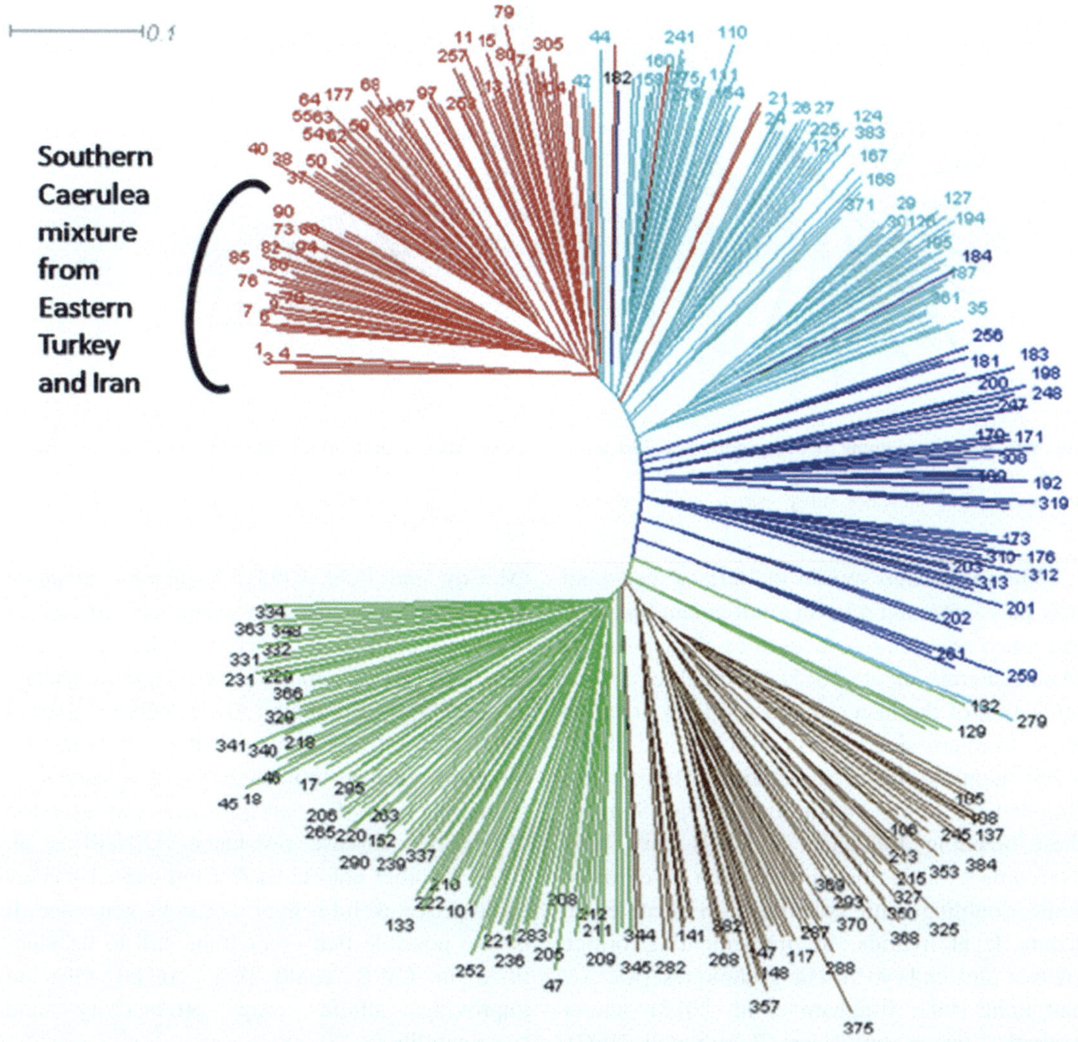

Fig. 4.5 Neighbor-joining dendrogram of 374 individual genotypes from 120 wild diploid accessions of *M. sativa*. *Falcata* A (lowland *falcata*) = brown; *Falcata B* (upland *falcata*) = green; *Caerulea* A (southern *caerulea*) = red; *Caerulea* B (northern *caerulea*) = light blue; and *Hemicycla* = dark blue (Şakiroğlu et al. 2010) *Subspecies *hemicycla* = *M. sativa* L. nothosubsp. *varia* in this book chapter. (*Reproduced with permission*)

have benefited from the introgression of traits from CWR (Table 4.4). Within the primary and secondary genepools, several different *M. sativa* subspecies have been used to introduce valuable traits. *Medicago s.* subsp. *falcata* germplasm has been used extensively to introgress cold hardiness into cultivated alfalfa (Lesins and Lesins 1979; Annicchiarico et al. 2015; Boe et al. 2020), with heterosis also being reported in subsp. *sativa* × subsp. *falcata* hybrids (Riday and Brummer 2002, 2005). Milić et al. (2018) identified aluminum tolerance in *M. s.* subsp. *caerulea* germplasm and proposed incorporating the trait into cultivars adapted to contaminated soils. Potato leafhopper (*Empoasca fabae* [Harris]) resistance in alfalfa, conferred by the glandular tipped hair trait, was introgressed from *M. s.* subsp. *glomerata* and *M. s.* subsp. *falcata* var. *viscosa* NPGS germplasm accessions (Sorensen et al. 1985, 1986, 1994; Shade and Kitch 1986; Bauchan and Greene 2001).

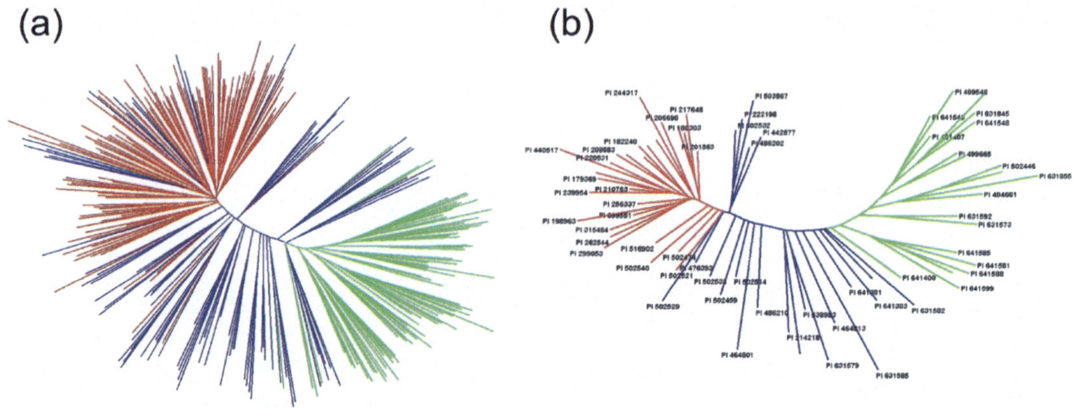

Fig. 4.6 Neighbor-joining dendrogram of **a** 280 tetraploid individuals and **b** 70 accessions from three subspecies. Red indicates subsp. *sativa*, green indicates subsp. *falcata*, and blue indicates nothosubsp. *varia*. (Ilhan et al. 2016) (*Reproduced with permission*)

Other *Medicago* spp. in the tertiary genepool may play important roles in alfalfa plant breeding and much research has focused on their potential (Annicchiarico et al. 2015). Difficulties in the utilization of the many alfalfa CWR in breeding arises from crossing barriers with unequal ploidy levels being a major constraint (McCoy and Bingham 1988; McCoy and Echt 1993). Some of these breeding difficulties in CWR have been overcome by using techniques such as chromosome doubling with colchicine to form equal ploidy level hybrids, by implementing bridge crosses and embryo rescue techniques (McCoy and Echt 1993; Bingham et al. 2013), and/or protoplast fusion approaches (Pupilli et al. 1992).

Examples of introduced traits, or of potential use, and interspecific hybrids in alfalfa CWR species in the tertiary genepool are listed in Table 4.4. Hybrids between *M. s.* subsp. *sativa* and *M. arborea* have been developed (Irwin et al. 2010, 2015; Bingham et al. 2013) with reported high forage yields and larger seed sizes (Humphries et al. 2020). Armour et al. (2008) described introgressing anthracnose disease resistance and the pod coiling trait from *M. arborea* into alfalfa. *Medicago daghestanica* Rupr. ex Boiss. and *M. pironae* Vis. were used in the development of interspecific hybrids at the diploid level via bridge crosses with *M. s.* subsp. *sativa* × *M. rupestris* M. Bieb. hybrids (McCoy and Echt 1993). *Medicago ruthenica* (L.) Trautv. was used in reciprocal crosses to introgress alkalinity, cold tolerance, and increased yield into cultivated alfalfa (Wang et al. 2008). In Bingham (2013), authors reported successful crosses between the autotetraploid (doubled using colchicine) *M. truncatula* 'Jemalong' and *M. s.* subsp. *sativa.* Introgression of reported disease resistance (O'Neill et al. 2003) or other abiotic traits from annual medics could prove useful if hybrids can be generated. It is also possible that other traits still to be identified in CWR could play crucial roles in improving alfalfa crop productivity and sustainability.

4.10 Alfalfa PGR in Breeding

Most cultivars of alfalfa grown today have been derived from and are the product of directed crosses and improved selections from original sources of germplasm. The history of cultivated alfalfa in the U.S. has been documented, with the first production fields established in the state of Utah (Clayton et al. 1997). Since then, nine "original" germplasm sources, "Falcata," "Varia," "Turkestan," "Flemish," "Ladak," "Chilean," "Peruvian," "Indian," and "African" have all been involved in the development of

Table 4.4 Primary, secondary, and tertiary genepools for alfalfa (*Medicago sativa* L.) and its crop wild relatives, estimated number of accessions in worldwide collections, and associated traits used or of potential use in crop breeding

Genepool	Taxon[a]	Accessions[b]	Trait[c]
Primary	*M. s.* subsp. *falcata* (L.) Arcang.	1,390	Readily hybridizes, with many traits introgressed. Examples include cold, drought, and grazing tolerance, which have been incorporated from subsp. *falcata* (Boe et al. 2020), and insect resistance from the glandular hair trait in subsp. *viscosa/glomerata* (Sorensen et al. 1985, 1986, 1994)
	M. s. subsp. *falcata* var. *viscosa* (Rchb.) Posp.	30	
	M. s. subsp. *glomerata* (Balb.) Rouy	98	
	M. s. nothosubsp. *varia* (Martyn) Arcang.	2,280	
Secondary	*M. s.* subsp. *caerulea* (Less. ex Ledeb.) Schmalh.	369	Cold, drought, and salt tolerance (Small 2011), and aluminum tolerance (Milić et al. 2018)
	M. s. nothosubsp. *tunetana* Murb.	28	–
Tertiary	*M. arborea* L.	158	Disease resistance (Renfro and Sprague 1959; Armour et al. 2008), yield (Irwin et al. 2010)
	M. cancellata M. Bieb.	58	Disease resistance (Borges et al. 1975)
	M. daghestanica Rupr. ex Boiss.	36	Disease resistance, fertility, cold tolerance (McCoy and Bingham 1988)
	M. hybrida (Pourr.) Trautv.	24	Disease resistance
	M. littoralis Rohde ex Loisel.	36	Disease resistance
	M. marina L.	238	Drought and salt tolerance (McCoy and Bingham 1988; Scippa et al. 2011)
	M. murex Willd.	619	Disease resistance (Elgin and Ostazeski 1982)
	M. papillosa[d] Boiss.	62	Disease resistance, drought tolerance (Quiros and Bauchan 1988)
	M. pironae Vis.	44	Fertility, cold tolerance, disease resistance (McCoy and Bingham 1988)
	M. rigidula (L.) All.	3,084	Insect resistance (Quiros and Bauchan 1988)
	M. rhodopeai Velen.	40	–
	M. rupestris M. Bieb.	35	–
	M. ruthenica (L.) Trautv.	167	Alkalinity, drought, and salt tolerance (Wang et al. 2008)
	M. saxatilis M. Bieb.	38	–
	M. sphaerocarpos Bertol.	35	Drought tolerance, disease resistance
	M. suffruticosa Ramond ex DC.	76	Disease resistance
	M. tenoreana Ser.	59	Disease resistance
	M. truncatula (L.) Mill.	9,571	Disease resistance (O'Neill and Bauchan 2000)

[a]Based on GRIN Taxonomy nomenclature with data derived from (Small 2011) and data derived from NPGS GRIN-Global https://npgsweb.ars-grin.gov/; APG GRIN-Global https://apg.pir.sa.gov.au/gringlobal/; Genesys https://www.genesys-pgr.org/ and the VIR Plant Genetic Resources Database http://db.vir.nw.ru/virdb/—*accessed* January, 2021
[b]Estimated number of accessions by taxon held in worldwide collections
[c]Examples of traits that have been or have the potential to be used in alfalfa breeding
[d]Includes accessions for both *M. papillosa* subspp. *macrocarpa* and *papillosa*

commercial cultivars in the U.S. (Barnes et al. 1977). Most of these germplasm genetic background (excluding "Falcata" and "Chilean") PGR were early introductions of the USDA's Plant Introduction Office (a predecessor of the NPGS) and were used widely for cultivar development (Bauchan and Greene 2001).

Over the years, breeding efforts in alfalfa have focused mostly on introgressing traits from germplasm sources for cold hardiness from *M. s.* subsp. *falcata* (Lesins and Lesins 1979; Annicchiarico et al. 2015; Boe et al. 2020) and disease (Havey and Maxwell 1988) and insect pest resistance (Sorensen et al. 1985; Shade and Kitch 1986; Sorensen et al. 1986, 1994; Bauchan and Greene 2001). More recent efforts have focused on identifying PGR traits for improving yield (Bhandari et al. 2007), forage quality, and abiotic stress tolerances (Zhang et al. 2018). Alfalfa PGR sourced from the NPGS, often identified in publications by their PI or plant introduction number, have been used widely in the development of cultivars (Caddel et al. 2000; Bauchan and Greene 2001; Viands et al. 2012). Many of these cultivars eventually are deposited and incorporated into the permanent NPGS collections. It is also true that un-adapted, unimproved, wild-collected, or landrace PGR are often not extensively utilized in cultivar development because of detrimental effects associated with linkage drag of undesirable traits including low yield and growth habit (Annicchiarico et al. 2015). Ongoing projects to evaluate and select from these underutilized PGR and to prebreed alfalfa providing sources for cultivar development are underway (Humphries et al. 2020). Although not directly used in breeding, alfalfa and *Medicago* spp. are often used in developing tools that aid in the plant breeding process. Research projects have used alfalfa PGR for developing trait associations and tools for marker-assisted selection (Yu et al. 2016; Yu 2017; Lin et al. 2020). Cazenave et al. (2019) used alfalfa PGR to develop high-throughput phenotyping approaches and tools to aid in data collection associated with breeding.

4.11 Summary

Important PGR collections for alfalfa and many of its wild relatives (i.e., *Medicago* spp.) are conserved in national and international genebanks. Managing these collections is a complicated process made particularly difficult by the extensive number of accessions and to the diversity in species/taxa. For major cultivated species like alfalfa (*M. sativa*), genetic diversity in collections is expansive, with specific gaps limited to unique ecogeographic regions of the world or to unique traits (e.g., drought tolerance). Plant genetic resources in secondary and tertiary genepools are less represented in ex situ collections, yet may be sources of useful traits, especially for adaptation to climate change. Access to information associated with accessions continues to amass, with the advent of larger genotyping datasets complicating hosting and data linkages. Germplasm of alfalfa, its CWR, and other *Medicago* spp. as well as their associated information are readily available for use in basic and applied research, with plant breeding benefiting especially. Ardent support and advocacy for alfalfa and other *Medicago* spp. PGR from stakeholders will be key in efforts to continue conservation and to provide long-term access to these invaluable resources.

References

Alercia A, Diulghero S, Mackay M (2015) FAO/Bioversity Multi-Crop. Passport Descriptors, V.2.1 (MCPD V.2.1). Bioversity International, Rome, Italy

Annicchiarico P, Barrett B, Brummer EC, Julier B, Marshall AH (2015) Achievements and challenges in improving temperate perennial forage legumes. Crit Rev Plant Sci 34:327–380

Annicchiarico P, Pecetti L, Proietti S (2004) Seed drying rates for long-term germplasm storage through a simple drying facility. Sementi Elette 50:17–22

Armour DJ, Mackie JM, Musial JM, Irwin JA (2008) Transfer of anthracnose resistance and pod coiling traits from *Medicago arborea* to *M. sativa* by sexual reproduction. Theor Appl Genet 117:149–156

Barnes DK, Bingham ET, Murphy RP, Hunt OJ, Beard DF, Skrdla WH, Teuber LR (1977) Alfalfa germplasm in the United States: Genetic vulnerability, use, improvement, and maintenance. USDA Tech Bull 171

Basigalup DH, Barnes DK, Stucker RE (1995) Development of a core collection for perennial *Medicago* plant introductions. Crop Sci 35:1163–1168

Bauchan GR, Greene SL (2001) Status *Medicago* germplasm. Plant Genet Resour Newsl 129:1–8

Belzile F, Abed A, Torkamaneh D (2020) Time for a paradigm shift in the use of plant genetic resources. Genome 63:189–194

Bhandari HS, Pierce CA, Murray LW, Ray IM (2007) Combining abilities and heterosis for forage yield among high-yielding accessions of the alfalfa core collection. Crop Sci 47:665–671

Bingham E (2013) Results using tetraploid *Medicago truncatula* cv. Jemalong in crosses with alfalfa. Medicago Genet Rep 13:1–16

Bingham E, Armour D, Irwin J (2013) The hybridization barrier between herbaceous *Medicago sativa* and woody *M. arborea* is weakened by selection of seed parents. Plants (Basel) 2:343–353

Bioversity International (2007) Developing crop descriptor lists. Bioversity Technical Bulletin 13:71

Boe A, Kephart KD, Berdahl JD, Peel MD, Brummer EC, Xu L, Wu Y (2020) Breeding alfalfa for semiarid regions in the northern great plains: history and additional genetic evaluations of novel germplasm. Agronomy 10:40–61

Boller B, Greene SL (2010) Genetic resources. In: Boller B, Posselt UK, Veronesi F (eds) Handbook of plant breeding, vol 5. Fodder and Amenity Grasses. Springer, New York, pp 13–38

Borges OL, Stanford EH, Webster RK (1975) The host-pathogen interaction of alfalfa and *Stemphylium botryosum*. Phytopathology 66:749–753

Bouton JH (1996) Screening the alfalfa core collection for acid soil tolerance. Crop Sci 36:198–200

Brink M, van Hintum T (2019) Genebank operation in the arena of access and benefit-sharing policies. Front Plant Sci 10:1712

Brummer EC, Cazcarro PM, Luth D (1999) Ploidy determination of alfalfa germplasm accessions using flow cytometry. Crop Sci 39:1202–1207

Brunet J, Zhao Y, Clayton MK (2019a) Linking the foraging behavior of three bee species to pollen dispersal and gene flow. PLoS ONE 14:

Brunet J, Ziobro R, Osvatic J, Clayton MK (2019b) The effects of time, temperature and plant variety on pollen viability and its implications for gene flow risk. Plant Biol 21:715–722

Byrne PF, Volk GM, Gardner C, Gore MA, Simon PW, Smith S (2018) Sustaining the future of plant breeding: the critical role of the USDA-ARS National plant germplasm system. Crop Sci 58:451–468

Caddel JL, Zarrabi AA, Berberet RC, Prater JD (2000) Registration of OK 163, OK 164, OK 187, OK 188, OK 189, and OK 208, enhanced alfalfa world collection germplasm. Crop Sci 42:316–317

Cazenave A-B, Shah K, Trammell T, Komp M, Hoffman J, Motes CM, Monteros MJ (2019) High-throughput approaches for phenotyping alfalfa germplasm under abiotic stress in the field. Plant Phenome J 2:1–13

Chen H, Zeng Y, Yang Y, Huang L, Tang B, Zhang H, Hao F, Liu W, Li Y, Liu Y, Zhang X, Zhang R, Zhang Y, Li Y, Wang K, He H, Wang Z, Fan G, Yang H, Bao A, Shang Z, Chen J, Wang W, Qiu Q (2020) Allele-aware chromosome-level genome assembly and efficient transgene-free genome editing for the autotetraploid cultivated alfalfa. Nat Commun 11:2494

Chen J, Wu G, Shrestha N, Wu S, Guo W, Yin M, Li A, Liu J, Ren G (2021) Phylogeny and species delimitation of chinese *Medicago* (Leguminosae) and its relatives based on molecular and morphological evidence. Front Plant Sci 11:1–10

Clark RL, Shands HL, Bretting PK, Eberhart SA (1997) Managing large diverse germplasm collections. Crop Sci 37:1–6

Clayton RB, Robison LR, Jackson RH (1997) The historical diffusion of alfalfa. J Agron Edu 6:13–19

Dempewolf H, Eastwood RJ, Guarino L, Khoury CK, Müller JV, Toll J (2014) Adapting agriculture to climate change: a global initiative to collect, conserve, and use crop wild relatives. Agroecol Sust Food 38:369–377

Diwan N, Bauchan GR, McIntosh MS (1994) A core collection for the United States annual *Medicago* germplasm collection. Crop Sci 34:279–285

Diwan N, McIntosh MS, Bauchan GR (1995) Methods of developing a core collection of annual *Medicago* species. Theor Appl Genet 90:755–761

Elgin JH Jr, Ostazeski SA (1982) Evaluation of selected alfalfa cultivars and related *Medicago* species for resistance to race 1 and race 2 *Anthracnose*. Crop Sci 22:39–42

Ellwood SR, D'Souza NK, Kamphuis LG, Burgess TI, Nair RM, Oliver RP (2006) SSR analysis of the *Medicago truncatula* SARDI core collection reveals substantial diversity and unusual genotype dispersal throughout the Mediterranean basin. Theor Appl Genet 112:977–983

Elsherif N, Ibrahim M (2020) Implications of *rbcL* and *rpoC1* DNA Barcoding in phylogenetic relationships of some Egyptian *Medicago sativa* L. cultivars. Egypt J Bot 60:451–460

Engels JMM, Visser L (eds) (2003) A guide to effective management of germplasm collections. IPGRI, Rome, Italy

FAO (2013) Genebank standards for plant genetic resources for food and agriculture. Rome

Frankel OH (1984) Genetic perspectives of germplasm conservation. In: Arber WK, Llimensee K, Peacock WJ, Starlinger P (eds) Genetic manipulation: impact on man and society. Cambridge University Press, Cambridge, Enlgand, pp 161–170

Greene SL, Afonin A, Dzyubenko E, Dzyubenko N (2011) Crop wild relatives of *Medicago* in Russia and neighboring countries: gap analysis for effective conservation. In: Maxted N, Dullo ME, Ford-Lloyd BV, Frese L, Iriondo J, Pinheiro de Carvalho MA (eds) Agrobiodiversity conservation: securing the diversity of crop wild relatives and landraces. Cab International, United Kingdom, pp 82–90

Greene SL, Hannan R, Afonin A, Dzyubenko NI, Khusainov A (2005) Collecting wild crop relatives in the northwestern steppes of Kazakhstan. Plant Genet Resour Newsl 141:1–6

Greene SL, Kesoju SR, Martin RC, Kramer M (2015) Occurrence of transgenic feral alfalfa (*Medicago sativa* subsp. *sativa* L.) in alfalfa seed production areas in the United States. PLoS One 10:e0143296

Greene SL, Kisha TJ, Dzyubenko NI (2008) Conserving alfalfa wild relatives: is past introgression with Russian varieties evident today? Crop Sci 48:1853–1864

Havey MJ, Maxwell DP (1988) Transfer of disease resistance from diploid to tetraploid alfalfa by unreduced female gametes. Plant Dis 72:603–604

Humphries AW, Ovalle C, Hughes S, del Pozo A, Inostroza L, Barahona V, Yu L, Yerzhanova S, Rowe T, Hill J, Meiirman G, Abayev S, Brummer EC, Peck DM, Toktarbekova S, Kalibayev B, Espinoza S, Ivelic-Saez J, Bingham E, Small E, Kilian B (2020) Characterization and pre-breeding of diverse alfalfa wild relatives originating from drought-stressed environments. Crop Sci 61:69–88

IBPGR (1984) Descriptor list for forage legumes. In: Andersen S, Davies WE (eds) Descriptor list for forage legumes. IBPGR Secretariat, Rome; Italy, pp 29–29

Ilhan D, Brummer EC, Li X, Şakiroğlu M (2016) Genetic diversity and population structure of tetraploid accessions of the *Medicago sativa–falcata* complex. Crop Sci 56:1146–1156

Irwin J, Sewell J, Woodfield D, Bingham E (2015) Restructuring lucerne (*Medicago sativa*) through introgression of the *Medicago arborea* genome. Agri Sci 28:40–46

Irwin JAG, Armour DJ, Pepper PM, Lowe KF (2010) Heterosis in lucerne testcrosses with *Medicago arborea* introgressions and Omani landraces and their performance in synthetics. Crop Pasture Sci 61:450–463

Ivanov AI, Brezhnev DD (1988) Alfalfa. Balkema, A.A

Jung HG, Sheaffer CC, Barnes DK, Halgerson JL (1997) Forage quality variation in the U.S. alfalfa core collection. Crop Sci 37:1361–1366

Kesoju SR, Greene SL, Martin RC, Kramer M, Walsh DB, Boydston RA (2020) Isolation distances for transgenic alfalfa seed production in the Pacific Northwest. Crop Sci 59:1701–1708

Kumar PL, Cuervo M, Kreuze JF, Muller G, Kulkarni G, Kumari SG, Massart S, Mezzalama M, Alakonya A, Muchugi A, Graziosi I, Ndjiondjop M-N, Sharma R, Negawo AT (2021) Phytosanitary interventions for safe global germplasm exchange and the prevention of transboundary pest spread: the role of CGIAR germplasm health units. Plants 10:328

Lesins KA, Lesins I (1979) Genus *Medicago* (Leguminosae). A taxongenic study. Junk, The Hague, Netherlands

Li A, Liu A, Du X, Chen JY, Yin M, Hu HY, Shrestha N, Wu SD, Wang HQ, Dou QW, Liu ZP, Liu JQ, Yang YZ, Ren GP (2020) A chromosome-scale genome assembly of a diploid alfalfa, the progenitor of autotetraploid alfalfa. Hortic Res 7:194

Lin S, Medina CA, Boge B, Hu J, Fransen S, Norberg S, Yu LX (2020) Identification of genetic loci associated with forage quality in response to water deficit in autotetraploid alfalfa (*Medicago sativa* L.). BMC Plant Biol 20:303

Lopez-Noriega I, Galluzzi G, Halewood M, Vernooy R, Bertacchini E, Gauchan D, Welch E (2012) Flows under stress: availability of plant genetic resources in times of climate and policy change. Working Paper 18, Copenhagen, Denmark: CCAFS

Loumerem M, Ferchichi A, Haddad M, Rahim MAA, Hajjaji H (2007) Collection and evaluation of lucerne (*Medicago sativa* L.) germplasm from oases of Tunisia. Genet Resour Crop Evol 54:1645–1651

Mascher M, Schreiber M, Scholz U, Graner A, Reif JC, Stein N (2019) Genebank genomics bridges the gap between the conservation of crop diversity and plant breeding. Nat Genet 51:1076–1081

McCoy TJ, Bingham ET (1988) Cytology and cytogenetics of alfalfa. Alfalfa and Alfalfa Improvement. ASA-CSSA-SSSA, Madison, WI, pp 737–766

McCoy TJ, Echt CS (1993) Potential of trispecies bridge crosses and random amplified polymorphic DNA markers for introgression of *Medicago daghestanica* and *M. pironae* germplasm into alfalfa (*M. sativa*). Genome 36:594–601

Meglič V, Pelikán J, Babinec J (2003) Current state of the alfalfa genetic resources in Europe. Czech J Genet Plant 39:20–22

Mezzalama M, Crouch JH, Ortiz R (2010) Monitoring the threat of unintentional transgene flow into maize gene banks and breeding materials. Electron J Biotechnol 13:1–6

Michaud RLWF, Rumbaugh MD (1988) World distribution and historical development. In: Hanson AA, Barnes DK, Hill RR Jr (eds) Alfalfa and alfalfa improvement. Wiley, Hoboken, NJ, pp 25–91

Milić D, Taški-Ajduković K, Nagl N, Atlagić J, Karagić D (2018) Utilization of *M. sativa* ssp. *caerulea* × *M. sativa* ssp. *sativa* hybridization in improvement of alfalfa aluminium tolerance. Plant Genet Resour-C 16:68–73

O'Neill NR, Bauchan GR (2000) Sources of resistance to anthracnose in the annual *Medicago* core collection. Plant Dis 84:261–267

O'Neill NR, Bauchan GR, Samac DA (2003) Reactions in the annual *Medicago* spp. core germplasm collection to *Phoma medicaginis*. Plant Dis 87:557–562

Odong TL, Jansen J, van Eeuwijk FA, van Hintum TJL (2013) Quality of core collections for effective

utilisation of genetic resources review, discussion and interpretation. Theor Appl Genet 126:289–305
Peterson SS, Baird CR (1994) The alfalfa seed chalcid: an important alfalfa seed pest. Agricultural Experiment & UI Extension Publications, Digital Initiatives, University of Idaho Library, Moscow, ID
Piano E, Valentini P, Pecetti L, Romani M (1996) Evaluation of a lucerne germplasm collection in relation to traits conferring grazing tolerance. Euphytica 89:279–288
Postman JD, Bretting PK, Kinard GR, Cyr PD, Weaver B, Millard MJ, Gardner CA, Bohning MA, Emberland GP, Sinnott QP, Ayala Silva T, Hummer KE, Franco T, Mackay M, Guarino L (2010) GRIN-Global: an international project to develop a global plant genebank information management system. Acta Hort 859:49–55
Prosperi J-M, Eric J, Joëlle R, Michel A (2006) Morphologic and agronomic diversity of wild genetic resources of *Medicago sativa* L. collected in Spain. Genet Resour Crop Evol 53:843–856
Prosperi J-M, Jenczewski E, Muller M-H, Fourtier S, Sampoux J-P, Ronfort J (2014) Alfalfa domestication history, genetic diversity and genetic resources. Legume Perspec 4:13–14
Pupilli E, Scarpa GM, Damiani F, Arcioni S (1992) Production of interspecific somatic hybrid plants in the genus *Medicago* through protoplast fusion. Theor Appl Genet 84:792–797
Qiang H, Chen Z, Zhang Z, Wang X, Gao H, Wang Z (2015) Molecular diversity and population structure of a worldwide collection of cultivated tetraploid alfalfa (*Medicago sativa* subsp. *sativa* L.) germplasm as revealed by microsatellite markers. PLOS ONE 10: e0124592
Quiros CF, Bauchan GR (1988) The genus *Medicago* and the origin of the *Medicago sativa* Comp. In: Hanson AA, Barnes DK, Hill RR Jr (eds) Alfalfa and alfalfa improvement. Wiley, Hoboken, NJ, pp 93–124
Renfro BL, Sprague EW (1959) Reaction of *Medicago* species to eight alfalfa pathogens. Agron J 51:481–483
Riday H, Brummer EC (2002) Forage yield heterosis in alfalfa. Crop Sci 42:716–723
Riday H, Brummer EC (2005) Heterosis in a broad range of alfalfa germplasm. Crop Sci 45:8–17
Rowe DE (1986) Effects and control of genetic drift in the autotetraploid population. Crop Sci 26:89–92
Rubenstein K, Smale M, Widrlechner MP (2006) Demand for genetic resources and the U.S. National Plant Germplasm System. Crop Sci 46:1021–1031
Rumbaugh MD, Graves WL, Caddel JL, Mohammad RM (1988) Variability in a collection of alfalfa germplasm from Morocco. Crop Sci 28:605–609
Şakiroğlu M, Doyle JJ, Brummer CE (2010) Inferring population structure and genetic diversity of broad range of wild diploid alfalfa (*Medicago sativa* L.) accessions using SSR markers. Theor Appl Genet 121:403–415
Samac DA, Rhodes LH, Lamp WO (2016) Compendium of alfalfa diseases and pests, 3rd edn. APS Press, St. Paul, MN
Scippa GS, Petrollini E, Trupiano D, Rocco M, Falco G, Di Michele M, Chiatante D (2011) Dormancy of *Medicago marina* (L.) seed. Environ Exp Bot 72:320–329
Shade RE, Kitch LW (1986) Registration of 81IND-2 germplasm. Crop Sci 26:205
Sinskaya EN (1961) Flora of cultivated plants of the USSR: Perenial leguminous plants: Part I, Medic, Sweetclover, Fenugreek. Israel Program for Scientific Translations, Jerusalem, Israel
Skinner DZ, Bauchan GR, Auricht G, Hughes S (1999) A method for the efficient management and utilization of large germplasm collections. Crop Sci 39:1237–1242
Smale M, Day-Rubenstein K (2002) The demand for crop genetic resources: international use of the US National Plant Germplasm System. World Dev 30:1639–1655
Small E (2011) Alfalfa and relatives: evloution and classification of *Medicago*. NRC Research Press, Ottawa, Ontario, Canada
Smýkal P, Coyne CJ, Ambrose MJ, Maxted N, Schaefer H, Blair MW, Berger J, Greene SL, Nelson MN, Besharat N, Vymyslický T, Toker C, Saxena RK, Roorkiwal M, Pandey MK, Hu J, Li YH, Wang LX, Guo Y, Qiu LJ, Redden RJ, Varshney RK (2015) Legume crops phylogeny and genetic diversity for science and breeding. Crit Rev Plant Sci 34:43–104
Solberg SØ, Yndgaard F, Andreasen C, von Bothmer R, Loskutov IG, Asdal Å (2020) Long-term storage and longevity of orthodox seeds: a systematic review. Front Plant Sci 11
Sorensen EL, Horber EK, Stuteville DL (1985) Registration of KS108GH5 germplasm. Crop Sci 25:1132
Sorensen EL, Horber EK, Stuteville DL (1986) Registration of KS94GH6 germplasm. Crop Sci 26:1088
Sorensen EL, Stuteville DL, Horber EK, Skinner DZ (1994) Registration of KS224 glandular-haired alfalfa germplasm with multiple pest resistance. Crop Sci 34:544–544
Steele K, Ickert-Bond S, Zarre S, Wojciechowski M (2010) Phylogeny and character evolution in *Medicago* (Leguminosae): Evidence from analyses of plastid *trnK/matK* and nuclear *GA3ox1* sequences. Am J Bot 97:1142–1155
Tang H, Krishnakumar V, Bidwell S, Rosen B, Chan A, Zhou S, Gentzbittel L, Childs KL, Yandell M, Gundlach H, Mayer KFX, Schwartz DC, Town CD (2014) An improved genome release (version Mt4.0) for the model legume *Medicago truncatula*. BMC Genom 312
Tiwari SP, Randhawa GJ (2010) Strategies to monitor the adventitious presence of transgenes in *ex situ* collections. Indian J Agric Sci 80:351–356
Tucak M, Popović S, Cupic T, Šimić G, Gantner R, Meglic V (2009) Evaluation of alfalfa germplasm collection by multivariate analysis based on phenotypic traits. Rom Agric Res 26:47–52

Vaitsis T (2000) Collecting and breeding *Medicago* perennial species in Greece. In: Veronesi F, Rosellini D (eds) Lucerne and medics for the XXI Century Proceedings XIII Eucarpia *Medicago* spp Group Meeting, Perugia, Italy, 13-16 September 1999. Universita di Perugia, Perugia; Italy, pp 19–27

Van Deynze AE, Fitzpatrick S, Hammon B, McCaslin MH, Putnam DH, Teuber LR, Undersander DJ (2008) Gene flow in alfalfa: biology, mitigation, and potential impact on production. 28 edn, United States

van Treuren R, Engels JMM, Hoekstra R, van Hintum TJL (2009) Optimization of the composition of crop collections for *ex situ* conservation. Plant Genet Resour-C 7:185–193

Viands DR, Hansen JL, Crawford JL (2012) Registration of 'Ezra' alfalfa. J Plant Regist 6:225–228

Volk GM, Bretting PK, Byrne PF (2019) Survey identifies essential plant genetic resources training program components. Crop Sci 59:2308–2316

Walters C, Wheeler LM, Grotenhuis JM (2007) Longevity of seeds stored in a genebank: species characteristics. Seed Sci Res 15:1–20

Wang DK, Li H, Luo XY (2008) Crossbreeding of Melilotoides ruthenicus and *Medicago sativa*. Acta Agrestia Sinica 16:458–465

Wang Z, Wang X, Zhang H, Ma L, Zhao H, Jones CS, Chen J, Liu G (2020) A genome-wide association study approach to the identification of candidate genes underlying agronomic traits in alfalfa (*Medicago sativa* L.). Plant Biotechnol J 18:611–613

Weise S, Lohwasser U, Oppermann M (2020) Document or lose it-on the importance of information management for genetic resources conservation in genebanks. Plants (Basel) 9:1050

Yaege JR, Stuteville DL (2000) Reactions in the annual *Medicago* core germplasm collection to two isolates of *Peronospora trifoliorum* from alfalfa. Plant Dis 84:521–524

Yaege JR, Stuteville DL (2002) Reactions of accessions in the annual *Medicago* core germplasm collection to *Erysiphe pisi*. Plant Dis 86:312–315

Yu LX (2017) Identification of single-nucleotide polymorphic loci associated with biomass yield under water deficit in alfalfa (*Medicago sativa* L.) using genome-wide sequencing and association mapping. Front Plant Sci 8:1152

Yu LX, Liu X, Boge W, Liu XP (2016) Genome-wide association study identifies loci for salt tolerance during germination in autotetraploid alfalfa (*Medicago sativa* L.) using genotyping-by-sequencing. Front Plant Sci 7:956

Zhang T, Kesoju S, Greene SL, Fransen S, Hu J, Yu L-X (2018) Genetic diversity and phenotypic variation for drought resistance in alfalfa (*Medicago sativa* L.) germplasm collected for drought tolerance. Genet Resour Crop Evol 65:471–484

Biotechnology Advances in Alfalfa

5

Deborah A. Samac
and Stephen J. Temple

Abstract

Unique traits not found in alfalfa (*Medicago sativa*) germplasm can be introduced through genetic engineering for crop improvement. *Agrobacterium*-mediated transformation by co-cultivation of plant tissue pieces followed by somatic embryogenesis and plant regeneration or transformation of cotyledonary node meristems is now routine for generating transgenic alfalfa plants. Numerous genes have been introduced into alfalfa for understanding gene function, characterizing promoters, and for introducing a wide array of agronomic traits: tolerance to aluminum toxicity, salt, and drought; resistance to herbicides, diseases, and insects; bioremediation of heavy metals; increased plant biomass, improved nutrient uptake, and improved forage quality and nutritional content. Genetically modified alfalfa was also explored for large-scale production of enzymes, biodegradable plastics, and pharmaceuticals including antigens for veterinary applications. The most widely used promoter for transgenic alfalfa research is the CaMV 35S promoter, although expression in alfalfa is lower than in other plants. This chapter presents a summary of constitutive, tissue-specific, and inducible promoters tested in alfalfa and recent advances in developing transgenic alfalfa for desired agronomic traits. The development of the two commercialized traits, glyphosate resistance (Roundup Ready® alfalfa) and reduced lignin (HarvXtra® alfalfa) is detailed including challenges encountered in breeding and deregulation of these genetically modified traits. The emerging use of gene editing will likely have a large impact on alfalfa improvement and commercialization of new traits.

5.1 Introduction

Alfalfa was among the first plants for which methods of regeneration of plants in tissue culture were developed. Nonetheless, commercialization and adoption of genetically modified (GM) alfalfa lagged behind soybean, canola, corn, and cotton. Up to the present time, only two transgenic traits have been commercialized in alfalfa, tolerance to the herbicide glyphosate (Roundup Ready®) and reduced lignin (HarvXtra®). The small number of commercialized traits is in stark contrast to a large number of reports of successful genetic engineering in alfalfa for crop improvement. Primarily this is due to the

D. A. Samac (✉)
USDA-ARS-Plant Science Research Unit,
St. Paul, MN 55108, USA
e-mail: debby.samac@usda.gov

S. J. Temple
Forage Genetics International, West Salem,
WI 54669, USA
e-mail: stemple@foragegenetics.com

L.-X. Yu and C. Kole (eds.), *The Alfalfa Genome*, Compendium of Plant Genomes,
https://doi.org/10.1007/978-3-030-74466-3_5

significant regulatory and economic hurdles to commercialization of transgenic traits (Wang and Brummer 2012) with the result that most GM traits do not offer a sufficient return on investment for commercialization. Additionally, some transgenic traits fail to perform well when transferred to other genetic backgrounds or when grown and tested under field conditions. Additional hurdles to commercialization are intellectual property (IP) issues in which genes, promoters, or technologies may need to be licensed for commercialization in alfalfa. For example, early research in alfalfa biotechnology successfully developed transgenic alfalfa lines producing high amounts of phytase, an enzyme that releases phosphorus from phytates found in plant seeds (Ullah et al. 2002). Phytase is added to animal feeds to enhance nutritional value and improve animal growth and health. The enzyme could be extracted efficiently from transgenic alfalfa or the dried alfalfa leaf meal added directly to feed to promote phosphorus nutrition (Austin-Phillips et al. 1999). However, this promising product never came to market because issues surrounding IP could not be resolved.

Several reviews on different aspects of alfalfa biotechnology have been published previously. A thorough review of alfalfa tissue culture and somatic embryogenesis was published by Bingham et al. (1988). As reviewed by these authors, alfalfa was a model for tissue culture research for some time and many studies were done to optimize methods and media to improve the frequency of regeneration and explore genotype specificity of regeneration. Most alfalfa cultivars have a low number of plants that will regenerate in culture, although some germplasm such as creeping rooted types have a high frequency of plants capable of regeneration. Bingham et al. (1975) found that the frequency of regeneration could be improved by breeding. Selection within the cultivar Saranac produced RegenS (Bingham 1989) and selection within *M. sativa* ssp. *falcata* produced RegenY (Bingham et al. 1975) with high regeneration potential. A cross of these two genotypes resulted in the RegenSYgermplasm (Bingham 1991), which has been used by academic and government scientists worldwide for alfalfa biotechnology research. This chapter will focus primarily on the recent advances in alfalfa biotechnology not covered in previous reviews (Samac and Temple 2004; Tesfaye et al. 2008; Wang and Brummer 2012; Singer et al. 2017; Liu et al. 2018).

5.2 Transformation Methods

Previous reviews have summarized research on *Agrobacterium*-mediated and particle bombardment transformation methods and optimizing transformation efficiency in alfalfa (Atkins and Smith 1997; Samac and Temple 2004; Liu et al. 2018). The most widely used method of gene transfer is *A. tumefaciens* co-cultivation. A number of strains have been used, although some strain–alfalfa genotype interaction has been noted (Samac 1995; Ziauddin et al. 2004). Both herbicide resistance (phosphinothricin, glyphosate) and antibiotic resistance (kanamycin, hygromycin) selectable marker genes are used successfully in alfalfa transformation. Use of gabaculine, which has herbicidal activity, has also been reported for selecting transgenic alfalfa (Rosellini et al. 2007; Ferradini et al. 2011) and tolerance to atrazine has also been obtained, although not used in selection (Wang et al. 2005; Vail et al. 2014). Two detailed protocols for *Agrobacterium*-mediated transformation and regeneration of alfalfa using the RegenSY germplasm have been published (Samac and Austin-Phillips 2006; Fu et al. 2015).

5.2.1 In Planta *Transformation*

A limitation of most *Agrobacterium*-mediated transformation methods is a need to go through a callus phase and induce the formation of somatic embryos. This step may be genotype-specific and require optimization of culture media for higher efficiency of embryogenesis. One means to eliminate these steps is the transformation of plant meristematic cells. Ding et al. (2003) reported efficient regeneration of cultivars from cotyledonary nodes of alfalfa seedlings, but transformation was reported only for *Trifolium* species. Successful transformation was obtained using cotyledonary nodes of seedlings after additional wounding (Weeks et al. 2008). Several advantages of this method were noted: no antibiotic resistance marker was required, seed production occurred within 5 months of the initial transformation, and a commercial cultivar could be used for transformation. Several groups have reported using the method successfully (Liu et al. 2013; Duan et al. 2015; Bao et al. 2016; Zhang et al. 2016; Wang et al. 2016a) and a similar method was described by Wang et al. (2019a). A disadvantage of this method is that the genetic background of transgenic plants is not uniform since transgenes are inserted into a genetically and phenotypically heterogeneous population of seedlings. Also, the frequency of transformation appears to be lower than those utilizing somatic embryogenesis.

5.2.2 Chloroplast Transformation

Chloroplast transformation has been pursued in a wide range of crop plants as a means of increasing the expression of transgenes and limiting gene flow in the environment (Adem et al. 2017). Expression of genes in chloroplasts is not subject to the gene silencing sometimes found with nuclear transformation. Gene products are compartmentalized in plastids, which may increase the accumulation of some products such as polyhydroxyalkanoates, polymers that can be used to manufacture biodegradable plastic items (Nawrath et al. 1994; McQualter et al. 2014). Wei et al. (2011) reported successful chloroplast transformation of alfalfa at a low frequency from particle bombardment of callus and leaf tissue and demonstrated green fluorescent protein expression in chloroplasts of regenerated plants. Plastid inheritance in alfalfa is biparental with a strong paternal bias (Smith et al. 1986) making the development and maintenance of homoplasmic lines a challenge. Nevertheless, chloroplast transformation would greatly facilitate projects in which the expression of a large amount of a target protein is desired.

5.3 Promoters for Gene Expression in Alfalfa

The most widely used promoter in alfalfa biotechnology experiments is the 35S promoter from the cauliflower mosaic virus (CaMV). Due to its wide use in the transformation of other plants and generally high constitutive expression, the 35S promoter is assumed to have strong activity in alfalfa. However, several reports found that the 35S promoter has lower activity in alfalfa than in model plants (Tabe et al. 1995; Khoudi et al. 1999; Samac et al. 2004b; D'Aoust et al. 2005). When 35S activity was measured with the β-glucuronidase (*gusA*) reporter gene, the amount of GUS activity in alfalfa leaves is considerably lower than the GUS activity in tobacco (Samac et al. 2004b). In stems, activity is found in the epidermis, chlorenchyma, phloem, and cambium, but rarely in xylem or pith cells. In roots and nodules, activity is limited in most plants to vascular tissues. Alternative constitutive promoters with higher expression and less tissue specificity include the Mac promoter, a chimeric promoter with elements from the 35S promoter, and mannopine synthetase promoter (Comai et al. 1990) which was used for the production of proteins that accumulated to high levels in alfalfa (Austin-Phillips and Ziegelhoffer 2001). The cassava vein mosaic virus promoter also was shown to have strong constitutive activity in alfalfa (Samac et al. 2004b). For commercialization of Roundup Ready® alfalfa, the enhanced figwort mosaic virus (FMV) sequence, which consists of the FMV promoter with a duplicated enhancer region (Richins et al. 1987), was used as a strong constitutive promoter (Rogan and Fitzpatrick 2004).

The promoters Act2 (Barone et al. 2008) and MtHP (Reyno et al. 2013) were reported to be constitutively expressed in alfalfa, but this has not been confirmed with reporter gene data, which is needed to identify expression patterns and compare expression levels with other promoters. The activity of promoters from alfalfa and from other plant species are listed in Tables 5.1 and 5.2, although not all promoters have been examined using a visible reporter gene.

Table 5.1 Expression of promoters from alfalfa

Gene promoter	Primary expression pattern	References
Asparagine synthase (AS)	Infected and uninfected nodule cells	Shi et al. (1997)
Aspartate amino transferase 1 (AAT-1)	Uninfected nodule cells	Yoshioka et al. (1999)
Aspartate amino transferase 2 (AAT-2)	Infected nodule cells	Yoshioka et al. (1999)
Early nodulin 40 (ENOD40)	Root vascular tissue, root tip, nodule primordium, vascular bundles	Fang and Hirsch (1998)
Glutamate synthase (GOGAT)	Infected nodule cells	Trepp et al. (1999)
Isoflavone reductase (IFR)	Root meristem and cortex, nodules; fungal induced	Oommen et al. (1994)
Mitogen-activated protein kinase (MAPK)	Meristematic cells, glandular hairs	Schoenbeck et al. (1999)
Pathogenesis Related -10 (PR10)	Roots, pathogen inducible in leaves	Sathoff et al. (2020)
Phosphoenolpyruvate carboxylase (PEPC)	Infected nodule cells, pulvinar cells	Pathirana et al. (1997)
Phosphoenolpyruvate carboxylase-4 (PEPC-4)	Vascular tissue, xylem cells	Pathirana et al. (1997)
Plastocyanin	Leaves	Weeks et al. (2008)
PO22, PO149	Mature pollen grains	Wu et al. (1998)
Proline rich protein 2 (MsPRP2)	Root cells, callus cells	Winicov et al. (2004)
Ribulose-1,5-bisphosphate carboxylase-oxygenase, small subunit (ssuRuBisCo)	Light inducible, leaf tissue	Khoudi et al. (1999), Samac et al. (2004b), Weeks et al. (2008)

5.4 Transgenic Traits in Alfalfa

An impressive amount of research in alfalfa biotechnology has been accomplished to introduce novel value-added traits, enhance biotic and abiotic stress tolerance, increase the nutritive value of foliage, and improve biomass production. The earliest biotechnology research focused on expanding the use of alfalfa from an animal feed to an industrial feedstock by introducing novel value-added traits. Genes for production of industrial enzymes; phytase, manganese-dependent lignin peroxidase, alpha-amylase, cellulase, and endochitinase; were selected with the goal of increasing alfalfa acreage (reviewed by Tesfaye et al. 2008). Additionally, alfalfa plants were engineered for the production of a biodegradable plastic polymer, polyhydroxybutyrate (Saruul et al. 2002). The press residue remaining after wet fractionation of these alfalfa feedstocks could be used for biomass energy, providing multiple streams of revenue for the producer (Koegel and Straub 1996). Additional value-added traits introduced into alfalfa include genes for bioremediation of the herbicide atrazine (Wang et al. 2005; Vail et al. 2014) and removing heavy metals from soil (Watrud et al. 2006; Zhang et al. 2013; Kumar et al. 2019).

Significant work has also been done developing alfalfa to produce viral antigens and biopharmaceuticals in alfalfa, so-called molecular farming (Khoudi et al. 1999; Bardor et al. 2003; D'Aoust et al. 2005; Stefanova et al. 2013). Although these traits would not necessarily increase alfalfa acreage, they could provide high revenue (Khoudi et al. 1999). However, there would be significant costs and complexity associated with the commercialization of edible vaccines. The utilization of alfalfa has the advantage of protein stability (Busse et al. 2001) and the ability to generate homogenous N-glycan

Table 5.2 Heterologous promoters expressed in alfalfa

Gene promoter	Origin	Expression pattern	References
Act2	Arabidopsis	Constitutive	Barone et al. (2008)
Blec4	*Pisum sativum*	Epidermal cells	Mandaci and Dobres (1997)
Class III chitinase	Arabidopsis	Vascular tissue	Samac et al. (2004a)
Glutamate-ammonia ligase (GS3A)	*Pisum sativum*	Phloem; nodule primordia, meristem, symbiotic zone, vascular tissue	Brears et al. (1991)
MtHP	*Medicago truncatula*	Constitutive	Reyno et al. (2013)
MtPT1	*Medicago truncatula*	Roots	Ma et al. (2012)
PAL2	*Phaseolus vulgaris*	Vascular tissue	Guo et al. (2001a)
pin2	*Solanum tuberosum*	Vascular tissue, mesophyll	Samac and Smigocki (2003)
PR5	*Medicago truncatula*	Roots	Sathoff et al. (2020)
PR10	*Medicago truncatula*	Roots	Sathoff et al. (2020)
PsUGT-1	*Pisum sativum*	Root meristem	Woo et al. (1999)
RB7	*Nicotiana tabacum*	Roots	Barone et al. (2008)
rd29A	Arabidopsis	Stress induced	Suárez et al. (2009), Jin et al. (2010)
SAG12	Arabidopsis	Senescence induced	Calderini et al. (2007)
SWPA2	*Ipomoea batatas*	Stress induced	Li et al. (2014), Wang et al. (2014)
TA29	*Nicotiana tabacum*	Anther tapetum	Rosellini et al. (2001)

structures (Bardor et al. 2003; D'Aoust et al. 2005). An edible plant producing a vaccine for enteric pathogens is particularly attractive because oral immunization may be able to elicit appropriate immune mechanisms for the induction of protective responses in livestock. Research on edible vaccines for the protection of cattle include expression of antigens for bovine viral diarrhea virus (Aguirreburualde et al. 2013), foot and mouth disease virus (Wigdorovitz et al. 1999, 2004; Dus Santos et al. 2005), bovine rotavirus (Wigdorovitz et al. 2004; Dus Santos and Wigdorovitz 2005; Dong et al. 2005), and bovine pneumonic pasteurellosis (Ziauddin et al. 2004; Lee et al. 2008). Similarly, an antigen for the avian reovirus was expressed in alfalfa (Huang et al. 2006). Commercialization of alfalfa for the production of antigens and biopharmaceuticals was pursued by the Canadian-based company Medicago Inc., but their current efforts focus on the use of tobacco for the production of biopharmaceuticals.

Early efforts in alfalfa biotechnology also focused on improving tolerance to abiotic stresses. Winterhardiness is a critical trait in a perennial crop for maintaining stand density and dry matter yields. It is a complex trait, involving many different stresses, but common to many stresses is the production of superoxide radicals. Expression of genes for Mn-superoxide dismutase (SOD) and Fe-SOD increased SOD activity in transgenic plants and increased survival and shoot dry matter yield compared to control non-transgenic lines (McKersie et al. 1999, 2000). Poor alfalfa growth in mineral acid soils is primarily due to aluminum (Al) toxicity. Tolerance in some crops is due to the production of organic acids that chelate Al, preventing uptake by roots. Tesfaye et al. (2001) reported increased malate secretion from alfalfa plants overexpressing the alfalfa nodule-enhanced malate dehydrogenase, which was associated with increased Al tolerance in a hydroponic system. Expression of bacterial citrate synthase from either a constitutive or root-specific promoter increased tolerance of alfalfa plants to acidic soil with toxic concentrations of Al (Barone et al. 2008). Most recently, Reyno et al. (2013) confirmed that the *Pseudomonas aeruginosa* citrate synthase gene confers Al tolerance in unlimed acidic soil and that DcPA1, an organic acid transporter, enhances Al tolerance. Acidic soils are increasing worldwide and although liming can ameliorate acidity in the plow layer, there is an increasing need for acid soil-tolerant crops. There is limited acid soil tolerance in alfalfa germplasm; thus, additional research under field conditions to determine if this is a trait suitable for commercialization is warranted. A significant advance to enhance drought tolerance was the overexpression of *WXP1*, an AP2 domain-containing transcription factor gene from *M. truncatula*, which increases the waxy layer on alfalfa leaves (Zhang et al. 2005), although field trials on this material has yet to be reported.

5.4.1 Glyphosate Tolerance: Development of Roundup Ready®Alfalfa (RRA)

Glyphosate (N-phosphonomethyl-glycine) is the active ingredient in the herbicide Roundup. It binds to the enzyme 5-enolpyruvylshikimate-3-phosphate synthase (EPSPS) blocking the biosynthesis of aromatic amino acids. Sources of tolerance to glyphosate were identified in the EPSPS enzyme from bacterial sources. The concept of conferring glyphosate tolerance in plants was tested starting with an EPSPS identified in the *Agrobacterium* sp. strain CP4 (Barry et al. 1992). The EPSPS gene from CP4, optimized for plant expression, fused to a chloroplast transit sequence, and under the control of the CaMV35S promoter, was used to transform soybean, which led to the development, deregulation, and subsequent commercialization of Roundup Ready® soybean in 1996 (Padgette et al. 1995). Roundup Ready® soybeans were grown on 91.9 million hectares in 2019 reaching near saturation planting levels in the US, Brazil, Argentina, and Canada (ISAAA 2018).

Starting in 1997, scientists with Forage Genetics International, in collaboration with Monsanto, transformed alfalfa with constructs similar to the one used to develop Roundup Ready®soybean. Commercial-scale alfalfa transformation produced 212 initial transformed (T0) plants, which were evaluated for glyphosate tolerance under greenhouse conditions (Samac and Temple 2004). In 1999, event sorting based on agronomic performance under glyphosate selection was carried out under greenhouse and field conditions. Concurrent molecular analysis identified four lead events to be used for trait introgression into elite germplasm and cultivar development.

In diploid plant species such as soybean, it is a relatively simple task to introgress the hemizygous (i.e. A-) single locus transgene into superior agronomic types following several cycles of backcrossing and achieve seed purities approaching 100% homozygosity (AA). By contrast, alfalfa is an insect-pollinated, outcrossing autotetraploid, and producing high trait purity seed (>90%) requires an intercross of plants that are duplex (AA–), triplex (AAA-), and/or quadraplex (AAAA) for the transgene. While this can be achieved by phenotypic recurrent selection, it requires genotyping large numbers of plants to accurately predict copy number or gene dosage. Techniques such as TaqMan PCR assays were either not available or cost-prohibitive in the early 2000s. Thus, the early generation of RRA varieties were developed using two independent transgenic events. Using line ID or event-specific PCR assays coupled to a gel-based detection platform, it was possible to identify populations of plants containing the two events (A—B—). Seed produced from these "dihomogenic" plants were predicted to have a trait purity of 93.7%. With glyphosate selection, the Syn3 commercial seed was predicted to have a trait purity of 93.8% (Samac and Temple 2004). Early varieties of RRA were developed using this two-event breeding system. This breeding method was later replaced with a more conventional single event strategy utilizing event J101-1 as genotyping technologies improved and glyphosate tolerance was stacked with the reduced lignin trait.

In the early 2000s, prominent alfalfa researchers at US universities began testing RRA in different geographic regions and under different crop management systems to evaluate Roundup application rates and timing of application to determine the spectrum of weed control and to develop take out options when terminating an alfalfa stand (Undersander et al. 2009). These studies showed excellent weed control across many environments and weed species. The concern that the introduction of another herbicide-resistant crop into the corn-soybean-alfalfa rotation would increase the rate of development of resistant weeds was also considered. In general, it was determined that most weeds do not tolerate the frequent harvests employed in a forage production system. Possible shifts in weed types were also considered (Undersander et al. 2009).

The deregulation process for RRA began in April 2004 when Forage Genetics International and Monsanto submitted a petition to the US Department of Agriculture-Animal and Plant Health Inspection Service (USDA-APHIS) requesting a determination of the non-regulated status of RRA lines J101 and J163 under 7 C.F.R. part 340 (Rogan and Fitzpatrick 2004). The submission provided the agency a rationale for the benefits proposed for RRA and extensive background on alfalfa as well as the plant transformation process, detailed information of the transformation vectors and the genetic elements they contained, and detailed molecular characterization of both events including generational stability. Also included were studies on the EPSPS protein produced in the alfalfa events including confirmation of the amino acid sequence and quantitation of EPSPS protein. The report documented extensive phenotypic comparison of the events to suitable controls. The final section of the application covered studies of agronomic practices, any environmental consequences caused by the cultivation of RRA, and proposed stewardship of the trait (Rogan and Fitzpatrick 2004). A petition was also made to the US Environmental Protection Agency (EPA) that provided glyphosate residue data and a proposed labeling rate for the use of Roundup on RRA. Petitions were also made to the US Food and Drug Administration (FDA) and the regulatory agencies of key export countries.

USDA-APHIS reviewed the submission and prepared an Environmental Assessment (EA) that considered the plant pest risk that RRA posed and in June 2005 RRA was deregulated in the US. Commercial sales of RRA followed shortly thereafter. However, a lawsuit filed by the Center for Food Safety and an organic alfalfa grower challenged the APHIS decision to grant non-regulated status to events J101 and J163. In February 2007 the Northern District of California ruled that the EA had failed to adequately consider environmental and economic impacts. The ruling returned events J101 and J163 to regulated status thus preventing further seed sales and planting of RRA. The court also ordered that USDA prepare an Environmental Impact Statement (EIS). Additional information on the court-ordered restrictions, its implications, and the EIS process have been discussed by Wang and Brummer (2012).

The Roundup Ready® alfalfa case went to the US Supreme Court that ruled 7 to 1 in favor of Monsanto in June 2010 (Suprement Court of the United States 2010). The draft EIS was published in December 2009 and a public comment period followed. APHIS published the final EIS on December 16, 2010, which concluded that transgenic alfalfa is safe for food and feed purposes and is unlikely to pose plant pest risks

(USDA-APHIS 2010). The agency proposed two possible actions: to approve the transgenic alfalfa fully or to approve the crop in part with restrictions on isolation distances and geographical locations. The latter option caused strong reactions from both the scientific and political communities (Waltz 2011; Wang and Brummer 2012). On January 27, 2011, the USDA announced it would fully deregulate RRA without restrictions allowing the resumption of sales after a delay of almost 4 years. In 2018, the growers in the US grew 1.14 million hectares of RRA (ISAAA 2018).

Production of seed for RRA provided unique challenges. Alfalfa seed production is mediated by leafcutter bees, alkali bees, or honeybees. Potential pollen flow by pollinators created concerns about seed production in neighboring conventional seed fields and trait purity for growers producing RRA seed. Adventitious presence (AP) of the transgene was of particular concern for the organic market, and hay and seed destined for export. Scientifically, RRA provided an excellent tool for gene and pollen flow studies. A study using predominantly leafcutter bees showed 0.32% gene flow at 305 meters, 0.003% at 805 meters, and no detectable gene flow at 1.6 km (Fitzpatrick et al. 2003). In a study using honeybees, which are known to have a longer flight range, gene flow (<0.06%) was detected out to 4.1 km (Teuber et al. 2004). The results of these studies have allowed the development of stewardship guidelines for alfalfa seed production and the development of grower opportunity zones for seed production (National Alfalfa and Forage Alliance 2014). These studies took advantage of inexpensive and highly accurate antibody-based lateral flow strips that can detect the CP4 EPSPS protein in alfalfa forage and seed.

Cross-pollination from RRA seed fields to feral alfalfa in alfalfa seed production areas in the Western United States has been the subject of several studies. From 4,580 sites surveyed, feral alfalfa plants were observed at 404 sites, and at only 27% of these sites were transgenic plants detected (Greene et al. 2015). An earlier study across wider US geographies found feral plants at only 3% of the sites surveyed (Kendrick et al. 2005; USDA-APHIS 2010). The authors noted that the potential for gene flow from RRA to feral alfalfa was low because the Roundup Ready® trait offers no increased fitness to the alfalfa plants. It was also noted that herbicides are not typically used to control feral plants and when employed they are typically used in combination to increase effectiveness.

Since the commercialization of RRA, there have been reports of additional glyphosate-resistant alfalfa being developed. Yi et al. (2018) used a new synthetic glyphosate-resistant gene tracing to a novel EPSPS gene from glyphosate contaminated soil. Nicolia et al. (2014) utilized a glycine oxidase gene optimized for plant expression. Although both groups demonstrated glyphosate tolerance, no events from either study appear to have been evaluated under field conditions. Although never commercialized, transgenic alfalfa plants resistant to glufosinate were first developed in 1990 (D'Halluin et al. 1990). A recent report from Argentina described the development of and field testing of a transgenic variety of alfalfa containing the *bar* gene for resistance to glufosinate (Jozefkowicz et al. 2018). Commercialization of alfalfa varieties with the second mode of herbicide action may help reduce future problems of management of glyphosate resistant weeds.

5.4.2 Forage Quality

Alfalfa forage provides a high concentration of crude protein; however, foliar proteins are deficient in sulfur-containing amino acids and most proteins are rapidly degraded by rumen microbes resulting in loss of nitrogen to the environment. Addressing these problems were goals of early alfalfa biotechnology research through overexpression of storage proteins from other plants or animal proteins (Schroeder et al. 1991; Wandelt et al. 1992; Bagga et al. 1992; Tabe et al. 1995; Bellucci et al. 2005; Bagga et al. 2005). Although these proteins accumulated in alfalfa herbage, concentrations were generally not sufficient to significantly improve amino acid nutrition. More recently, approaches to introduce enzymes to modify amino acid content have shown some success. Expression of Arabidopsis cystathionine gamma-synthase, the enzyme controlling the synthesis of the first intermediate metabolite in the methionine biosynthetic pathway, increased the contents of methionine up to 32-fold compare to the wild type plants (Avraham et al. 2004). Modest increases in methionine and cysteine were obtained by expression of bacterial aspartate kinase and adenylylsulfate reductase genes to provide increased amounts of reduced sulfur for amino acid synthesis (Tong et al. 2014). Overexpression of sucrose phosphate synthase increased nodule number and nodule mass with a concomitant increase in nitrogenase activity and total protein content (Gebril et al. 2015).

Forage quality may also be improved by increasing the leaf to stem ratio. A modest increase in leaf retention by delaying senescence was obtained by controlled expression of isopentenyl transferase (Calderini et al. 2007), the rate-liming step in cytokinin synthesis. A productive approach to increase plant biomass has been to identify *M. truncatula* mutants with desired characteristics. Mutation of the *M. truncatula* STAY-GREEN gene results in the maintenance of green tissue during plant senescence. Silencing the homologous alfalfa gene resulted in the retention of chlorophyll and an increase in crude protein in alfalfa hay (Zhou et al. 2011). A recent report using CRISPR-Cas9 based gene editing targeted the alfalfa STAY-GREEN gene. The editing efficiency was high enough that homozygous mutants with a complete knockout of the four allelic copies in the T0 generation were recovered (Wolabu et al. 2020). The gene-edited plants showed a stronger STAY-GREEN phenotype than the RNAi lines. Similarly, mutation of SQUAMOSA PROMOTER BINDING PROTEIN-LIKE 8 (SPL8) in *M. truncatula* increased stem branching and leafy biomass. Downregulation of the alfalfa SPL8 gene increased stem branching and increased biomass yield by 43–86%. Alfalfa plants also showed increased salt and drought tolerance (Gou et al. 2018). In a separate study, downregulation of SPL13 increased branching of transgenic alfalfa plants and delayed flowering (Gao et al. 2018). SPL13 was shown to bind to the promoter of MTB112, a transcription factor involved in plant development. Downregulation of the MYB112 transcription factor also increased shoot branching, suggesting that SPL13 binding negatively regulates MYB112. These genes are attractive targets for gene editing to enhance biomass production and forage quality.

Ensiling is a popular method for preserving forage crops. However, wounding of plants during harvest and ensiling releases protein degrading enzymes that can lead to significant losses of crude protein. In efforts to reduce these losses in alfalfa, the polyphenol oxidase (PPO) gene from red clover was cloned and expressed in alfalfa (Sullivan et al. 2004). In red clover, this enzyme acts on endogenous *o*-diphenol compounds, which prevents degradation of proteins during ensiling. Transgenic alfalfa expressed the red clover PPO and when substrate was added, protein degradation was inhibited. However, alfalfa lacks endogenous enzymatic substrates for red clover PPO. Recently, to re-create the red clover PPO/*o*-diphenol system in alfalfa, Sullivan et al. (2021) expressed the red clover gene for hydroxycinnamoyl-coenzyme A (CoA):malate hydroxycinnamoyl transferase (HMT) in alfalfa. The activity of this enzyme in red clover results in the accumulation of hydroxycinnamoyl-malate esters, which are substrates for PPO. Accumulation of substrates in alfalfa was enhanced by super-transformation of plants with an RNAi construct to silence caffeoyl-CoA 3-*O*-methytransferase (CCOMT). When extracts from the HMT and CCOMT RNAi plants were mixed with extracts from alfalfa plants expressing PPO, there was sufficient oxidation of substrates to inhibit proteolysis by approximately 50%. These experiments are encouraging that protein losses and conversion to non-protein nitrogen during ensiling of alfalfa can be reduced, although additional experiments are needed to evaluate field performance including resistance to biotic and abiotic stresses.

5.4.3 Reduced Lignin Alfalfa

Natural mutations of genes of the lignin biosynthetic pathway were discovered in the 1920s. These so-called brown mid-rib (BMR) corn mutants confer enhanced cell wall digestibility due to a reduction in lignin concentration. With the widespread production of corn silage starting in the 1980s, corn hybrids containing the BMR mutation become synonymous with high-quality corn silage. Early efforts using genetic engineering reduced lignin levels in tobacco plants by suppression of the caffeic acid 3-*O*-methyltransferase (COMT) gene (Sewalt et al. 1997). Alfalfa knockdowns in most of the key genes in the lignin biosynthetic pathway have been evaluated for their effects on alfalfa forage composition, fiber digestibility, lignin concentration and composition, and agronomic performance. Knockouts of COMT and caffeoyl-CoA 3-*O*-methytransferase (CCOMT) under the control of the vascular specific phenylalanine ammonia-lyase (Pal2) promoter resulted in the desired lignin reduction (Guo et al. 2001a) and improvement in rumen digestibility (Guo et al. 2001b). Subsequent field evaluation of some of these lines and others generated using RNAi technology (Chen et al. 2006) confirmed that lines downregulated in CCOMT had improved forage quality without negative impacts on yield and without the increase in lodging observed in downregulated COMT lines. Hay from these trials was used in lamb and dairy cow feeding studies and for in vitro gas production analysis, which demonstrated improved fiber digestibility with gene downregulation (Mertens and McCaslin 2008; Weakley et al. 2008; Getachew et al. 2011). Based on the incidence of lodging seen with COMT downregulation, the CCOMT gene was advanced for commercialization.

A recent publication detailed the development and characterization of the reduced lignin event which was subsequently used to develop the commercial HarvXtra® alfalfa varieties (Barros et al. 2019). Alfalfa events were developed via *Agrobacterium tumefaciens*-mediated transformation using a binary vector containing two T-DNAs. The first contained a cassette to suppress CCOMT expression; the second contained an NPTII selection cassette. During the transformation process, either or both T-DNA regions were inserted into the alfalfa genome. Primary transformation events containing the CCOMT suppression cassette were identified and used to generate F1 populations which were screened by PCR to identify plants with unlinked insertions of the CCOMT suppression cassette. This process identified small populations of marker-free plants representing 74 independent insertional events. These plants were evaluated for general agronomic traits, biomass accumulation, and lignin concentration in field trials at the Forage Genetics research station located in West Salem, Wisconsin during the summer of 2007. Extensive molecular characterization of the best events followed, and multi-location field trials were established in 2008 that focused on agronomic evaluation to meet yield and forage quality trait goals and allowed identification of the eight and subsequently four lead events. In 2009, Syn1 seed (75% trait purity) was produced under isolation cages from small populations of plants tracing to the four lead events, which were subsequently used for regulatory studies and trait integration into the reduced lignin breeding program. Event KK179 was identified as the lead event and was the subject of regulatory studies. Submissions were made to the FDA and USDA-APHIS in August and November 2012, respectively (Hall and Whalen 2012; USDA-APHIS 2014). The regulatory packages included detailed molecular characterization of the event, generational stability data, compositional analysis of alfalfa forage containing event KK179 compared to conventional alfalfa controls looking at key nutrients, anti-nutrients, and secondary metabolites based on the guidelines for alfalfa set by the Organization for Economic Cooperation and Development. The conclusions drawn from these studies was that the genetic modification in KK179 does not result in meaningful changes in composition other than the intended reduction in guaiacyl lignin (19%) and total lignin (22%) and that the feed, food safety, and nutritional quality are comparable to those of conventional alfalfa. Following extensive review and public comment,

USDA-APHIS published their assessment in November 2014 that KK179 was no longer considered a regulated article under the regulations governing the introduction of genetically engineered organisms (USDA-APHIS 2014). Forage Genetics commercialized reduced lignin alfalfa varieties starting in 2015 under the trade name HarvXtra®. The HarvXtra® alfalfa varieties are a breeding stack of reduced lignin event KK179 and the previously developed Roundup Ready® event J101. Event KK179 has also received deregulated status allowing cultivation in Canada and more recently Argentina.

When compared to commercial checks, the first generation of HarvXtra®cultivars consistently showed a ~15% reduction in whole plant lignin which results in a 10–15% increase in neutral detergent fiber digestibility (NDFD) and relative feed quality (RFQ) (Barros et al. 2019). Forage yield, crude protein, and neutral detergent fiber (NDF) in HarvXtra®cultivars are comparable to appropriate reference cultivars (Barros et al. 2019). On-farm feeding studies showed an average daily increase of 4 lb of milk per cow although this varied by location based on the concentration of alfalfa in the diet and exact growth stage at which the alfalfa was harvested.

Alfalfa producers have long faced a dilemma on when to harvest alfalfa because quality decreases with increasing maturity. The slower rate of lignification seen in reduced lignin plants has provided growers management flexibility tools (Barros et al. 2019). By delaying harvest by 5–10 days, growers can achieve a ~20% gain in yield with little reduction in RFQ compared to conventional cultivars harvested on a more typical harvest regime. Reducing the cutting frequency has the benefit of lowering harvest costs by reducing the number of harvests per year by one or more in many regions while improving the persistence of the stand. Alternatively, harvesting HarvXtra® on a 30-day window significantly increases the chances of producing dairy quality forage (Barros et al. 2019). Attempts have been made to calculate the value of HarvXtra® alfalfa for producers. Based on current midwestern United States figures estimates range from $113–154 per year per acre based on cultivation practices to maximize yield and quality, respectively, or $1550–2125 per bag of seed (https://www.pioneer.com/us/tools-services/calculate-harvxtra-value.html).

Since the commercialization of HarvXtra® alfalfa, several independent studies carried out by University and government scientists have evaluated its performance and made recommendations to growers. The largest study was conducted at sites in six states in the northern United States thus providing testing data across a broad geographic range (Arnold et al. 2019). Plots were seeded in the spring of 2015 using fall dormant varieties with data collection over two production years. Harvest intervals of 28, 33, and 38 days were employed. Overall, an 8.4% reduction in lignin concentration and 5.3–7.7% increase in NDFD was measured. Small reductions in NDF and an increase in crude protein were also reported. Dry matter yields were reduced by 4.8–7.0% in the HarvXtra®cultivar in this study. The authors concluded that HarvXtra® alfalfa can be used to extend the time interval when it is possible to harvest forage with adequate fiber digestibility for animals with high energy requirements. It also validated the management flexibility concept. The results of this large study support the earlier expectations for reduced lignin alfalfa (McCaslin et al. 2014) and one that evaluated production from four sites in Minnesota (Grev et al. 2017). This study measured an 8% reduction in lignin, 10% increase in NDFD but no changes in NDF or crude protein, and no yield penalty. Under a delayed harvest schedule a 21% yield gain was obtained with only a 3% reduction in RFQ (Grev et al. 2017). In a third study, Getachew et al. (2018) found significant reductions in lignin content resulting in increase in in vitro dry gas production. The authors also concluded that the harvest window can be extended for higher yield without compromising the nutritional quality of the alfalfa forage.

5.4.4 Salt Tolerance

An increasing amount of alfalfa biotechnology research has focused on tolerance to saline soil, alkaline soil, and drought. Alfalfa is considered to be moderately sensitive to soil salinity which inhibits seed germination and plant growth. Extensive research with model plants such as Arabidopsis and several halophytes has described the responses of plants to salt treatment and identified tolerance mechanisms (Parihar et al. 2015; Ismail and Horie 2017; van Zelm et al. 2020). However, to increase salt tolerance in alfalfa, an understanding of the adaptations that are beneficial is needed in order to optimize the responses that increase biomass yield. The Na^{2+} in saline soil reduces root water uptake, causing water deficit effects on plants. Thus, salt tolerance often increases drought tolerance. Salt also causes injury to plant cells and competes with mineral nutrient uptake. The Na^{2+} ion competes with K^{+}, leading to potassium deficiencies. In response to the perception of Na^{2+} ions or osmotic stress, a number of early signaling events occur and reactive oxygen species (ROS) are rapidly generated. Downstream signaling leads to changes in plant hormones and gene expression, which reduces growth in sensitive plants or can increase tolerance. Tolerance mechanisms include restricting Na uptake or transport to shoots, compartmentalization of Na in the vacuole, and/or production of compatible solutes that provide osmotic protection to retain cell volume and turgor pressure. Biotechnology approaches to salt tolerance in alfalfa are summarized in Table 5.3. Overexpression of each gene is reported to increase salt tolerance in controlled conditions. Approaches include single genes to reduce damage from ROS, reduce ion transport, compartmentalization of Na in vacuoles, and production of solutes. Expression of transcription factors generally leads to multiple tolerance responses. An increased understanding of responses of alfalfa selected for tolerance, and identification of salt responsive genes, may help inform the most productive path forward for a biotechnology solution to salt sensitivity in alfalfa.

5.4.5 Alkaline Soil Tolerance

In contrast to the understanding of salt stress in plants, much less is known about alkaline or high pH stress. Saline–alkaline soils occur in arid and semiarid regions and stress from these soil conditions severely reduces alfalfa production in these areas. In such soils, plants suffer from saline toxicity and damage from alkaline conditions which affect mineral uptake, ion balance, and pH homeostasis. To address these problems, the research identified saline–alkali-tolerant *Glycine soja* plants and genes induced in these plants with alkali treatment. Several genes from alkali tolerant accessions of *G. soja* were constitutively expressed in alfalfa, glutathione S-transferase (Wang et al. 2012; Jia et al. 2016), phosphoenolpyruvate carboxylase kinase (Sun et al. 2014), and a basic leucine zipper transcription factor (Wu et al. 2018). Similarly, genes from the halophyte *Suaedacorniculata*, a vacuolar proton pump ATPase (Wang et al. 2016a), vacuolar Na^{+}/H^{+} antiporter, and vacuolar proton phosphatase (Liu et al. 2013) were constitutively expressed. Each transgene improved alkali or salt tolerance over non-transgenic plants. However, treatments were made to established plants for a short period. Information is needed on performance from seedling stages in alkaline field soil. Additionally, root-specific or constitutive promoters with stronger root expression could potentially enhance tolerance.

5.4.6 Insect Resistance

One aspect of alfalfa biotechnology that has not received much attention in the past is engineering plants for insect resistance. This is somewhat surprising given the tremendous effort and success of engineering insect resistance into other crop plants. Corn and cotton expressing genes for *Bacillus thuringiensis* (Bt) endotoxin genes comprise a major part of the crop biotechnology sector. With the removal of important alfalfa insecticides from the market and resistance to insecticides in the alfalfa weevil (*Hypera postica*) populations in the western US, further

Table 5.3 Transgenes for salt tolerance

Target	Gene	References
Early signaling: Restrict Na^{2+} uptake or movement, minimize ROS damage	*AtNDPK2*	Wang et al. (2014)
	AtSOS1,2,3	Wang et al. (2019a)
	Os ascorbate peroxidase	Guan et al. (2012), Zhang et al. (2014)
	rstB	Zhang and Wang (2015)
Downstream signaling	*Alfin1*	Winicov (2000)
	AtABF3	Wang et al. (2016b)
	CsALDH12A1	Duan et al. (2015)
	CsLEA	Zhang et al. (2016)
	GsCBRLK	Bai et al. (2013)
	GsSRK	Sun et al. (2018)
	GsWRKY20	Tang et al. (2014)
	GsZFP1	Tang et al. (2013)
Compatible solutes	Betaine aldehyde dehydrogenase	Liu et al. (2011)
	choline oxidase (*codA*)	Li et al. (2014)
	GmDREB1	Jin et al. (2010)
	Yeast trehalose	Suárez et al. (2009)
Ion homeostasis	*AtAVP1*	Bao et al. (2009)
	AtNHX1	Stritzler et al. (2018)
Na^{2+} sequestration	*TaNHX2*	Zhang et al. (2015)
	Zx NHX	Bao et al. (2016)

research into engineering insect resistance into alfalfa is warranted. Additionally, damage from clover root curculio (CRC; *Sitona hispidulus*) has often been overlooked as a pest problem in alfalfa, but significant outbreaks cause losses in forage yield and crude protein, reduces nitrogen fixation, and results in premature stand decline from directly killing plants and from winter heaving and winter kill of plants with weakened root systems. There are currently no cultivars with resistance, no insecticides labeled for use of the larval stages, and few management methods that can be used to reduce damage from the insect. Blister beetles (*Epicauta* spp.) are also significant pests of alfalfa that reduce forage yield can cause toxicosis in horses fed contaminated hay.

Over 700 insecticidal protein toxins have been identified in *B. thuringiensis,* many of which remain to be characterized for insecticidal activity, and additional novel genes are being discovered through next-generation sequencing technologies (Palma et al. 2014). Once ingested by insects, the proteins bind to specific receptors located in the insect cell membrane leading to cell disruption and insect death. The family of toxins in the Cry3A and Vip1/2 families are known to have activity against Coleopteran insects, which include alfalfa weevil, CRC, and blister beetles, and also have activity against Hemipteran insects which include Lygus bugs, aphids, and leafhoppers, all significant alfalfa pests. Tohidfar et al. (2013) reported that constitutive expression of a synthetic *cry3a* gene in alfalfa resulted in 73–90% mortality of alfalfa weevil larvae. Additional research in this area is needed to identify the most effective toxins, optimal expression levels, and performance under field conditions. Based on issues of insects developing resistance to Bt toxins deployed in other crops, it will be essential to develop best practices for stewardship of alfalfa cultivars expressing Bt genes. The extensive research on Bt genes and their use in other crops should facilitate their utilization in alfalfa. The need for sustainable and effective insect management across the US should drive the economics of developing a Bt alfalfa.

An alternative approach to reducing damage from aphids was taken by Wang et al. (2019b) to engineer alfalfa plants to synthesize the aphid repellant (E)–farnesene (EF), an aphid alarm pheromone. Expression of the EF synthase gene from pea in transgenic alfalfa resulted in the production of volatile EF that repelled pea aphids under lab conditions. If results can be confirmed under field conditions, this type of approach has promise for managing aphid pests in alfalfa.

5.4.7 Disease Resistance

Early alfalfa biotechnology research investigated if disease resistance could be enhanced through the expression of single genes encoding for hydrolytic enzymes to attack pathogen cell walls (Masoud et al. 1996; Samac and Smigocki 2003; Samac et al. 2004b) and to modify production of phytoalexins with antimicrobial activity (Hipskind and Paiva 2000; He and Dixon 2000). Such single gene "silver bullet" approaches did not significantly increase disease resistance with the exception of expression of an anthracnose resistance gene from *Medicago truncatula*. Expression of *RCT1* in alfalfa provides broad nonrace-specific resistance to at least three races of *Colletotrichum trifolii* (Yang et al. 2008). Also, some promising results have been reported for the expression of defensins, antimicrobial peptides, some of which have broad antimicrobial and antifungal activity (García et al. 2014; Sathoff et al. 2019). Interestingly, expression of human lactoferrin, an iron-binding glycoprotein, increased resistance to bacterial pathogens of alfalfa under controlled conditions (Stefanova et al. 2013).

A major mechanism of disease resistance in legumes is the production of phytoalexins, antimicrobial phenolic compounds that can inhibit the growth of fungal, oomycete, and nematode pathogens of alfalfa (Vaziri et al. 1981; Blount et al. 1992; Baldridge et al. 1998). Research leading to the development of reduced lignin alfalfa (HarvXtra® trait) generated transgenic alfalfa plants downregulated for specific genes in the lignin biosynthetic pathway. Downregulation of hydroxycinnamoyl CoA: shikimate hydroxycinnamoyl transferase (HCT) resulted in accumulation of several phenolic aglycones, increased anthocyanin accumulation, and expression of pathogenesis-related proteins involved in plant defense against pathogens. Downregulated plants showed increased resistance to the anthracnose pathogen *C. trifolii*, but not to *Phytophthora medicaginis, Phoma medicaginis,* or *Sclerotinia sclerotiorum* (Gallego-Giraldo et al. 2011). Strong downregulation of CCOMTresulted in lignin pathway intermediates being shunted into the isoflavonoid pathway with an accumulation of the phytoalexins medicarpin and 7,4'-dihydroxyflavone. Both compounds inhibit the growth of *Fusarium oxysporum* f. sp. *medicaginis*, a vascular wilt pathogen of alfalfa, and downregulated plants showed increased resistance to Fusarium wilt (Gill et al. 2017). Although these plants had poor agronomic characteristics, these studies provide data on the relative importance of isoflavonoid phytoalexins in resistance toward specific alfalfa pathogens.

5.5 Future Aspects

What are the prospects for commercializing additional transgenic traits for alfalfa? Developing non-bloating alfalfa with the potential for enhanced protein utilization by the production of tannins in foliage has been a long-sought-after goal. Much has been learned about the genes in the condensed tannin or proanthocyanidin biosynthetic pathway and their regulation (Dixon and Sarnalac 2020). A novel transcription factor was identified in the foliar proanthocyanidin accumulating legume *Trifolium arvense* (Hancock et al. 2012). Its constitutive overexpression in alfalfa resulted in the accumulation of condensed tannins from 0.9 to 1.6% dry weight in alfalfa under greenhouse conditions (Hancock et al. 2014). However, transgenic approaches to increase tannins in alfalfa forage have so far led to unacceptable agronomic traits.

Recently, the burden for developing some organisms through genetic engineering was reduced with the first comprehensive revision of the biotechnology regulations administered by APHIS. The SECURE Rule makes plants developed through biotechnology exempt from regulation if they could have been developed through conventional breeding. Thus, gene mutations or introduction of a gene from the plant's gene pool will not be regulated. Genome editing in alfalfa has been reported using TALENS (Luo and Baltes 2020) and CRISPR/Cas9 (Wolabu et al. 2020). Alfalfa plants developed with these technologies would have substantially lower regulatory costs and face lower hurdles to commercialization. Reductions in lignin through COMT editing are in development (Luo and Baltes 2020). As noted above, a number of known mutations in *M. truncatula* could be targeted for genome editing to improve forage quality and biomass accumulation. Additional candidate genes for genome editing for crop improvement may be mined from the extensive research on mutations in Arabidopsis and other crop model species.

References

Adem M, Beyene D, Feyissa T (2017) Recent achievements obtained by chloroplast transformation. Plant Methods 13:30. https://doi.org/10.1186/s13007-017-0179-1

Aguirreburualde MSP, Gómez MC, Ostachuk A, Wolman F, Albanesi G et al (2013) Efficacy of a BVDV subunit vaccine produced in alfalfa transgenic plants. Vet Immunol Immunopathol 151:315–324. https://doi.org/10.1016/j.vetimm.2012.12.004

Arnold AM, Cassida KA, Albrecht KA, Hall MH, Min D et al (2019) Multistate evaluation of reduced-lignin alfalfa harvested at different intervals. Crop Sci 59:1799–1807. https://doi.org/10.2135/cropsci2019.01.0023

Atkins CA, MC Smith P (1997) Genetic transformation and regeneration of legumes. In: Legocki, A'Bothe H, Puhler A (eds) Biological fixation of nitrogen for ecology and sustainable agriculture. Springer, Berlin, Heidelberg, pp 283–304

Austin-Phillips S, Koegel R, Straub R, Cook M (1999) Animal feed compositions containing phytase derived from transgenic alfalfa and method of use thereof. United States Patent 5(900):525

Austin-Phillips S, Ziegelhoffer T (2001) The production of value-added proteins in transgenic alfalfa. In: Spangenberg G (ed) Molcular Breeding of Forage Crops. Kluwer Academic, Dordrecht, pp 285–301

Avraham T, Badani H, Galili S, Amir R (2004) Enhanced levels of methionine and cysteine in transgenic alfalfa (*Medicago sativa* L.) plants over-expressing the Arabidopsis cystathionine γ-synthase gene. Plant Biotechnol J 3:71–79. https://doi.org/10.1111/j.1467-7652.2004.00102.x

Bagga S, Potenza C, Ross J, Martin MN, Leustek T et al (2005) A transgene for high methionine protein is posttranscriptionally regulated by methionine. Vitro Cell Dev Biol- Plant 41:731–741. https://doi.org/10.1079/IVP2005709

Bagga S, Sutton D, Kemp JD, Sengupta-Gopalan C (1992) Constitutive expression of the β-phaseolin gene in different tissues of transgenic alfalfa does not ensure phaseolin accumulation in non-seed tissue. Plant Mol Biol 19:951–958. https://doi.org/10.1007/BF00040527

Bai X, Liu J, Tang L, Cai H, Chen M et al (2013) Overexpression of GsCBRLK from *Glycine soja* enhances tolerance to salt stress in transgenic alfalfa (*Medicago sativa*). Funct Plant Biol 40:1048. https://doi.org/10.1071/FP12377

Baldridge GD, O'Neill NR, Samac DA (1998) Alfalfa (*Medicago sativa* L.) resistance to the root-lesion nematode, *Pratylenchus penetrans*: Defense-response gene mRNA and isoflavonoid phytoalexin levels in roots. Plant Mol Biol 38:999–1010. https://doi.org/10.1023/A:1006182908528

Bao A-K, Du B-Q, Touil L, Kang P, Wang Q-L et al (2016) Co-expression of tonoplast Cation/H + antiporter and H + -pyrophosphatase from xerophyte *Zygophyllum xanthoxylum* improves alfalfa plant growth under salinity, drought and field conditions. Plant Biotechnol J 14:964–975. https://doi.org/10.1111/pbi.12451

Bao A-K, Wang S-M, Wu G-Q, Xi J-J, Zhang J-L et al (2009) Overexpression of the Arabidopsis H + -PPase enhanced resistance to salt and drought stress in transgenic alfalfa (*Medicago sativa* L.). Plant Sci 176:232–240. https://doi.org/10.1016/j.plantsci.2008.10.009

Bardor M, Loutelier-Bourhis C, Paccalet T, Cosette P, Fitchette A-C et al (2003) Monoclonal C5-1 antibody produced in transgenic alfalfa plants exhibits a N-glycosylation that is homogenous and suitable for glyco-engineering into human-compatible structures. Plant Biotechnol J 1:451–462. https://doi.org/10.1046/j.1467-7652.2003.00041.x

Barone P, Rosellini D, LaFayette P, Bouton J, Veronesi F et al (2008) Bacterial citrate synthase expression and soil aluminum tolerance in transgenic alfalfa. Plant Cell Rep 27:893–901. https://doi.org/10.1007/s00299-008-0517-x

Barros J, Temple S, Dixon RA (2019) Development and commercialization of reduced lignin alfalfa. Curr Opin Biotechnol 56:48–54. https://doi.org/10.1016/j.copbio.2018.09.003

Barry G, Kishore G, Padgette S, Taylor M, Kolacz K et al (1992) Inhibitors of amino acid biosynthesis: strategies for imparting glyphosate tolerance to crop plants. Curr Top Plant Physiol 7:139–145

Bellucci M, De Marchis F, Mannucci R, Bock R, Arcioni S (2005) Cytoplasm and chloroplasts are not suitable subcellular locations for -zein accumulation in transgenic plants. J Exp Bot 56:1205–1212. https://doi.org/10.1093/jxb/eri114

Bingham ET (1989) Registration of Regen-S alfalfa germplasm useful in tissue culture and transformation research. Crop Sci 29:1095–1096. https://doi.org/10.2135/cropsci1989.0011183X002900040071x

Bingham ET (1991) Registration of alfalfa hybrid Regen-SY germplasm for tissue culture and transformation research. Crop Sci 31:1098–1098. https://doi.org/10.2135/cropsci1991.0011183x003100040075x

Bingham ET, Hurley LV, Kaatz DM, Saunders JW (1975) Breeding alfalfa which regenerates from callus tissue in culture. Crop Sci 15:719–721. https://doi.org/10.2135/cropsci1975.0011183X001500050032x

Bingham ET, McCoy TJ, Walker KA (1988) Alfalfa tissue culture. In: Hanson A, Barnes DK, Hill RJ (eds) Alfalfa and alfalfa improvement. ASA-CSSA-SSSA, Madison, WI, pp 903–929

Blount JW, Dixon RA, Paiva NL (1992) Stress responses in alfalfa (*Medicago sativa* L.) XVI. Antifungal activity of medicarpin and its biosynthetic precursors; implications for the genetic manipulation of stress metabolites. Physiol Mol Plant Pathol 41:333–349. https://doi.org/10.1016/0885-5765(92)90020-V

Brears T, Walker EL, Coruzzi GM (1991) A promoter sequence involved in cell-specific expression of the pea glutamine synthetase GS3A gene in organs of transgenic tobacco and alfalfa. Plant J 1:235–244. https://doi.org/10.1111/j.1365-313X.1991.00235.x

Busse U, Levée V, Trépanier S, Vézina L (2001) Production of antibodies in alfalfa (*Medicago sativa*). In: Erickson L (ed) Molecular Farming of Plants and Animal for Human and Veterinary Medicine. J. Wiley & Sons, New York, pp 199–219

Calderini O, Bovone T, Scotti C, Pupilli F, Piano E et al (2007) Delay of leaf senescence in *Medicago sativa* transformed with the ipt gene controlled by the senescence-specific promoter SAG12. Plant Cell Rep 26:611–615. https://doi.org/10.1007/s00299-006-0262-y

Chen F, Srinivasa Reddy MS, Temple S, Jackson L, Shadle G et al (2006) Multi-site genetic modulation of monolignol biosynthesis suggests new routes for formation of syringyl lignin and wallbound ferulic acid in alfalfa (*Medicago sativa* L.). Plant J. 48:113–124. https://doi.org/10.1111/j.1365-313X.2006.02857.x

Comai L, Moran P, Maslyar D (1990) Novel and useful properties of a chimeric plant promoter combining CaMV 35S and MAS elements. Plant Mol Biol 15:373–381

D'Aoust M-A, Lerouge P, Busse U, Bilodeau P, Trépanier S et al (2005) Efficient and reliable production of pharmaceuticals in alfalfa. In: Fischer R, Schillberg S (eds) Molecular Farming. Wiley-VCH Verlag GmbH & Co, KGaA, Weinheim, FRG, pp 1–12

D'Halluin K, Botterman J, De Greef W (1990) Engineering of herbicide-resistant alfalfa and evaluation under field conditions. Crop Sci 30:866–871. https://doi.org/10.2135/cropsci1990.0011183x003000040020x

Ding Y-L, Aldao-Humble G, Ludlow E, Drayton M, Lin Y-H et al (2003) Efficient plant regeneration and *Agrobacterium*-mediated transformation in *Medicago* and *Trifolium* species. Plant Sci 165:1419–1427. https://doi.org/10.1016/j.plantsci.2003.08.013

Dixon RA, Sarnalac S (2020) Proanthocyanidin biosynthesis—a matter of protection. Plant Physiol 184:579–591. https://doi.org/10.1104/pp.20.00973

Dong J-L, Liang B-G, Jin Y-S, Zhang W-J, Wang T (2005) Oral immunization with pBsVP6-transgenic alfalfa protects mice against rotavirus infection. Virology 339:153–163. https://doi.org/10.1016/j.virol.2005.06.004

Duan Z, Zhang D, Zhang J, Di H, Wu F, et al (2015) Co-transforming bar and CsALDH genes enhanced resistance to herbicide and drought and salt stress in transgenic alfalfa (*Medicago sativa* L.). Front Plant Sci 6:1115. https://doi.org/10.3389/fpls.2015.01115

Dus Santos MJ, Carrillo C, Ardila F, Ríos RD, Franzone P et al (2005) Development of transgenic alfalfa plants containing the foot and mouth disease virus structural polyprotein gene P1 and its utilization as an experimental immunogen. Vaccine 23:1838–1843. https://doi.org/10.1016/j.vaccine.2004.11.014

Dus Santos MJ, Wigdorovitz A (2005) Transgenic plants for the production of veterinary vaccines. Immunol Cell Biol 83:229–238. https://doi.org/10.1111/j.1440-1711.2005.01338.x

Fang Y, Hirsch AM (1998) Studying early nodulin gene ENOD40 expression and induction by nodulation factor and cytokinin in transgenic alfalfa. Plant Physiol 116:53–68. https://doi.org/10.1104/pp.116.1.53

Ferradini N, Nicolia A, Capomaccio S, Veronesi F, Rosellini D (2011) Assessment of simple marker-free genetic transformation techniques in alfalfa. Plant Cell Rep 30:1991–2000. https://doi.org/10.1007/s00299-011-1107-x

Fitzpatrick S, Reisen P, McCaslin M (2003) Pollen-mediated gene flow in alfalfa : a three-year summary of field research. In: Proceedings of the central alfalfa improvement conference virtual meeting, July 21–25, 2003

Fu C, Hernandez T, Zhou C, Wang Z-Y (2015) Alfalfa (*Medicago sativa* L.). In: Wang K (ed) *Agrobacterium* protocols: volume 1, methods in molecular biology, vol 1223. https://doi.org/10.1007/978-1-4939-1695-5_17, pp 213–221

Gallego-Giraldo L, Jikumaru Y, Kamiya Y, Tang Y, Dixon RA (2011) Selective lignin downregulation leads to constitutive defense response expression in alfalfa (*Medicago sativa* L.). New Phytol 190:627–639. https://doi.org/10.1111/j.1469-8137.2010.03621.x

Gao R, Gruber MY, Amyot L, Hannoufa A (2018) SPL13 regulates shoot branching and flowering time in *Medicago sativa*. Plant Mol Biol 96:119–133. https://doi.org/10.1007/s11103-017-0683-8

García AN, Ayub ND, Fox AR, Gómez MC, Diéguez MJ et al (2014) Alfalfa snakin-1 prevents fungal colonization and probably coevolved with rhizobia. BMC Plant Biol 14:248. https://doi.org/10.1186/s12870-014-0248-9

Gebril S, Seger M, Villanueva FM, Ortega JL, Bagga S et al (2015) Transgenic alfalfa (*Medicago sativa*) with increased sucrose phosphate synthase activity shows enhanced growth when grown under N_2-fixing conditions. Planta 242:1009–1024. https://doi.org/10.1007/s00425-015-2342-0

Getachew G, Ibáñez AM, Pittroff W, Dandekar AM, McCaslin M et al (2011) A comparative study between lignin down regulated alfalfa lines and their respective unmodified controls on the nutritional characteristics of hay. Anim Feed Sci Technol 170:192–200. https://doi.org/10.1016/j.anifeedsci.2011.09.009

Getachew G, Laca EA, Putnam DH, Witte D, McCaslin M et al (2018) The impact of lignin downregulation on alfalfa yield, chemical composition, and in vitro gas production. J Sci Food Agri 98:4205–4215. https://doi.org/10.1002/jsfa.8942

Gill US, Uppalapati SR, Gallego-Giraldo L, Ishiga Y, Dixon RA et al (2017) Metabolic flux towards the (iso)flavonoid pathway in lignin modified alfalfa lines induces resistance against *Fusarium oxysporum* f. sp. *medicaginis*. Plant Cell Environ 41:1997–2007. https://doi.org/10.1111/pce.13093

Gou J, Debnath S, Sun L, Flanagan A, Tang Y et al (2018) From model to crop: functional characterization of SPL8 in *M. truncatula* led to genetic improvement of biomass yield and abiotic stress tolerance in alfalfa. Plant Biotechnol J 16:951–962. https://doi.org/10.1111/pbi.12841

Greene SL, Kesoju SR, Martin RC, Kramer M (2015) Occurrence of transgenic feral alfalfa (*Medicago sativa* subsp. *sativa* L.) in alfalfa seed production areas in the United States. PLoS One 10:e0143296. https://doi.org/10.1371/journal.pone.0143296

Grev AM, Scott Wells M, Samac DA, Martinson KL, Sheaffer CC (2017) Forage accumulation and nutritive value of reduced lignin and reference alfalfa cultivars. Agron J 109:2749–2761. https://doi.org/10.2134/agronj2017.04.0237

Guan Q, Takano T, Liu S (2012) Genetic transformation and analysis of rice OsAPx2 gene in *Medicago sativa*. PLoS One 7: https://doi.org/10.1371/journal.pone.0041233

Guo D, Chen F, Inoue K, Blount JW, Dixon RA (2001a) Downregulation of caffeic acid 3-O-methyltransferase and caffeoyl CoA 3-O-methyltransferase in transgenic alfalfa: Impacts on lignin structure and implications for the biosynthesis of G and S lignin. Plant Cell 13:73–88. https://doi.org/10.1105/tpc.13.1.73

Guo D, Chen F, Wheeler J, Winder J, Selman S et al (2001b) Improvement of in-rumen digestibility of alfalfa forage by genetic manipulation of lignin O-methyltransferases. Transgen Res 10:457–464. https://doi.org/10.1023/A:1012278106147

Hall M, Whalen D (2012) Petition for the determination of nonregulated status for reduced lignin alfalfa KK179. United States Department of Agriculture, Animal and Plant Health Inspection Service. https://www.aphis.usda.gov/brs/aphisdocs/12_32101p.pdf. Accessed 28 Jan 2021

Hancock KR, Collette V, Fraser K, Greig M, Xue H et al (2012) Expression of the R2R3-MYB transcription factor TaMYB14 from *Trifolium arvense* activates proanthocyanidin biosynthesis in the legumes *Trifolium repens* and *Medicago sativa*. Plant Physiol 159:1204–1220. https://doi.org/10.1104/pp.112.195420

Hancock K, Collette V, Chapman E, Hanson K, Temple SJ et al (2014) Progress towards developing bloat-safe legumes for the farming industry. Crop Pasture Sci 65:1107–1113. https://doi.org/10.1071/CP13308

He X-Z, Dixon RA (2000) Genetic manipulation of isoflavone 7-O-methyltransferase enhances biosynthesis of 4′-O-methylated isoflavonoid phytoalexins and disease resistance in alfalfa. Plant Cell 12:1689–1702. https://doi.org/10.1105/tpc.12.9.1689

Hipskind JD, Paiva NL (2000) Constitutive accumulation of a resveratrol-glucoside in transgenic alfalfa increases resistance to *Phoma medicaginis*. Mol Plant-Microbe Interact 13:551–562. https://doi.org/10.1094/MPMI.2000.13.5.551

Huang L-K, Liao S-C, Chang C-C, Liu H-J (2006) Expression of avian reovirus σC protein in transgenic plants. J Virol Methods 134:217–222. https://doi.org/10.1016/j.jviromet.2006.01.013

ISAAA (2018) Global status of commercialized Biotech/GM Crops in 2018: biotech crops continue to help meet the challenges of increased population and climate change. In: ISAAA Brief No. 54. ISAAA, Ithaca

Ismail AM, Horie T (2017) Genomics, physiology, and molecular breeding approaches for improving salt tolerance. Annu Rev Plant Biol 68:405–434. https://doi.org/10.1146/annurev-arplant-042916-040936

Jia B, Sun M, Sun X, Li R, Wang Z et al (2016) Overexpression of GsGSTU13 and SCMRP in *Medicago sativa* confers increased salt-alkaline tolerance and methionine content. Physiol Plant 156:176–189. https://doi.org/10.1111/ppl.12350

Jin T, Chang Q, Li W, Yin D, Li Z et al (2010) Stress-inducible expression of GmDREB1 conferred salt tolerance in transgenic alfalfa. Plant Cell Tissue Organ Cult 100:219–227. https://doi.org/10.1007/s11240-009-9628-5

Jozefkowicz C, Frare R, Fox R, Odorizzi A, Arolfo V et al (2018) Maximizing the expression of transgenic traits into elite alfalfa germplasm using a supertransgene configuration in heterozygous conditions. Theor Appl Genet 131:1111–1123. https://doi.org/10.1007/s00122-018-3062-1

Kendrick D, Pester T, Horak M, Rogan G, Nickson T (2005) Biogeographic survey of feral alfalfa populations in the U.S. during 2001 and 2002 as a component of an ecological risk assessment of Roundup Ready Alfalfa®. ASA-CSSA-SSSA Abstracts

Khoudi H, Laberge S, Ferullo J-M, Bazin R, Darveau A et al (1999) Production of a diagnostic monoclonal antibody in perennial alfalfa plants. Biotechnol Bioeng 64:135–143. https://doi.org/10.1002/(SICI)1097-0290(19990720)64:2%3c135:AID-BIT2%3e3.0.CO;2-Q

Koegel RG, Straub RJ (1996) Fractionation of alfalfa for food, feed, biomass, and enzymes. Trans Amer Soc Agric Eng 39:769–774. https://doi.org/10.13031/2013.27559

Kumar V, AlMomin S, Al-Shatti A, Al-Aqeel H, Al-Salameen F et al (2019) Enhancement of heavy metal tolerance and accumulation efficiency by expressing Arabidopsis ATP sulfurylase gene in alfalfa. Intl J Phytoremed 21:1112–1121. https://doi.org/10.1080/15226514.2019.1606784

Lee R, Cornelisse M, Ziauddin A, Slack P, Hodgins D et al (2008) Expression of a modified *Mannheimia haemolytica* GS60 outer membrane lipoprotein in transgenic alfalfa for the development of an edible vaccine against bovine pneumonic pasteurellosis. J Biotechnol 135:224–231. https://doi.org/10.1016/j.jbiotec.2008.03.006

Li H, Wang Z, Ke Q, Ji CY, Jeong JC et al (2014) Overexpression of codA gene confers enhanced tolerance to abiotic stresses in alfalfa. Plant Physiol Biochem 85:31–40. https://doi.org/10.1016/j.plaphy.2014.10.010

Liu L, Fan X-D, Wang F-W, Wang N, Dong Y-Y et al (2013) Coexpression of ScNHX1 and ScVP in transgenic hybrids improves salt and saline-alkali tolerance in alfalfa (*Medicago sativa* L.). J Plant Growth Regul 32:1–8. https://doi.org/10.1007/s00344-012-9270-z

Liu W, Aung B, Hannoufa A, Xing T, Tian L (2018) Recent progress of transgenic technology development for alfalfa. Am J Plant Sci 09:467–482. https://doi.org/10.4236/ajps.2018.93035

Liu Z-H, Zhang H-M, Li G-L, Guo X-L, Chen S-Y et al (2011) Enhancement of salt tolerance in alfalfa transformed with the gene encoding for betaine aldehyde dehydrogenase. Euphytica 178:363–372. https://doi.org/10.1007/s10681-010-0316-7

Luo S, Baltes NJ (2020) Alfalfa with reduced lignin composition. United States Patent Application US2020/0048648 A1

Ma X-F, Tudor S, Butler T, Ge Y, Xi Y et al (2012) Transgenic expression of phytase and acid phosphatase genes in alfalfa (*Medicago sativa*) leads to improved phosphate uptake in natural soils. Mol Breed 30:377–391. https://doi.org/10.1007/s11032-011-9628-0

Mandaci S, Dobres MS (1997) A promoter directing epidermal expression in transgenic alfalfa. Plant Mol Biol 34:961–965. https://doi.org/10.1023/A:1005804514854

Masoud SA, Zhu Q, Lamb C, Dixon RA (1996) Constitutive expression of an inducible β-1,3-glucanase in alfalfa reduces disease severity caused by the oomycete pathogen*Phytophthora megasperma* f. sp. *medicaginis*, but does not reduce disease severity of chitin-containing fungi. Transgen Res 5:313–323. https://doi.org/10.1007/BF01968941

McCaslin M, Resisen P, Ho J (2014) New strategies for forage quality improvement in alfalfa. In: Proceedings of the 44th California alfalfa, forage, and grain symposium. University of California Cooperative Extension, Long Beach

McKersie BD, Bowley SR, Jones KS (1999) Winter survival of transgenic alfalfa overexpressing superoxide dismutase. Plant Physiol 119:839–848. https://doi.org/10.1104/pp.119.3.839

McKersie BD, Murnaghan J, Jones KS, Bowley SR (2000) Iron-superoxide dismutase expression in transgenic alfalfa increases winter survival without a detectable increase in photosynthetic oxidative stress tolerance. Plant Physiol 122:1427–1438. https://doi.org/10.1104/pp.122.4.1427

McQualter RB, Somleva MN, Gebbie LK, Li X, Petrasovits LA et al (2014) Factors affecting polyhydroxybutyrate accumulation in mesophyll cells of sugarcane and switchgrass. BMC Biotechnol 14:83. https://doi.org/10.1186/1472-6750-14-83

Mertens D, McCaslin M (2008) Evaluation of alfalfa hays with down-regulated lignin biosynthesis. J Anim Sci 86, E-Suppl 2:170

National Alfalfa and Forage Alliance (2014) Coexistence for alfalfa seed export markets. https://www.alfalfa.org/pdf/ExportSeed.pdf. Accessed 28 Jan 2021

Nawrath C, Poirier Y, Somerville C (1994) Targeting of the polyhydroxybutyrate biosynthetic pathway to the plastids of *Arabidopsis thaliana* results in high levels of polymer accumulation. Proc Natl Acad Sci USA 91:12760–12764. https://doi.org/10.1073/pnas.91.26.12760

Nicolia A, Ferradini N, Molla G, Biagetti E, Pollegioni L et al (2014) Expression of an evolved engineered variant of a bacterial glycine oxidase leads to glyphosate resistance in alfalfa. J Biotechnol 184:201–208. https://doi.org/10.1016/j.jbiotec.2014.05.020

Oommen A, Dixon RA, Paiva NL (1994) The elicitor-inducible alfalfa isoflavone reductase promoter confers different patterns of developmental expression in homologous and heterologous transgenic plants. Plant Cell 6:1789–1803. https://doi.org/10.1105/tpc.6.12.1789

Padgette SR, Kolacz KH, Delannay X, Re DB, LaVallee BJ et al (1995) Development, identification, and characterization of a glyphosate-tolerant soybean line. Crop Sci. 35:145101461. https://doi.org/10.2135/cropsci1995.0011183X003500050032x

Palma L, Muñoz D, Berry C, Murillo J, Caballero P (2014) *Bacillus thuringiensis*toxins: an overview of their biocidal activity. Toxins (Basel) 6:3296–3325. https://doi.org/10.3390/toxins6123296

Parihar P, Singh S, Singh R, Singh VP, Prasad SM (2015) Effect of salinity stress on plants and its tolerance strategies: a review. Environ Sci Pollut Res 22:4056–4075. https://doi.org/10.1007/s11356-014-3739-1

Pathirana S, Samac DA, Roeven R, Yoshioka H, Vance CP et al (1997) Analyses of phosphoenolpyruvate carboxylase gene structure and expression in alfalfa nodules. Plant J 12:293–304. https://doi.org/10.1046/j.1365-313X.1997.12020293.x

Reyno R, Khu D-M, Monteros MJ, Bouton JH, Parrott W et al (2013) Evaluation of two transgenes for aluminum tolerance in alfalfa. Crop Sci 53:1581–1588. https://doi.org/10.2135/cropsci2012.12.0676

Richins RD, Scholthof HB, Shepherd RJ (1987) Sequence of figwort mosaic virus DNA (caulimovirus group). Nucl Acids Res 15:8451–8466. https://doi.org/10.1093/nar/15.20.8451

Rogan G, Fitzpatrick S (2004) Petition for the determination of nonregulated status: Roundup Ready alfalfa (*Medicago sativa* L.) events J101 and J163. USDA-APHIS. https://www.aphis.usda.gov/brs/aphisdocs/04_11001p.pdf

Rosellini D, Capomaccio S, Ferradini N, Savo Sardaro ML, Nicolia A et al (2007) Non-antibiotic, efficient selection for alfalfa genetic engineering. Plant Cell Rep 26:1035–1044. https://doi.org/10.1007/s00299-007-0321-z

Rosellini D, Pezzotti M, Veronesi F (2001) Characterization of transgenic male sterility in alfalfa. Euphytica 118:313–319. https://doi.org/10.1023/A:1017568201732

Samac D, Temple SJ (2004) Development and utilization of transformation in *Medicago*species. In: Liang G, Skinner DZ (eds) Genetically modified crops: their development, uses, and risks. The Haworth Press, Binghampton, pp 165–202

Samac DA (1995) Strain specificity in transformation of alfalfa by *Agrobacterium tumefaciens*. Plant Cell Tissue Organ Cult 43:271–277. https://doi.org/10.1007/BF00039955

Samac DA, Austin-Phillips S (2006) Alfalfa (*Medicago sativa* L.). In: Wang K (ed) *Agrobacterium* protocols, 2nd edn, Methods in molecular biology, vol 343. https://doi-org.ezp3.lib.umn.edu/10.1385/1-59745-130-4:301, pp 301–311

Samac DA, Litterer L, Temple G, Jung H-JG, Somers DA (2004a) Expression of UDP-glucose dehydrogenase reduces cell-wall polysaccharide concentration and increases xylose content in alfalfa stems. Appl Biochem Biotechnol Part A Enzym Eng Biotechnol 116:1167–1182. https://doi.org/10.1385/ABAB:116:1-3:1167

Samac DA, Smigocki AC (2003) Expression of oryzacystatin I and II in alfalfa increases resistance to the root-lesion nematode. Phytopathology 93:799–804. https://doi.org/10.1094/PHYTO.2003.93.7.799

Samac DA, Tesfaye M, Dornbusch M, Saruul P, Temple SJ (2004b) A comparison of constitutive promoters for expression of transgenes in alfalfa (*Medicago sativa*). Transgen Res 13:349–361. https://doi.org/10.1023/B:TRAG.0000040022.84253.12

Saruul P, Srienc F, Somers DA, Samac DA (2002) Production of a biodegradable plastic polymer, poly-β-hydroxybutyrate, in transgenic alfalfa. Crop Sci 42:919–927. https://doi.org/10.2135/cropsci2002.9190

Sathoff AE, Dornbusch MR, Miller SS, Samac DA (2020) Functional analysis of *Medicago*-derived pathogen-induced gene promoters for usage in transgenic alfalfa. Mol Breed 40:66. https://doi.org/10.1007/s11032-020-01144-6

Sathoff AE, Velivelli S, Shah DM, Samac DA (2019) Plant defensin peptides have antifungal and antibacterial activity against human and plant pathogens. Phytopathology 109:402–408. https://doi.org/10.1094/PHYTO-09-18-0331-R

Schoenbeck MA, Samac DA, Fedorova M, Gregerson RG, Gantt JS, Vance CP (1999) The alfalfa (*Medicago sativa*) TDY1 gene encodes a mitogen-activated protein kinase homolog. Mol Plant-Microbe Interact 12:882–893. https://doi.org/10.1094/MPMI.1999.12.10.882

Schroeder H, Khan M, Knibb W, Spencer D, Higgins T (1991) Expression of a chicken ovalbumin gene in three lucerne cultivars. Austral J Plant Physiol 18:495–505. https://doi.org/10.1071/PP9910495

Sewalt VJH, Ni W, Jung HG, Dixon RA (1997) Lignin impact on fiber degradation: increased enzymatic digestibility of genetically engineered tobacco (*Nicotiana tabacum*) stems reduced in lignin content. J Agric Food Chem 45:1977–1983. https://doi.org/10.1021/jf9609690

Shi L, Twary SN, Yoshioka H, Gregerson RG, Miller SS et al (1997) Nitrogen assimilation in alfalfa: isolation and characterization of an asparagine synthetase gene showing enhanced expression in root nodules and dark-adapted leaves. Plant Cell 9:1339–1356. https://doi.org/10.1105/tpc.9.8.1339

Singer SD, Hannoufa A, Acharya S (2017) Molecular improvement of alfalfa for enhanced productivity and adaptability in a changing environment. Plant Cell Environ 41:1955–1971. https://doi.org/10.1111/pce.13090

Smith S, Bingham ET, Fulton R (1986) Transmission of chlorophyll deficiencies in *Medicago sativa*. Evidence for biparental inheritance of plastids. J Hered 77:35–38. https://doi.org/10.1093/oxfordjournals.jhered.a110163

Stefanova G, Slavov S, Gecheff K, Vlahova M, Atanassov A (2013) Expression of recombinant human lactoferrin in transgenic alfalfa plants. Biol Plant 57:457–464. https://doi.org/10.1007/s10535-013-0305-5

Stritzler M, Elba P, Berini C, Gomez C, Ayub N et al (2018) High-quality forage production under salinity by using a salt-tolerant AtNXH1-expressing transgenic alfalfa combined with a natural stress-resistant nitrogen-fixing bacterium. J Biotechnol 276–277:42–45. https://doi.org/10.1016/j.jbiotec.2018.04.013

Suárez R, Calderón C, Iturriaga G (2009) Enhanced tolerance to multiple abiotic stresses in transgenic alfalfa accumulating trehalose. Crop Sci 49:1791–1799. https://doi.org/10.2135/cropsci2008.09.0573

Sullivan ML, Green HA, Verdonk JC (2021) Engineering alfalfa to produce 2-*O*-caffeoyl-L-malate (phaselic acid) for preventing post-harvest protein loss via oxidation by polyphenol oxidase. Front Plant Sci 11: https://doi.org/10.3389/fpls.2020.610399

Sullivan ML, Hatfield RD, Thoma SL, Samac DA (2004) Cloning and characterization of red clover polyphenol oxidase cDNAs and expression of active protein in *Escherichia coli* and transgenic alfalfa. Plant Physiol 136:3234–3244. https://doi.org/10.1104/pp.104.047449

Sun M, Qian X, Chen C, Cheng S, Jia B, Zhu Y et al (2018) Ectopic expression of GsSRK in *Medicago sativa* reveals its involvement in plant architecture and salt stress responses. Front Plant Sci 9:226. https://doi.org/10.3389/fpls.2018.00226

Sun M, Sun X, Zhao Y, Zhao C, DuanMu H et al (2014) Ectopic expression of GsPPCK3 and SCMRP in *Medicago sativa* enhances plant alkaline stress tolerance and methionine content. PLoS One 9: https://doi.org/10.1371/journal.pone.0089578

Supreme Court of the United States (2010) Monsanto Co. v. Geertson Seed Farms, 561 U.S. 139 (2010). https://supreme.justia.com/cases/federal/us/561/139/

Tabe LM, Wardley-Richardson T, Ceriotti A, Aryan A, McNabb W et al (1995) A biotechnological approach to improving the nutritive value of alfalfa. J Anim Sci 73:2752–2759. https://doi.org/10.2527/1995.7392752x

Tang L, Cai H, Ji W, Luo X, Wang Z et al (2013) Overexpression of GsZFP1 enhances salt and drought tolerance in transgenic alfalfa (*Medicago sativa* L.). Plant Physiol Biochem 71:22–30. https://doi.org/10.1016/j.plaphy.2013.06.024

Tang L, Cai H, Zhai H, Luo X, Wang Z et al (2014) Overexpression of *Glycine soja* WRKY20 enhances both drought and salt tolerance in transgenic alfalfa (*Medicago sativa* L.). Plant Cell Tissue Organ Cult 118:77–86. https://doi.org/10.1007/s11240-014-0463-y

Tesfaye M, Samac DA, Lamb JFS (2008) Alfalfa. In: Kole C, Hall TC (eds) A compendium of transgenic crop plants: transgenic legume grains and forages. Blackwell Publishing, Oxford, pp 199–210

Tesfaye M, Temple SJ, Allan DL, Vance CP, Samac DA (2001) Overexpression of malate dehydrogenase in transgenic alfalfa enhances organic acid synthesis and confers tolerance to aluminum. Plant Physiol 127:1836–1844. https://doi.org/10.1104/pp.127.4.1836

Teuber LR, van Deynze A, Mueller S, McCaslin M, Fitzpatrick S, et al (2004) Gene flow in alfalfa under honey bee (*Apis mellifera*) pollination. In: Proceedings of the 39th North American alfalfa improvement conference

Tohidfar M, Zare N, Jouzani GS, Eftekhari SM (2013) *Agrobacterium*-mediated transformation of alfalfa (*Medicago sativa*) using a synthetic *cry3a* gene to enhance resistance against alfalfa weevil. Plant Cell Tiss Org Cult 113:227–235. https://doi.org/10.1007/s11240-012-0262-2

Tong Z, Xie C, Ma L, Liu L, Jin Y, et al (2014) Co-expression of bacterial aspartate kinase and adenylylsulfate reductase genes substantially increases sulfur amino acid levels in transgenic alfalfa (*Medicago sativa* L.). PLoS One 9:e88310. https://doi.org/10.1371/journal.pone.0088310

Trepp GB, Van De Mortel MV, Yoshioka H, Miller SS, Samac DA et al (1999) NADH-glutamate synthase in alfalfa root nodules. Genetic regulation and cellular expression. Plant Physiol 119:817–828. https://doi.org/10.1104/pp.119.3.817

Ullah AH, Sethumadhavan K, Mullaney EJ, Ziegelhoffer T, Austin-Phillips S (2002) Cloned and expressed fungal phya gene in alfalfa produces a stable phytase. Biochem Biophys Res Commun 290:1343–1348. https://doi.org/10.1006/bbrc.2002.6361

Undersander D, Martin N, Hall M, Mueller S (2009) Review of roundup ready alfalfa. Forage and Grazinglands. https://doi.org/10.1094/fg-2009-1019-01-rv

USDA-APHIS (2010) Glyphosate-tolerant alfalfa events J101 and J163: request for nonregulated status final environmental impact statement. https://www.aphis.usda.gov/biotechnology/downloads/alfalfa/gt_alfalfa_feis.pdf. Accessed 28 Jan 2021

USDA-APHIS (2014) Determination of nonregulated status for GE alfalfa (APHIS Petition No. 12-321-01p). https://www.aphis.usda.gov/aphis/ourfocus/biotechnology/sa_environmental_documents/sa_environmental_assessments/ct_det_nonreg_alfalfa_kk179. Accessed 28 Jan 2021

Vail AW, Wang P, Uefuji H, Samac DA, Vance CP et al (2014) Biodegradation of atrazine by three transgenic grasses and alfalfa expressing a modified bacterial atrazine chlorohydrolase gene. Transgenic Res 24:475–488. https://doi.org/10.1007/s11248-014-9851-7

van Zelm E, Zhang Y, Testerink C (2020) Salt tolerance mechanisms of plants. Annu Rev Plant Biol 71:403–433. https://doi.org/10.1146/annurev-arplant-050718-100005

Vaziri A, Keen NT, Erwin DC (1981) Correlation of medicarpin production with resistance to *Phytophthora megasperma* f. sp. *medicaginis* in alfalfa seedlings. Phytopathology 71:1235–1238. https://doi.org/10.1094/Phyto-71-1235

Waltz E (2011) Industry exhales as USDA okays glyphosate-resistant alfalfa. Nat. Biotechnol. 29:179–181. https://doi.org/10.1038/nbt0311-179a

Wandelt C, Khan M, Craig S, Schroeder H, Spencer D et al (1992) Vicilin with carboxy-terminal KDEL is retained in the endoplasmic reticulum and accumulates to high levels in the leaves of transgenic plants. Plant J 2:181–192. https://doi.org/10.1046/j.1365-313X.1992.t01-41-00999.x

Wang F, Wang C, Sun Y, Wang N, Li X et al (2016a) Overexpression of vacuolar proton pump ATPase (V-H + -ATPase) subunits B, C and H confers tolerance to salt and saline-alkali stresses in transgenic alfalfa (*Medicago sativa* L.). J Integr Agri 15:2279–2289. https://doi.org/10.1016/S2095-3119(16)61399-0

Wang J, Xu X, Ma D (2019a) Establishment of an efficient alfalfa regeneration system via organogenesis and the co-expression of Arabidopsis SOS genes improves salt tolerance in transgenic alfalfa (*Medicago sativa* L.). Can J Plant Sci 99:348–363. https://doi.org/10.1139/cjps-2018-0105

Wang L, Samac DA, Shapir N, Wackett LP, Vance CP et al (2005) Biodegradation of atrazine in transgenic plants expressing a modified bacterial atrazine chlorohydrolase (*atzA*) gene. Plant Biotechnol J 3:475–486. https://doi.org/10.1111/j.1467-7652.2005.00138.x

Wang X, Gao Y, Chen Z, Li J, Huang J et al (2019b) (E)-β-farnesene synthase gene affects aphid behavior in transgenic *Medicago sativa*. Pest Manag Sci 75:622–631. https://doi.org/10.1002/ps.5153

Wang Z-Y, Brummer EC (2012) Is genetic engineering ever going to take off in forage, turf and bioenergy crop breeding? Ann Bot 110:1317–1325. https://doi.org/10.1093/aob/mcs027

Wang Z, Li H, Ke Q, Jeong JC, Lee HS et al (2014) Transgenic alfalfa plants expressing AtNDPK2 exhibit increased growth and tolerance to abiotic stresses. Plant Physiol Biochem 84:67–77. https://doi.org/10.1016/j.plaphy.2014.08.025

Wang Z, Su G, Li M, Ke Q, Kim SY et al (2016b) Overexpressing Arabidopsis ABF3 increases tolerance to multiple abiotic stresses and reduces leaf size in alfalfa. Plant Physiol Biochem 109:199–208. https://doi.org/10.1016/j.plaphy.2016.09.020

Wang ZY, Song FB, Cai H, Zhu YM, Bai X et al (2012) Over-expressing GsGST14 from *Glycine soja* enhances alkaline tolerance of transgenic *Medicago sativa*. Biol Plant 56:516–520. https://doi.org/10.1007/s10535-012-0075-5

Watrud LS, Misra S, Gedamu L, Shiroyama T, Maggard S et al (2006) Ecological risk assessment of alfalfa (*Medicago varia* L.) genetically engineered to express a human metallothionein (hMT) gene. Water Air Soil Pollut 176:329–349. https://doi.org/10.1007/s11270-006-9171-5

Weakley D, Mertens D, McCaslin M (2008) Lactating cow responses to alfalfa hays with down-regulated lignin biosynthesis. J Anim Sci 86, E-Suppl 2:170

Weeks JT, Ye J, Rommens CM (2008) Development of an *in planta* method for transformation of alfalfa (*Medicago sativa*). Transgen Res 17:587–597. https://doi.org/10.1007/s11248-007-9132-9

Wei Z, Liu Y, Lin C, Wang Y, Cai Q et al (2011) Transformation of alfalfa chloroplasts and expression of green fluorescent protein in a forage crop. Biotechnol Lett 33:2487–2494. https://doi.org/10.1007/s10529-011-0709-2

Wigdorovitz A, Carrillo C, Dus Santos MJ, Trono K, Peralta A et al (1999) Induction of a protective antibody response to foot and mouth disease virus in mice following oral or parenteral immunization with alfalfa transgenic plants expressing the viral structural protein VP1. Virology 255:347–353. https://doi.org/10.1006/viro.1998.9590

Wigdorovitz A, Mozgovoj M, Santos MJD, Parreño V, Gómez C et al (2004) Protective lactogenic immunity conferred by an edible peptide vaccine to bovine rotavirus produced in transgenic plants. J Gen Virol 85:1825–1832. https://doi.org/10.1099/vir.0.19659-0

Winicov I (2000) Alfin1 transcription factor overexpression enhances plant root growth under normal and saline conditions and improves salt tolerance in alfalfa. Planta 210:416–422. https://doi.org/10.1007/PL00008150

Winicov I, Valliyodan B, Xue L, Hoober JK (2004) The MsPRP2 promoter enables strong heterologous gene expression in a root-specific manner and is enhanced by overexpression of Alfin-1. Planta 219:925–935. https://doi.org/10.1007/s00425-004-1296-4

Wolabu TW, Cong L, Park J-J, Bao Q, Chen M et al (2020) Development of a highly efficient multiplex genome editing system in outcrossing tetraploid alfalfa (*Medicago sativa*). Front Plant Sci 11:1063. https://doi.org/10.3389/fpls.2020.01063

Woo H-H, Orbach MJ, Hirsch AM, Hawes MC (1999) Meristem-localized inducible expression of a UDP-glycosyltransferase gene is essential for growth and development in pea and alfalfa. Plant Cell 11:2303–2315. https://doi.org/10.1105/tpc.11.12.2303

Wu S, Zhu P, Jia B, Yang J, Shen Y et al (2018) A *Glycine soja* group S2 bZIP transcription factor GsbZIP67 conferred bicarbonate alkaline tolerance in *Medicago sativa*. BMC Plant Biol 18:234. https://doi.org/10.1186/s12870-018-1466-3

Wu Y, Pereira F, Qiu X, Erickson L (1998) A comparison of the promoter regions of three pollen-specific genes in alfalfa. Sex Plant Reprod 11:181–182. https://doi.org/10.1007/s004970050139

Yang S, Gao M, Xu C, Gao J, Deshpande S et al (2008) Alfalfa benefits from *Medicago truncatula*: The *RCT1* gene from *M. truncatula* confers broad-spectrum resistance to anthracnose in alfalfa. Proc Natl Acad Sci U S A 105:12164–12169. https://doi.org/10.1073/pnas.0802518105

Yi D, Ma L, Lin M, Li C (2018) Development of glyphosate-resistant alfalfa (*Medicago sativa* L.) upon transformation with the GR79Ms gene encoding 5-enolpyruvylshikimate-3-phosphate synthase. Planta 248:211–219. https://doi.org/10.1007/s00425-018-2898-6

Yoshioka H, Gregerson RG, Samac DA, Hoevens KCM, Trepp G et al (1999) Aspartate aminotransferase in alfalfa nodules: localization of mRNA during effective and ineffective nodule development and promoter analysis. Mol Plant-Microbe Interact 12:263–274. https://doi.org/10.1094/MPMI.1999.12.4.263

Zhang J-Y, Broeckling CD, Blancaflor EB, Sledge MK, Sumner LW et al (2005) Overexpression of *WXP1*, a putative *Medicago truncatula* AP2 domain-containing transcription factor gene, increases cuticular wax accumulation and enhances drought tolerance in transgenic alfalfa (*Medicago sativa*). Plant J 42:689–707. https://doi.org/10.1111/j.1365-313X.2005.02405.x

Zhang J, Duan Z, Zhang D, Zhang J, Di H et al (2016) Co-transforming bar and CsLEA enhanced tolerance to drought and salt stress in transgenic alfalfa (*Medicago sativa* L.). Biochem Biophys Res Commun 472:75–82. https://doi.org/10.1016/j.bbrc.2016.02.067

Zhang Q, Ma C, Xue X, Xu M, Li J et al (2014) Overexpression of a cytosolic ascorbate peroxidase gene, OsAPX2, increases salt tolerance in transgenic alfalfa. J Integr Agric 13:2500–2507. https://doi.org/10.1016/S2095-3119(13)60691-7

Zhang W-J, Wang T (2015) Enhanced salt tolerance of alfalfa (*Medicago sativa*) by *rstB* gene transformation. Plant Sci 234:110–118. https://doi.org/10.1016/j.plantsci.2014.11.016

Zhang Y-M, Zhang H-M, Liu Z-H, Li H-C, Guo X-L et al (2015) The wheat NHX antiporter gene *TaNHX2* confers salt tolerance in transgenic alfalfa by increasing the retention capacity of intracellular potassium. Plant Mol Biol 87:317–327. https://doi.org/10.1007/s11103-014-0278-6

Zhang Y, Liu J, Zhou Y, Gong T, Wang J et al (2013) Enhanced phytoremediation of mixed heavy metal (mercury)–organic pollutants (trichloroethylene) with transgenic alfalfa co-expressing glutathione S-transferase and human P450 2E1. J Hazard Mater 260:1100–1107. https://doi.org/10.1016/j.jhazmat.2013.06.065

Zhou C, Han L, Pislariu C, Nakashima J, Fu C et al (2011) From model to crop: functional analysis of a STAY-GREEN gene in the model legume *Medicago truncatula* and effective use of the gene for alfalfa improvement. Plant Physiol 157:1483–1496. https://doi.org/10.1104/pp.111.185140

Ziauddin A, Lee RWH, Lo R, Shewen P, Strommer J (2004) Transformation of alfalfa with a bacterial fusion gene, *Mannheimia haemolytica* A1 leukotoxin50-gfp: Response with *Agrobacterium tumefaciens* strains LBA4404 and C58. Plant Cell Tissue Organ Cult 79:271–278. https://doi.org/10.1007/s11240-004-4540-5

Sequencing, Assembly, and Annotation of the Alfalfa Genome

6

Joann Mudge and Andrew D. Farmer

Abstract

While the alfalfa community originally relied on *Medicago truncatula* (especially the reference assembly, A17) for genomic resources, recent changes in sequencing and scaffolding technologies and algorithms have enabled the sequencing and assembly of five different alfalfa accessions, to date. These assemblies include two diploid assemblies, CADL and PI464715, as well as three tetraploid assemblies, NECS-141, Zhongmu No. 1, and XinJiangDaYe. Technological changes within the approximately half a decade over which these assemblies were produced, have allowed for increasingly contiguous assemblies and improved scaffolding resulting in chromosome level assemblies that allow for the detection of large-scale structural rearrangements. They have also made possible the assembly of all four subgenomes of the tetraploid in the XinJiangDaYe assembly. While subgenome haplotypes were very similar and sometimes indistinguishable, nevertheless, structural differences between haplotypes were uncovered. These included local differential gene content between subgenome haplotypes as well as larger structural variants such as inversions. Compared to the *M. truncatula* assembly and annotation, the approximately 75% increase in genome size in alfalfa is mainly due to the expansion of repeats. The availability of five different annotated alfalfa genome assemblies, including those of both diploid and tetraploid accessions, will be a significant asset to the alfalfa community.

6.1 Introduction

6.1.1 The Alfalfa Genome

Plants have very dynamic genomes making plant genome assembly especially challenging. Flexibility and instability in plant genomes is reflected in genome size expansion and contraction and higher rates of polyploidy, heterozygosity, repeats, and pseudogenes compared to eukaryotic organisms from other kingdoms (Schatz et al. 2012; Jiao and Schneeberger 2017).

The alfalfa genome is no exception. Obvious sources of alfalfa genome complexity include autopolyploidy and high rates of heterozygosity. Both heterozygosity and polyploidy can lead to diverging haplotypes that complicate assembly.

J. Mudge (✉) · A. D. Farmer
Department of Bioinformatics, National Center for Genome Resources, Santa Fe, NM, USA
e-mail: jm@ncgr.org

A. D. Farmer
e-mail: adf@ncgr.org

L.-X. Yu and C. Kole (eds.), *The Alfalfa Genome*, Compendium of Plant Genomes,
https://doi.org/10.1007/978-3-030-74466-3_6

The obligate outcrossing reproductive mechanism of alfalfa ensures that heterozygosity rates remain high and polypoidy provides further opportunities for haplotype diversity. While the *Medicago sativa* complex includes both diploid and tetraploid forms, cultivated alfalfa is tetraploid. Most cultivars belong to the *sativa* subspecies or the *varia* subspecies, which represents introgressions of the *falcata* subspecies into the *sativa* subspecies. However, a few cultivars, especially those harboring cold tolerance, are from the subspecies *falcata* (Veronesi et al. 2010).

The autotetraploid genome of cultivated alfalfa allows for up to four different subgenome haplotypes, with the number of distinguishable haplotypes at a locus varying across the genome. In contrast to allopolyploids, whose subgenomes originate from different progenitor species' genomes that are typically relatively divergent, autopolyploids have chromosomes doubled from genomes within the same species and may allow recombination among the homoeologues. Tetraploidy also results in a large genome size requiring an increased sequencing volume to achieve the same genome coverage. While alfalfa has a base (haploid) chromosome number of 8 and a base genome size of ~800 Mb, cultivated and some wild alfalfa species have 32 chromosomes and ~3.2 Gb in its tetraploid genome (Blondon et al. 1994).

Original genomic analyses used the congeneric *Medicago truncatula* as a model (Yang et al. 2008; Young et al. 2011). *M. truncatula* has a smaller genome size (~450 Mb). In addition, this diploid plant has a high rate of selfing (Barker et al. 1990; Cook 1999) resulting in a low heterozygosity rate that makes assembly easier and lowers coverage requirements. Recent advances in sequencing and scaffolding technologies have lowered cost and increased throughput. These advances, along with improved assembly algorithms, and have recently made directly sequencing and assembling plant genomes, including polyploid genomes, more feasible (Mishra et al. 2017; Kyriakidou et al. 2018; Jung et al. 2019; Michael and VanBuren 2020).

6.1.2 Changing Technologies

Long-read sequencing technologies, including Pacific Biosciences Single Molecule Real-time or SMRT sequencing (PacBio) and Oxford Nanopore sequencing (ONT), have vastly improved our ability to generate reference-quality plant genomes, with relative ease and low cost compared to Sanger sequencing. These technologies produce higher quality assemblies compared to short-read or short and long-read hybrid assemblies (Eid et al. 2009; Deamer et al. 2016; Jiao and Schneeberger 2017). Plant assemblies generated based on long-read sequences first began to appear in 2015, with *M. truncatula* having some of the earliest PacBio-based plant assemblies (Berlin et al. 2015; VanBuren et al. 2015; Moll et al. 2017). PacBio-based plant assemblies showed much improved continuity, fewer gaps, and captured more of the genome compared to assemblies based on short reads (VanBuren et al. 2015; Moll et al. 2017; Jiao and Schneeberger 2017). But these assemblies still struggled to span long, closely related repeats or efficiently navigate differing levels of haplotype divergence. This is in part due to the fact that sequence error rates were higher than the divergence levels that needed to be discriminated in order to resolve these types of elements. The use of correction strategies based on consensus among reads taken from different molecules made it difficult to discriminate between closely related repeats or haplotypes, as they would often be lumped together during correction, forming a single chimeric consensus sequence.

But the recent transition from PacBio CLR (continuous long read) to HiFi (high fidelity) reads (Wenger et al. 2019), in which high accuracy consensus (>99%) is achieved by utilizing correction based on multiple sequencing passes on the same template molecule, has improved our ability to discriminate between closely related repeats or slightly diverged haplotypes. PacBio HiFi reads, because of their accuracy, require reduced consensus read coverage for assembly. Reduced coverage combined with increased throughput on the Sequel II, mean that higher quality plant assemblies can be obtained for a lower cost with reduced computational requirements and time compared to the original PacBio-based assemblies. In addition, the high read accuracy makes it possible to distinguish alternate haplotypes and similar but not identical repeat sequences, increasing continuity and improving our ability to phase haplotypes.

Oxford Nanopore Technology (ONT) is another long read technology. Since first conceptualized in the late 1980s, ONT has made recent advances in both length and accuracy (Deamer et al. 2016). In just a few short years, ONT-based assemblies have gone from bacteria (Deschamps et al. 2016) to higher organisms, including plants (Michael et al. 2018; Belser et al. 2018; Deschamps et al. 2018). A recent study on the comparison of PacBio HiFi and ONT in rice (Lang et al. 2020) indicated that while PacBio's high read accuracy enabled higher accuracy at the nucleotide level, including fewer artificial SNPs and small indels in the assembly, the longer ONT read length (up to 2 Mb) enabled higher assembly continuity and better spanning of repetitive genomic sections and resolution of gene family copy number. While both technologies were able to assemble some rice chromosomes in a single contig, ONT technology captured more chromosome length contigs (10 compared to 3 with PacBio HiFi) and 7 of these appeared to be gapless assemblies extending into telomeres on either end. The two technologies appear to complement each other with PacBio delivering high continuity and accuracy and ONT delivering even higher continuity tempered by a small reduction in accuracy.

Whichever long-read technology is used for plant genome assembly, additional scaffolding technologies are often applied to improve contiguity of the assembly. Recently, new technologies have replaced more expensive and cumbersome methods of scaffolding such as physical and genetic maps. Long-range, whole genome scaffolding technologies, including optical mapping and chromatin conformation technologies provide high-throughput and relatively inexpensive methods to scaffold contigs together, improving contiguity, often to the pseudo-chromosome level (Burton et al. 2013; O'Bleness et al. 2014; Steinberg et al. 2014; Mostovoy et al. 2016; Staňková et al. 2016).

In less than a decade, technology improvements have allowed the alfalfa community to move from reliance on *M. truncatula* genome assemblies to sequencing the alfalfa genome directly. To date, the community has generated five publicly available alfalfa genome assemblies. The sequencing and assembling of these genomes have been pursued at different times across a rapidly changing technological background, providing an interesting view into not only how changes in technology and strategies affect genome assemblies, but also elucidating structural challenges inherent in the alfalfa genome.

6.2 Genome Assemblies

6.2.1 Diploid Assemblies

6.2.1.1 Cultivated Alfalfa at the Diploid Level

The first genome assembly was generated from a plant from the Cultivated Alfalfa at the Diploid Level (CADL) population. This population is a stable diploid alfalfa population that is able to reproduce by seed (Bingham and McCoy 1979). It took advantage of diploid cultivated alfalfa germplasm generated by the $4x - 2x$ cross method (Bingham 1969). Though fertility of the diploid lines was low, crossing them as females to diploid *M. sativa* subspecies *falcata* lines improved fertility of the F_1, allowing backcrossing to the $2x$ diploids. This resulted in a stable diploid, fertile population whose germplasm is estimated to be derived from at least 98% cultivated germplasm. A single, clonally propagated plant was chosen from the CADL population for genome sequencing and assembly to avoid any interplant variability.

Using a diploid plant for sequencing and assembly reduced the amount of sequence data needed and the complexity of the assembly, which should lead to a high-quality assembly while requiring fewer computational resources.

The eventual goal was to use this assembly as a scaffold for assembling a tetraploid genome. Whole-genome PacBio continuous long read (CLR) sequencing was begun in early 2015 with a preliminary version of the assembly publicly released in mid-2016. Just over 100X subread coverage (based on an 800 Mb genome size) or just over 50X coverage per haplotype was generated. Subreads, sequencing reads resulting from each of the multiple sequencing passes of a DNA fragment, had a mean length of 8.0 kb and an N50 length of 13.1 kb (Table 6.1).

Several assembly iterations were tried, though computational constraints made it difficult to test assembly strategies extensively. The current assembly version (version 1.0) was generated as follows. DAligner (Myers 2014) was used to align reads. Using these alignments, Falcon (Chin et al. 2016) was used to assemble the reads. The resulting assembly was polished using Quiver (https://github.com/PacificBiosciences/GenomicConsensus). The polished assembly was scaffolded with long-distance maps generated from chromatin conformation Dovetail libraries using the HiRise algorithm (Putnam et al. 2016). A final polish with Quiver completed the assembly.

The resulting assembly was fragmented (5,753 pieces) but, nevertheless, contained most of the genespace (96.7%) based on capture of single-copy eukaryotic orthologous genes with Benchmarking Universal Single-Copy Orthologs (BUSCO) (Simão et al. 2015) (Tables 6.1 and 6.2), which is a surrogate for overall gene capture. Even with the fragmentation, much of the assembly was in megabase-sized pieces with a contig N50 of 1.27 Mb (half of the assembly is in pieces of 1.27 Mb or larger). While still far short of expected chromosome sizes, this is, nevertheless, an important improvement over the short-read plant genome assemblies that had previously dominated.

The total assembly size of 1,200 Mb is approximately 50% larger than the expected 800 Mb base genome size. This is due to the assembly of multiple haplotypes in some, but not all regions of the genome. In comparing haplotypes that were divergent enough to assemble separately, it became clear that different haplotypes of this diploid genome were often missing genes from the syntenic haplotype (Fig. 6.1a). Therefore, the full gene complement was not present in a single haplotype. This might be an artifact of creating a diploid from cultivated autotetraploid germplasm. Upon plant whole-genome duplication that results in an autopolyploid, differential gene loss between haplotypes can occur (Doyle et al. 2008; Hufton and Panopoulou 2009). The assembly of multiple haplotypes in about half of the genome is confirmed by the BUSCO results, with 57.4% of the typical single-copy orthologs duplicated (Table 6.2 and Fig. 6.2). The CADL assembly shows good coverage of the *M. truncatula* genome (Fig. 6.3). Regions of the assembly showing one or two haplotypes assembled are visible as double versus single diagonals (Fig. 6.4).

Table 6.1 Sequence read and assembly statistics for the five alfalfa genome assemblies

Accession	Ploidy	Sequencing technology	Read N50 (kb)	Scaffolding technology	Scaffold length (Mb)	Contig N50 (Mb)	Scaffold N50 (Mb)
CADL	*2x*	PacBio	13.1	Dovetail	1,251	1.27	1.27
PI464715	*2x*	Oxford Nanopore	27.9	Hi-C	793	3.86	102.49
NECS-141	*4x*	PacBio	17.4	BioNano	2,698	0.22	2.21
Zhongmu No. 1	*4x*	PacBio	12.2	BioNano and Hi-C	817	3.92	102.29
XinJiangDaYe	*4x*	PacBio HiFi	12.6	Hi-C	3158	0.46	84.27

Table 6.2 Gene statistics for the five alfalfa genome assemblies

Accession	Ploidy of source	Assembly ploidy[a]	Protein coding genes (thousands)	BUSCO database	Complete BUSCO Genes (%)	Complete and duplicated BUSCO Genes (%)
CADL	*2x*	*2x*	111	eudicotyledons_odb10	96.7	57.4
PI464715	*2x*	*1x*	47	embryophyta_odb10	97.7	8.7
NECS-141	*4x*	*4x*	103	eudicotyledons_odb10	95.7	75.8
Zhongmu No. 1	*4x*	*1x*	50	embryophyta_odb9	93.3	5.5
XinJiangDaYe	*4x*	*4x*	165	unavailable	97.2	90.1

[a]Upper estimate as some haplotypes were collapsed in assemblies at more than 1X

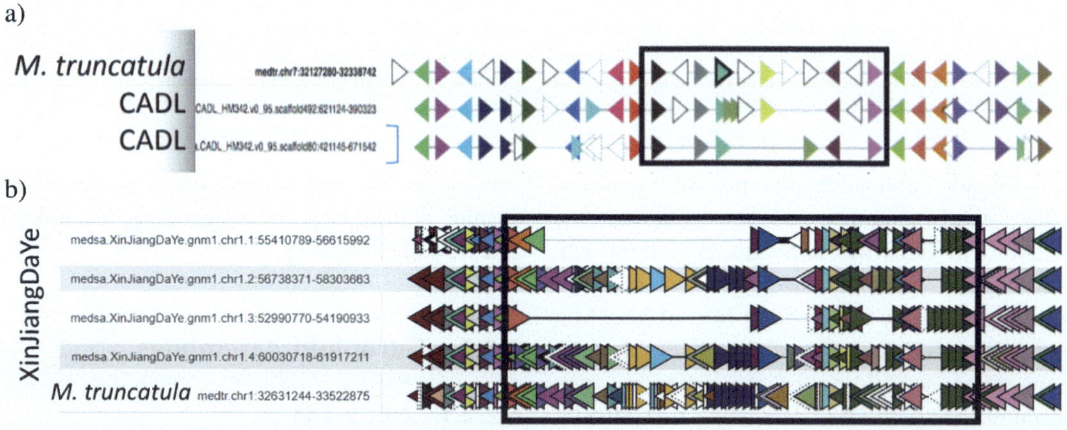

Fig. 6.1 Comparison of alfalfa genomic regions to syntenic *Medicago truncatula* regions. Triangles represent genes with orientation indicated by the direction of the pointed side. Genes are colored by gene family. The boxed regions show synteny breaking down through the differential loss of genes between haplotypes in alfalfa assembly. **a** *M. truncatula* (top) chromosome 7 compared to CADL's two syntenic haplotypes. **b** *M. truncatula* (bottom) chromosome 1 compared to XinJiangDaYe's four syntenic haplotypes

6.2.1.2 Pi464715

In 2020, another diploid alfalfa genome assembly was published (Li et al. 2020). This germplasm with plant introduction (PI) 464715, belongs to *Medicago sativa* subsp. *caerulea* and is thought to be the diploid progenitor of the autotetraploid alfalfa (Small and Jomphe 1989). This wild diploid provides an important contrast to CADL, a diploid derived from a tetraploid.

PI464715 was sequenced and assembled using ONT reads, currently the only alfalfa genome assembly based on ONT technology. ONT reads were corrected with Illumina sequences. With ~145X read coverage of the 800 Mb haploid genome size, the sequencing coverage is higher than that in CADL (just over 100X) and read lengths are longer. The mean read length of 19.7 kb and N50 read length of 27.9 kb are both more than twice those seen in the CADL data (Table 6.1). PI464715 sequence was corrected, assembled, and polished with NextDenovo with additional rounds of correction with Illumina and ONT reads. This resulted in an assembly of 1.35 Gb in length, with part of the genome likely assembled into two haplotypes, as in CADL. Indeed BUSCO duplication rates and alignments to *M. truncatula* for this intermediate assembly support this conclusion (Figs. 6.2 and

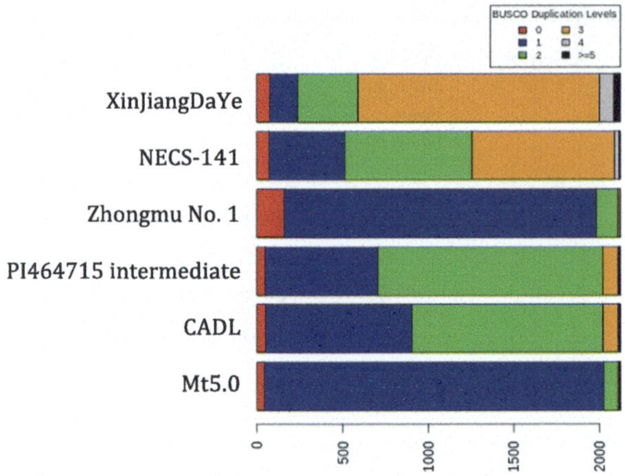

Fig. 6.2 A modified BUSCO analysis was run on each of the five alfalfa genomes and *M. truncatula* that enabled counting of duplication number for each captured BUSCO. To facilitate haplotype analyses, the version of the PI464715 assembly before haplotypes were collapsed and before scaffolding was used, which the authors kindly made available, rather than the final version of the assembly. The analysis was run with BUSCO 3.1.0 in genome mode using the eudicotyledons_odb10 database

6.5). Finally, duplicate haplotypes were removed using purge_haplotigs, which collapsed the assembly to 793.2 Mb in length, consistent with the haploid genome size (Table 6.1). The resulting assembly had a contig N50 of 3.86 Mb, approximately 3-fold that of CADL (Table 6.1). Long-range scaffolding of the assembly was accomplished with Hi-C data using LACHESIS. The final assembly consisted of 355 contigs scaffolded into 8 pseudo-chromosomes that cover 98.5% of the assembly and captured 97.7% of BUSCO gene orthologs (Table 6.2).

The final assembly covered the *M. truncatula* genome well (Fig. 6.6). Assembly contiguity is high enough that it is easy to see the chromosome 4/8 translocation, known to have occurred in the A17 accession of *M. truncatula,* as well as several small inversions (Figs. 6.6 and 6.7). The presence of only one haplotype in the final assembly is supported by the lack of double diagonals when compared to *M. truncatula* (Figs. 6.6 and 6.7) as well as by the low number of duplicate BUSCO genes (8.7%) (Table 6.2). It is interesting to note that the initial assembly length (1.35 Gb) before haplotypes were collapsed and scaffolding was run was very close in size to that of CADL (1.25 Gb), indicating that a similar proportion of the genome had diverged enough to assemble haplotypes independently despite a difference in sequencing technology and assembly strategy.

6.2.2 Tetraploid Assemblies

6.2.2.1 NECS-141

The first tetraploid alfalfa genome to be sequenced was that of NECS-141, a semi-dormant breeding line developed in Iowa (Khu et al. 2010). While originally meant to be a hybrid PacBio and Illumina assembly, because of the complications of the genome such as the repeat and ploidy structure, additional PacBio data was obtained as costs of the technology came down and accuracy, read length, and assembly algorithms improved.

The PacBio CLR sequencing reads were obtained in 2014 and 2015, around the time that fully PacBio plant genome assemblies were first being contemplated. It was sequenced around the same time as CADL with slightly higher coverage overall ($\sim$115X per haplotype vs $\sim$100X in CADL) but lower coverage per haplotype ($\sim$29X vs $\sim$50X for CADL). It is not surprising that it, too, is fragmented, with lack of continuity exacerbated by the increase in ploidy compared to CADL. However, scaffolding with BioNano

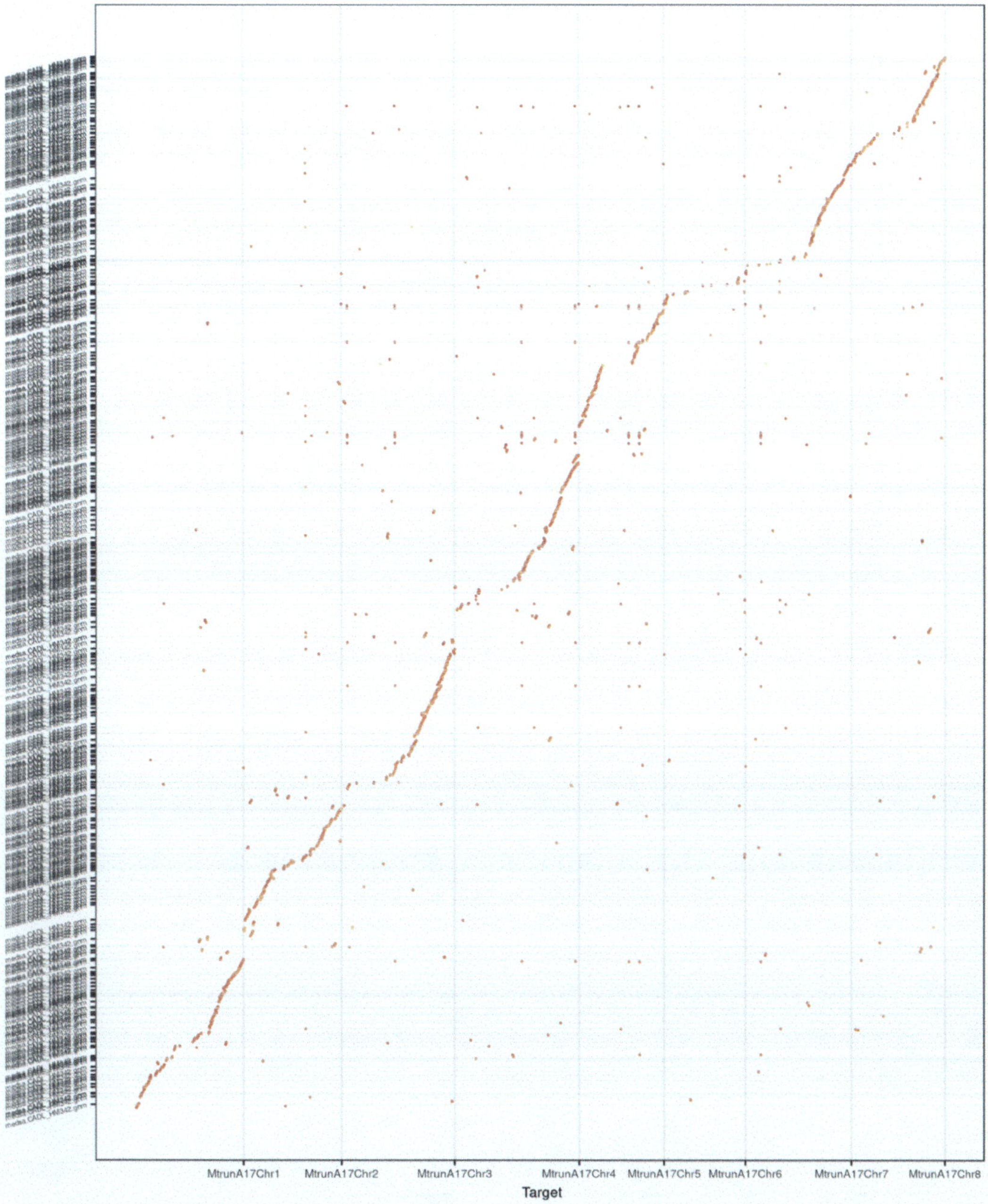

Fig. 6.3 Dotplot comparing the CADL assembly (y-axis) to the eight *Medicago truncatula* v. 5.0 chromosomes (x axis). Nucleotide level alignments were generated with minimap2 (Li 2018) using the asm20 preset, which allows up to 5% sequence divergence. Dotplots were generated using dotplotly with a minimum query length of 50 kb and a minimum alignment length of 10 kb (https://github.com/tpoorten/dotPlotly)

optical maps merged it into fewer pieces, though it still contained approximately twice the number of pieces as CADL, but with a scaffold N50 that exceeded that of CADL. More specifically, BioNano scaffolding was able to collapse the approximately 67 k contigs (N50 = 221 kb) into

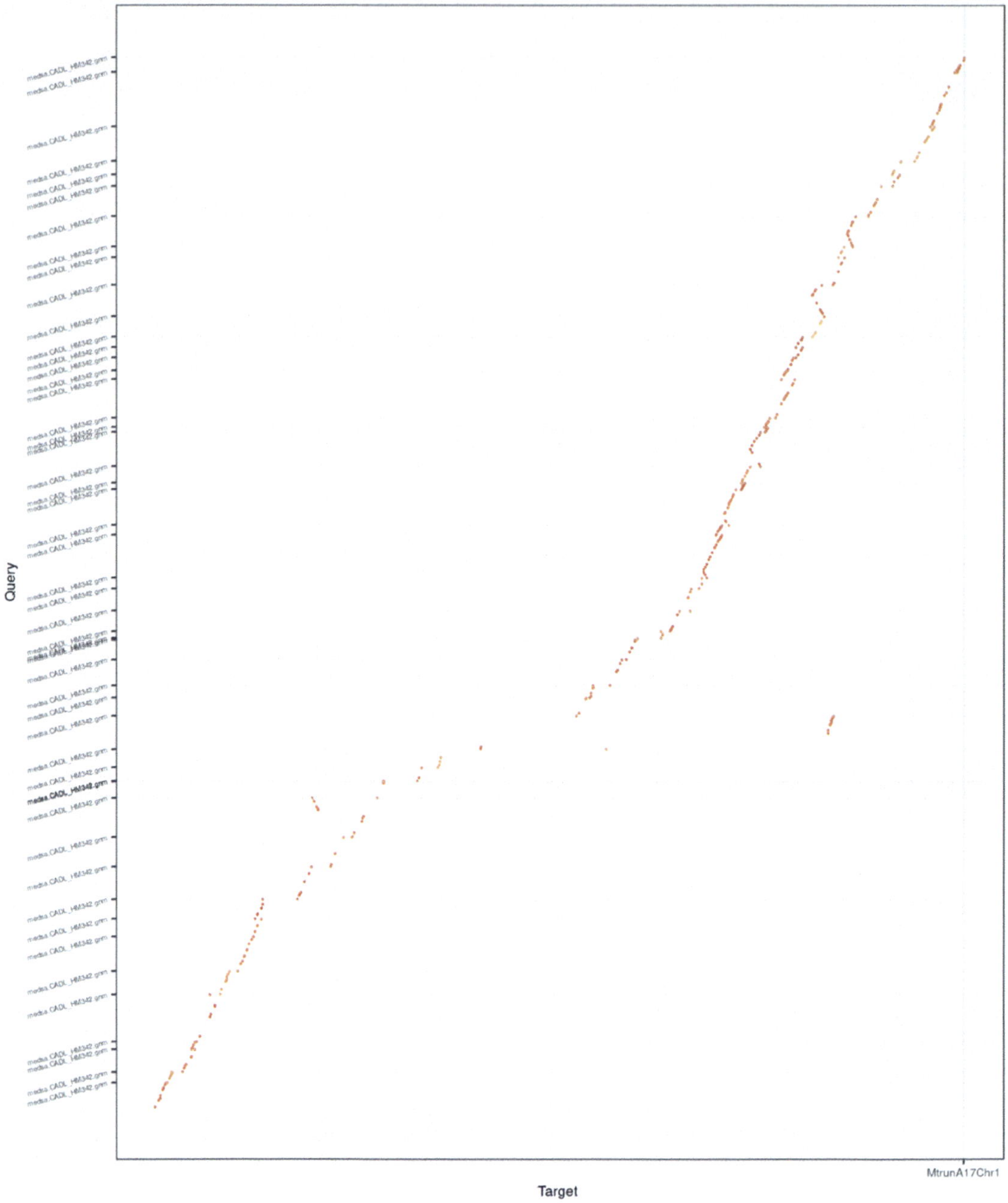

Fig. 6.4 Dotplot comparing CADL (y-axis) to *Medicago truncatula* v. 5.0 chromosome 1 (x-axis) showing capture of differing numbers of haplotypes across the genome, likely due to differing levels of haplotype divergence. Dotplots were generated as described in Fig. 6.3

just under 10 k scaffolds (N50 = 2.2 Mb), increasing the N50 by 10-fold (Table 6.1). Hi-C was also obtained but pieces were small enough that the Hi-C assembly was not able to separate out the haplotypes nor resolve local ordering so it was left out of the final assembly (Unpublished). The assembly covers the *M. truncatula* genome well, indicating that it is largely complete (Fig. 6.8).

Fig. 6.5 Dotplot comparing a preliminary version of the PI464715 assembly before haplotypes were collapsed and before scaffolding (y-axis) to *Medicago truncatula* v. 5.0 chromosome 1 (x-axis). This shows capture of differing numbers of haplotypes across the genome, likely due to differing levels of haplotype divergence. Dotplots were generated as described in Fig. 6.3

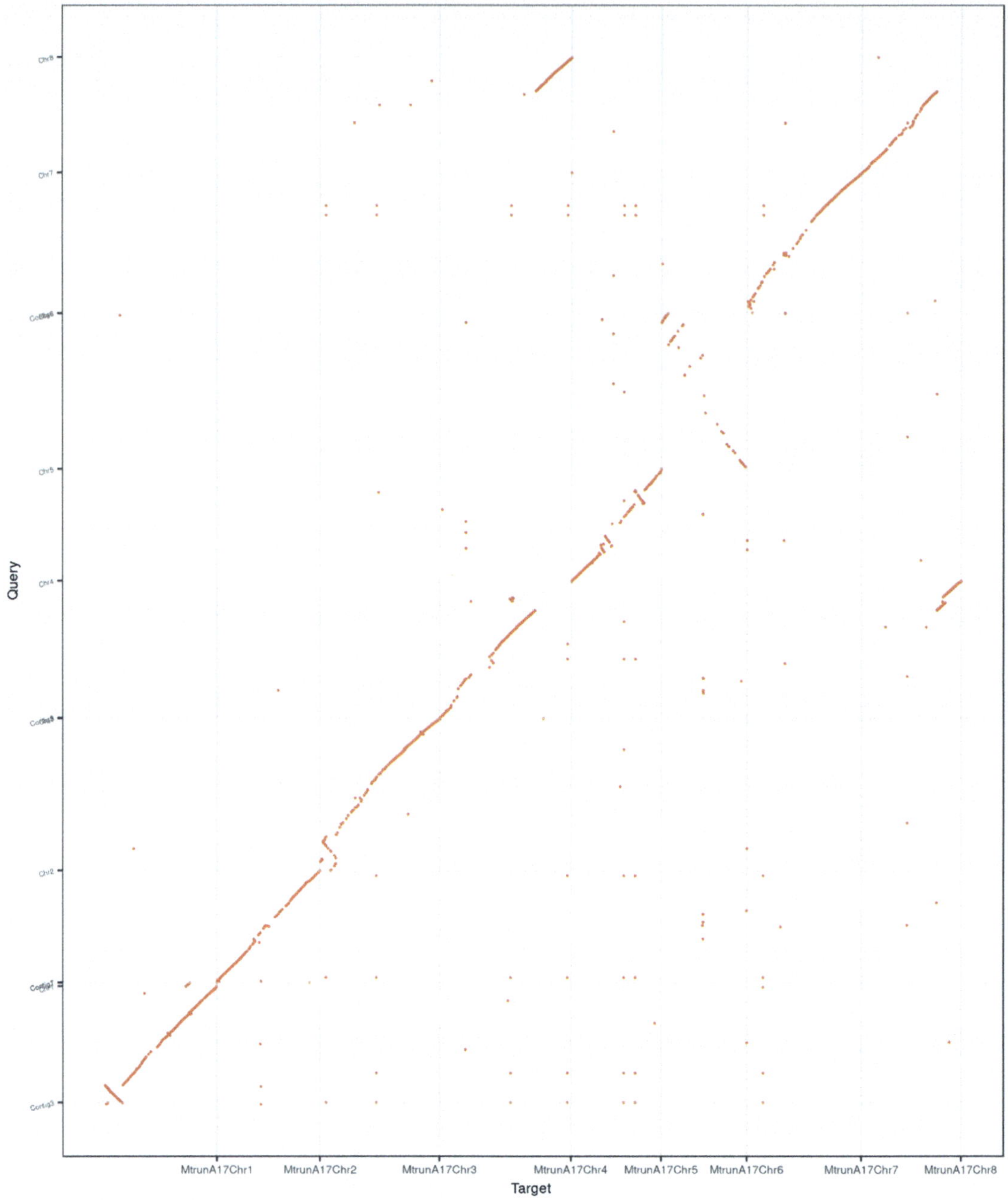

Fig. 6.6 Dotplot comparing the final PI464715 assembly (y-axis) to the eight *Medicago truncatula* v. 5.0 chromosomes (x-axis). Dotplots were generated as described in Fig. 6.3

BUSCO results on the percentage of core eukaryotic orthologs captured reveal some interesting insights (Table 6.2 and Fig. 6.2). As with the diploid genome assemblies, NECS-141 captured the vast majority of BUSCO genes (95.7%) despite its fragmentation. Approximately 76% of these genes were duplicated in the assembly, likely indicating that about three-quarters of the genome had multiple haplotypes assembled. Unlike the diploid genomes, these

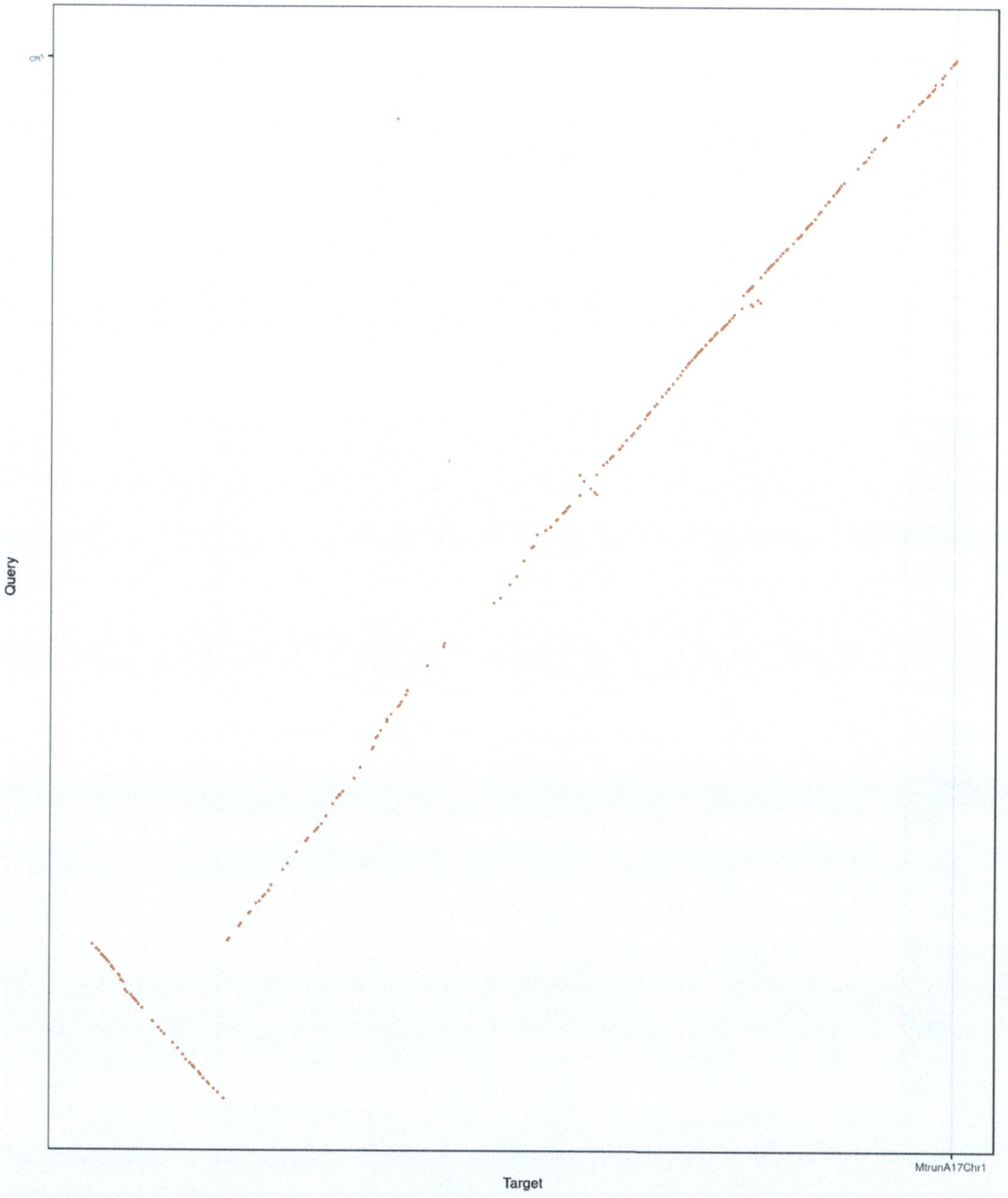

Fig. 6.7 Inversion shown in PI464715 final assembly chromosome 1 (y-axis) compared to *Medicago truncatula* chromosome 1 (x-axis). Dotplots were generated as described in Fig. 6.3

duplicated genes not only included those with two assembled haplotypes but also included slightly more genes with 3 assembled haplotypes, though 4 or more assembled haplotypes were rare (Fig. 6.2). In addition, duplicate or even triplicate haplotypes are visible when aligning to *M. truncatula* (Fig. 6.9). The assembly of multiple haplotypes in some but not genomic regions is further supported by the assembly length (2.70 Gb including 2.35 Gb of

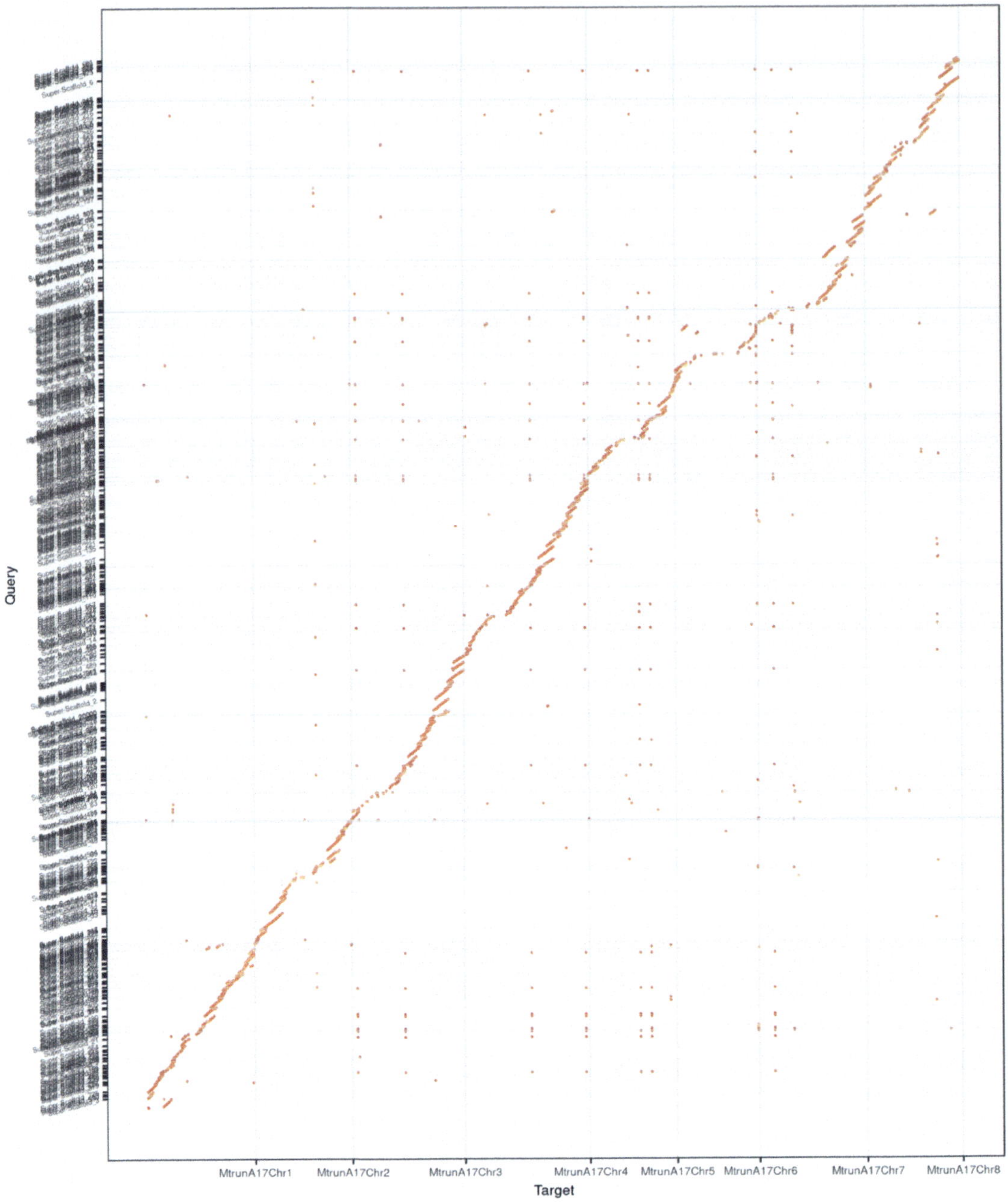

Fig. 6.8 Dotplot comparing the NECS-141 assembly (y-axis) to the eight *Medicago truncatula* v. 5.0 chromosomes (x-axis). Dotplots were generated as described in Fig. 6.3

non-gap sequence), representing approximately 84% and 73% of the expected 3.2 Gb genome covered and captured, respectively, by the assembly (Table 6.1). It appears, therefore, that haplotypes were collapsed to a single version in roughly one-quarter of the genome.

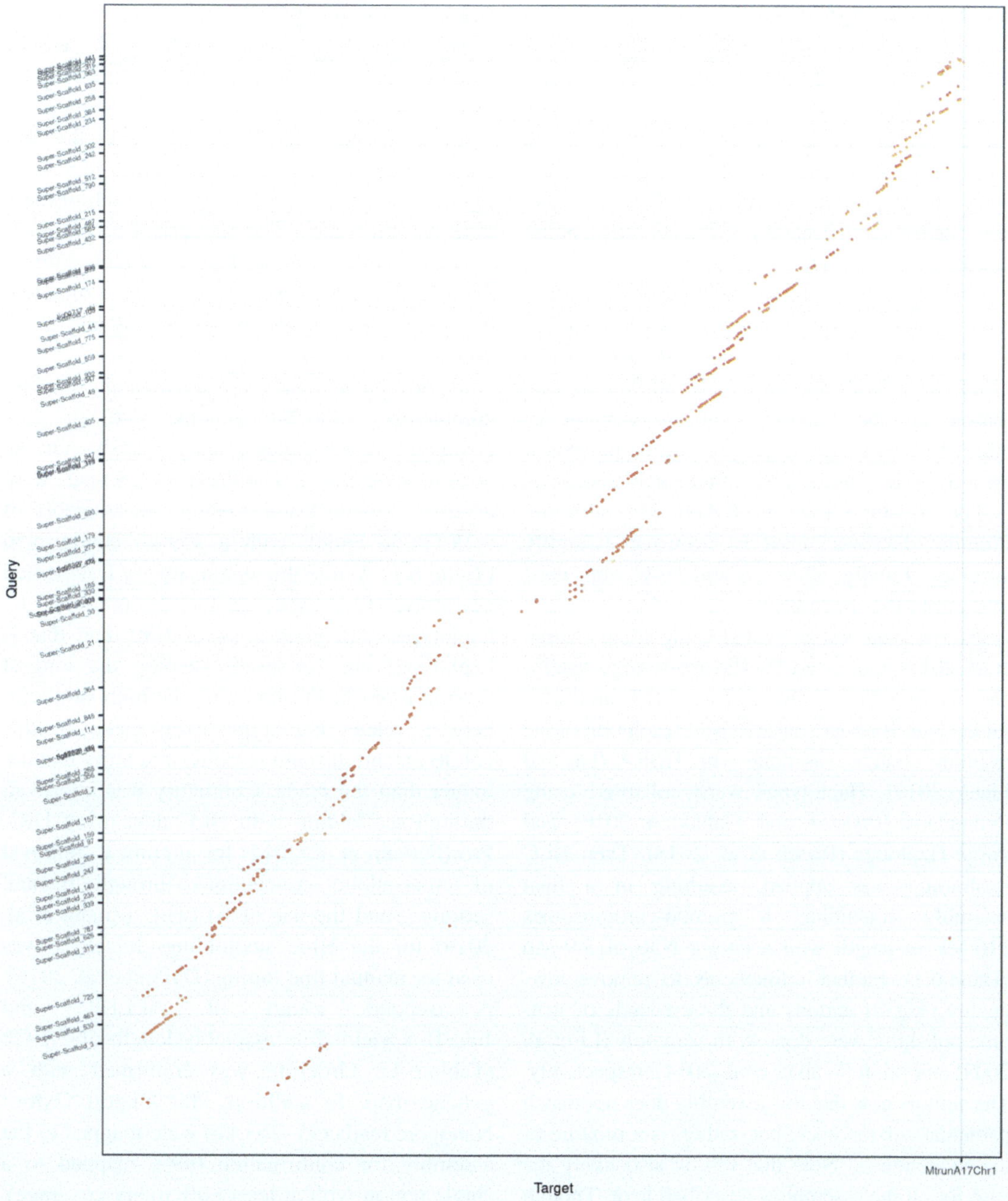

Fig. 6.9 Dotplot comparing NECS-141 (y-axis) to *Medicago truncatula* v. 5.0 chromosome 1 (x-axis) showing capture of differing numbers of haplotypes across the genome, likely due to differing levels of haplotype divergence. Dotplots were generated as described in Fig. 6.3

6.2.2.2 Zhongmu No. 1

A pseduo-chromosome level assembly of the tetraploid cultivar Zhongmu No. 1 was recently published (Shen et al. 2020). Zhongmu No. 1 is a subspecies *sativa* cultivar from Northern China that is salt tolerant (Shi et al. 2017). The continuity at the pseudo-chromosome level is made possible by a combination of PacBio sequencing and BioNano and Hi-C scaffolding.

This assembly is based on PacBio long reads with subread lengths comparable to that of CADL (N50 = 12.1 kb vs 13.1 kb for CADL) (Table 6.1). With approximately 300X coverage (based on the haploid genomic content of 800 Mb) or 75X coverage per haplotype (based on a 3.2 Gb genome size), there was about 3X higher coverage than in CADL. In addition, Illumina data was used to improve assembly accuracy. Finally, BioNano and Hi-C were used to scaffold the assembly.

PacBio data was corrected using Canu (Koren et al. 2017). Corrected PacBio reads were assembled using MECAT (Xiao et al. 2017) and scaffolded with BioNano data. Repeat resolution of the resulting contigs was done with HERA (Du and Liang 2019). Haplotypes were collapsed using Redundans (Pryszcz and Gabaldón 2016) and Purge Haplotigs (Roach et al. 2018). Then Hi-C scaffolding was applied, resulting in a final assembly containing 8 pseudo-chromosomes 816 Mb in length with a contig N50 of 3.9 Mb (Table 6.1). Further refinements to remove low-quality (<Q30) regions and three rounds of genome polishing were done with samtools (Li et al. 2009) and pilon (Walker et al. 2014), respectively. The authors note that this assembly does not match particular subgenomes, but, rather, is a mixture of the subgenomes. Note that this is also likely the case for all the assemblies described here. Though slightly lower than the other assemblies, the Zhongmu No. 1 assembly captured the majority of genes as estimated by BUSCO (93.3%), with most genes captured as single copy (Table 6.2 and Fig. 6.2), reflecting the deredundification step. Dotplots show good coverage of *M. truncatula* and structural variation including the chromosome 4/8 translocation and inversions compared to *M. truncatula* (Figs. 6.10 and 6.11).

6.2.2.3 XinJiangDaYe

Finally, an "allele-aware" tetraploid genome assembly has been published (Chen et al. 2020), using PacBio Circular Consensus Sequencing (CCS) technology and Hi-C scaffolding, that was able to assemble all 32 chromosomes representing the four different haplotypes of each of the 8 base chromosomes. The sequenced accession is XinJiangDaYe, a large-leaved alfalfa cultivar from Xinjiang Provence of northwest China that has good regeneration properties (Zhang et al. 2010; Shi et al. 2017).

Approximately 88X coverage of the haploid complement (800 Mb genome size) or 22X coverage of each haplotype (3.2 Gb genome size) was generated. The PacBio CCS reads were assembled using Canu, yielding an assembly of 3.15 Gb in length with a 459 kb contig N50 (Table 6.1). While this is a relatively fragmented assembly, it is expected that keeping all the haplotypes will yield a lower N50 than that if haplotypes are collapsed, keeping the longest version of each. Furthermore, the high similarity between related haplotypes likely makes it difficult to extend through regions of identity that are longer than the reads. Continuity was improved through scaffolding with Hi-C data using HiC-Pro (Servant et al. 2015) for alignment, removal of cross-allelic connections through manual scripting, and the use of ALLHiC (Zhang et al. 2019) for the Hi-C scaffolding. JuiceBox was used for manual fine-tuning (Durand et al. 2016). A second round of ALLHiC and JuiceBox yielded an assembly length of 2.7 Gb (Table 6.1). Ordering was confirmed with a genetic map. In addition, 200 longest Oxford Nanopore reads (95–263 kb) were mapped to the assembly for confirmation (89% mapped to a single region with at least 80% query coverage).

The assembly length (2.738 Gb) is slightly short of the expected ~3.2 Gb full genome size and only 40 Mb larger than that of NECS-141 (Table 6.1). XinJiangDaYe's assembly length is approximately 86% of the expected genome size and matches up well with estimates of captured core eukaryotic conserved orthologs (BUSCO) in each haplotype (88.50, 88.30, 87.50, and 87.20%), with 97.2% captured in at least one haplotype

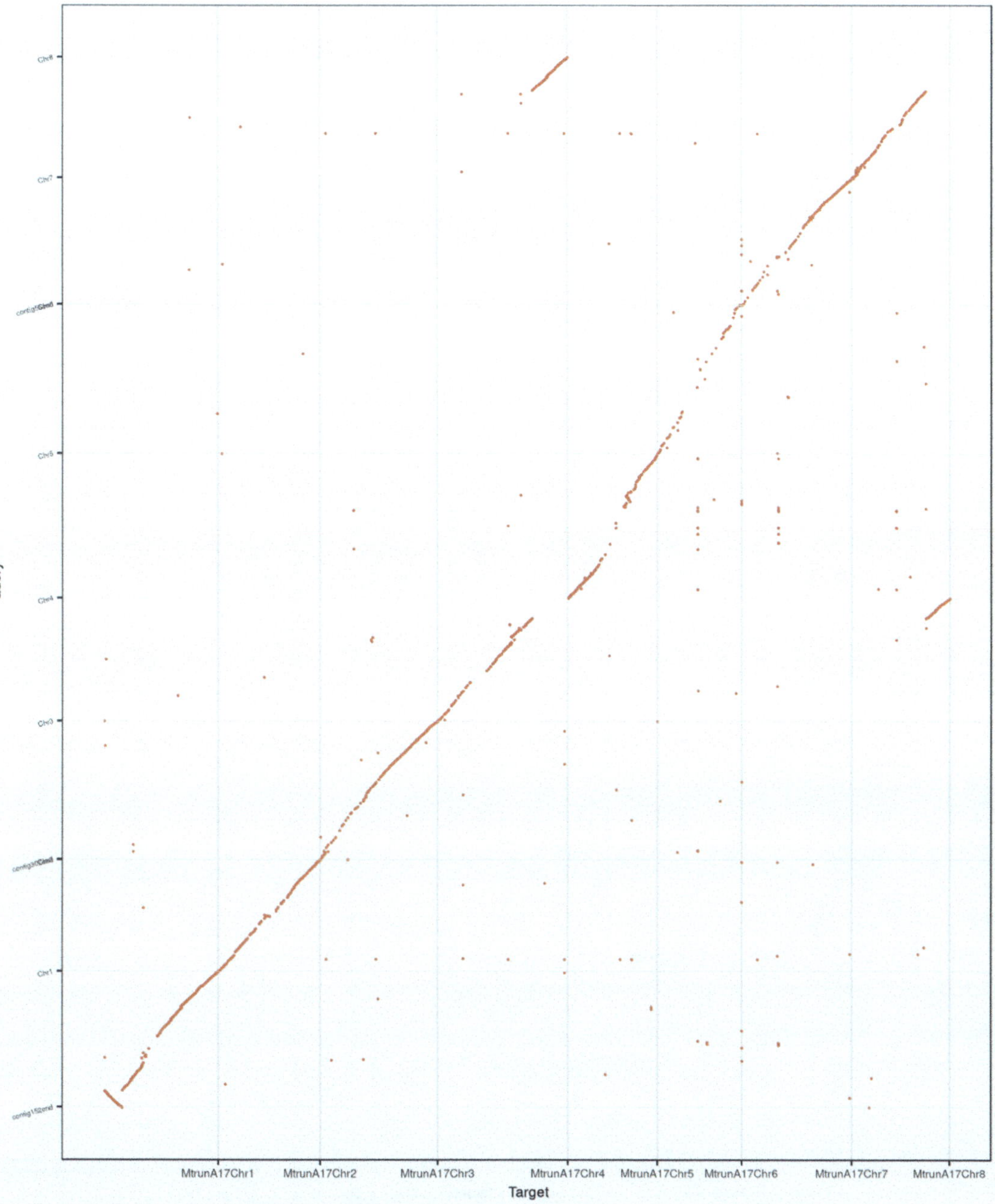

Fig. 6.10 Dotplot comparing the Zhongmu No. 1 assembly (y-axis) to the eight *Medicago truncatula* v. 5.0 chromosomes (x-axis). Dotplots were generated as described in Fig. 6.3

(Table 6.2). The majority of duplicated BUSCO genes were captured three times, though some were captured once or twice, and a small number were captured 4 or more times (Fig. 6.2). While the differential gene capture seen between haplotypes may reflect actual differential gene content as seen in CADL, the authors found that the number of genes is similar between haplotypes, retained synteny is high, and evolutionary pressures on genes are similar between haplotypes.

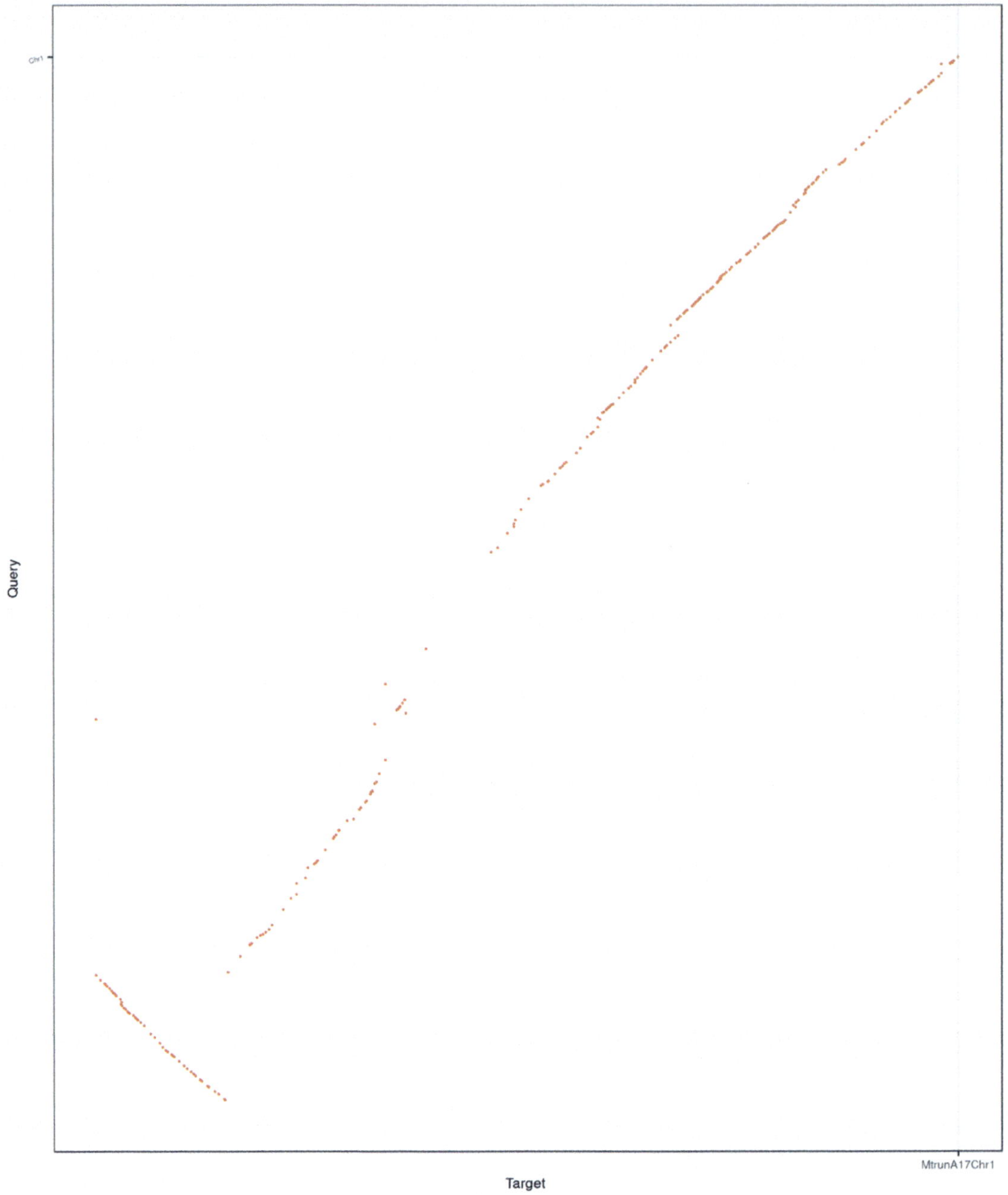

Fig. 6.11 Inversion shown in Zhongmu No. 1 chromosome 1 (y-axis) compared to *Medicago truncatula* chromosome 1 (x-axis). Dotplots were generated as described in Fig. 6.3

Indeed, the percentage of duplicated BUSCO genes in the genome is just over 90%, indicating that most genes are in multiple haplotypes. Furthermore, the percentage of the genome that remains uncaptured is sufficient to explain missing genes in each haplotype. Nevertheless, it is clear that some differential gene loss occurs between the XinJiangDaYe haplotypes (Fig. 6.1b).

At least some of the uncaptured portion of the genome is likely reflected in haplotypes that were collapsed due to strong similarity across chromosome distances that exceed PacBio subread lengths. There was approximately double coverage on 3.2% of the genome, reflecting possible collapse of haplotypes, though collapse of repeats or tandem duplications within a haplotype could also account for some of the double coverage regions. Nevertheless, there is good evidence that, overall, haplotypes were highly similar. Sequence divergence wasn't always sufficient to distinguish haplotypes and the consensus genetic map that was used couldn't distinguish the subgenomes, leading to some possible phasing errors. Nevertheless, it is interesting that, in addition to structural variation identified between XinJiangDaYe and *M. truncatula* (Fig. 6.12), some structural differences between subgenomes were uncovered. For instance, Hi–C data supports the two inversions that occur in only one of the chromosome 1 haplotypes (Fig. 6.13).

6.3 Annotation

6.3.1 Genes

Gene counts vary between the assemblies but there are some interesting patterns (Table 6.2). PI464715, Zhongmu No. 1, and XinJiangDaYe all have between 47 and 50 k genes per 800 Mb haploid genome complement. CADL, on the other hand, has considerably higher at ~71 k genes per haploid genome complement. About ¼ of these are redundant at >98% identity, indicating they might be from alternate haplotypes. Surprisingly, CADL has more protein-coding gene annotations than NECS-141 despite being less than half as long. This not only reflects the high gene count in CADL but also a low gene count in NECS-141 (~30 k genes per haploid genome complement). Differences in annotation pipelines likely account for the differing gene counts. For example, the CADL annotation was the only one that used the SPADA pipeline (Zhou et al. 2013), adding about 8,000 small peptides to the annotation.

BUSCO was originally generated for assessing completeness of genome assembly and annotation (Simão et al. 2015). While it only assays "near-universal single-copy" genes, BUSCO analyses are a reasonable surrogate for overall gene capture. All five genome assemblies had complete gene capture of more than 93% of genes, ranging from 93.3% in Zhongmu No. 1 to 97.7% in PI464715 (Table 6.2). This indicates that, at least in the gene space, all of these assemblies are nearly complete. The capture of duplicate copies of the BUSCO genes mirrors well estimates of duplication based on extra genome length beyond the 800 Mb base genome size and through alignments to *M. truncatula* (Table 6.2, Figs. 6.3, 6.4, 6.5, 6.6, 6.7, 6.8, 6.9, 6.10, 6.11, 6.12 and 6.13). The diploid PI464715 and the tetraploid Zhongmu No. 1 have both been compressed into a haploid genome complement. CADL and NECS-141, along with the intermediate PI464715 assembly version, have retained diverged duplicate haplotypes while collapsing highly similar or identical ones, and XinJiangDaYe has separately assembled all four subgenomes.

A customized BUSCO analysis that identifies copy number of captured genes reflects differing levels of haplotype capture in different assemblies (Fig. 6.2). The *Medicago truncatula* (A17) and Zhongmu No. 1 assemblies capture mostly single copy BUSCO genes reflecting the haploid assembly strategy. Based on the small percentage of duplicated genes in the final PI464715 genome (Table 6.2), the final PI464715 genome would have looked similar had it been included in this figure. The diploid assemblies, CADL and the intermediate assembly of PI464715, which both captured some haplotype variation, show similar profiles with the largest fraction of BUSCO genes captured with two copies but with a significant fraction collapsed into a single haplotype. The two uncollapsed tetraploid assemblies, NECS-141 and XinJiangDaYe, show BUSCO gene counts that vary. The biggest fraction has a count of three, with that fraction being larger in XinJiangDaYe where a concerted effort to capture all four haplotypes was

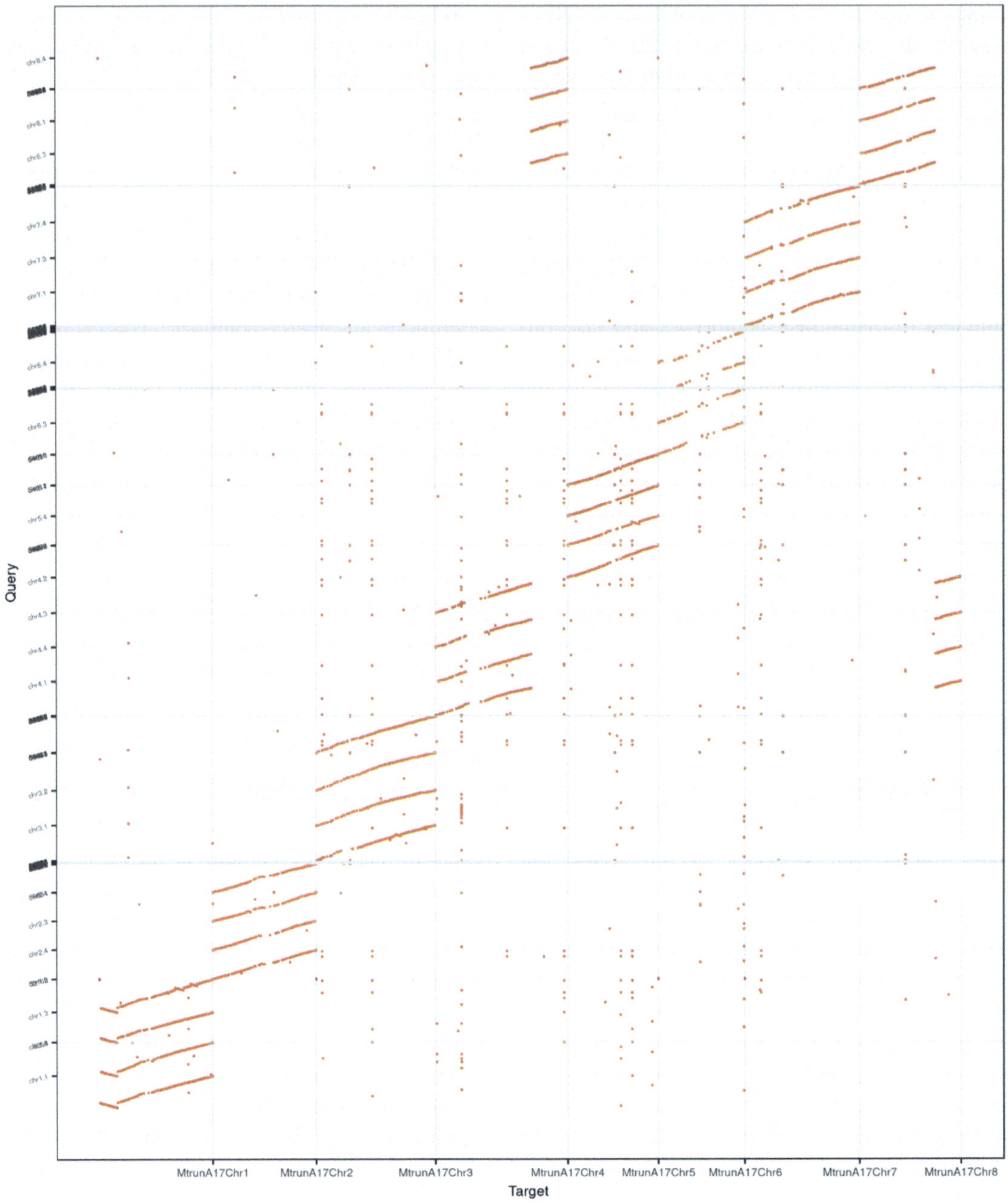

Fig. 6.12 Dotplot comparing the XinJiangDaYe assembly (y-axis) to the eight *Medicago truncatula* v. 5.0 chromosomes (x-axis). Dotplots were generated as described in Fig. 6.3

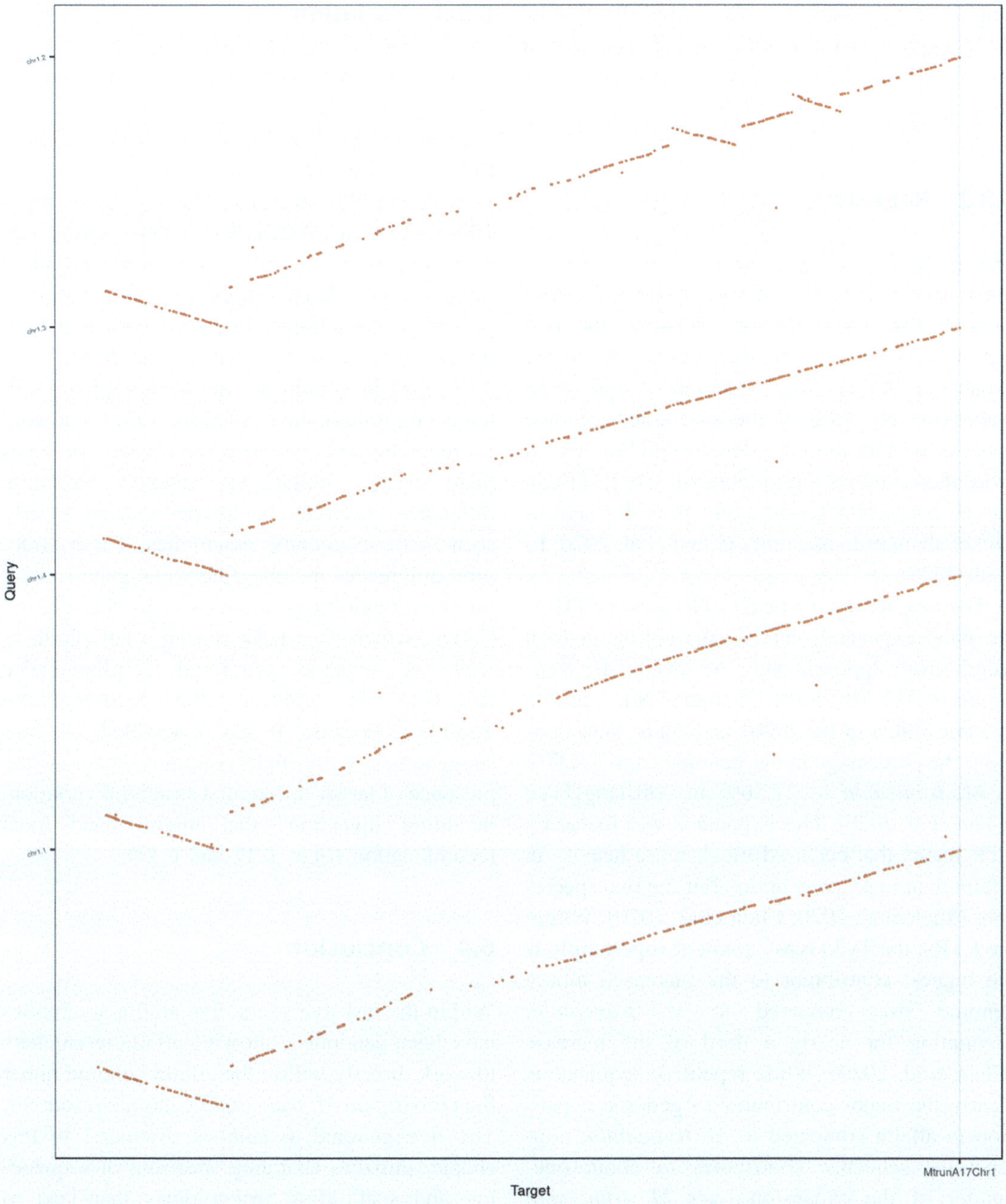

Fig. 6.13 Inversion shown in four XinJiangDaYe chromosome 1 subgenomes (y-axis) compared to *Medicago truncatula* chromosome 1 (x-axis). Dotplots were generated as described in Fig. 6.3

employed. Surprisingly, only a small fraction of BUSCO genes with a count of 4 were captured in the assembly, though more were captured in XinJiangDaYe than in NECS-141.

6.3.2 Repeats

The alfalfa haploid genome size (~800 Mb) is much larger than that of *Medicago truncatula* (~450 Mb). The difference between the two genomes appears to be due mainly to repeat expansion rather than genome duplication. Approximately 55% of the assembled genome consists of transposable elements (TEs), which more than doubles the number of Mb of TEs in the *M. truncatula* genome and provides significant challenges to assembly (Chen et al. 2020; Li et al. 2020).

The long terminal repeat (LTR) class of TEs is the most expanded, nearly quintupling in total length from approximately 65 Mb in *M. truncatula* to 315 Mb in the Zhongmu No. 1 alfalfa genome (Shen et al. 2020) and more than doubling the percentage in the genome from 13.37% in *M. truncatula* to 27.36% in XinJiangDaYe (Chen et al. 2020). This expansion was fueled by LTR bursts that occurred much more heavily in alfalfa than in *M. truncatula* after the two species split (Shen et al. 2020; Chen et al. 2020). Within the LTRs, the Ty3/Gypsy element superfamily is the biggest contributor to the increased alfalfa genome size compared to *M. truncatula*, accounting for nearly a third of the increase (Chen et al. 2020). While repetitive sequence is clearly the major contributor to genome expansion in alfalfa compared to *M. truncatula*, non-repetitive sequence contributes to about one-quarter of the expansion over *M. truncatula* (Chen et al. 2020). Further evidence that large-scale duplications do not appear to have contributed significantly to genome expansion in alfalfa is confirmed by comparisons of the alfalfa genomes to *M. truncatula* (Figs. 6.3, 6.4, 6.5, 6.6, 6.7, 6.8, 6.9, 6.10, 6.11, 6.12 and 6.13).

6.3.3 Variation

Alfalfa is an outcrossing species, and so heterozygosity is expected to be high. Genome sequencing data confirms this. In the diploid PI464715, the average heterozygosity rate estimate is ~1.9%, or nearly 2 heterozygous nt per 100 nt (Li et al. 2020). The heterozygosity rate estimate is nearly double (3.7%) in the tetraploid XinjiangDaye, reflecting the increased variation present in the tetraploid genome with four haplotypes rather than two (Chen et al. 2020).

Tetraploid alfalfa is an autotetraploid with tetrasomic inheritance (Stanford 1951), allowing for recombination between haplotypes that keeps them highly similar. Nevertheless, structural differences between haplotypes can be clearly seen in these genome assemblies. These structural differences include differential gene content between haplotypes as shown in the diploid CADL, which was derived from a tetraploid, as well as in the tetraploid XinJiangDaYe (Fig. 6.1). In addition, the XinJiangDaYe assembly, because it has assembled all four subgenomes with high continuity, shows the presence of larger differential structural variation, including inversions that might affect local recombination (Figs. 6.12 and 6.13).

6.4 Conclusion

Within the last five years, five alfalfa assemblies have been generated, allowing alfalfa researchers to work directly within the alfalfa genome rather than relying on *M. truncatula* genomic resources. The five genome assemblies discussed in this chapter utilize a changing spectrum of sequencing and scaffolding technologies that lead to improved assembly continuity and an increased ability to distinguish between repeats and subgenome haplotypes. This ability to distinguish nearly identical sequences is critical in alfalfa genome assembly because transposable repeats alone comprise more than half of the alfalfa genome. Furthermore, haplotypes present in the different subgenomes are highly similar, as evidenced by insufficient sequence divergence to distinguish some haplotypes in the CADL, NECS-141, and XinJiangDaYe assemblies, even with genetic map support, as well as in preliminary versions of the PI464715 assembly. Nevertheless, these assemblies uncover local haplotype differences in gene content as well as larger structural rearrangements that distinguish some of the subgenomes. The increase in repeat content compared to *M. truncatula*, as well as heterozygosity and ploidy challenges, are now more easily navigable with improved, highly accurate PacBio HiFi long reads or even longer, moderately accurate ONT reads, as well as rapid, inexpensive whole-genome scaffolding technologies such as BioNano and chromatin conformation technologies. Given these technological breakthroughs, we fully expect to see additional alfalfa genome assemblies released in upcoming years as alfalfa researchers sequence additional alfalfa germplasm that has important scientific or breeding applications.

6.5 Assembly Availability

CADL is available under a Ft. Lauderdale usage agreement at https://legumeinfo.org/data/index/public/Medicago_sativa/CADL_HM342.gnm1.rVNY/. NECS-141 is available under MTA from the Noble Research Institute in Ardmore, Oklahoma. All other assemblies are available as described in their manuscripts (Shen et al. 2020; Chen et al. 2020; Li et al. 2020).

Acknowledgements We would like to thank the following people for helping with the CADL and/or NECS-141 assembly or annotation: Connor T. Cameron, Xinbin Dai, Nicholas P. Devitt, Diego A. Fajardo, Joseph Guhlin, Jason R. Miller, Maria J. Monteros, Thiru Ramaraj, Kevin A. T. Silverstein, Haibao Tang, Christopher D. Town, Nevin D. Young, Patrick X. Zhao, Peng Zhou, Zhaohong Zhuang. We would like to thank the *Medicago* HapMap project (NSF Project IOS-1237993) and Noble Research Institute for funding for the CADL and NECS-141 assemblies. Acknowledgements for the other genome assemblies can be found in their respective manuscripts.

References

Barker DG, Bianchi S, Blondon F, Dattée Y, Duc G, Essad S, Flament P, Gallusci P, Génier G, Guy P, Muel X, Tourneur J, Dénarié J, Huguet T (1990) *Medicago truncatula*, a model plant for studying the molecular genetics of the Rhizobium-legume symbiosis. Plant Mol Biol Rep 8:40–49

Belser C, Istace B, Denis E, Dubarry M, Baurens F-C, Falentin C, Genete M, Berrabah W, Chèvre A-M, Delourme R, Deniot G, Denoeud F, Duffé P, Engelen S, Lemainque A, Manzanares-Dauleux M, Martin G, Morice J, Noel B, Vekemans X, D'Hont A, Rousseau-Gueutin M, Barbe V, Cruaud C, Wincker P, Aury J-M (2018) Chromosome-scale assemblies of plant genomes using nanopore long reads and optical maps. Nat Plants 4:879–887

Berlin K, Koren S, Chin C-S, Drake JP, Landolin JM, Phillippy AM (2015) Assembling large genomes with single-molecule sequencing and locality-sensitive hashing. Nat Biotechnol 33:623–630

Bingham ET (1969) Haploids from cultivated alfalfa, *Medicago sativa L.* Nature 221:865–866

Bingham ET, McCoy TJ (1979) Cultivated Alfalfa at the diploid level: origin, reproductive stability, and yield of seed and forage 1. Crop Sci 19:97–100

Blondon F, Marie D, Brown S, Kondorosi A (1994) Genome size and base composition in *Medicago sativa* and *M. truncatula* species. Genome 37:264–270

Burton JN, Adey A, Patwardhan RP, Qiu R, Kitzman JO, Shendure J (2013) Chromosome-scale scaffolding of de novo genome assemblies based on chromatin interactions. Nat Biotechnol 31:1119–1125

Chen H, Zeng Y, Yang Y, Huang L, Tang B, Zhang H, Hao F, Liu W, Li Y, Liu Y, Zhang X, Zhang R, Zhang Y, Li Y, Wang K, He H, Wang Z, Fan G, Yang H, Bao A, Shang Z, Chen J, Wang W, Qiu Q (2020) Allele-aware chromosome-level genome assembly and efficient transgene-free genome editing for the autotetraploid cultivated alfalfa. Nat Commun 11:2494

Chin C-S, Peluso P, Sedlazeck FJ, Nattestad M, Concepcion GT, Clum A, Dunn C, O'Malley R, Figueroa-Balderas R, Morales-Cruz A, Cramer GR, Delledonne M, Luo C, Ecker JR, Cantu D, Rank DR, Schatz MC (2016) Phased diploid genome assembly with single-molecule real-time sequencing. Nat Methods 13:1050–1054

Cook DR (1999) *Medicago truncatula*—a model in the making! Curr Opin Plant Biol 2:301–304

Deamer D, Akeson M, Branton D (2016) Three decades of nanopore sequencing. Nat Biotechnol 34:518–524

Deschamps S, Mudge J, Cameron C, Ramaraj T, Anand A, Fengler K, Hayes K, Llaca V, Jones TJ, May G (2016) Characterization, correction and de novo assembly of an Oxford Nanopore genomic dataset from Agrobacterium tumefaciens. Sci Rep 6:28625

Deschamps S, Zhang Y, Llaca V, Ye L, Sanyal A, King M, May G, Lin H (2018) A chromosome-scale assembly of the sorghum genome using nanopore sequencing and optical mapping. Nat Commun 9:4844

Doyle JJ, Flagel LE, Paterson AH, Rapp RA, Soltis DE, Soltis PS, Wendel JF (2008) Evolutionary genetics of genome merger and doubling in plants. Annu Rev Genet 42:443–461

Du H, Liang C (2019) Assembly of chromosome-scale contigs by efficiently resolving repetitive sequences with long reads. Nat Commun 10:5360

Durand NC, Robinson JT, Shamim MS, Machol I, Mesirov JP, Lander ES, Aiden EL (2016) Juicebox provides a visualization system for Hi-C contact maps with unlimited zoom. Cell Syst 3:99–101

Eid J, Fehr A, Gray J, Luong K, Lyle J, Otto G, Peluso P, Rank D, Baybayan P, Bettman B, Bibillo A, Bjornson K, Chaudhuri B, Christians F, Cicero R, Clark S, Dalal R, Dewinter A, Dixon J, Foquet M, Gaertner A, Hardenbol P, Heiner C, Hester K, Holden D, Kearns G, Kong X, Kuse R, Lacroix Y, Lin S, Lundquist P, Ma C, Marks P, Maxham M, Murphy D, Park I, Pham T, Phillips M, Roy J, Sebra R, Shen G, Sorenson J, Tomaney A, Travers K, Trulson M, Vieceli J, Wegener J, Wu D, Yang A, Zaccarin D, Zhao P, Zhong F, Korlach J, Turner S (2009) Real-time DNA sequencing from single polymerase molecules. Science 323:133–138

Hufton AL, Panopoulou G (2009) Polyploidy and genome restructuring: a variety of outcomes. Curr Opin Genet Dev 19:600–606

Jiao W-B, Schneeberger K (2017) The impact of third generation genomic technologies on plant genome assembly. Curr Opin Plant Biol 36:64–70

Jung H, Winefield C, Bombarely A, Prentis P, Waterhouse P (2019) Tools and strategies for long-read sequencing and De Novo assembly of plant genomes. Trends Plant Sci 24:700–724

Khu D-M, Reyno R, Charles Brummer E, Bouton JH, Han Y, Monteros MJ (2010) QTL mapping of aluminum tolerance in tetraploid alfalfa. In: Sustainable use of genetic diversity in forage and turf breeding, pp 437–442

Koren S, Walenz BP, Berlin K, Miller JR, Bergman NH, Phillippy AM (2017) Canu: scalable and accurate long-read assembly via adaptive k-mer weighting and repeat separation. Genome Res 27:722–736

Kyriakidou M, Tai HH, Anglin NL, Ellis D, Strömvik MV (2018) Current strategies of polyploid plant genome sequence assembly. Front Plant Sci 9:1660

Lang D, Zhang S, Ren P, Liang F, Sun Z, Meng G, Tan Y, Li X, Lai Q, Han L, Wang D, Hu F, Wang W, Liu S (2020) Comparison of the two up-to-date sequencing technologies for genome assembly: HiFi reads of Pacific Biosciences Sequel II system and ultralong reads of Oxford Nanopore. Gigascience 9:giaa123. https://doi.org/10.1093/gigascience/giaa123

Li A, Liu A, Du X, Chen J-Y, Yin M, Hu H-Y, Shrestha N, Wu S-D, Wang H-Q, Dou Q-W, Liu Z-P, Liu J-Q, Yang Y-Z, Ren G-P (2020) A chromosome-scale genome assembly of a diploid alfalfa, the progenitor of autotetraploid alfalfa. Hortic Res 7:194

Li H (2018) Minimap2: pairwise alignment for nucleotide sequences. Bioinformatics 34:3094–3100

Li H, Handsaker B, Wysoker A, Fennell T, Ruan J, Homer N, Marth G, Abecasis G, Durbin R, 1000 Genome Project Data Processing Subgroup (2009) The sequence alignment/map format and SAMtools. Bioinformatics 25:2078–2079

Michael TP, Jupe F, Bemm F, Motley ST, Sandoval JP, Lanz C, Loudet O, Weigel D, Ecker JR (2018) High contiguity Arabidopsis thaliana genome assembly with a single nanopore flow cell. Nat Commun 9:541

Michael TP, VanBuren R (2020) Building near-complete plant genomes. Curr Opin Plant Biol 54:26–33

Mishra DC, Lal SB, Sharma A, Kumar S, Budhlakoti N, Rai A (2017) Strategies and tools for sequencing and assembly of plant genomes. Compend Plant Genomes 81–93

Moll KM, Zhou P, Ramaraj T, Fajardo D, Devitt NP, Sadowsky MJ, Stupar RM, Tiffin P, Miller JR, Yound ND, Silverstein KAT, Mudge J (2017) Strategies for optimizing BioNano and Dovetail explored through a second reference quality assembly for the legume model *Medicago truncatula*. BMC Genomics 18:578

Mostovoy Y, Levy-Sakin M, Lam J, Lam ET, Hastie AR, Marks P, Lee J, Chu C, Lin C, Džakula Ž, Cao H, Schlebusch SA, Giorda K, Schnall-Levin M, Wall JD, Kwok P-Y (2016) A hybrid approach for de novo human genome sequence assembly and phasing. Nat Methods 13:587–590

Myers G (2014) Daligner: fast and sensitive detection of all pairwise local alignments

O'Bleness M, Searles VB, Dickens CM, Astling D, Albracht D, Mak ACY, Lai YYY, Lin C, Chu C, Graves T, Kwok P-Y, Wilson RK, Sikela JM (2014) Finished sequence and assembly of the DUF1220-rich 1q21 region using a haploid human genome. BMC Genom 15:387

Pryszcz LP, Gabaldón T (2016) Redundans: an assembly pipeline for highly heterozygous genomes. Nucleic Acids Res 44:

Putnam NH, O'Connell BL, Stites JC, Rice BJ, Blanchette M, Calef R, Troll CJ, Fields A, Hartley PD, Sugnet CW, Haussler D, Rokhsar DS, Green RE (2016) Chromosome-scale shotgun assembly using an in vitro method for long-range linkage. Genome Res 26:342–350

Roach MJ, Schmidt SA, Borneman AR (2018) Purge Haplotigs: allelic contig reassignment for third-gen diploid genome assemblies. BMC Bioinform 19

Schatz MC, Witkowski J, McCombie WR (2012) Current challenges in de novo plant genome sequencing and assembly. Genome Biol 13:243

Servant N, Varoquaux N, Lajoie BR, Viara E, Chen C-J, Vert J-P, Heard E, Dekker J, Barillot E (2015) HiC-Pro: an optimized and flexible pipeline for Hi-C data processing. Genome Biol 16

Shen C, Du H, Chen Z, Lu H, Zhu F, Chen H, Meng X, Liu Q, Liu P, Zheng L, Li X, Dong J, Liang C, Wang T (2020) The chromosome-level genome

sequence of the autotetraploid alfalfa and resequencing of core germplasms provide genomic resources for alfalfa research. Mol Plant 13:1250–1261
Shi S, Nan L, Smith KF (2017) The current status, problems, and prospects of alfalfa (*Medicago sativa L.*) breeding in China. Agronomy 7:1
Simão FA, Waterhouse RM, Ioannidis P, Kriventseva EV, Zdobnov EM (2015) BUSCO: assessing genome assembly and annotation completeness with single-copy orthologs. Bioinformatics 31:3210–3212
Small E, Jomphe M (1989) A synopsis of the genus Medicago (Leguminosae). Can J Bot 67:3260–3294
Stanford EH (1951) Tetrasomic inheritance in alfalfa 1. Agron J 43:222–225
Staňková H, Hastie AR, Chan S, Vrána J, Tulpová Z, Kubaláková M, Visendi P, Hayashi S, Luo M, Batley J, Edwards D, Doležel J, Šimková H (2016) BioNano genome mapping of individual chromosomes supports physical mapping and sequence assembly in complex plant genomes. Plant Biotechnol J 14:1523–1531
Steinberg KM, Schneider VA, Graves-Lindsay TA, Fulton RS, Agarwala R, Huddleston J, Shiryev SA, Morgulis A, Surti U, Warren WC, Church DM, Eichler EE, Wilson RK (2014) Single haplotype assembly of the human genome from a hydatidiform mole. Genome Res 24:2066–2076
VanBuren R, Bryant D, Edger PP, Tang H, Burgess D, Challabathula D, Spittle K, Hall R, Gu J, Lyons E, Freeling M, Bartels D, Ten Hallers B, Hastie A, Michael TP, Mockler TC (2015) Single-molecule sequencing of the desiccation-tolerant grass Oropetium thomaeum. Nature 527:508–511
Veronesi F, Brummer EC, Huyghe C (2010) Alfalfa. In: Boller B, Posselt UK, Veronesi F (eds) Fodder crops and amenity grasses. Springer, New York, New York, NY, pp 395–437
Walker BJ, Abeel T, Shea T, Priest M, Abouelliel A, Sakthikumar S, Cuomo CA, Zeng Q, Wortman J, Young SK, Earl AM (2014) Pilon: an integrated tool for comprehensive microbial variant detection and genome assembly improvement. PLoS ONE 9:
Wenger AM, Peluso P, Rowell WJ, Chang P-C, Hall RJ, Concepcion GT, Ebler J, Fungtammasan A, Kolesnikov A, Olson ND, Töpfer A, Alonge M, Mahmoud M, Qian Y, Chin C-S, Phillippy AM, Schatz MC, Myers G, DePristo MA, Ruan J, Marschall T, Sedlazeck FJ, Zook JM, Li H, Koren S, Carroll A, Rank DR, Hunkapiller MW (2019) Accurate circular consensus long-read sequencing improves variant detection and assembly of a human genome. Nat Biotechnol 37:1155–1162
Xiao C-L, Chen Y, Xie S-Q, Chen K-N, Wang Y, Han Y, Luo F, Xie Z (2017) MECAT: fast mapping, error correction, and de novo assembly for single-molecule sequencing reads. Nat Methods 14:1072–1074
Yang S, Gao M, Xu C, Gao J, Deshpande S, Lin S, Roe BA, Zhu H (2008) Alfalfa benefits from *Medicago truncatula*: the RCT1 gene from *M. truncatula* confers broad-spectrum resistance to anthracnose in alfalfa. Proc Natl Acad Sci U S A 105:12164–12169
Young ND, Debellé F, Oldroyd GED, Geurts R, Cannon SB, Udvardi MK, Benedito VA, Mayer KFX, Gouzy J, Schoof H, Van de Peer Y, Proost S, Cook DR, Meyers BC, Spannagl M, Cheung F, De Mita S, Krishnakumar V, Gundlach H, Zhou S, Mudge J, Bharti AK, Murray JD, Naoumkina MA, Rosen B, Silverstein KAT, Tang H, Rombauts S, Zhao PX, Zhou P, Barbe V, Bardou P, Bechner M, Bellec A, Berger A, Bergès H, Bidwell S, Bisseling T, Choisne N, Couloux A, Denny R, Deshpande S, Dai X, Doyle JJ, Dudez A-M, Farmer AD, Fouteau S, Franken C, Gibelin C, Gish J, Goldstein S, González AJ, Green PJ, Hallab A, Hartog M, Hua A, Humphray SJ, Jeong D-H, Jing Y, Jöcker A, Kenton SM, Kim D-J, Klee K, Lai H, Lang C, Lin S, Macmil SL, Magdelenat G, Matthews L, McCorrison J, Monaghan EL, Mun J-H, Najar FZ, Nicholson C, Noirot C, O'Bleness M, Paule CR, Poulain J, Prion F, Qin B, Qu C, Retzel EF, Riddle C, Sallet E, Samain S, Samson N, Sanders I, Saurat O, Scarpelli C, Schiex T, Segurens B, Severin AJ, Sherrier DJ, Shi R, Sims S, Singer SR, Sinharoy S, Sterck L, Viollet A, Wang B-B, Wang K, Wang M, Wang X, Warfsmann J, Weissenbach J, White DD, White JD, Wiley GB, Wincker P, Xing Y, Yang L, Yao Z, Ying F, Zhai J, Zhou L, Zuber A, Dénarié J, Dixon RA, May GD, Schwartz DC, Rogers J, Quétier F, Town CD, Roe BA (2011) The Medicago genome provides insight into the evolution of rhizobial symbioses. Nature 480:520–524
Zhang H, Huang Q-M, Su J (2010) Development of alfalfa (*Medicago sativa L.*) regeneration system and agrobacterium-mediated genetic transformation. Agric Sci China 9:170–178
Zhang X, Zhang S, Zhao Q, Ming R, Tang H (2019) Assembly of allele-aware, chromosomal-scale autopolyploid genomes based on Hi-C data. Nat Plants 5:833–845
Zhou P, Silverstein KA, Gao L et al (2013) Detecting small plant peptides using SPADA (Small Peptide Alignment Discovery Application). BMC Bioinform 14:335

7 Transcription Factors in Alfalfa (*Medicago sativa* L.): Genome-Wide Identification and a Web Resource Center AlfalfaTFDB

Lev G. Nemchinov, Jonathan Shao, Samuel Grinstead, and Olga A. Postnikova

Abstract

Research on the genomics of cultivated alfalfa (*Medicago sativa* L.), over many years hindered by the lack of the reference genome, has substantially progressed during the last decade. Cutting-edge high-throughput sequencing technologies increased our understanding of transcriptomic processes in the plant in response to a variety of environmental factors, advanced whole-genome sequencing, and transformed plant genotyping and breeding strategies. This newly gained information presents unique opportunities for the discovery of novel sequences, identification of genes, functional DNA elements, and genetic markers. It also provides critical tools to decipher molecular mechanisms involved in crop adaptation to the environment, evolution, plant productivity, and biomass increase. This chapter pertains to transcription factors, the regulatory components of the alfalfa genome that coordinate gene expression and activity.

Electronic supplementary material The online version of this chapter (https://doi.org/10.1007/978-3-030-74466-3_7) contains supplementary material, which is available to authorized users.

L. G. Nemchinov (✉) · J. Shao
United States Department of Agriculture, Beltsville Agricultural Research Center, Molecular Plant Pathology Laboratory, 10300 Baltimore Avenue, Beltsville, MD 20705, USA
e-mail: Lev.Nemchinov@ars.usda.gov

S. Grinstead
United States Department of Agriculture, Beltsville Agricultural Research Center, National Germplasm Resources Laboratory, 10300 Baltimore Avenue, Beltsville, MD 20705, USA

O. A. Postnikova
National Institute of Health, 9000 Rockville Pike, Bethesda, MD 20892, USA

7.1 Introduction

Transcription factors (TFs) are proteins that govern organismal development and response to the environment by regulating gene expression (Postnikova et al. 2014). The number, distribution, and diversity of TFs directly correlate with the complexity of organisms (De Mendoza et al. 2013). The information on the specific TF repertoires within individual plant species is critical for the understanding of their biological roles and evolutionary history across the plant kingdom.

For years, only a scattered knowledge on a few individual TFs was available for alfalfa, one of the most widely planted forage legumes in the

L.-X. Yu and C. Kole (eds.), *The Alfalfa Genome*, Compendium of Plant Genomes,
https://doi.org/10.1007/978-3-030-74466-3_7

world (Li and Brummer 2012). Meanwhile, a large number of TFs have been discovered in other plants, including legume species with sequenced genomes (Zhang et al. 2011; Udvardi et al. 2007; Libault et al. 2009). In 2014, an in silico analysis of transcriptome data generated in our laboratory and publicly acquirable from other sources enabled prediction of nearly a thousand alfalfa TFs along with their sequence features and putative phylogenies of the largest families (Postnikova et al. 2014). All data were integrated into a simple open-access database named AlfalfaTFDB, offering an essential tool for studying the regulation of gene expression in alfalfa.

In this study, taking advantage of the most recent developments in the field of alfalfa genomics, we have re-analyzed and significantly expanded the alfalfa TFs repertoire. It is now comprised of 8465 TFs, distributed among 53 families. We have also updated the AlfalfaTFDB database to include new transcription factors identified since 2014. This chapter reports general information on the identified TFs and brief descriptions of the five major TF families in alfalfa.

7.2 Data Mining and Analysis

7.2.1 Identification of Alfalfa Transcription Factors

The allele-aware chromosome-level alfalfa genome assembly was downloaded from https://figshare.com/projects/whole_genome_sequencing_and_assembly_of_Medicago_sativa/66380/ (Chen et al. 2020). The putative encoded protein sequences were submitted to the PlantTFcat (http://plantgrn.noble.org/PlantTFcat/) and PlantTFDB (http://planttfdb.gao-lab.org/prediction_result.php) web servers. Transcription factor domains were identified using InterProScan v. 5.44-79.0 (Jones et al. 2014) and ARS-SCINet (https://scinet.usda.gov). The results were consolidated and filtered for putative predicted transcription factors and agreement between prediction programs. For proteins with multiple domains, only the top domain was extracted for each sequence. Five TF families—SAP, NF-X, GRF, DBB, and CPP—were extracted using the same bioinformatics software as the other alfalfa TF's, except that they were identified as transcription factors by either PlantTFDB or PlantTFcat, but not necessarily both web servers.

7.2.2 Development of the TF Database

A relational MySQL database was created storing the open reading frames (ORF), (nucleotide and protein), relevant domain, and literature description of the transcription factor. This database is available online and delineated into a table by transcription factors (https://plantpathology.ba.ars.usda.gov/alfalfatfdb.html).

7.2.3 Phylogenetic Analysis

Phylogenetic trees were deduced from domain alignments of *M. sativa* and *M. truncatula* TF families with Arabidopsis TFs. The *A. thaliana* TFs were retrieved from TAIR (The Arabidopsis Information Resource) and *M. truncatula* sequences were downloaded from PlantTFDB (http://planttfdb.cbi.pku.edu.cn/). Alfalfa TFs were named according to the pattern adopted by Chen et al. (2020) for genome assembly of the cultivated alfalfa. Sequences of DNA binding domains from *A. thaliana*, *M. trunatula,* and *M. sativa* were aligned with the Geneious software v. 2020.2.4 and the resulting alignments were used to construct UPGMA consensus trees based on the bootstrap analysis of 1000 replicates. Trees were customized and visualized with CLC Main Workbench v. 20.0.4. Only complete alfalfa DNA-binding domains (DBD) domains identified via InterPro and CDD (Marchler-Bauer et al. 2011) were used for alignments. The confidence level was estimated. Phylogenetic trees were not rooted so that evolutionary information could be derived within each class. Sequences that were not clearly clustered within the same TF family and formed minor subtrees with five or fewer leaf nodes, were not shown.

7.3 Identification and Characterization of Transcription Factors

7.3.1 Identification of Putative Alfalfa TFs

Since our last study on alfalfa TF (Postnikova et al. 2014), the genome of autotetraploid cultivated alfalfa has been assembled (Chen et al. 2020). The availability of the genome has significantly facilitated the identification of alfalfa TFs. Here, using this genome assembly and two prediction programs, PlantTFcat (Dai et al. 2013) and PlantTFDB (Jin et al. 2017) as well as InterProScan (Jones et al. 2014), we were able to predict 8465 TFs in alfalfa. This quantity exceeds the one in a close alfalfa relative, *Medicago truncatula*, more than threefold. Whereas these two species share a high degree of sequence similarity and genome synteny (Yang et al. 2008), observations gathered on *M. truncatula*, an annual legume with a diploid genome, may not always apply to alfalfa, a perennial and an outcrossing tetraploid.

The predicted 8465 TFs were tentatively classified into 53 families according to the types of DNA binding domains. (Table 7.1; Supplementary Table 7.1). The Nuclear Factor Y TF complex, composed of three subunits (NF-YA, NF-YB, and NF-YC), was counted as one TF family, NF-Y (Siefers et al. 2009). The individual numbers for each of the subunit families are shown in Table 7.2 and Supplementary Table 7.1.

Five TF families—SAP, NF-X, GRF, DBB, and CPP—did not have the term “transcription factor” as a standalone annotation and instead have been annotated as “transcription factor interactors and regulators”. The alfalfa SAP TF family appeared to have domains similar to C3H or C3H-like TFs. It was previously reported that the C3H TF family proteins can carry several other known functional domains, including ANK, KH, RRM, SAP, WD-40, B-box, DEXDc, HELICc, PHD, SWIB, Plus3, GYF, G-patch, and ZF-Ring (Liu et al. 2020). Two TFs, HRT-like and VOZ (Raventós et al. 1998; Mitsuda et al. 2004), were not detected. This could be related either to the methodology of analysis (software prediction programs annotated them with another description) or to the missing contigs in the genome assembly.

When all identified TFs were mapped to the eight homologous groups, each containing four allelic chromosomes, the distribution of the TF families among allelic sets was uneven (Table 7.2). Chromosomal sets contained different numbers of TFs; some TF families have more members preferentially localized to specific chromosomes and several families were either uniquely confined or absent from particular chromosomal groups. Divergent clustering of TF families on alfalfa chromosomes suggests a relationship between functional roles associated with particular gene groups and the structural organization of the alfalfa genome.

The polypoid nature of alfalfa makes it particularly difficult to suggest the precise number of TFs since it increases with the number of genes in the genome: more complex species need more elaborate regulation of plant response to the environment. While polyploidization may cause changes in TF family sizes, it can also apply selective pressure on duplicated genes so that they could develop new functional roles or different expression patterns. Meanwhile, the size of TF superfamilies could be limited, depending on the number of DNA bases recognized by the binding mechanism (Itzkovitz et al. 2006).

TF superfamilies in alfalfa, such as WKRY, bHLH, NAC, C2H2, and ERF, were often substantially larger than those in *M. truncatula* (Table 7.1). These TFs are implicated in a variety of developmental processes and are known to control several useful agronomic traits, including responses to stresses and the symbiotic relationship with nitrogen-fixing rhizobacteria (Vernié et al. 2008; Libault et al. 2009; Moreau et al. 2011; Godiard et al. 2011; De Zélicourt et al. 2012). Our earlier studies corroborated the importance of these TFs in the regulation of

Table 7.1 Transcription factors computationally predicted in *Medicago sativa* L

AP2 (106)	ARF (148)	ARR-B (123)	B3 (520)
BBR-BPC (19)	BES (25)	bHLH (637)	bZIP (320)
C2H2 (508)	C3H (228)	CAMTA (38)	CO-like (35)
CPP (31)	DBB (24)	DOF (141)	E2F_DB (33)
EIL (34)	ERF (686)	FAR1 (604)	G2-like (164)
GATA (166)	GeBP (25)	GRAS (255)	GRF (139)
HB-other (30)	HB-phd (8)	HD-ZIP (193)	HSF (115)
LBD (196)	LFY (4)	LSD (17)	M-type_MADS (462)
MIKC_MADS (52)	MYB (357)	MYB-rel (429)	NAC (445)
NF-X (12)	NF-Y(A, B, C) (163)	Nin-like (55)	RAV (7)
S1Fa-like (10)	SAP (11)	SBP (83)	SRS (29)
STAT (4)	TALE (78)	TCP (93)	Trihelix (70)
Whirly (10)	WOX (70)	WRKY (362)	YABBY (39)
ZF-HD (52)			

stress-related genes and signaling networks in alfalfa. Increased amounts of mRNA transcribed from the genes of all five TF families were recorded in alfalfa plants under different biotic and abiotic stresses, such as salinity (Postnikova et al. 2013), root-knot nematode (Postnikova et al. 2015), bacterial infection (Nemchinov et al. 2017) and root-lesion nematode (Vieira et al. 2019). It may be assumed that they are also involved in conferring resistance against environmental changes that cause the decline in alfalfa fitness. Considering that the size of the TF families reflects, through transcriptional control, the scale and diversity of their involvement in many key aspects of plant development, these superfamilies in alfalfa are likely to define distinct characteristics of the plant as a valuable agricultural crop.

7.3.2 Phylogenetic Analysis of Major TF Families

The TF superfamilies, bHLH, ERF, MYB, NAC, and WRKY, represent one-third of all alfalfa TFs. We examined their phylogenetic relationships with the respective proteins of *M. truncatula* and *A. thaliana*. The latter species served as a core dataset needed to depict the ancestral lineage of alfalfa TFs. The resulting phylogenetic clustering into subfamilies within each TF family is tentative and influenced by the alignment algorithms, methods used for the phylogenetic analysis, and topology of the tree. Nevertheless, it is expected that phylogenetic trees, whose reliability was estimated by bootstrap methods, provided a generally accurate representation of the relationships, diversity, and composition of alfalfa TFs.

7.3.3 The BHLH TF Family

Land plants have large numbers of bHLH proteins that are implicated in a wide range of processes related to growth, development, and stress responses (Toledo-Ortiz et al. 2003; Pires and Dolan 2010; Goossens et al. 2017). In *M. truncatula*, bHLH TFs control the development of root nodules (Godiard et al. 2011) and in soybean, they are engaged in response to biotic and abiotic stress (Osorio et al. 2012).

A bHLH signature domain comprises 50–60 amino acids and incorporates two functional

Table 7.2 Distribution of TFs among 8 homologous groups of alfalfa chromosomes

TF ID	chr1	chr2	chr3	chr4	chr5	chr6	chr7	chr8	Unplaced
AP2	9	12	4	10	14	2	17	38	0
ARF	15	35	6	25	43	0	11	13	0
ARR-B	19	8	17	31	6	6	9	23	4
B3	173	15	94	48	25	12	98	20	35
BBR-BPC	4	2	0	8	0	4	0	1	0
BES1	3	7	3	0	8	0	4	0	0
bHLH	111	112	53	93	67	13	65	106	17
bZIP	41	42	54	40	34	14	49	37	9
C2H2	117	48	78	88	41	7	67	47	15
C3H	38	19	52	22	14	4	24	53	2
CAMTA	0	4	3	15	0	0	4	7	5
CO-like	8	0	6	3	4	0	9	4	1
Dof	8	26	12	23	18	10	16	28	0
E2F/DP	3	0	4	15	0	0	5	4	2
EIL	0	4	12	4	4	6	0	4	0
ERF	123	74	59	78	87	79	77	94	15
FAR1	78	98	91	61	72	67	49	68	20
G2-like	19	30	10	22	21	12	30	17	3
GATA	58	12	20	28	13	2	9	18	6
GeBP	9	4	3	0	0	6	3	0	0
GRAS	20	57	25	45	27	4	39	26	12
HB-other	0	2	7	2	4	3	2	10	0
HB-PHD	4	4	0	0	0	0	0	0	0
HD-ZIP	18	37	19	9	30	12	22	40	6
HSF	15	11	7	22	17	23	8	9	3
LBD	16	13	29	19	41	17	26	31	4
LFY	0	0	4	0	0	0	0	0	0
LSD	0	5	0	8	0	0	0	4	0
MIKC_MADS	0	0	9	17	9	2	1	14	0
M-type_MADS	112	39	74	81	28	18	41	55	14
MYB	59	22	41	32	63	33	57	44	6
MYB_related	60	27	69	41	54	40	51	68	19
NAC	77	53	49	60	40	40	60	63	3
NF-YA	3	11	5	0	0	0	3	6	2
NF-YB	36	10	5	8	9	0	4	14	1
NF-YC	4	12	10	1	4	4	7	4	0
Nin-like	4	4	24	4	11	4	0	4	0
RAV	4	0	0	0	3	0	0	0	0
S1Fa-like	0	0	0	2	0	0	8	0	0
SBP	11	20	7	13	3	0	16	11	2
SRS	4	0	8	1	8	0	0	8	0
STAT	0	0	0	0	0	0	0	4	0

(continued)

Table 7.2 (continued)

TF ID	chr1	chr2	chr3	chr4	chr5	chr6	chr7	chr8	Unplaced
TALE	21	14	12	11	12	0	7	1	0
TCP	18	10	4	8	4	18	13	18	0
Trihelix	22	7	11	6	6	4	12	2	0
Whirly	0	0	0	0	3	0	4	3	0
WOX	7	8	5	17	8	7	17	0	1
WRKY	35	47	44	53	55	8	59	50	11
YABBY	3	4	0	8	13	0	0	11	0
ZF-HD	4	0	5	0	12	1	4	25	1
CPP	11	0	4	4	5	5	0	0	2
DBB	1	6	5	8	4	0	0	0	0
GRF	11	16	28	13	19	7	21	17	7
NF-X1	0	0	4	5	0	3	0	0	0
SAP	4	0	3	0	4	0	0	0	0
Total	**1420**	**991**	**1098**	**1112**	**967**	**497**	**1028**	**1124**	**228**

regions: the basic region, located at the N-terminal end of the domain and involved in DNA binding, and the helix-loop-helix region, located at the C-terminal end and involved in dimerization between two HLH proteins (Pires and Dolan 2010). The core DNA sequence motif recognized by the bHLH TFs is a consensus E-box sequence 5-CANNTG-3 (Heim et al. 2003; Toledo-Ortiz et al. 2003; Carretero-Paulet et al. 2010).

Genome-wide profiling identified 637 bHLH TFs in alfalfa (Table 7.1). This is more than twice the currently known number of bHLH proteins in a model legume, *M. truncatula* ($n = 259$). Alignment of the *M. sativa* bHLH genes with the complete domains revealed a consensus amino acid sequence and the structure of the conserved domain regions (Supplementary Fig. 7.1). Phylogenetic analysis demonstrated that alfalfa bHLH TFs formed at least 36 tentative subfamilies, arbitrarily numbered 1–36 (Fig. 7.1). Many of the alfalfa proteins, along with selected bHLH TFs of *M. truncatula*, clustered into monophyletic groups proposed for *A. thaliana* (Heim et al. 2003: Carretero-Paulet et al. 2010; Pires and Dolan 2010), therefore indicating conservation of bHLH genes between these three species and a likely resemblance of their biological roles to those suggested for *A. thaliana*. However, a large number of alfalfa TFs ($n = 363$) formed their own clades ($n = 20$), different from the Arabidopsis nomenclature, signifying the possible divergence of their functions in alfalfa.

7.3.4 The ERF TF Family

The ERF family is part of the larger AP2/ERF superfamily named after the 40–70 conserved amino acid-long AP2/ERF DNA-binding domain (Nakano et al. 2006; Yamasaki et al. 2013; Xie et al. 2019). The ERF TFs control diverse biological processes such as hormonal signal transduction, response to biotic and abiotic stresses, regulation of developmental pathways, and programmed cell death (Nakano et al. 2006; Mase et al. 2013; Müller and Munné-Bosch 2015; Heyman et al. 2018). In *M. truncatula*, the ERF

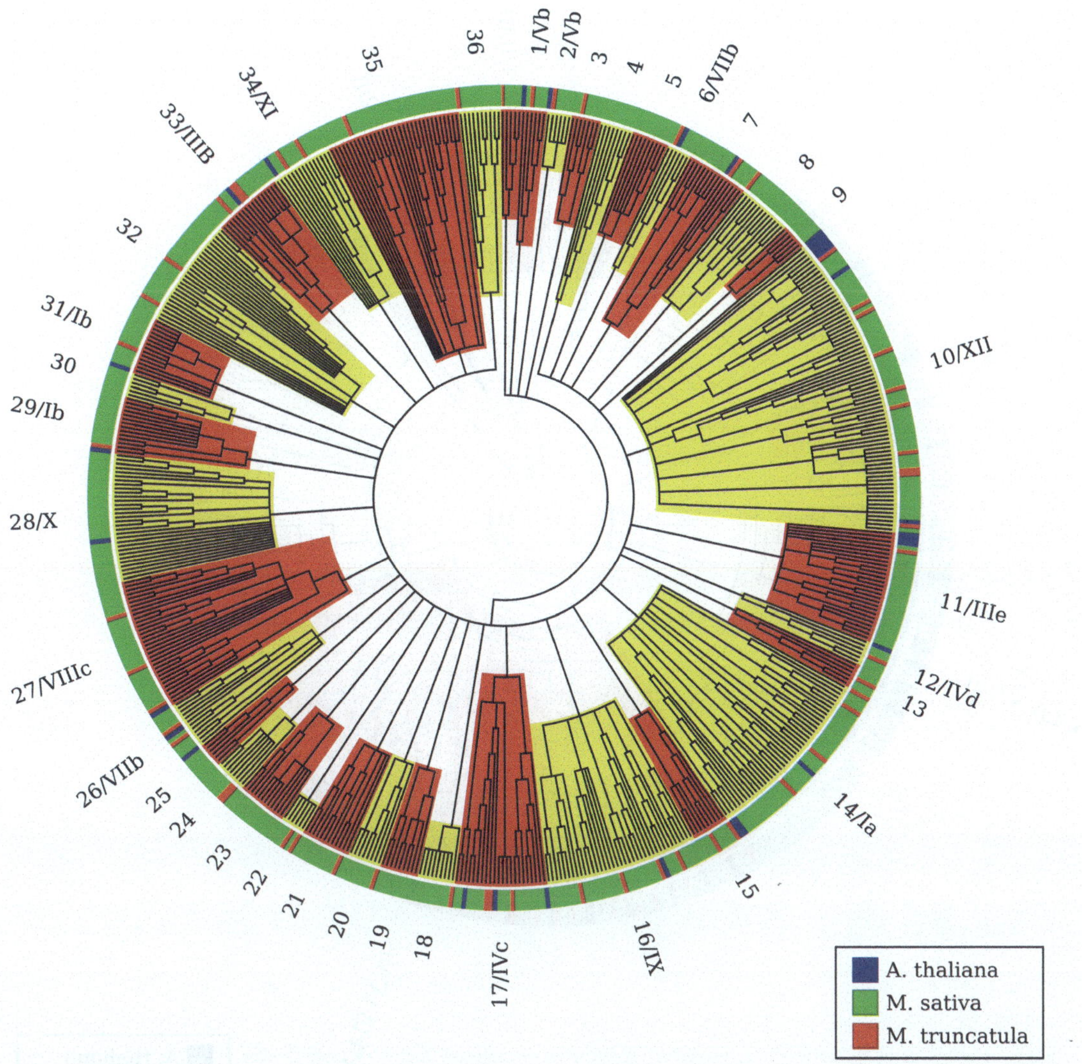

Fig. 7.1 Unrooted phylogenetic tree of the bHLH proteins in alfalfa. Tentative subfamilies are shown in alternating red and yellow backgrounds and numbered with Arabic numerals. *M. sativa*, *M. truncatula,* and *A. thaliana* TFs are shown in green, red, and blue rectangles, respectively. bHLH subfamilies previously classified in Arabidopsis are depicted with Roman numerals

TFs were shown to be involved in nodulation (Middleton et al. 2007), hormonal induction of somatic embryogenesis (Mantiri et al. 2008), and resistance to a subset of root pathogens (Anderson et al. 2010). Jin et al. (2019) have recently reported identification, phylogenetic analysis, and expression patterns under different environmental conditions of 155 ERF TFs from alfalfa.

We were able to identify 686 ERF TFs with complete DBD in alfalfa (3.5-fold increase vs *M. truncatula*), (Table 7.1). Phylogenetic analysis of *M. sativa* ERF TFs revealed at least 8 distinct monophyletic groups arbitrarily named 1–8 (Fig. 7.2). Majority of the TFs clustered together with the ERF members from *A. thaliana* and *M. truncatula* implying general conservation of the ERF domains in alfalfa. This conservation was also noticeable in the alignment of ERF TFs originated from these three species (Supplementary Fig. 7.2).

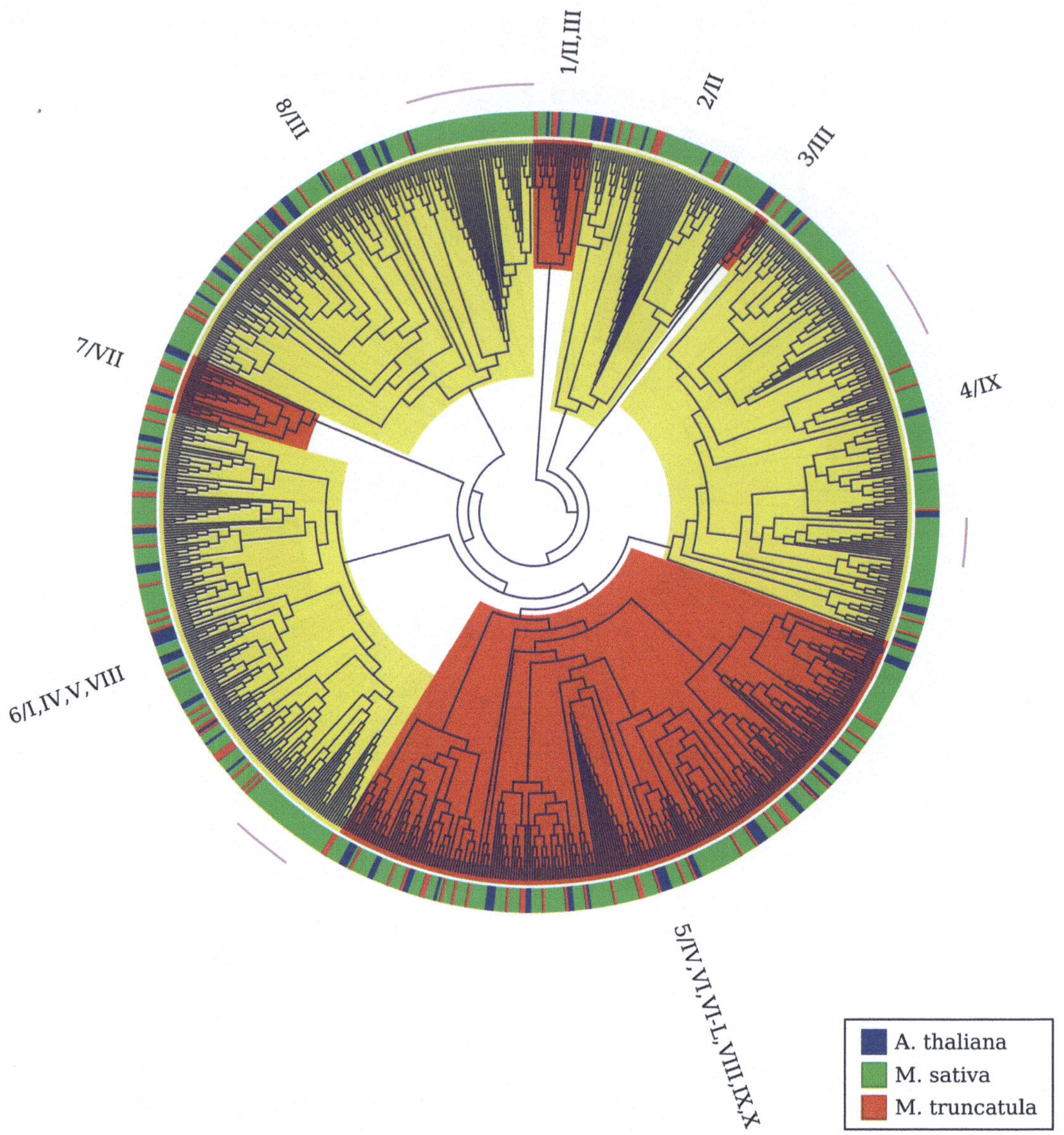

Fig. 7.2 Unrooted phylogenetic tree of alfalfa ERF TFs. Tentative subfamilies are shown in alternating red and yellow backgrounds and numbered with Arabic numerals. *M. sativa*, *M. truncatula*, and *A. thaliana* TFs are shown in green, red, and blue rectangles, respectively. ERF subfamilies previously classified in Arabidopsis are depicted with Roman numerals. Subtrees within groups 4, 6, and 8 of alfalfa ERF TFs that did not congregate with *A. thaliana* and *M. truncatula* TFs, are marked by red lines

Nevertheless, provisional subfamilies of the alfalfa ERF TFs only loosely corresponded to the clades I–X previously described for *A. thaliana* (Nakano et al. 2006). Large alfalfa ERF subfamilies included several groups defined for *A. thaliana*. This suggested that further breakdown of 8 phylogenetically classified subfamilies based on the domain structure would likely be necessary to determine their exact organization and evolutionary similarities. Additionally, several subtrees within larger groups 4, 6, and 8 of alfalfa ERF, did not assemble with *A. thaliana* and *M. truncatula* TFs, assuming that these proteins could be distinct from other members of the family.

Therefore, despite a broad similarity of the *M. sativa* ERF proteins to those of *A. thaliana* and *M. truncatula*, their composition, organization into subfamilies and evolutionary trajectory could be substantially different, considering complicated autopolyploid evolution in alfalfa (Havananda et al. 2011).

7.3.5 The NAC TF Family in Alfalfa

NAC is one of the largest families of transcription factors specific to plants (Olsen et al. 2005; Puranik et al. 2012). The NAC TFs contain a conserved N-terminal region, or NAC domain, and a highly variable C-terminal part that presumably functions as a transcriptional activation domain (Ernst et al. 2004; Olsen et al. 2005). The core DNA motif, recognized by the N-terminal NAC domain in the promoter region, is a 5-CACG-3 sequence. Ling et al. (2017) identified 97 putative NAC transcription factors in the genome of the *M. truncatula* Several NAC proteins in this species are known to be involved in cell wall development and adaptive root responses (Zhao et al. 2010; De Zélicourt et al. 2012).

According to the PlantTFDB database, 138 NAC TFs are currently known in *A. thaliana* and 123 in *M. truncatul*a (database accessed on 01/11/2021). Although NAC TFs present in many land plants and play diverse biological roles (embryonic, floral, and vegetative development, root formation, disease resistance, stress tolerance), information on the composition of the NAC family in alfalfa until recently has been very limited. Earlier, we found 67 NAC TFs in alfalfa out of which 52 contained complete DBD (Postnikova et al. 2014). Min et al. (2020) reported the identification of 113 *M. sativa* NAC, which they divided into 15 distinct subgroups.

In this work, we have updated these lists to reveal 445 NAC TFs in *M. sativa* (Table 7.1). Phylogenetic analysis of alfalfa NAC TFs with complete DBD motif identified at least 16 subfamilies arbitrarily numbered 1–16 (Fig. 7.3; Supplementary Fig. 7.3). Ten NAC polytomies were proposed for *A. thaliana* (Jensen et al. 2010) based on domain motifs and 14 subfamilies, clustered into two large groups, for *M. truncatula* (Ling et al. 2017). Provisional grouping of alfalfa NAC TFs only partially corresponded to the categorization of NAC TFs in *A. thaliana* and *M. truncatula*. Alfalfa subfamilies 5, 6, and 12 did not cluster together with any Arabidopsis and *M. truncatula* TFs. The phylogenetic tree also appeared different from the one reported for *M. sativa* (Min et al. 2020). This could be explained by the fact that the latter study seemingly used CADL resources (cultivated alfalfa at diploid level) for its identification of TFs (http://www.medicagohapmap.org/downloads/cadl), while in this research we have conducted the genome-wide profiling of tetraploid alfalfa. This resulted in a larger number of NAC TFs and the emergence of novel clades/subfamilies revealed by phylogenetic analysis, implying diversification of NAC TF family in alfalfa.

7.3.6 The WRKY TF Family

WRKY is a large plant-specific family of transcription factors defined by the presence of the conserved WRKYGQK amino acid sequence at the N-terminus of the protein and zinc finger motif ($CX_{4-7}CX_{22-23}HXH/C$) in the C-terminus (Pandey and Somssich 2009; Rushton et al. 2010). All known proteins of this family have either one or two WRKY domains (Eulgem et al. 2000). The DNA motif recognized by WRKY TFs is a conserved W-Box sequence (TTGACC/T), the minimum consensus needed for binding (Rushton et al. 2010).

Similar to other TF superfamilies, WRKY proteins have numerous biological functions among which the most noticeable and well-described is their role in plant defense responses (Zhang and Wang 2005; Pandey and Somssich 2009; Phukan et al. 2016; Chen et al. 2019). It was recently shown that WRKY TFs maintain pith cells in their parenchymatous state in *M. truncatula* and that their mutation results in a

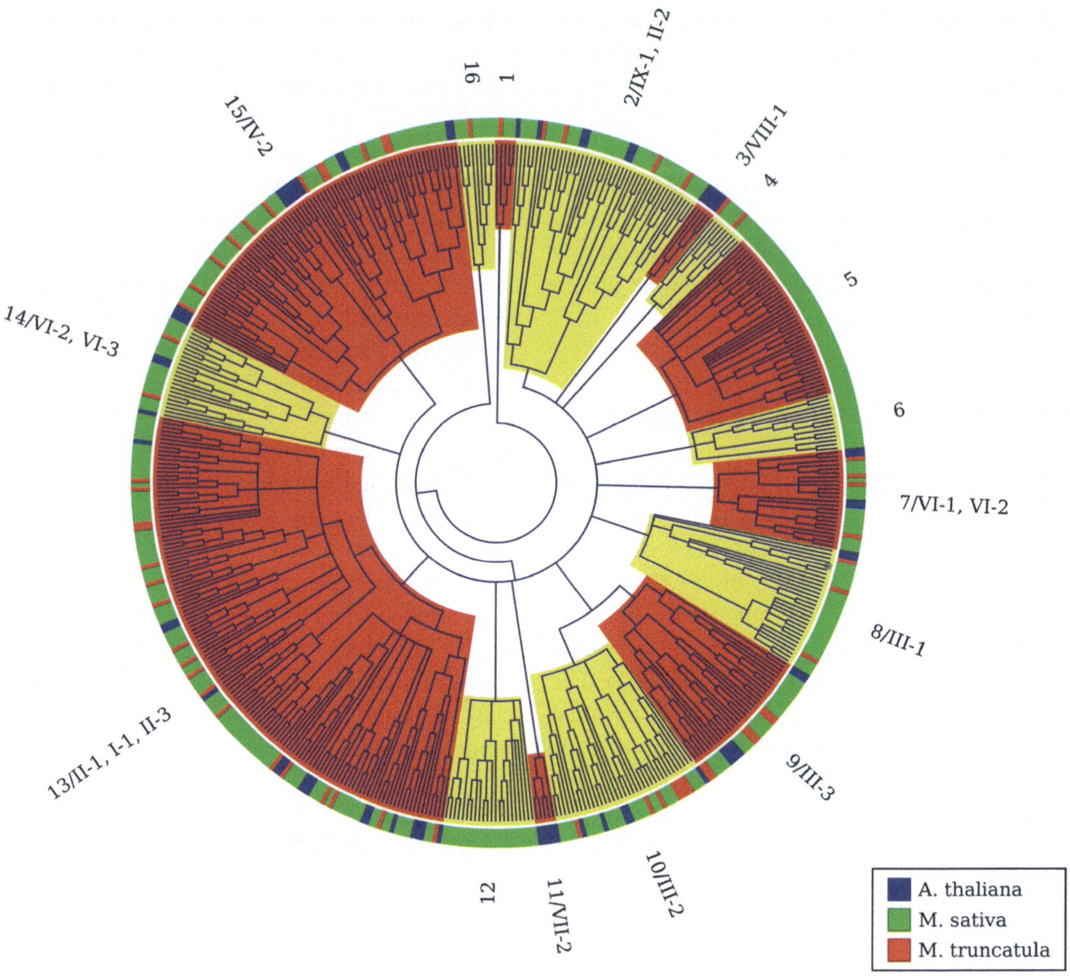

Fig. 7.3 Unrooted phylogenetic tree of alfalfa NAC TFs. Tentative subfamilies are shown in alternating red and yellow backgrounds and numbered with Arabic numerals. *M. sativa*, *M. truncatula,* and *A. thaliana* TFs are shown in green, red, and blue rectangles, respectively. NAC subfamilies previously classified in Arabidopsis are depicted with Roman numerals

significant increase in stem biomass (Wang et al. 2010). In the previous study (Postnikova et al. 2014), we have identified 71 WRKY TFs in alfalfa, among which 65 have complete DBD. More recently, Mao et al. (2020) reported 107 WRKY TFs in *M. sativa* that were partitioned by the authors into three main groups, similar to those proposed for this family in *A. thaliana* (Eulgem et al. 2000).

We have further expanded the identified WRKY TFs in alfalfa to 362 members (Table 7.1). Phylogenetic analysis of the alfalfa WRKY TFs identified six tentative subfamilies (Fig. 7.4; Supplementary Fig. 7.4) which, in general, corresponds to the nomenclature of the WRKY TFs established for *A. thaliana* (Eulgem et al. 2000). Alfalfa proteins clustered together with known WRKY TFs from *A. thaliana* and *M. truncatula*. All the indications are that alfalfa WRKY TFs are mainly conserved and the composition of this family resembles that found in other higher plants, aside from the considerable expansion of this family in alfalfa.

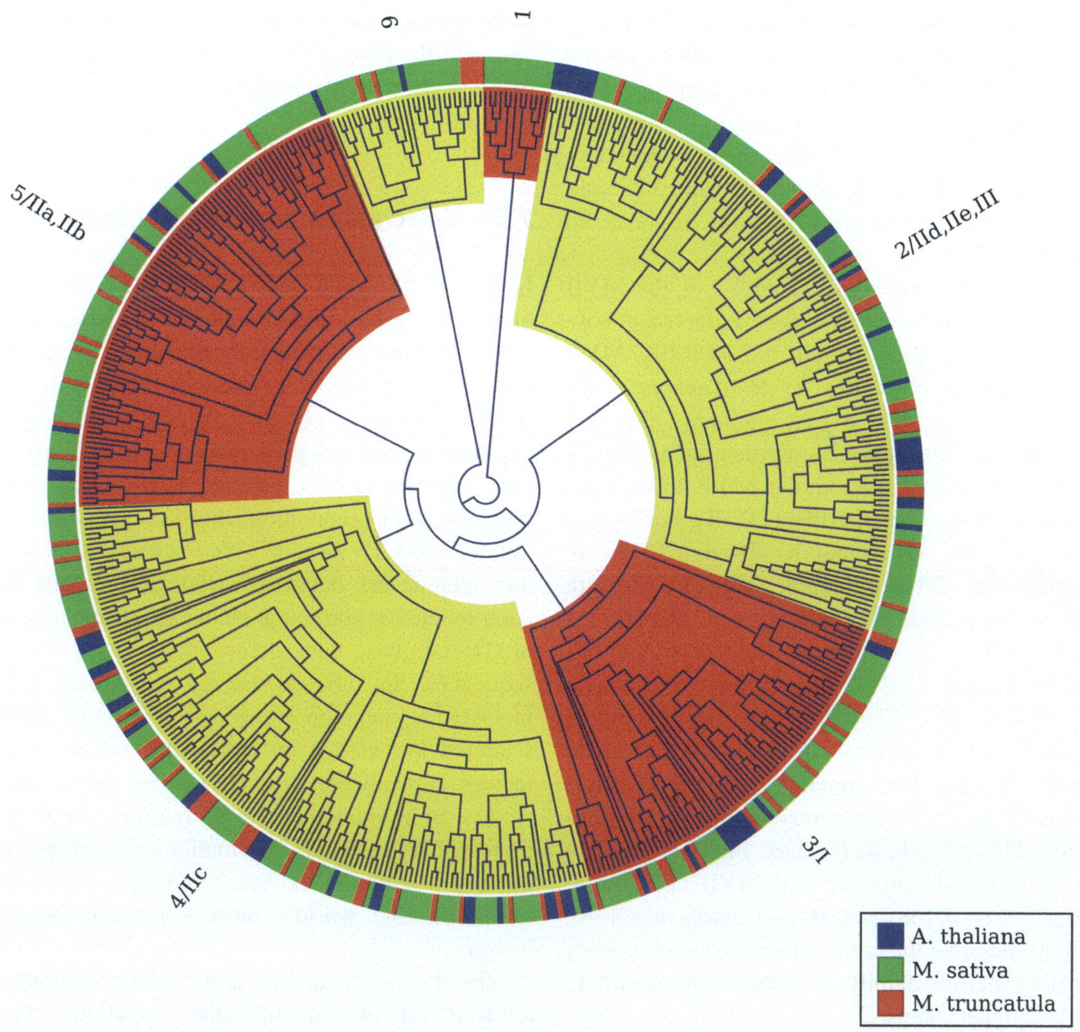

Fig. 7.4 Unrooted phylogenetic tree of alfalfa WRKY TFs. Tentative subfamilies are shown in alternating red and yellow backgrounds and numbered with Arabic numerals. *M. sativa*, *M. truncatula*,and *A. thaliana* TFs are shown in green, red, and blue rectangles, respectively. WRKY subfamilies previously classified in Arabidopsis are depicted with Roman numerals

7.3.7 The MYB TF Family

Plants encode a large number of MYB TFs with a highly conserved MYB DBD consisting of ~50 amino acids and a variable activation domain located at the C-terminal region of the protein. The MYB gene was first identified in the avian myeloblastosis virus and the first plant MYB-like TF was discovered in *Zea mays* (Klempnauer et al. 1982; Lata et al. 2011). MYB proteins are classified depending on the number of up to four imperfect amino acid repeats present in their DBD, named R1–R4 (Dubos et al. 2010; Feller et al. 2011; Katiyar et al. 2012). MYB genes encoding proteins of the R2R3 type are prevalent in plants (Stracke et al. 2001).

The core DNA motif recognized by R2R3 MYB TFs is enriched in adenosine (A) and cytosine (C) residues and is referred to as AC elements (Katiyar et al. 2012; Prouse and Campbell 2013). The MYB-like proteins with a single or partial repeat bind the consensus sequence of plant telomeric DNA TTTAGGG (Stracke et al. 2001; Yu et al. 2000) and are members of the MYB-related family. MYB transcription factors play important roles in plant growth and organ development, secondary metabolism, hormone signal transduction, and response to environmental stresses (Lata et al. 2011).

The first genome-wide analysis of the MYB family in legume species was conducted in soybean and identified a total of 244 R2R3 MYB genes (Du et al. 2012). In *M. truncatula*, the MYB TF served as a key regulator of proanthocyanidin biosynthesis (Verdier et al. 2012). There are currently 185 known MYB TFs for this close relative of alfalfa (PlantTFDB). Earlier, we identified 48 MYB and MYB-related TFs in alfalfa that did not include a SANT domain (Boyer et al 2002; Postnikova et al. 2014). This number was recently updated to 265 MYB TFs by Zhou et al. (2019).

We found 357 MYB TFs in *M sativa* (Table 7.1). Phylogenetic analysis performed with 357 complete domains of MYB proteins, revealed at least 10 arbitrary-numbered subfamilies (Fig. 7.5; Supplementary Fig. 7.5). Alfalfa proteins broadly grouped with MYB TFs from *A. thaliana* (n = 168) and *M. truncatula* (n = 178), suggesting that the composition and evolutionary history of this family in alfalfa correspond to other higher plants.

7.4 AlfalfaTFDB: A Comprehensive Resource of Alfalfa Transcription Factors

All the new data on the alfalfa TFs obtained in this study, were used to update and improve an open-access database called AlfalfaTFDB that we originally developed in 2014 (Postnikova et al. 2014). The database depicts a chart of identified 8465 identified alfalfa TFs with active links to their descriptions in the literature, conserved protein domains, ORF, and protein sequences (Fig. 7.6). At the moment, the AlfalfaTFDB represents the only comprehensive resource for all currently identified alfalfa transcription factors, offering an essential tool for studying the regulation of gene expression in the plant (https://plantpathology.ba.ars.usda.gov/alfalfatfdb.html).

7.5 Conclusion and Perspectives

In this study, using a newly available draft genome of tetraploid alfalfa (Chen et al. 2020), we have substantially expanded previously defined lists of computationally predicted alfalfa transcription factors, proteins that regulate gene expression, and assigned them to known structural families. As of January 2021, when this article was in preparation, this list comprises a total of 8465 TF proteins. We have also updated our open-access database, AlfalfaTFDB, into a comprehensive portal that includes classifications of TF families, their descriptions, and coding sequences. In the absence of the broad and inclusive resource on *M. sativa*'s TFs, when only scattered knowledge on individual transcription factors is available through random published studies on individual alfalfa TF families, our data represent an essential information tool, addressing key proteins in the gene regulatory network of alfalfa, the world's most important forage legume.

The newly available draft alfalfa genome allowed reliable identification of alfalfa TF domains that were not previously possible using transcriptome resources due to the presence of only fragmentary sequences/contigs, thus leading to incorrect estimates of the TF distribution. Our data on the composition of TFs in alfalfa are broadly comparable with the makeup of TFs in a close alfalfa relative, *M. truncatula* TFs, bearing in mind a difference in their genome sizes.

Phylogenetic analysis of the largest alfalfa TF families, bHLH, ERF, MYB, NAC, and WRKY, showed that their composition, although similar to *M. truncatula* and a model plant *A. thaliana*, is substantially more diverse. The implication is that, despite the general conservation of alfalfa TFs as compared to other higher plants, the spectrum of their functions and biological roles in response to environmental stimuli can be

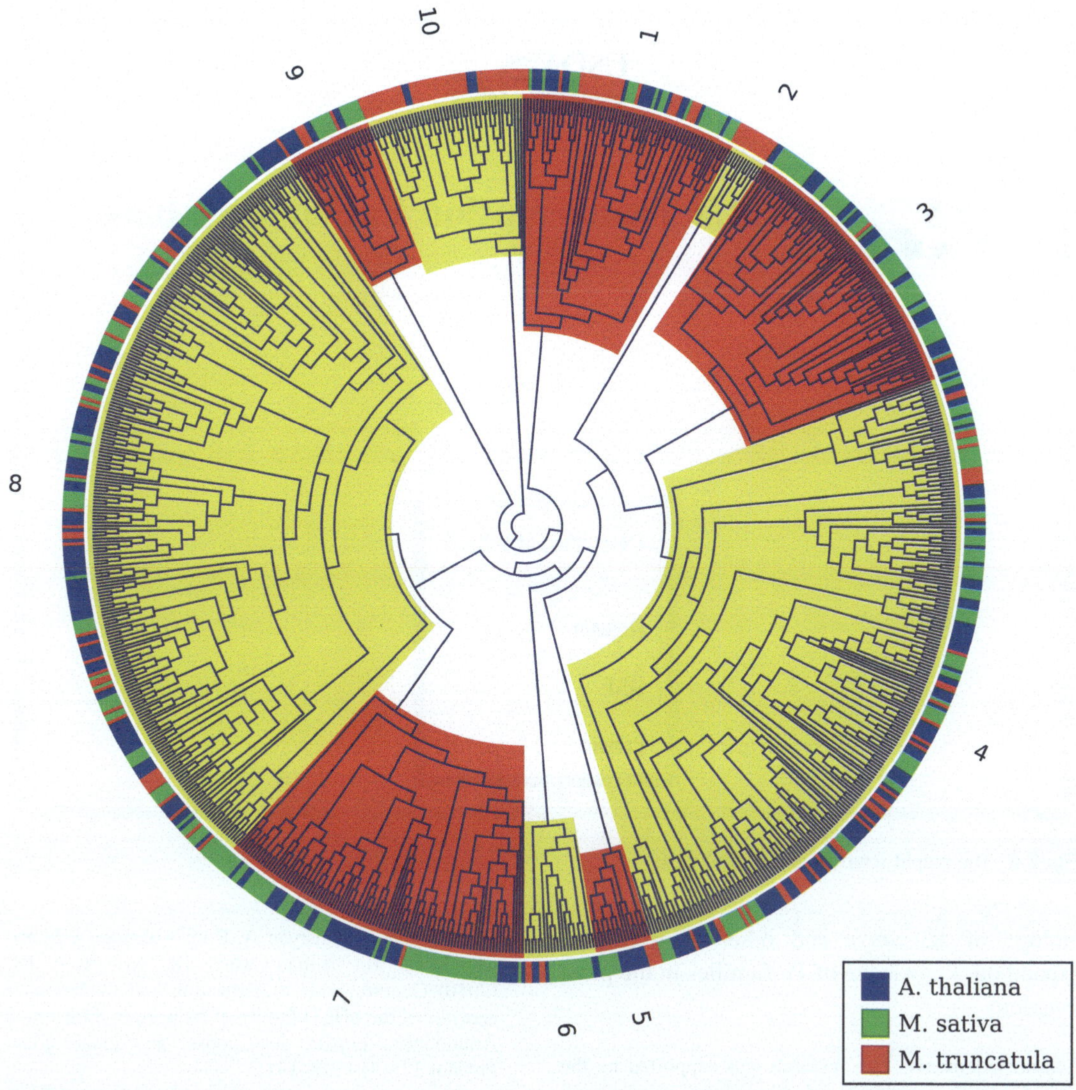

Fig. 7.5 Unrooted phylogenetic tree of alfalfa MYB TFs. Tentative subfamilies are shown in alternating red and yellow backgrounds and numbered with Arabic numerals. *M. sativa*, *M. truncatula,* and *A. thaliana* TFs are shown in green, red and blue red, and green rectangles, respectively

distinct and species-specific. Functional divergence due to expansion of TF families would not be unusual in autotetraploid cultivated alfalfa, whose complicated evolution may have followed routes different from other polyploidy complexes (Havanada et al. 2011).

Overall, the results of this study demonstrate that TF families in alfalfa sustained a pronounced expansion that may indicate sophisticated mechanisms of adaptive responses to environmental stresses (Shiu et al. 2005). The data reported here will further contribute to the

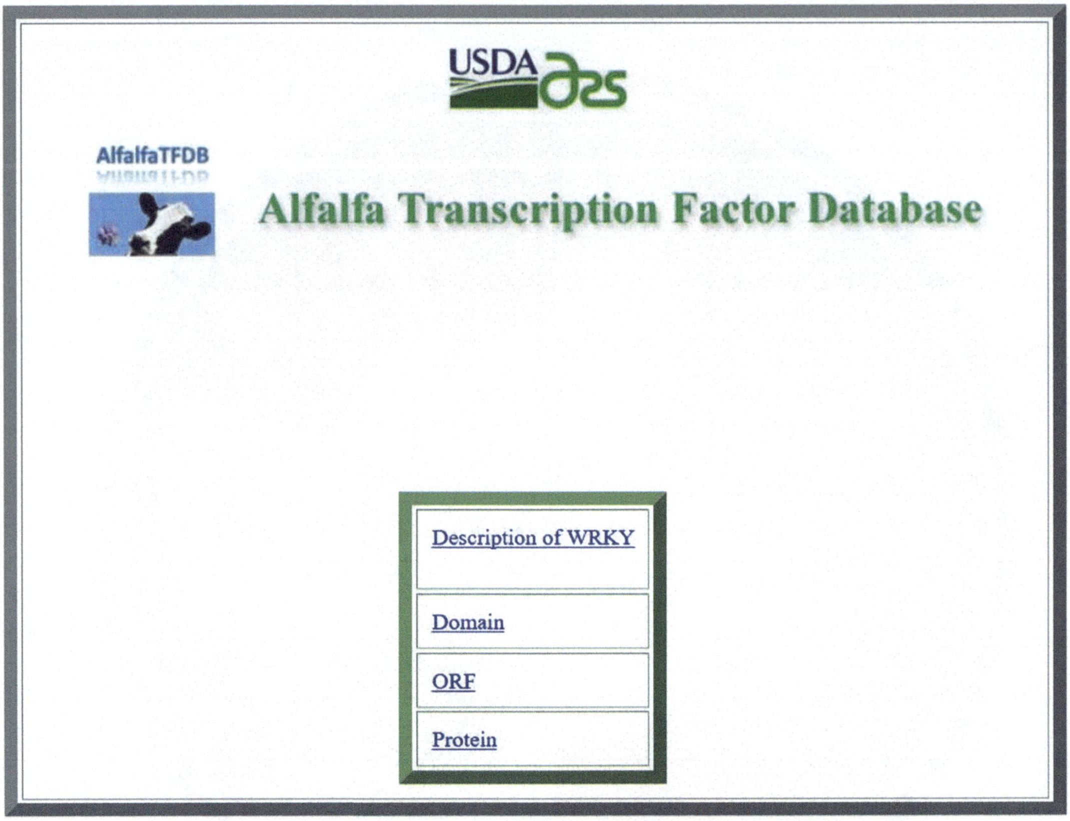

Fig. 7.6 The output interface of AlfalfaTFDB

biology of *M. sativa* and other legumes and expand our knowledge of TF families in the plant kingdom.

Acknowledgments This research was supported by the United States Department of Agriculture, Agricultural Research Service, CRIS project number 8042-21000-300-00D. SCINet resources were provided by the ARS CRIS project number 0500-00093-001-00-D.

References

Anderson JP, Lichtenzveig J, Gleason C, Oliver RP, Singh KB (2010) The B-3 ethylene response factor MtERF1-1 mediates resistance to a subset of root pathogens in *Medicago truncatula* without adversely affecting symbiosis with rhizobia. Plant Physiol 154 (2):861–873

Boyer LA, Langer MR, Crowley KA, Tan S, Denu JM, Peterson CL (2002) Essential role for the SANT domain in the functioning of multiple chromatin remodeling enzymes. Mol Cell 10(4):935–942

Carretero-Paulet L, Galstyan A, Roig-Villanova I, Martínez-García JF, Bilbao-Castro JR, Robertson DL (2010) Genome-wide classification and evolutionary analysis of the bHLH family of transcription factors in Arabidopsis, poplar, rice, moss, and algae. Plant Physiol 153(3):1398–1412

Chen X, Li C, Wang H, Guo Z (2019) WRKY transcription factors: evolution, binding, and action. Phytopathol Res 1(13)

Chen H, Zeng Y, Yang Y, Huang L, Tang B et al (2020) Allele-aware chromosome-level genome assembly and efficient transgene-free genome editing for the autotetraploid cultivated alfalfa. Nat Comm 11(1):2494

Dai X, Sinharoy S, Udvardi M, Zhao PX (2013) PlantTFcat: an online plant transcription factor and transcriptional regulator categorization and analysis tool. BMC Bioinform 14:321

De Mendoza A, Sebé-Pedrós A, Sestak MS, Matejcic M, Torruella G, Domazet-Loso T, Ruiz-Trillo I (2013) Transcription factor evolution in eukaryotes and the assembly of the regulatory toolkit in multicellular lineages. Proc Natl Acad Sci USA 110(50):E4858–E4866

De Zélicourt A, Diet A, Marion J, Laffont C, Ariel F, Moison M, Zahaf O, Crespi M, Gruber V, Frugier F

(2012) Dual involvement of a *Medicago truncatula* NAC transcription factor in root abiotic stress response and symbiotic nodule senescence. Plant J 70(2):220–230

Du H, Yang SS, Liang Z, Feng BR, Liu L, Huang YB, Tang YX (2012) Genome-wide analysis of the MYB transcription factor superfamily in soybean. BMC Plant Biol 12:106

Dubos C, Stracke R, Grotewold E, Weisshaar B, Martin C, Lepiniec L (2010) MYB transcription factors in Arabidopsis. Trends Plant Sci 15(10):573–581

Ernst HA, Olsen AN, Larsen S, Lo Leggio L (2004) Structure of the conserved domain of ANAC, a member of the NAC family of transcription factors. EMBO Rep 5(3):297–303

Eulgem T, Rushton PJ, Robatzek S, Somssich IE (2000) The WRKY superfamily of plant transcription factors. Trends Plant Sci 5(5):199–206

Feller A, Machemer K, Braun EL, Grotewold E, Plant J (2011) Evolutionary and comparative analysis of MYB and bHLH plant transcription factors. 66 (1):94–116

Godiard L, Lepage A, Moreau S, Laporte D, Verdenaud M, Timmers T, Gamas P (2011) MtbHLH1, a bHLH transcription factor involved in *Medicago truncatula* nodule vascular patterning and nodule to plant metabolic exchanges. New Phytol 191(2):391–404

Goossens J, Mertens J, Goossens A (2017) Role and functioning of bHLH transcription factors in jasmonate signalling. J Exp Bot 68(6):1333–1347

Havananda T, Brummer EC, Doyle JJ (2011) Complex patterns of autopolyploid evolution in alfalfa and allies (*Medicago sativa;* Leguminosae). Am J Bot 98 (10):1633–1646

Heim MA, Jakoby M, Werber M, Martin C, Weisshaar B, Bailey PC (2003) The basic helix-loop-helix transcription factor family in plants: a genome-wide study of protein structure and functional diversity. Mol Biol Evol 20(5):735–747

Heyman J, Canher B, Bisht A, Christiaens F, De Veylder L (2018) Emerging role of the plant ERF transcription factors in coordinating wound defense responses and repair. J Cell Sci 131(2):jcs208215

Itzkovitz S, Tlusty T, Alon U (2006) Coding limits on the number of transcription factors. BMC Genom 7:239

Jensen MK, Kjaersgaard T, Nielsen MM, Galberg P, Petersen K, O'Shea C, Skriver K (2010) The Arabidopsis thaliana NAC transcription factor family: structure-function relationships and determinants of ANAC019 stress signalling. Biochem J 426(2):183–196

Jin JP, Tian F, Yang DC, Meng YQ, Kong L, Luo JC and Gao G (2017) PlantTFDB 4.0: toward a central hub for transcription factors and regulatory interactions in plants. Nucl Acids Res 45(D1):D1040–D1045

Jin X, Yin X, Ndayambaza B, Zhang Z, Min X, Lin X, Wang Y, Liu W (2019) Genome-wide identification and expression profiling of the ERF gene family in *Medicago sativa* L. under various abiotic stresses. DNA Cell Biol 38(10):1056–1068

Jones P, Binns D, Chang HY, Fraser M, Li W et al (2014) InterProScan 5: genome-scale protein function classification. Bioinformatics 30(9):1236–1240

Katiyar A, Smita S, Lenka SK, Rajwanshi R, Chinnusamy V, Bansal KC (2012) Genome-wide classification and expression analysis of MYB transcription factor families in rice and Arabidopsis. BMC Genom 13:544

Klempnauer KH, Gonda TJ, Bishop JM (1982) Nucleotide sequence of the retroviral leukemia gene v-myb and its cellular progenitor c-myb: The architecture of a transduced oncogene. The Cell 31(2):

Lata C, Yadav A, and Prasad M (2011) Role of plant transcription factors in abiotic stress tolerance, abiotic stress response in plants. In: Shanker A, Venkateswarlu B (eds) Physiological, biochemical and genetic perspectives. IntechOpen. https://doi.org/10.5772/23172

Li X, Brummer EC (2012) Applied genetics and genomics in alfalfa breeding. Agronomy 2(1):40–61

Libault M, Joshi T, Benedito VA, Xu D, Udvardi MK, Stacey G (2009) Legume transcription factor genes: what makes legumes so special? Plant Physiol 151 (3):991–1001

Ling L, Song L, Wang Y, Guo C (2017) Genome-wide analysis and expression patterns of the NAC transcription factor family in *Medicago truncatula*. Physiol Mol Biol Plants 23(2):343–356

Liu C, Xu X, Kan J, Cheng ZM, Chang Y, Lin J, Li H (2020) Genome-wide analysis of the C3H zinc finger family reveals its functions in salt stress responses of *Pyrus betulaefolia*. PeerJ 8:

Mantiri FR, Kurdyukov S, Lohar DP, Sharopova N, Saeed NA, Wang XD, Vandenbosch KA, Rose RJ (2008) The transcription factor MtSERF1 of the ERF subfamily identified by transcriptional profiling is required for somatic embryogenesis induced by auxin plus cytokinin in *Medicago truncatula*. Plant Physiol 146(4):1622–1636

Mao P, Jin X, Bao Q, Mei C, Zhou Q, Min X, Liu Z (2020). WRKY transcription factors in *Medicago sativa L.*: Genome-wide identification and expression analysis under abiotic stress. DNA Cell Biol. https://doi.org/10.1089/dna.2020.5726. Online ahead of print

Marchler-Bauer A et al (2011) CDD: a Conserved Domain Database for the functional annotation of proteins. Nucleic Acids Res 39(database issue):225–229

Mase K, Ishihama N, Mori H, Takahashi H, Kaminaka H, Kodama M, Yoshioka H (2013) Ethylene-responsive AP2/ERF transcription factor MACD1 participates in phytotoxin-triggered programmed cell death. Mol Plant Microbe Interact 26(8):868–879

Middleton PH, Jakab J, Penmetsa RV, Starker CG, Doll J, Kaló P, Prabhu R, Marsh JF, Mitra RM, Kereszt A, Dudas B, VandenBosch K, Long SR, Cook DR, Kiss GB, Oldroyd GE (2007) An ERF transcription factor in *Medicago truncatula* that is essential for Nod factor signal transduction. Plant Cell 19(4):1221–1234

Min X, Jin X, Zhang Z, Wei X, Ndayambaza B, Wang Y, Liu W (2020) Genome-wide identification of NAC transcription factor family and functional analysis of the abiotic stress-responsive genes in *Medicago sativa* L. J Plant Growth Regul 39:324–337

Mitsuda N, Hisabori T, Takeyasu K, Sato MH (2004) Isolation and characterization of novel vascular plant transcription factors with a one-zinc finger from Arabidopsis thaliana. Plant Cell Physiol 45(7):845–854

Moreau S, Verdenaud M, Ott T, Letort S, de Billy F, Niebel A, Gouzy J, de Carvalho-Niebel F, Gamas P (2011) Transcription reprogramming during root nodule development in *Medicago truncatula*. PLoS One 6:e16463

Müller M, Munné-Bosch S (2015) Ethylene response factors: a key regulatory hub in hormone and stress signaling. Plant Physiol 169(1):32–41

Nakano T, Suzuki K, Fujimura T, Shinshi H (2006) Genome-wide analysis of the ERF gene family in Arabidopsis and rice. Plant Physiol 140(2):411–432

Nemchinov LG, Shao J, Lee MN, Postnikova OA, Samac DA (2017) Resistant and susceptible responses in alfalfa (*Medicago sativa*) to bacterial stem blight caused by Pseudomonas syringae pv. syringae. PLoS One 12(12):e0189781

Olsen AN, Ernst HA, Leggio LL, Skriver K (2005) NAC transcription factors: structurally distinct, functionally diverse. Trends Plant Sci 10(2):79–87

Osorio MB, Bücker-Neto L, Castilhos G, Turchetto-Zolet AC, Wiebke-Strohm B, Bodanese-Zanettini MH, Margis-Pinheiro M (2012) Identification and in silico characterization of soybean trihelix-GT and bHLH transcription factors involved in stress responses. Genet Mol Biol 35(1):233–246

Pandey SP, Somssich IE (2009) The role of WRKY transcription factors in plant immunity. Plant Physiol 150(4):1648–1655

Phukan UJ, Jeena GS, Shukla RK (2016) WRKY transcription factors: molecular regulation and stress responses in plants. Front Plant Sci 7:760

Pires N, Dolan L (2010) Origin and diversification of basic-helix-loop-helix proteins in plants. Mol Biol Evol 27(4):862–874

Postnikova OA, Shao J, Nemchinov LG (2013) Analysis of the alfalfa root transcriptome in response to salinity stress. Plant Cell Physiol 54(7):1041–1055

Postnikova OA, Shao J, Nemchinov LG (2014) In silico identification of transcription factors in *Medicago sativa* using available transcriptomic resources. Mol Genet Genomics 289(3):457–468

Postnikova OA, Hult M, Shao J (2015) Skantar A and Nemchinov LG (2015) Transcriptome analysis of resistant and susceptible alfalfa cultivars infected with root-knot nematode *Meloidogyne incognita*. PLoS One 10(2): https://doi.org/10.1371/journal.pone.0118269

Prouse MB, Campbell MM (2013) Interactions between the R2R3-MYB transcription factor, AtMYB61, and target DNA binding sites. PLoS One 8:e65132

Puranik S, Sahu PP, Srivastava PS, Prasad M (2012) NAC proteins: regulation and role in stress tolerance. Trends Plant Sci 17(6):369–381

Raventós D, Skriver K, Schlein M, Karnahl K, Rogers SW, Rogers JC, Mundy J (1998) HRT, a novel zinc finger, transcriptional repressor from barley. J Biol Chem 273(36):23313–23320

Rushton PJ, Somssich IE, Ringler P, Shen QJ (2010) WRKY transcription factors. Trends Plant Sci 15 (5):247–258

Shiu SH, Shih MC, Li WH (2005) Transcription factor families have much higher expansion rates in plants than in animals. Plant Physiol 139(1):18–26

Siefers N, Dang KK, Kumimoto RW, Bynum WE, Tayrose G, Holt BF (2009) Tissue-specific expression patterns of Arabidopsis NF-Y transcription factors suggest potential for extensive combinatorial complexity. Plant Physiol 149(2):625–641

Stracke R, Werber M, Weisshaar B (2001) The R2R3-MYB gene family in Arabidopsis thaliana. Curr Opin Plant Biol 4(5):447–456

Toledo-Ortiz G, Huq E, Quail PH (2003) The Arabidopsis basic/helix-loop-helix transcription factor family. Plant Cell 15(8):1749–1770

Udvardi MK, Kakar K, Wandrey M, Montanari O, Murray J, Andriankaja A, Zhang JY, Benedito V, Hofer JM, Chueng F, Town CD (2007) Legume transcription factors: global regulators of plant development and response to the environment. Plant Physiol 144(2):538–549

Verdier J, Zhao J, Torres-Jerez I, Ge S, Liu C, He X, Mysore KS, Dixon RA, Udvardi MK (2012) MtPAR MYB transcription factor acts as an on switch for proanthocyanidin biosynthesis in *Medicago truncatula*. Proc Natl Acad Sci USA 109(5):1766–1771

Vernié T, Moreau S, de Billy F, Plet J, Combier JP, Rogers C, Oldroyd G, Frugier F, Niebel A, Gamas P (2008) EFD is an ERF transcription factor involved in the control of nodule number and differentiation in *Medicago truncatula*. Plant Cell 20(10):2696–2713

Vieira P, Mowery J, Eisenback JD, Shao J, Nemchinov LG (2019) Cellular and transcriptional responses of resistant and susceptible cultivars of alfalfa to the root lesion nematode, Pratylenchus penetrans. Front Plant Sci 10:971

Wang H, Avci U, Nakashima J, Hahn MG, Chen F, Dixon RA (2010) Mutation of WRKY transcription factors initiates pith secondary wall formation and increases stem biomass in dicotyledonous plants. Proc Natl Acad Sci USA 107(51):22338–22343

Xie Z, Nolan TM, Jiang H, Yin Y (2019) AP2/ERF transcription factor regulatory networks in hormone and abiotic stress responses in Arabidopsis. Front Plant Sci 10:228

Yamasaki K, Kigawa T, Seki M, Shinozaki K, Yokoyama S (2013) DNA-binding domains of plant-specific transcription factors: structure, function, and evolution. Trends Plant Sci 18(5):267–276

Yang S, Gao M, Xu C, Gao J, Deshpande S, Lin S, Roe BA, Zhu H (2008) Alfalfa benefits from *Medicago truncatula*: The RCT1 gene from *M. truncatula* confers broad-spectrum resistance to anthracnose in alfalfa. Proc Natl Acad Sci USA 105(34):12164–12169

Yu EY, Kim SE, Kim JH, Ko JH, Cho MH, Chung IK (2000) Sequence-specific DNA recognition by the Myb-like domain of plant telomeric protein RTBP1. J Biol Chem 275(31):24208–24214

Zhang Y, Wang L (2005) The WRKY transcription factor superfamily: its origin in eukaryotes and expansion in plants. BMC Evol Biol 5:1

Zhang H, Jin JP, Tang L, Zhao Y, Gu XC, Gao G, Luo JC (2011) PlantTFDB 2.0: update and improvement of the comprehensive plant transcription factor database. Nucl Acids Res 39(database issue):D1114–D1117

Zhao Q, Gallego-Giraldo L, Wang H, Zeng Y, Ding SY, Chen F, Dixon RA (2010) An NAC transcription factor orchestrates multiple features of cell wall development in *Medicago truncatula*. Plant J 63 (1):100–114

Zhou Q, Jia C, Ma W, Cui Y, Jin X, Luo D, Min X, Liu Z (2019) MYB transcription factors in alfalfa (*Medicago sativa*): genome-wide identification and expression analysis under abiotic stresses. PeerJ 7:e7714

Genomics of Forage Quality in Alfalfa

8

Sen Lin, Steve Norberg, and David Combs

Abstract

Alfalfa is called the "Queen of Forages", due to its adaptability in a changing environment, high digestibility and enrichment of multiple nutrients that are beneficial for animal performance and dairy products. However, the forage quality can be quite different among cultivars. In addition, alfalfa quality is affected by various environmental factors. Different methods and a series of parameters have been developed to evaluate alfalfa quality. These methods and parameters can help us quantitatively analyze and compare forage quality from digestibility, protein content, fiber content, energy, and minerals among different cultivars. Given the complexity of factors affecting forage quality, it is very difficult and time-consuming for breeding with traditional methods. Recently developed sequencing techniques combined with bioinformatic software provide the new platform to identify markers and genes associated with traits of interests. This review is focused on studies on genomics of forage quality of alfalfa to provide helpful information for breeding alfalfa with improved forage quality.

S. Lin
USDA-ARS Plant Germplasm Introduction Testing and Research, 24016 North Bunn Road, Prosser, WA, USA

S. Norberg (✉)
Washington State University, 404 West Clark Street, Pasco, WA, USA
e-mail: s.norberg@wsu.edu

D. Combs
University of Wisconsin, 1675 Observatory Drive, Madison, WI, USA

8.1 Introduction

Alfalfa is a highly desirable forage crop because of its high density of nutrients such as protein, vitamins and minerals, and high digestibility of fiber and non-structural carbohydrates (NSC). The definition of "high-quality alfalfa" differs depending on the end users. For horses, alfalfa quality is positively related to palatability, crude protein and NSC and negatively related to neutral detergent fiber (NDF). Plant genetic factors related to fine stems, bright green color, and high protein content are most important to horse owners (Shewmaker et al. 2005). For ruminant sufficient intake of dietary protein and energy is essential for high production of milk, high rate of gain and efficient reproduction. Therefore, the definition of "high-quality alfalfa" in ruminants is less focused on palatability and more on supply of protein and energy to support low diet costs and high levels of production. Beef cattle fed alfalfa as the primary source of feed can support as much as 2.16 lb gain per day, whereas cattle fed diets composed of primarily tall fescue, Sudan grass, *Sericea lespedeza*, orchard grass,

L.-X. Yu and C. Kole (eds.), *The Alfalfa Genome*, Compendium of Plant Genomes,
https://doi.org/10.1007/978-3-030-74466-3_8

white clover, or annual ryegrass support average daily gains lower than 2 lb per day (Ball and Prevatt 2009).

The protein content in late vegetative alfalfa is generally between 18 and 24%, although this number varies among cultivars. Alfalfa contains higher concentrations of protein than grasses or corn silage and ruminant diets formulated with high proportions of alfalfa require far less supplemental protein than diets with high proportions of corn silage, especially for lactating dairy cows.

While protein is an important factor that is defining alfalfa quality from a feed cost perspective, genetic improvement of alfalfa is usually focused on developing cultivars that provide necessary energy to support high levels of milk production and/or rapid daily weight gains. Around 25% of the dry matter of late vegetative alfalfa consists of non-structural carbohydrates (NSC) such as starch, simple sugars, fructan, soluble fiber and in ensiled material, fermentation acids. The NSC feed fraction is calculated as [100 − (CP,% + aNDF,% + fat,% +ash,%)]. The NSC component of alfalfa represents carbohydrates that are rapidly and almost completely digested in the rumen (Robinson 1998).

Plant fiber is found in the cell wall and is composed of cellulose, hemicellulose, and lignin. Total plant fiber is measured as neutral detergent fiber (NDF). The physical and chemical structure of NDF varies due to alfalfa maturity and environmental conditions under which the forage was grown. Alfalfa fiber is digested much more slowly and less completely in the rumen than the NSC fraction. The digestibility of alfalfa fiber can be very different even at the immature stage suggesting that the chemical and physical structure of fiber can vary due to environment and also by genetic variance due to cultivar (Goeser and Combs 2009).

Alfalfa quality decreases with increasing maturity. The content of protein, energy, vitamin, and minerals goes down with aging of alfalfa. After the flowering period, the nutrient loss is even faster. Leaf is the major part contributing the feed value compared with other tissues of alfalfa. Leaves contain 2–3 times the protein content of stems. However, leaf-to-stem ratio decreases gradually during alfalfa maturation. On the other hand, lignification level and indigestible fiber increase with advancing maturity, particularly in the stem fraction. In addition, the crude protein content decreases by about 0.25–0.5% per day beyond the bud stage (Buse 2018). Besides the impact from alfalfa maturity, other environmental factors such as climate, irrigation, and storage conditions can also affect alfalfa quality. Alfalfa grown in higher temperatures usually has a lower forage quality than those grown in cooler conditions (Vough and Marten 1971). Alfalfa has lower ADF, NDF, and acid detergent lignin (ADL) when it is grown in a drought environment, although the values of the three parameters could be still greater than those in other legume species (Peterson et al. 1992).

Alfalfa quality can vary markedly among different cultivars. HarvXtra 118 showed significantly lower ADL and higher neutral detergent fiber digestibility (NDFD) value after 20 days from last harvest, compared with LegenDairy XHD, Hi-Gest 360, and Pioneer 54R02. Lignin content in HarvXtra 118 was over 20% lower than the other 3 check varieties (Smith 2016). In another study that compared forage yield and crude protein content of several alfalfa cultivars, two varieties (Prista 3 and Roxana) had higher crude protein (CP) content than Catinca, Magnat, Madalina, Sandra and Daniela, when grown in the same conditions as Prista 3 and Roxana (Marinova et al. 2018).

8.2 Parameters for Evaluation of Forage Quality in Ruminants

For high producing ruminants, alfalfa quality is generally defined as the ability of the forage to provide digestible energy and support high voluntary feed intake. Around 20–30% of the digestible energy comes from NDF, which accounts for 35–55% of the DM content of alfalfa forage. Digested NFC, CP, and fat provide the remaining digestible energy. Voluntary intake is primarily related to NDF, and is associated with slow and incomplete digestion of fiber in the rumen, which increases gut fill.

8.2.1 Methods for Evaluation of Forage Quality

The quality of alfalfa can be evaluated in different ways. One approach to assessing alfalfa quality is based on measurement of the concentration of feed components that are related to livestock performance, such as fiber, crude protein, carbohydrate, and fat. Unfortunately, animal performance is generally poorly correlated with the concentration of any individual feed component (REF).

Another way to compare quality between alfalfa cultivars is to calculate a feeding value or index which is directly related to the animal performance. Measurement of fiber-related parameters, such as acid detergent fiber (ADF) and neutral detergent fiber (NDF), is useful for indexing feeding values of alfalfa. Relative feed value (RFV) is a quality-related parameter based on ADF and NDF concentrations of alfalfa. Calculation of RFV accounts for the supply of digestible dry matter in alfalfa (DDM), which is an empirical equation calculated from ADF (DDM = 88.9 – (ADF × 0.779), and the dry matter intake (DMI) potential, which is correlated to NDF (DMI = 120 ÷ NDF). The DDM and DMI values are multiplied and the product is divided by 1.29 to create the RFV index value. The RFV is widely used to index and compare alfalfas in hay marketing. The RFV is not used in formulating diets and cannot be used to index or compare quality across different types of forages, such as grass or corn silage.

The RFV index also does not account for how variation in NDF digestibility affects the energy content or intake potential of alfalfa. The digestibility of NDF from forage is positively correlated with voluntary dry matter intake and with increased milk yield in dairy cattle (Oba and Allen 1999). In 2002, scientists at the University of Wisconsin designed another index, relative forage quality (RFQ) for estimating forage quality. The RFQ index accounts for the concentration of NDF and fiber digestibility on intake and uses a summative equation to predict total digestible nutrients (TDN) content of the forage instead of the empirical DDM equation (Undersander and Moore 2002). The RFQ index is an improvement over the RFV index because it uses a summative equation to predict TDN content rather than an empirical ADF-based equation to predict DDM (Weiss et al. 1992). This summative equation is also used to measure TDN concentrations of forages and concentrates and to define energy requirements for dairy cattle (National Research Council 2001).

Fiber digestibility is influenced by forage characteristics such as lignification, leaf-to-stem ratio and plant maturity as well as by diet characteristics and feed intake. Selecting cultivars for high fiber digestibility requires a system of measurement that focuses on plant characteristics related to digestibility. One of the primary plant factors affecting NDF digestibility is lignification. Lignin concentration in forages is negatively correlated to NDF digestibility.

There are two traditional plant breeding methods that have been used to alter the NDF digestibility of alfalfa: altering leaf:stem ratio or reducing the lignin content of the stem material (Buxton 1996; Lee et al. 2011). Alfalfa leaves maintain high NDF digestibility throughout the growth cycle, while the stem material becomes increasingly lignified as the plant approaches full bloom and pod set (Buxton and Hornstein 1986). Lignification occurs in the secondary cell wall and has several different phases of formation in the various cell tissues (Vallet et al. 1996). Fiber digestibility of alfalfa declines as the stem lignifies with advancing maturity and the leaves fall off, as well as when leaf loss occurs during harvest (Albrecht 1983).

Reduced lignin traits have been selected for corn to improve NDF digestibility. The natural recessive brown-midrib (BMR) trait in corn and sorghums reduce the lignin content in leaves and stalks and significantly reduce the proportion of indigestible NDF (**iNDF**) compared to isogenic corn hybrids without the trait (Oba and Allen 2000a). The rate of digestion (**Kd**) of potentially digestible NDF does not appear to differ greatly between BMR and isogenic hybrids of silage corn (Oba and Allen 2000b; Taylor and Allen 2005).

Plant geneticists have attempted to improve alfalfa quality by reducing lignification by either selecting for lines of alfalfa with lower lignin or by genetically modifying lignin biosynthesis (Guo et al. 2001; Lee et al. 2011). Grev et al. (2017) and Getachew et al. (2018) found that NDF digestibility was higher in transgenic reduced-lignin alfalfas as compared to non-transgenic lines of alfalfa.

8.2.2 Assessing Fiber Quality and Digestibility

Alfalfa fiber quality is difficult to assess because many factors affect digestion and utilization of fiber in ruminants. From a plant genetics perspective, the many animal factors that affect fiber digestion (intake, rumen pH, forage particle length, associative effects due to diet) should be considered separately from the chemical/physical factors related to the growth and development of the plant. In vivo measures of plant digestion in ruminants are poorly correlated to fiber digestion, unless animal factors such as intake, diet composition, and rumen environment are carefully defined. In vitro or in situ assays of NDF degradation, however, can be a useful means of comparing NDF digestibility of plant cultivars because the many animal-related factors are standardized. In vitro and in situ analyses are also much cheaper and less time-consuming and lend themselves well to simultaneously evaluating many samples at one time.

There are several in vitro methods that are used to index or compare NDF digestibility among alfalfa lines. The most commonly used approach to compare fiber digestibility is to digest a feed sample under anaerobic conditions in a buffered-rumen fluid solution for a specific period of time (in vitro). These tests are typically described as NDFD (in vitro NDF digestibility as a percentage of total plant NDF, where x denotes the time in hours, of incubation). Residual NDF after long term in vitro incubations of 240 h or more describe the proportion of fiber that cannot be digested in the rumen. An $NDFD_{240}$ value in theory represents the indigestible fiber fraction and would be related to gut fill in ruminants. From a plant genetics perspective, the iNDF fraction is thought to be related to total lignin and how the lignin cross-links with hemicellulose in the secondary cell wall. Low $NDFD_{240}$ values are indicative of higher fiber digestibility and higher quality.

$NDFD_{30}$ and $NDFD_{48}$ values are typically used to index or compare how quickly fiber degrades. Higher $NDFD_{30}$ or $NDFD_{48}$ values suggest higher fiber quality because of faster disappearance of fiber due to microbial digestion. It is not clear why or how rates of NDF digestion vary. Potentially digestible NDF in alfalfa degrades nearly twice as fast as potentially digestible NDF in corn silage. Rates of NDF degradation of potentially digestible NDF in alfalfa samples that contain approximately 40% NDF are on average about 6.5% per hour with a standard deviation of ±1.6% (Combs, unpublished). It is not clear how much the rates of fiber digestion in alfalfa vary due to genotype and how much variation is due to growing conditions and environment.

Differences in fiber digestibility between alfalfa cultivars are due to both the proportion of fiber that is iNDF and the rate at which the potentially digestible fraction of NDF is digested by rumen microbes. For example, on average about 40% (±6.5%) of the total NDF in late vegetative alfalfa is indigestible. High-quality fiber is composed of a relatively low proportion of iNDF and a potentially digestible fiber fraction that is rapidly degraded. It is important to emphasize that $NDFD_{30}$ or $NDFD_{48}$ residues contain both the indigestible fiber fraction and undigested potentially digested NDF, so $NDFD_{30}$ and $NDFD_{48}$ values are not reliable measures of the rate of digestion of potentially digestible NDF. The rate of degradation of the potentially digestible fiber fraction cannot be quantified from a single measure such as $NDFD_{30}$ or $NDFD_{48}$. A quantitative estimate of kd can be measured, however, if the iNDF fraction is measured and the amounts of potentially digestible NDF remaining after a series of short term in vitro assays are measured. Quantifying the iNDF fraction and the rate of degradation of the potentially digestible NDF in forage could potentially provide insights to different genetically controlled biochemical pathways that influence fiber digestibility.

Lopes et al. (2015) published an approach to integrating iNDF and kd of the potentially digestible NDF into a single term and to describe fiber digestion in ruminants. They proposed a method to estimate total tract NDF digestibility (TTNDFD) in lactating dairy cows. A single estimate of NDF digestibility that accounts for the rate of digestion of NDF, the proportion of NDF that is potentially digestible was developed from the in vitro measures of iNDF and kd. These parameters were then incorporated into an animal model to account for how intake and rate of feed passage would impact digestion of NDF. The model provides a means for validating in vitro estimates of fiber digestion with controlled feeding experiments. The TTNDFD of alfalfa can be estimated from the in vitro or in situ measurements of iNDF and kd of alfalfa NDF. RFQ involves fiber digestibility, whereas RFV emphasizes digestible dry matter intake. The RFV index is still widely used as an index to assess quality and compare forage quality among varieties. Other indicators such as daily gains, milk production, wool production can also indirectly reflect the quality of alfalfa. In Table 8.1, we listed several parameters for evaluation of alfalfa quality.

8.3 Genetic Bases of Alfalfa Quality Factors

Given that alfalfa quality is affected by many factors, and forage quality is related to multiple traits, it is a challenging task to genetically enhance alfalfa quality. But recently developed high-throughput sequencing techniques and bioinformatic analysis software provide us new tools to localize the gene locus associated with traits we are interested in. Identification of molecular markers is an important step in marker-assisted breeding for genetic improvement of crops. Single nucleotide polymorphism (SNP) is a type of marker that widely exists among different varieties. SNPs can represent the difference in a DNA region of different cultivars. This kind of difference may play an important role on gene function or expression. High density of SNPs can help us to better understand population structure, genetic diversity and discover locus associated with traits of interest. Owing to the advanced sequencing technologies, a large number of SNPs can be obtained through NGS techniques. One of the high-throughput and highly efficient approaches to discover SNPs is GBS, which has been broadly used for phylogenesis and GWAS in many plant species. Based on the GBS data from *Medicago sativa* and *Medicago truncatula*, a high synteny of genomes between these two close relatives were discovered (Li et al. 2014). Typically, GBS requires a digestion of the whole genome using a frequent-cutter enzyme, such as ApeKI, followed by ligation of adaptors to the digested DNA fragments (Elshire et al. 2011). Different from the traditional genetic mapping, high-density markers obtained from NGS can be used for GWAS. The physical position of significant markers in the genome can be localized. Annotation of the candidate gene associated with specific traits can also be found by BLAST of sequence encompassing the significant marker in the reference genome (Schebenet al. 2017).

Compared to the commercial SNP arrays, sequencing withGBS is cost-saving, time-saving, and easy to automate (Poland and Rife 2012). High-quality SNP markers obtained from GBS can be used for downstream analysis, such as GWAS, which requires a large number of SNP markers to map trait-associated loci in whole-genome level. Although GWAS power can be affected by several factors, such as phenotypic variation, allele frequency, population size and structure, several linear mixed models have been developed and applied in GWAS to reduce false-positive rates of marker-trait association analysis (Yu et al. 2006; Zhang et al. 2010; Lippert et al. 2011; Runcie and Crawford 2019). To date, GWAS has been successfully applied in many crops (maize, rice, soybean, barley, etc.) for identification of loci associated with agronomic traits (Alqudahet al. 2020). In *M. truncatula*,

Table 8.1 Parameters for evaluation of forage quality

Category	Parameter	Full name	Annotation
Ash	Ash	Ash	Total mineral content of forage or diet, including K, Mg, Ca, P, Cu, etc.
Fat	Fat	Fat	Including fat, fat soluble vitamins, and different types of fatty acids
Proteins	ADICP	Acid Detergent Insoluble Crude Protein	Insoluble protein fraction remaining in ADF residual
	CP	Crude Protein	Protein and non-protein nitrogen in forage
	NDICP	Neutral Detergent Insoluble Crude Protein	Insoluble protein fraction remaining after extraction using neutral detergent solution
	RUP	Rumen UNDEGRADABLE PROTEIN	The portion of protein that escapes degradation by ruminal microorganisms and passed into small intestine (Hersom and Carter 2010)
Non-structural carbohydrates	NFC	Nonfibrous Carbohydrates	Starch, sugar, pectin
Structural carbohydrates	ADF	Acid Detergent Fiber	Percentage of highly indigestible material, containing lignin, cellulose, silica
	NDF	Neutral Detergent Fiber	Total plant fiber. Including cellulose, hemicellulose, and lignin
	aNDF	Neutral Detergent Fiber analyzed with amylase	The residue after boiling sample in neutral detergent solution. Amylase is used to remove starch and CP (Ball et al. 2001)
	aNDFom	Amylase Neutral Detergent Fiber organic matter	Contamination by minerals is removed
	ADL	Acid Detergent Lignin	Forage cellulose content can be estimated as ADF minus ADL
	$NDFD_x$	NDF Digestibility	Digestibility coefficient of NDF after a specified time of incubation in rumen fluid. $NDFD_{30}$, for example, would represent the proportion of total NDF that has been digested after 30 h in rumen fluid. Often measured by in situ or in vitro methods. Expressed as a % of the initial amount of NDF. Note that NDFD values include the iNDF and residual undigested pdNDF after a specific incubation time
	iNDF	Indigestible NDF	Indigestible part of NDF. In theory, the proportion of total plant NDF that remains after infinite time in rumen fluid. Typically estimated by incubation of fiber in rumen fluid for 240 h or longer
	RFQ	Relative Forage Quality	An index improved from RFV. RFQ emphasizes NDF digestibility and measures feed energy content as TDN instead of DDM (Undersander and Moore 2002)
	RFV	Relative Feed Value	A forage quality index based on energy. It takes digestibility and potential intake into account
	TTNDFD	Total Tract NDF Digestibility	TTNDFD can be used to compare fiber utilization across forage or fiber sources, and predict how the process of

(continued)

Table 8.1 (continued)

Category	Parameter	Full name	Annotation
			forage fiber digestion is expected to occur in high producing dairy cows (Combs 2014)
	uNDF	Undigested Neutral Detergent Fiber	The residue remaining after a given length of time (usually 30, 120, and 240 h) of digestion (Cotanch et al. 2014)
	NE	Net Energy	Referring to net energy for maintenance (NEM), gain (NEG), and lactation (NEL) (Henning et al. 1996)
	ME	Metabolizable Energy	Net energy remaining after fecal and urinary energy loss.
	NEM	Net Energy for Maintenance	Estimation of energy value to keep a stable weight of an animal
	NEG	Net Energy for Gain	The amount of energy available for animal growth, above that required for maintenance
	NEL	Net Energy for Lactation	Reflecting the available energy amount to support milk production. NEL is calculated from ADF
	TDN	Total Digestible Nutrients	Calculated from ADF, reflecting the percentage of digestible material in a forage (Newman et al. 2009)
	IVDDM	In Vitro Digestible Dry Matter	Determined by incubation of forage samples and rumen fluid at body temperature for a certain period of time (24 to 48 h) (Marten and Barnes 1979)
Moisture	DM	Dry Matter	Non-moisture portion of a feed ingredient or diet
	DDM	Digestible Dry Matter	Reflecting the digestible proportion of forage, calculated from ADF
	DMI	Dry Matter Intake	Reflecting how much dry matter is consumed by a cow per day

GWAS was performed to identify significant markers associated with biomass yield and cell wall biosynthesis using over 15,000 SNPs (Arruda et al. 2016). In alfalfa, GWAS has also been successfully applied to locate quantitative trait loci (QTL) associated with biomass yield and stress tolerance (Ray et al. 2015; Liu and Yu 2017; Yu 2017).

Unlike GWAS designed to discover genomic areas linked to traits of interest, genomic selection (GS) is a method to predict genetic value of selection candidates based on the genomic estimated breeding value (GEBV) (Newell and Jannink 2014). Genomic selection can help to infer the phenotype of crops based on the genotype information, as phenotype is affected by the environment and it requires resources to grow the plants and costs time to measure. For predicting the phenotype of one population, a training population with available genotype and phenotype is required. The data of variations in loci of genotypes and observed phenotypes is used to create a statistical model, which embodies the relation between genotype and phenotype. Then, the model is used to predict the breeding value of individuals with genotypic data available. The feasibility and accuracy of genomic selection has been verified in different crops (Lorenzana and Bernardo 2009; Asoro et al. 2011; Lorenz et al. 2012). There are also studies using genomic selection to predict phenotypes of alfalfa. Annicchiarico et al. (2015b) used genomic selection to predict alfalfa yield in different reference populations. Medina et al. (2020) estimated the breeding value of training populations under salt stress using eight GS models. Understanding how well genomic selection works in alfalfa will help us to predict the performance of alfalfa individuals and accelerate breeding progress.

Alfalfa is an autotetraploid species and highly heterozygous. It is challenging to genotype individuals and develop markers with allele dosage in such a complex genome (Hawkins and Yu 2018). Thanks to recently advanced sequencing technologies and bioinformatic tools, a large progress has been made on understanding the genetic basis of alfalfa and its relationship with desired traits. In this chapter, we will review recent discoveries on alfalfa quality improvement using different genomic tools. These results are helpful for future studies of marker-assisted selection (MAS) and genomic selection. Furthermore, these findings will help us to better understand the genetic basis of forage quality and expedite the breeding progress of high-quality alfalfa.

8.4 Forage Quality Associated Markers Identified by GWAS

Alfalfa has a large genome size which is estimated between 800 and 1000 Mbps (Blondon et al. 1994). It is difficult and complicated to perform genomic and genetic studies in alfalfa due to its out-crossing and autotetraploid features. Diploid *Medicago truncatula* is a self-fertile plant species and close relative to alfalfa. The genome size of *M. truncatula* was around 550 Mbps. Given the high synteny and similarity between these 2 *Medicago* genomes, *M. truncatula* is a model legume species for study of alfalfa (Zhou et al. 2011; Li et al. 2014; Choi et al. 2004). The *M. truncatula* genome can be used to identify the key genomic loci in alfalfa for improvement of alfalfa quality.

Biazzi et al. (2017) sequenced 154 alfalfa genotypes by using GBS. In total 8494 markers were obtained for GWAS by aligning GBS data with *M. truncatul*a genome. Their results demonstrated that alfalfa quality traits were polygenically controlled. In total, 83 significant markers associated with forage quality were mapped on the reference genome. The forage quality traits were measured and compared using alfalfa grown in 3 different conditions including 2 conditions in summer (C1 and C2) and 1 condition in autumn (C3). The quality of alfalfa harvested in C3 was markedly higher than those harvested in C1 and C2. This resulted from higher protein content and NDFD values but lower NDF and ADL values in alfalfa stems grown in autumn. The leaf-to-stem ratio of alfalfa grown in C3 was 1.43, whereas the ratios were 1.06 and 1.09 of those grown in C1 and C2, respectively, demonstrating a better forage quality in autumn compared to summer harvests. Three significant markers associated with leaf-to-stem ratio were identified by GWAS, with 2 on chromosome 3 and the other one on chromosome 4. Forty-two significant markers were identified in association with leaf NDF digestibility in 24 h with the significance threshold setting at 3.0. Two markers were associated with leaf protein content. Interestingly, the markers on the same target trait were different when phenotypic data collected from different tissues were used for GWAS. For example, 6 significant markers associated with CP in the stem were detected, whereas only one was associated with CP in the leaf with the threshold −Log10(P-value) > 4.0. These results implied the genetic control on forage quality traits could be highly variable in different tissues.

Since the genome of tetraploid alfalfa is complex, diploid alfalfa is an ideal material that can be used as an instead for identification of useful genetic markers. Sakiroglu and Brummer (2017) selected 120 accessions of diploid alfalfa including 362 genotypes for GBS sequencing and identification of candidate genes associated with forage quality-related traits. In this study, over 15,000 SNPs were obtained by GBS after genotype calling in Universal Network Enabled Analysis Kit (UNEAK) pipeline and imputation using Random Forest Regression Imputation Model. The results highlighted a significant marker associated with fall ADF, NDF, arabinose, and xylose content. Based on the annotation of the *M. truncatula* genome, this candidate gene is coding nucleotide-diphospho-sugar transferase, which is involved in cell wall biosynthesis. In addition, some of the significant markers (p-value < 0.001) identified in this study were in association with other nutritive value-related traits, such as ADL, total nonstructural carbohydrates (TNC), and glucose content.

High protein content is also a desired trait for alfalfa breeding. A recent study using 85 simple sequence repeat (SSR) markers from 336 genotypes for GWAS reported several marker-trait associations (MTAs) for protein content in alfalfa (Jia et al. 2017). The phenotypic data in this study was collected for a consecutive 3 years (2013, 2014, 2015). Three MTAs in 2013, two in 2014, and eight in 2015 were detected for crude protein content. Additionally, 6, 4, and 2 significant markers were associated with rumen undegraded protein (RUP) when using phenotype data of 2013–2015, respectively.

The content of different types of minerals in alfalfa also acts on the performance of animals. Potassium homeostasis is one of the factors that affect lactation performance of dairy cows (Jarrett et al. 2012). The requirement of potassium is higher than all the other minerals (Grant 1992). It was reported that increasing potassium concentration in the diet helped cows to avoid hypocalcemia, milk fever, and metabolic alkalosis (Goff et al. 2007). Moreover, optimizing concentration of potassium in dairy cows' diet contributes to higher quality of milk with lower trans-fatty acids included (Harrison et al. 2012). Calcium is the second-highest demanded mineral after potassium (Grant 1992). The need for calcium is greatly raised at the beginning of lactation period (Goff 2008). Normal blood calcium concentration in cows is usually between 9 and 10 mg/dL, but this number can drop below 5 mg/dl when cattle are suffering from hypocalcemia (Lamar 2013). Insufficient calcium intake causes hypocalcemia, which is harmful for muscle and nerve systems of cows and may cause some other disorders (Curtis et al. 1985). Jia et al. (2017) also identified 6 significant associations between markers and potassium accumulation in alfalfa using phenotypic data collected in 2013 and 2015. In total 19 significant MTAs were identified on calcium concentration, with 5 markers repeatedly detected in 2 years. Significant markers associated with phosphorus and magnesium were also reported in this study.

To date, a number of markers associated with forage quality have been detected, but very few markers are repeatedly observed in different studies, even on the same quality traits. It is because the genotypes of alfalfa used in each study were different, and each genotype may respond to the environment variously. Moreover, the alfalfa plants in each study were planted in different conditions. Environmental factors, such as light, temperature, and irrigation, can all influence forage quality in various degrees. Some traits are greatly affected by genetics, whereas the other traits are more influenced by environmental factors. Genetics and environment interact with each other and both impact on the phenotypic variation, which is called genotype-environment interaction (G $\times$ E). Studies on G $\times$ E can help to understand complex trait variations. Forage quality traits are also both affected by genetic and environmental factors. Wang et al. (2016) identified 124 MTAs with 5 traits related to fiber content in alfalfa in 4 environments. Of those significant MTAs, only 8 associations were repeatedly observed in 2 environments, indicating the activity of the marker genes in multiple environments. On the other hand, it was also suggested that fiber-related traits were hugely affected by the environment. Lin et al. (2020) identified over one hundred genetic loci associated with 26 forage quality traits using the phenotypic data collected from 198 alfalfa accessions grown in 3 water-deficit conditions (well-watered, mild drought, and severe drought). Genomic libraries of 198 accessions were sequenced on Illumina HiSeq 2000 platform. After filtering missing > 50% and minor allele frequency (MAF) < 0.05 loci, 10,327 SNPs were obtained for association analysis in this study. In the three water-deficit treatments, the mild drought condition intended to decrease fiber content and improve digestibility of alfalfa. Two times of markers were detected in mild drought treatment compared to those identified in severe drought conditions. The authors also compared significant markers identified in different drought environments. Although 133 markers were observed, only nine associated markers showed consistent effects in different water-deficit treatments, demonstrating that most of the associated markers were dependent on water-deficit levels. This study highlighted that genetic regulation on forage quality was influenced by surrounding environmental factors.

8.5 Genomic Selection on Forage Quality

Genetic gain of alfalfa is still low because of the low heritability, strong genotype-environment interaction, and complex genetic architecture (Annicchiarico et al. 2015a). Knowing the breeding value (BV) of an individual is quite helpful for the breeding programs. BV is determined by the average performance of a population and the values of the alleles that can be passed down to the progeny (Falconer and Mackay 1981). However, it is unfeasible to accurately estimate BV of complex traits by using phenotype data only. Thanks to the high-efficiency and low-cost high-throughput sequencing technologies, genomic selection (GS) is a promising strategy and showed great potential on improvement of prediction accuracy on BV (Hayes and Goddard 2001). The basic assumption of GS is that quantitative trait loci tend to be in strong linkage disequilibrium (LD) with the alleles affecting the traits. High-density markers across the genome are desirable for an accurate genomic selection. According to the published results, the accuracy of GS can reach as high as 0.85 (Hayes and Goddard 2001). However, the results of genomic selection can be affected by marker density, population size, statistical model, as well as heritability, minor allele frequency and the genetic architecture of traits (Habier et al. 2007; Heffner et al. 2011; Combs and Bernardo 2013; Zhang et al. 2019; Liu et al. 2018).

Bayesian is one of the frequently used statistical methods on GS. It can be represented by equation as follows:

$$y_i = \mu + \sum_{j=1}^{m} X_{ij}\beta_j + e_i$$

where y_i is the vector of adjusted phenotypic observations; $\mu$is the overall mean; m is the number of markers; B_j is the average effect of allele substitution for SNP j; X_{ij} is the jth SNP genotype of plant i; e_i is the residual with an assumed normal distribution N (0, σ_e^2). Jia et al. (2018) applied three Bayes models on prediction of 25 alfalfa agronomic traits including 15 quality traits and compared the prediction accuracy of the 3 models. But the three Bayes models used in this study did not show a significant difference in prediction accuracy on the quality traits. In this study, over 44,000 SNPs were obtained from 322 alfalfa genotypes for genomic selection after imputation. The prediction accuracy on some quality traits, such as CP, RUP, and ADL were all lower than 0.1. This might be caused by the low heritability of traits. The prediction accuracy on calcium was the highest with the value over 0.34, followed by dNDF48 with the predictability of around 0.25. These results demonstrate the feasibility of genomic selection on forage quality traits, although predictability still needs to be further improved.

Similar results were also reported by Biazzi et al. (2017). Prediction accuracy on NDFD in stem was above 0.3. Notably, it dropped below 0.2 when leaf samples were used for prediction on NDFD. The best prediction was observed on CP content in leaves with the value close to 0.4. However, the accuracy of CP in stem was lower than 0.2. Biazzi et al. (2017) adopted 5 different statistical models for genomic selection, including rrBLUP, SVR-lin, and 3 Bayes models. These models exhibited modest differences in prediction accuracy. SVR-lin model performed better than the other four when predicting leaf-to-stem ratio. For some of the traits, such as ADL in leaf, NDFD in leaf and NDFD in stem, the Bayes models displayed a bit higher capacity on prediction than the others. The model rrBLUP appeared to have a good prediction ability on overall traits averagely. These results suggested that prediction accuracy could be quite different on the same forage quality trait but in different tissues.

8.6 Functional Genes Involved in Regulation of Alfalfa Quality

Most forage digestibility-related traits are directly influenced by the lignin content. Therefore, repressing lignin biosynthesis in alfalfa can technically increase the forage quality. Lignin is a complex organic polymer of monolignol linked by oxidative coupling. It is enriched in cell walls and important in supporting tissues of vascular plants and keeping structural integrity of plants. However, lignin is a negative compound on alfalfa quality because of its indigestibility to livestock. Excess lignin content in alfalfa can devalue forage quality. There are three types of lignin in plants, monomethoxylatedguaiacyl (G), dimethyxylatedsyringyl (S), and p-hydroxyphenyl (H). G and S are two major monolignols existing in most angiosperms. The ratio of S/G in alfalfa increases during maturity of alfalfa (Jung and Vogel 1986; Buxton and Russell 1998). Given that both lignin content and S/G ratio are negatively correlated with forage quality, minimizing these factors is helpful for improvement of alfalfa quality.

Several enzymes have been verified on regulating the levels of substrate in the lignin biosynthesis pathway. Caffeic acid 3-O-methyltransferase (COMT) and caffeoyl CoA 3-O-methyltransferase (CCOMT) play important roles in the formation of monolignols and dimethoxylated lignin precursors. Repression either *COMT* or *CCOMT* result in reduction of lignin content. This was verified in transgenic alfalfa with *COMT* or *CCOMT* down regulated (Guo et al. 2001). Cinnamate 4-hydroxylase (C4H), coumaroyl shikimate 3-hydroxylase (C3H), and ferulate5-hydroxylase (F5H) are three cytochrome P450 enzymes. Knocking down the expression of *C4H, F5H,* or *C3H* significantly reduced lignin content in alfalfa (Reddy et al. 2005). However, the phenotypes of the knockdown lines were different when *C4H, C3H,* or *F5H* was repressed singly. The content of S lignin was strongly decreased in the transgenic lines with *F5H* down regulated, although lignin composition was not significantly impacted. Down-regulation of *C4H* induced an increased ratio of G lignin, meanwhile lower S lignin ratio was also observed in *C4H* down-regulated lines. The H units of lignin were dramatically increased in *C3H* knockdown lines, but the proportion of G lignin was decreased. Acid detergent fiber (ADF) and neutral detergent fiber (NDF) were reduced in the transgenic lines of *C4H, C3H* or *F5H* downregulated, suggesting that repression of these genes resulted in an improved digestibility of alfalfa. To date, only one commercial line (HarvXtra) in the market is lignin-reduced through transgenic technique (Barros et al. 2019). The findings on other genes implicated in lignin biosynthesis as introduced above are hopeful on breeding new forage cultivars with higher nutrition value.

Besides the genes regulating lignin synthesis, genes related to other forage nutrition traits were also reported. Overexpression of a *sucrose phosphate synthase* (*SPS*) gene from maize in alfalfa induced higher sucrose content, chlorophyll content, and photosynthesis rate (Gebril et al. 2015). Additionally, the crude protein in leaves was also raised. Zhou et al. (2011) reported another gene *STAY-GREEN (SGR)* related to CP content in alfalfa. This gene was identified from a mutant line in *M. truncatula. SGR* was involved in nodule development and senescence. Mutation of *SGR* helped to maintain the leaves, as well as anthers pods and seeds of *M. truncatula* green for a longer period. Silencing the expression of *SGR* homologs in alfalfa resulted in a more greenish appearance in the transgenic lines. Furthermore, the crude protein content was also increased by 2.3–5.5%. These results showed the negative effect of *SGR* on forage quality and implied a relationship between leaf senescence and nutrition quality in forage crops. Notably, the increase of crude protein along with chlorophyll increased were reported in both studies, indicating a correlation might exist between chlorophyll and crude protein contents in forage crops.

8.7 Conclusions and Prospects

Alfalfa is the major forage crop because of its high-nutrition feature and high resilience to multiple types of environments. The quality of alfalfa is not only influencing animal performance, but also human being's daily diet. However, tetraploid, self-incompatibility, and high heterozygosity features of alfalfa make it difficult to genetically improve alfalfa quality.

Recent studies have identified a number of markers and genes that are associated with different quality-related traits. These markers are promising for future breeding programs of forage to overall improve alfalfa quality. In addition, two versions of chromosome-level reference genome of alfalfa had been released recently (Chen et al. 2020; Shen et al. 2020). It is predictable that more markers and candidate genes associated with forage nutrition will be identified by application of alfalfa reference genome. These genomic resources will provide a new platform for facilitating the genetic improvement of alfalfa quality as well as other forage traits.

References

Albrecht KA (1983) Studies on nitrogen accumulation, fiber chemistry, and in vitro digestibility of alfalfa. Doctoral dissertation, Iowa State University

Alqudah AM, Sallam A, Baenziger PS, Börner A (2020) GWAS: fast-forwarding gene identification and characterization in temperate cereals: lessons from barley—a review. J Adv Res 22:119–135

Annicchiarico P, Barrett B, Brummer EC, Julier B, Marshall AH (2015a) Achievements and challenges in improving temperate perennial forage legumes. Crit Rev Plant Sci 34(1–3):327–380

Annicchiarico P, Nazzicari N, Li X, Wei Y, Pecetti L, Brummer EC (2015b) Accuracy of genomic selection for alfalfa biomass yield in different reference populations. BMC Genom 16(1):1020

Arruda MP, Brown P, Brown-Guedira G, Krill AM, Thurber C, Merrill KR, Foresman BJ, Kolb FL (2016) Genome-wide association mapping of Fusarium head blight resistance in wheat using genotyping-by-sequencing. Plant Genome 9(1):1–4

Asoro FG, Newell MA, Beavis WD, Scott MP, Jannink JL (2011) Accuracy and training population design for genomic selection on quantitative traits in elite North American oats. Plant Genome 4(2):132–144

Ball D, Prevatt W (2009) Stocker cattle performance and calculated pasture costs. Alabama Coop Ext Sys ANR-1348

Ball DM, Collins M, Lacefield GD, Martin NP, Mertens DA, Olson KE, Putnam DH, Undersander DJ, Wolf MW (2001) Understanding forage quality, vol 1, no 1. American Farm Bureau Federation Publication

Barros J, Temple S, Dixon RA (2019) Development and commercialization of reduced lignin alfalfa. Curr Opin Biotechnol 56:48–54

Biazzi E, Nazzicari N, Pecetti L, Brummer EC, Palmonari A, Tava A, Annicchiarico P (2017) Genome-wide association mapping and genomic selection for alfalfa (*Medicago sativa*) forage quality traits. PLoS One 12(1):

Blondon F, Marie D, Brown S, Kondorosi A (1994) Genome size and base composition in *Medicago sativa* and *M. truncatula* species. Genome 37(2):264–270

Buse K (2018) Widen margins with alfalfa protein. Hay and Forage Magazine

Buxton DR (1996) Quality-related characteristics of forages as influenced by plant environment and agronomic factors. Anim Feed Sci Technol 59(1–3):37–49

Buxton DR, Hornstein JS (1986) Cell-Wall concentration and components in stratified canopies of alfalfa, birdsfoot trefoil, and red clover. Crop Sci 26(1):180–184

Buxton DR, Russell JR (1998) Lignin constituents and cell-wall digestibility of grass and legume stems. Crop Sci 28(3):553–558

Chen H, Zeng Y, Yang Y, Huang L, Tang B, Zhang H, Hao F, Liu W, Li Y, Liu Y, Zhang X (2020) Allele-aware chromosome-level genome assembly and efficient transgene-free genome editing for the autotetraploid cultivated alfalfa. Nat Commun 11(1):1–11

Choi HK, Kim D, Uhm T, Limpens E, Lim H, Mun JH, Kalo P, Penmetsa RV, Seres A, Kulikova O, Roe BA (2004) A sequence-based genetic map of *Medicago truncatula* and comparison of marker colinearity with *M. sativa*. Genetics 166(3):1463–1502

Combs D (2014) Using in vitro total-tract NDF digestibility in forage evaluation. Focus Forage 15(2):1–3

Combs E, Bernardo R (2013) Accuracy of genome-wide selection for different traits with constant population size, heritability, and number of markers. Plant Genome 6(1):1–7

Cotanch KW, Grant RJ, Van Amburgh ME, Zontini A, Fustini M, Palmonari A, Formigoni A (2014) Applications of uNDF in ration modeling and formulation. eCommons, Cornell University, Ithaca, NY

Curtis CR, Erb HN, Sniffen CJ, Smith RD, Kronfeld DS (1985) Path analysis of dry period nutrition, postpartum metabolic and reproductive disorders, and mastitis in Holstein cows. J Dairy Sci 68(9):2347–2360

Elshire RJ, Glaubitz JC, Sun Q, Poland JA, Kawamoto K, Buckler ES, Mitchell SE (2011) A robust, simple genotyping-by-sequencing (GBS) approach for high diversity species. PLoS One 6(5):

Falconer DS, Mackay TF (1981) Introduction to quantitative genetics. Longman Inc., New York, p 104

Gebril S, Seger M, Villanueva FM, Ortega JL, Bagga S, Sengupta-Gopalan C (2015) Transgenic alfalfa (*Medicago sativa*) with increased sucrose phosphate synthase activity shows enhanced growth when grown under N 2-fixing conditions. Planta 242(4):1009–1024

Getachew G, Laca EA, Putnam DH, Witte D, McCaslin M, Ortega KP, DePeters EJ (2018) The impact of lignin downregulation on alfalfa yield, chemical composition, and in vitro gas production. J Sci Food Agric 98(11):4205–4215

Goeser JP, Combs DK (2009) An alternative method to assess 24-h ruminal in vitro neutral detergent fiber digestibility. J Dairy Sci 92(8):3833–3841

Goff JP (2008) The monitoring, prevention, and treatment of milk fever and subclinical hypocalcemia in dairy cows. Vet J 176(1):50–57

Goff JP, Brummer EC, Henning SJ, Doorenbos RK, Horst RL (2007) Effect of application of ammonium chloride and calcium chloride on alfalfa cation-anion content and yield. J Dairy Sci 90(11):5159–5164

Grant RJ (1992) G92-1111 Mineral and vitamin nutrition of dairy cattle

Grev AM, Wells MS, Samac DA, Martinson KL, Sheaffer CC (2017) Forage accumulation and nutritive value of reduced lignin and reference alfalfa cultivars. J Agron 109(6):2749–2761

Guo D, Chen F, Inoue K, Blount JW, Dixon RA (2001) Downregulation of caffeic acid 3-O-methyltransferase and caffeoyl CoA 3-O-methyltransferase in transgenic alfalfa: impacts on lignin structure and implications for the biosynthesis of G and S lignin. Plant Cell 13 (1):73–88

Habier D, Fernando RL, Dekkers JC (2007) The impact of genetic relationship information on genome-assisted breeding values. Genetics 177(4):2389–2397

Harrison J, White R, Kincaid R, Block E, Jenkins T, St-Pierre N (2012) Effectiveness of potassium carbonate sesquihydrate to increase dietary cation-anion difference in early lactation cows. J Dairy Sci 95(7):3919–3925

Hawkins C, Yu LX (2018) Recent progress in alfalfa (*Medicago sativa L.*) genomics and genomic selection. Crop J6(6):565–575

Hayes BJ, Goddard ME (2001) Prediction of total genetic value using genome-wide dense marker maps. Genetics 157(4):1819–1829

Heffner EL, Jannink JL, Iwata H, Souza E, Sorrells ME (2011) Genomic selection accuracy for grain quality traits in biparental wheat populations. Crop Sci 51 (6):2597–2606

Henning JC, Lacefield GD, Amaral-Phillips D (1996) Interpreting forage quality reports. Agron Publ ID-101. University of Kentucky, Lexington, KY

Hersom M, Carter JN (2010) Total protein requirement of beef cattle II: Protein components. EDIS. 2010(2)

Jarrett JP, Taylor MS, Nennich TD, Knowlton KF, Harrison J, Block E (2012) Effect of dietary calcium and stage of lactation on potassium balance in lactating Holstein cows through 20 weeks of lactation. Prof Anim Sci 28(5):502–506

Jia C, Wu X, Chen M, Wang Y, Liu X, Gong P, Xu Q, Wang X, Gao H, Wang Z (2017) Identification of genetic loci associated with crude protein and mineral concentrations in alfalfa (*Medicago sativa*) using association mapping. BMC Plant Biol 17(1):97

Jia C, Zhao F, Wang X, Han J, Zhao H, Liu G, Wang Z (2018) Genomic prediction for 25 agronomic and quality traits in alfalfa (*Medicago sativa*). Front Plant Sci 9:1220

Jung HG, Vogel KP (1986) Influence of lignin on digestibility of forage cell wall material. J Anim Sci 62(6):1703–1712

Lamar KC (2013) Effects of dietary potassium carbonate and fat concentration in high distiller grain diets fed to dairy cows. Doctoral dissertation, Ohio State University

Lee Y, Chen F, Gallego-Giraldo L, Dixon RA, Voit EO (2011) Integrative analysis of transgenic alfalfa (*Medicago sativa* L.) suggests new metabolic control mechanisms for monolignol biosynthesis. PLoS Comput Biol 7(5):e1002047

Li X, Wei Y, Acharya A, Jiang Q, Kang J, Brummer EC (2014) A saturated genetic linkage map of autotetraploid alfalfa (*Medicago sativa L.*) developed using genotyping-by-sequencing is highly syntenous with the *Medicago truncatula* genome. G3—Gene Genom Genet 4(10):1971–1979

Lin S, Medina CA, Boge B, Hu J, Fransen S, Norberg S, Yu LX (2020) Identification of genetic loci associated with forage quality in response to water deficit in autotetraploid alfalfa (*Medicago sativa L.*). BMC Plant Biol 20(1):1–8

Lippert C, Listgarten J, Liu Y, Kadie CM, Davidson RI, Heckerman D (2011) FaST linear mixed models for genome-wide association studies. Nat Methods 8 (10):833–835

Liu XP, Yu LX (2017) Genome-wide association mapping of loci associated with plant growth and forage production under salt stress in alfalfa (*Medicago sativa L.*). Front Plant Sci 8:853

Liu X, Wang H, Wang H, Guo Z, Xu X, Liu J, Wang S, Li WX, Zou C, Prasanna BM, Olsen MS (2018) Factors affecting genomic selection revealed by empirical evidence in maize. Crop J 6(4):341–352

Lopes F, Ruh K, Combs DK (2015) Validation of an approach to predict total-tract fiber digestibility using a standardized in vitro technique for different diets fed to high-producing dairy cows. J Dairy Sci 98(4):2596–2602

Lorenz AJ, Smith KP, Jannink JL (2012) Potential and optimization of genomic selection for Fusarium head blight resistance in six-row barley. Crop Sci 52 (4):1609–1621

Lorenzana RE, Bernardo R (2009) Accuracy of genotypic value predictions for marker-based selection in biparental plant populations. Theor Appl Genet 120(1):151–161

Marinova DH, Ivanova II, Zhekova ED (2018) Evaluation of Romanian alfalfa varieties under the agro-environmental conditions in northern Bulgaria. Banats J Biotechnol 9(18)

Marten GC, Barnes RF (1979) Prediction of energy digestibility of forages with in vitro rumen fermentation and fungal enzyme systems. Standardization of analytical methodology for feeds: proceedings. IDRC, Ottawa, ON, CA

Medina CA, Hawkins C, Liu XP, Peel M, Yu LX (2020) Genome-wide association and prediction of traits related to salt tolerance in autotetraploid alfalfa (*Medicago sativa L.*). Intl J Mol Sci 21(9):3361

National Research Council. Nutrient requirements of dairy cattle (2001) National Academies Press

Newell MA, Jannink JL (2014) Genomic selection in plant breeding. Methods Mol Biol 1145:117–130. https://doi.org/10.1007/978-1-4939-0446-4_10

Newman YC, Adesogan AT, Vendramini JM, Sollenberger LE (2009) Defining forage quality. EDIS 2009(5)

Oba M, Allen MS (1999) Evaluation of the importance of the digestibility of neutral detergent fiber from forage: effects on dry matter intake and milk yield of dairy cows. J Dairy Sci 82(3):589–596

Oba M, Allen MS (2000a) Effects of brown midrib 3 mutation in corn silage on productivity of dairy cows fed two concentrations of dietary neutral detergent fiber: 1. Feeding behavior and nutrient utilization. J Dairy Sci 83(6):1333–1341

Oba M, Allen MS (2000b) Effects of brown midrib 3 mutation in corn silage on productivity of dairy cows fed two concentrations of dietary neutral detergent fiber: 3. Digestibility and microbial efficiency. J Dairy Sci 83(6):1350–1358

Peterson PR, Sheaffer CC, Hall MH (1992) Drought effects on perennial forage legume yield and quality. Agron J 84(5):774–779

Poland JA, Rife TW (2012) Genotyping-by-sequencing for plant breeding and genetics. Plant Genome 5 (3):92–102

Ray IM, Han Y, Lei E, Meenach CD, Santantonio N, Sledge MK, Pierce CA, Sterling TM, Kersey RK, Bhandari HS, Monteros MJ (2015) Identification of quantitative trait loci for alfalfa forage biomass productivity during drought stress. Crop Sci 55 (5):2012–2033

Reddy MS, Chen F, Shadle G, Jackson L, Aljoe H, Dixon RA (2005) Targeted down-regulation of cytochrome P450 enzymes for forage quality improvement in alfalfa (*Medicago sativa L.*). Proc Natl Acad Sci USA 102(46):16573–16578

Robinson PH (1998) What are dairy nutritionists looking for in alfalfa hay. Proceedings, 28th California Alfalfa symposium, Reno, NV, USA

Runcie DE, Crawford L (2019) Fast and flexible linear mixed models for genome-wide genetics. PLoS Genet 15(2):

Sakiroglu M, Brummer EC (2017) Identification of loci controlling forage yield and nutritive value in diploid alfalfa using GBS-GWAS. TheorApplGenet 130 (2):261–268

Scheben A, Batley J, Edwards D (2017) Genotyping-by-sequencing approaches to characterize crop genomes: choosing the right tool for the right application. Plant Biotechnol J 15(2):149–161

Shen C, Du H, Chen Z, Lu H, Zhu F, Chen H, Meng X, Liu Q, Liu P, Zheng L, Li X (2020) The chromosome-level genome sequence of the autotetraploid alfalfa and resequencing of core germplasms provide genomic resources for alfalfa research. Mol Plant 13(9):1250–1261

Shewmaker GE, Undersander D, Lawrence LM, Lacefield GD (2005) Alfalfa: the high-quality hay for horses. Bulletin, National Alfalfa Alliance, Kennewick, WA, USA

Smith S (2016) Reduced or low lignin alfalfa: advantages for hay and grazing. uknowledge.uky.edu

Taylor CC, Allen MS (2005) Corn grain endosperm type and brown midrib 3 corn silage: site of digestion and ruminal digestion kinetics in lactating cows. J Dairy Sci 88(4):1413–1424

Undersander D, Moore JE (2002) Relative forage quality. Focus on Forage. 4(5):1–2

Vallet C, Chabbert B, Czaninski Y, Monties B (1996) Histochemistry of lignin deposition during sclerenchyma differentiation in alfalfa stems. Ann Bot 1996:625–632

Vough LR, Marten GC (1971) Influence of soil moisture and ambient temperature on yield and quality of alfalfa forage. Agron J 63(1):40–42

Wang Z, Qiang H, Zhao H, Xu R, Zhang Z, Gao H, Wang X, Liu G, Zhang Y (2016) Association mapping for fiber-related traits and digestibility in alfalfa (*Medicago sativa*). Front Plant Sci 7:331

Weiss WP, Conrad HR, Pierre NS (1992) A theoretically-based model for predicting total digestible nutrient values of forages and concentrates. Anim Feed Sci Technol 39(1–2):95–110

Yu LX (2017) Identification of single-nucleotide polymorphic loci associated with biomass yield under water deficit in alfalfa (*Medicago sativa L.*) using genome-wide sequencing and association mapping. Front Plant Sci 8:1152

Yu J, Pressoir G, Briggs WH, Bi IV, Yamasaki M, Doebley JF, McMullen MD, Gaut BS, Nielsen DM, Holland JB, Kresovich S (2006) A unified mixed-model method for association mapping that accounts for multiple levels of relatedness. Nat Genet 38 (2):203–208

Zhang Z, Ersoz E, Lai CQ, Todhunter RJ, Tiwari HK, Gore MA, Bradbury PJ, Yu J, Arnett DK, Ordovas JM, Buckler ES (2010) Mixed linear model approach adapted for genome-wide association studies. Nat Genet 42(4):355–360

Zhang H, Yin L, Wang M, Yuan X, Liu X (2019) Factors affecting the accuracy of genomic selection for agricultural economic traits in maize, cattle, and pig populations. Front Plant Sci 10:189

Zhou C, Han L, Pislariu C, Nakashima J, Fu C, Jiang Q, Quan L, Blancaflor EB, Tang Y, Bouton JH, Udvardi M (2011) From model to crop: functional analysis of a *STAY-GREEN* gene in the model legume *Medicago truncatula* and effective use of the gene for alfalfa improvement. Plant Physiol 157(3):1483–1496

Physiological, Morphological, Biochemical, and Genetic Responses of Alfalfa to Salinity

9

Biswa R. Acharya, Devinder Sandhu, and Jorge F. S. Ferreira

Abstract

Alfalfa is an important forage crop known for its perennial nature, high biomass yield, high nutritive value, and high water demand. Global reduction in the availability of clean water in arid and semiarid regions is forcing farmers to irrigate with low-quality degraded/recycled waters, which are generally higher in salts than freshwater. High salt concentrations in soils not only affect soil properties but also result in reduced plant growth and tissue ion toxicities. Plants employ a wide variety of morphological, anatomical, physiological, biochemical, and molecular mechanisms to cope with salinity stress. The self-incompatibility and polyploidy of alfalfa makes the genetic dissection of these complex mechanisms challenging. Understanding these mechanisms is critical to effectively manipulate underlying genetic determinants needed to develop salt-tolerant alfalfa genotypes. This chapter focuses on the effects of salinity on alfalfa growth and development, mechanisms of salinity tolerance in plants, and the current status of the knowledge related to salinity tolerance in alfalfa. Recent studies on the effects of salinity on growth, biomass yield, photosynthesis, water and ion relationships, nutritive value, and antioxidant capacity provide a clear picture of different component traits involved in the salinity response of alfalfa. Based on these studies and the molecular information generated on model plants, tremendous progress has been made to fill the major knowledge gaps in different pathways regulating salt stress responses in alfalfa. This knowledge will facilitate the manipulation of genetic components that would allow alfalfa to thrive when irrigated with recycled/highly saline waters and expand alfalfa cultivation to marginal lands.

9.1 Introduction

Alfalfa (*Medicago sativa* L.) is an important forage crop that is grown worldwide due to its perennial nature, wide adaptability, high biomass yield, resistance to multiple cuttings, and tolerance to various biotic and abiotic stresses (Riday and Brummer 2002; Singer et al. 2018). Additionally, alfalfa has high nutritive value, protein content, and palatability to ruminants. It can be used fresh for animal grazing or dehydrated to hay and meal pellets. Alfalfa is the 3rd most valuable field crop (US$10.8 billion) in the

B. R. Acharya · D. Sandhu (✉) · J. F. S. Ferreira
US Salinity Laboratory (USDA-ARS), 450 West Big Springs Rd., Riverside, CA 92507, USA
e-mail: devinder.sandhu@usda.gov

B. R. Acharya
College of Natural and Agricultural Sciences, University of California Riverside, Riverside, CA 92507, USA

L.-X. Yu and C. Kole (eds.), *The Alfalfa Genome*, Compendium of Plant Genomes,
https://doi.org/10.1007/978-3-030-74466-3_9

United States, behind corn grain (US$52.9) and soybean (US$31.2), according to the 2019 data (https://www.nass.usda.gov/).

Water is the most crucial resource for agriculture. With the reduced availability of good quality irrigation water, the demand for alternate/degraded waters, which have higher salt concentrations than freshwater, is growing. Salinity is a mounting threat to global agriculture. About one billion hectares of arable land is affected by salinity/sodicity (Shahid et al. 2018), and millions of hectares of land are destroyed each year due to salinity worldwide (Pimentel et al. 2004). It has been predicted that about 50% of the cultivated land may be affected by salinity by 2050 (Bartels and Sunkar 2005).

Earlier studies on alfalfa classify it as moderately sensitive to salinity, based on the electrical conductivity of soil-paste (EC_e) with a salinity threshold of 2.0 dS m^{-1} (Maas and Hoffman 1977). However, recent studies showed that alfalfa could produce high biomass yield with an EC_e of 11.0 dS m^{-1} (equivalent to $EC_{iw} = 24$ dS m^{-1}) while maintaining its nutrient composition and antioxidant capacity, and with a slight but significant increase in forage crude protein and quality (Cornacchione and Suarez 2017; Ferreira et al. 2015). These observations propose that alfalfa is moderately tolerant to salinity, and developing new salt-tolerant cultivars may lead to the expansion of alfalfa cultivation to marginal lands, unsuitable for producing staple crops. To realize this goal, the understanding of genetic determinants regulating morphological, physiological, and biochemical traits responsible for salinity tolerance in alfalfa is crucial. Genetics studies in alfalfa indicate considerable variation in salinity tolerance in alfalfa (Cornacchione and Suarez 2017; Sandhu et al. 2017; Yu et al. 2016).

9.2 Effects of Salinity on Alfalfa

9.2.1 Growth

Salinity typically induces osmotic stress (first phase) and ionic stress (second phase) that directly impact biochemical, physiological, and molecular plant responses. Osmotic stress limits plants' ability to uptake water from the soil, while the ionic stress results in the build-up of toxic levels of salt ions. These, together, lead to stunted plant growth and development (Munns and Tester 2008). In alfalfa, ionic toxicity is primarily caused by Na^+ and, to some extent, by Cl^-. For plant species sensitive to salinity, the salt concentration does not reach a level to cause osmotic stress (Sandhu and Kaundal 2018). In those cases, ionic stress is the key to suppressing plant growth. In alfalfa, it has been shown that some genotypes have a high salt tolerance index (STI) (performance under salt/performance under control conditions) for biomass yield resulting in almost no reduction when irrigated with saline water of high electrical conductivity ($EC_{iw} = 17$ dS m^{-1}) (Sandhu et al. 2017). Hence, for alfalfa, both the osmotic and ionic stresses are critical during salinity stress.

Roots are initially exposed to the high salinity of water or soil. Then, the stress signal is distributed throughout the plant. Leaves are ultimately affected, leading to several modifications under stress. Plants employ a wide variety of morphological, anatomical, physiological, molecular, and biochemical mechanisms to cope with salinity stress. Alfalfa responds to salinity stress by reducing young growth, leaf size and by accelerating the senescence of older leaves (Cornacchione and Suarez 2015). On the other hand, roots are more tolerant to salinity compared to shoots and suffer a smaller reduction in biomass than shoots. In alfalfa, biomass yield and forage quality are important traits that must be maintained in salinity-tolerant cultivars.

9.2.2 Shoot Biomass

Biomass yield is a complex trait regulated by several genes in a tetraploid species, such as alfalfa. On top of that, various biotic (e.g., insect or fungal attack) or abiotic stresses (e.g., drought, heat, salinity) stresses make it harder to dissect the genetic networks underlying this trait. High salinity of irrigation water or soil is known to reduce alfalfa shoot biomass, and that has been reported by several authors cited in a recent review (Al-Farsi et al. 2020b). Most researchers who evaluated several genotypes agree that there is a significant variability in alfalfa response to salinity (Al-Farsi et al. 2020a; Cornacchione and Suarez 2017; Sandhu et al. 2017). Researchers at the US Salinity Laboratory (USDA-ARS) in California evaluated 12 genotypes (G01-G12) under greenhouse conditions with EC_{iw} of 2.0 (control) and 17.0 dS m^{-1} (saline treatment) for 18 months (Sandhu et al. 2017). In this long-term study, biomass was determined based on the average of 11 harvests. Salinity impact on plant growth was genotype-dependent, and the maximum reduction recorded under salinity was 61% for G06, while G03 and G10 had no reduction in biomass. Interestingly, G02 had a significant reduction in biomass at 17.0 dS m^{-1} compared to control, but its biomass under salinity was slightly higher than G03 and G10. Thus, under salinity, G02, G03, and G10, respectively, were the most productive in shoot biomass (Sandhu et al. 2017). The genotypes G03 and G10 also had the highest STI. The performance of the best genotypes was consistent with their mother plants, with shoot biomass inversely correlated with their shoot accumulation of both Na and Cl (Sandhu et al. 2017). These authors also reported that the reduction in biomass under salinity was primarily attributed to the number of shoots per plant rather than plant height. Hence, high branching ability under salinity may be an important morphological trait to be considered while breeding for salinity tolerance in alfalfa.

When evaluating a perennial forage such as alfalfa, it is important to stress that long-term experiments of at least one year, preferably two, should be conducted to allow plants to establish and adjust to salinity. Also, one should use the same number of cuts used in the region where the genotype is cultivated. Although greenhouses in semiarid and arid areas usually get hot and dry during summer months, outdoor studies should be done to confirm that plants will perform as expected under extreme heat and evapotranspiration during summer months.

9.2.3 Photosynthesis

The efficiency of photosynthesis determines growth rate under stress conditions. The effect of salinity stress on photosynthesis in alfalfa is dependent on multiple factors like genotype, salinity level, developmental stage, and growth conditions. Plants irrigated with high-salinity water develop smaller leaves, darker in color due to the higher chlorophyll concentration (Sandhu et al. 2017). A greenhouse study was conducted with 15 commercial alfalfa genotypes from Argentina and the USA, with waters dominated by sulfate or chloride and salinities ranging from 0.85 to 24.0 dS m^{-1} (Cornacchione et al. 2018). This study revealed no difference between sulfate- or chloride-dominated waters for photosynthetic rate (*Pn*), stomatal conductance (*gs*), or transpiration rate (*Tr*). However, increased salinity significantly reduced leaf area (Cornacchione et al. 2018). Also, a highly positive correlation was found between the specific leaf weight and chlorophyll ($R^2 = 0.9653$), suggesting that, under high salinity, the leaves became thicker (with a thicker palisade layer that could have contained more chloroplasts per leaf unit area). Still, in this study, *Pn* increased with salinity while leaf area was reduced, suggesting that this reduction in *Pn* with salinity could be either "the cause or the effect" of the decreased biomass per plant.

9.2.4 Mineral Imbalance

Mineral balance in plant cells plays a critical role in plant growth and development. High concentrations of salt ions, including Na^+ and Cl^-, in irrigation water or soil, are known to affect the mineral balance of other nutrients necessary for plant growth and development, such as the previously reported Na^+ antagonism to K^+ and Ca^{2+} and Cl^- antagonism to NO_3^- uptake (Maas and Grattan 1999). Most alfalfa salinity studies show increased tissue concentrations of Na and decreased concentrations of K and Ca under salinity (Cornacchione and Suarez 2015; Ferreira et al. 2015; Cornacchione and Suarez 2017; Sandhu et al. 2017). K is an important nutrient and an important cell osmoticum, whereas Ca is critical for plasma membrane and cell wall stability. Although salinity may lead to reduced accumulation of some mineral nutrients, alfalfa irrigation with saline waters of $EC_{iw} = 24$ dS m^{-1} and mimicking the waters of the California Central Valley, and of ground-water-fed wells of the islands of Lanzarote and Fuerteventura, Spain ($EC_{iw} = 10$ dS m^{-1}), resulted in darker green shoots, which had a significantly higher N%, crude protein, and forage quality (Ferreira et al. 2015; Díaz et al. 2018). The salinity experiment performed in California lasted for 300 days, with plants being cut monthly with shoot mineral analysis and forage values performed at 84 days after sowing (DAS) and at 299 DAS (Ferreira et al. 2015). EC_{iw} ranged from 3.1 dS m^{-1} (control) to 30 dS m^{-1}. Although concentrations of N, P, Mg, and S increased significantly at $EC_{iw} = 18$ dS m^{-1}, those of K and Ca decreased (Ferreira et al. 2015). Interestingly, these authors also reported that Zn, Mo, and Mn increased significantly in shoots with salinity. The plants likely absorbed more micronutrients to boost their defenses against the salinity stress. None of the plants, even at 24 dS m^{-1}showed visual symptoms of mineral deficiency. For the salinity experiment performed in Spain, the relative biomass reduced by 50% under EC_{iw} of 5 dS m^{-1}, but forage crude protein increased, slightly increasing the relative forage value (RFV) and metabolizable energy (ME) (Díaz et al. 2018). Shoot mineral concentrations for N, Mg, S, and Na increased with salinity, while K and Ca decreased. The interplay of salt ions among themselves and with minerals is complex, and care should be taken to ensure that the plant has enough macro and micronutrients to sustain its growth and biomass production.

9.2.5 Forage Nutritional Value and Antioxidants

It is widely accepted that alfalfa is the most important forage crop used as a high protein source for dairy and beef cattle, sheep, horses, and birds, among other livestock (Radović et al. 2009). Besides protein, alfalfa silage contains higher concentrations of minerals compared to other cereals (Khorasani et al. 1997). Also, according to these authors, cows absorb more potassium (89%) from alfalfa silage than from cereal silages (74–83%). However, alfalfa demands a high volume of water in arid and semiarid lands to produce high forage biomass from several cuts a year. Among approximately 270 million hectares of irrigated land worldwide, close to 40% is located in arid/semiarid zones (Smedema and Shiati 2002). This irrigation, even if done with low-salinity water, is linked to increased salinity of irrigated lands. Although alfalfa is more tolerant to salinity than previously reported and can produce a reasonable amount of biomass under salinities of irrigation water up to 24 dS m^{-1} (Cornacchione and Suarez 2015, 2017; Ferreira et al. 2015), there are very few reports on the effect of salinity on mineral composition, forage parameters, and antioxidant capacity under salinity. We would like to emphasize that we are not using the term "nutritional value" because that implies evaluating the forage in livestock and their feed conversion into weight gain and milk production. These studies are more challenging to perform as they involve evaluating the forage through livestock parameters, and we found none with alfalfa in the literature. Thus, we cite the works that have been done evaluating mineral composition and/or antioxidant capacity of alfalfa biomass under salinity. Field-cultivated alfalfa in the San Joaquin Valley was reported to produce over 20 t/acre when irrigated with freshwater, but its yield dropped to a bit over 16 t/acre when the alfalfa cultivar (Salado/801S) was irrigated with drainage water of 4.7–6.9 dS m^{-1} (Suyama et al. 2007). These authors reported that alfalfa irrigated for five years with these drainage waters had 4–10 mg kg^{-1} of Se, highly present as selenite in the soil used for the experiment. The maximum limit of Se tolerated by most ruminants in forage is 2.0 mg kg^{-1} (NRC 2000).

Although mineral nutrients may be added to an animal diet through other feed sources, it is crucial to evaluate the mineral composition of alfalfa cultivated with water of elevated salinity. Four commercial alfalfa cultivars grown in sand tanks for ten months and irrigated with water salinities ranging from EC_{iw} = 3.0 (control) to 24.0 dS m^{-1} maintained sufficient concentrations of mineral nutrients for their growth and biomass accumulation (Ferreira et al. 2015). This study on the "potential nutritional value" of alfalfa reported that crude protein significantly increased at 18 dS m^{-1} in both harvest dates, while relative forage value (RFV) increased significantly for EC_{iw} = 18 dS m^{-1} at 84 DAS but remained constant at 299 DAS (Ferreira et al. 2015). However, after three years of field growth, "Salado/801S" alfalfa irrigated with either freshwater (EC_{iw} = 4.7 dS m^{-1}) or drainage water (EC_{iw} = 6.9 dS m^{-1}) had similar shoot concentrations of K, Ca, Mg, and P, and twice as much NO_3 nitrogen when irrigated with 6.9 dS m^{-1} drainage water (Suyama et al. 2007).

9.3 Mechanisms of Salinity Tolerance in Plants

Nonselective cation channels (NSSCs) are the primary route through which Na^+ enters roots. The mechanisms by which plants sense salt upon exposure are not well understood. Recently, an extracellular salt sensor, MONOCATION-INDUCED $[Ca^{2+}]$I INCREASES 1 (MOCA1), has been identified, which senses Na^+ and some other monovalent cations (Jiang et al. 2019). MOCA1 is responsible for the production of glycosyl inositol phosphorylceramide (GIPC) sphingolipids at the plasma membrane. Upon binding with monovalent cations, GIPCs can interact with the Ca^{2+} channel, increasing the intracellular cytosolic Ca^{2+} concentration. An increase of cytosolic Ca^{2+} level activates the Salt Overly Sensitive (SOS) pathway (Zhu 2000). SOS3 is a calcium-binding protein (calcineurin B-like protein 4) that senses increased calcium concentration in the cytosol of plant cells in response to salt stress. Binding of SOS3 to Ca^{2+} promotes interaction between SOS3 and SOS2 (calcium-induced protein kinase 24), and this process activates SOS2 protein kinase. Subsequently, active SOS2–SOS3 complex phosphorylates SOS1 at Ser^{1044} position that activates SOS1 by relieving from an autoinhibitory state of SOS1 (Quintero et al. 2011). Active SOS1 extrudes Na^+ from cytosol to outside. The SOS pathway is conserved in several plant species (Munns and Tester 2008; Zhao et al. 2020).

The movement of ions from the cytoplasm to the vacuole is critical in keeping ion concentration low in the cytoplasm. Tonoplast -localized Na^+/H^+ exchangers (NHX), including NHX1 and NHX2, mediate sequestration of Na^+ in the vacuole leading to enhanced salinity tolerance (Apse et al. 1999). It has been suggested that both NHX1 and NHX2 also play roles in K^+ homeostasis as they have equal affinity for both Na^+ and K^+ ions (Maathuis et al. 2014). These findings suggest that NHX1 and NHX2 would import Na^+ into the vacuole when the cytoplasmic concentration of Na^+ is higher than K^+. Na^+/K^+ homeostasis plays an important role in salinity tolerance in plants. Multiple potassium channels and transporters have been implicated in K^+ homeostasis during salinity stress, including AKT1, GORK, TPK1, HAK5, CHX17, NHX5, and NHX6 (van Zelm et al. 2020).

High-affinity potassium transporters (HKTs) play vital roles in ion transport during salinity stress in plants (Horie et al. 2009). Substitution of a single amino acid in the second pore-loop region determines affinity for Na^+ or K^+. These transporters belong to two subclasses: HKT1 has an affinity for Na^+, and HKT2 has an affinity for K^+. In Arabidopsis, HKT1 has been shown to inhibit root-to-shoot Na^+ transport but promotes Na^+ transport from shoot-to-root and provides tolerance to salt stress by keeping Na^+ content low in leaves (Kaundal et al. 2019).

There are many additional signaling pathways and elements that contribute to salinity tolerance like osmolytes, ROS signaling, hormone signaling, IP3 signaling, and sugar signaling (Park et al. 2016; van Zelm et al. 2020). Import of Na^+ also increases production of reactive oxygen species (ROS) and 3′, 5′-cyclic guanosine monophosphate (cGMP) levels. Ca^{2+}, cGMP, and ROS serve as important secondary messengers during salinity stress. cGMP inhibits the import of Na^+, reduces K^+ efflux, and increases Ca^{2+} import. Increase of Ca^{2+} promotes ROS production mediated by RbohD, RbohF, and RbohJ. Ca^{2+} and ROS form a positive feedback loop in the signaling pathway (van Zelm et al. 2020).

9.4 Progress Toward Understanding Salinity Tolerance in Alfalfa

Several genes have been identified that play critical roles in salinity tolerance in alfalfa, indicating that salinity stress tolerance is a multigenic trait (Bhattarai et al. 2020). Here, we discuss the importance of different component traits of the salinity tolerance mechanisms in alfalfa.

9.4.1 Transporters

Several transporter proteins are critical during salinity stress. The salt overly sensitive (SOS) pathway that consists of three important proteins (SOS1, SOS2, and SOS3) plays an important role in the removal of Na^+ from the cytoplasm in response to salinity stress (Zhu 2000). *MsSOS1*, *MsSOS2*, and *MsSOS3* were upregulated in response to salt stress in salt-tolerant genotypes (Sandhu et al. 2017). Coexpression of Arabidopsis SOS pathway genes in alfalfa has been shown to enhance salt tolerance in alfalfa by enhancing plant growth, lowering Na^+ accumulation, increasing K^+ accumulation in the leaves, increasing proline content, and enhancing activities of antioxidant enzymes (Wang et al. 2019). Additionally, the SOS core pathway in alfalfa was shown to be activated in response to ionic stress (Luo et al. 2019b). These findings indicate that the SOS pathway is also conserved in alfalfa.

Upregulation of *NHX1*, *HKT1*, and *AKT1* was observed in salt-tolerant genotypes of alfalfa, suggesting that these genes play a vital role in Na^+/K^+ homeostasis in alfalfa (Sandhu et al. 2017). The complementation of *NHX* genes from various plant species including *SeNHX1* (*Salicornia europaea NHX1*), *SsNHX1* (*Suaeda salsa NHX1*), and *TaNHX2* (*Triticum aestivum NHX2*) have been shown to provide tolerance to salt stress in alfalfa (Hrbáčková et al. 2020; Zhang et al. 2012). Arabidopsis vacuolar proton-pumping pyrophosphatase (H^+-PPase) gene *AVP1* functions with vacuolar H^+-ATPase (V-ATPases) to establish a difference in electrochemical potential for H^+ across the tonoplast, which helps in the sequestration of Na^+ into the vacuole (Gaxiola et al. 2001; Schilling et al. 2017). Constitutive expression of *AVP1* protects various plants from salinity stress and improves growth and yield. Overexpression of Arabidopsis *AVP1* in alfalfa enhanced salinity tolerance and membrane integrity in response to 200 mM NaCl (Bao et al. 2009) and resulted in the accumulation of more cations (Na^+, K^+, and Ca^{2+}) than wild-type plants in response to salinity stress (Schilling et al. 2017).

9.4.2 Osmoregulators

Salinity stress induces osmotic stress that leads to loss of turgor pressure, which in turn causes changes in membrane structure, leading to membrane leakage (Zhu 2002). To maintain turgor pressure in the cell during osmotic stress, plants synthesize compatible organic solutes (or osmolytes) and inorganic solutes for osmotic adjustment (Zhu 2002). Common solutes synthesized by plants include soluble sugars (fructose and sucrose), complex sugars (raffinose, trehalose, and fructans), polyols (pinitol, mannitol, glycerol, arabinitol, and sorbitol), charged metabolites (betaines, prolines, aspartate, glutamate, glycine, putrescine, choline, malate, and oxalate), specific proteins, and ions (mainly K^+). Some osmolytes also serve as osmoprotectants, antioxidants, and as signaling molecules (Park et al. 2016). Overexpression of *Glycine soja* S-locus LecRLK gene *GsSRK* in alfalfa enhanced salinity tolerance by controlling osmotic regulation, ion homeostasis, and ROS scavenging (Song et al. 2019; Sun et al. 2018). Salt-tolerant alfalfa genotypes are known to accumulate more proline than their corresponding sensitive cultivars (Torabi and Halim 2010). *P5CS1*, a gene required for proline biosynthesis, was upregulated in salt-tolerant alfalfa cultivars compared to salt-sensitive ones (Sandhu et al. 2017). Further, by employing a biotechnological approach, it was shown that coexpression of Arabidopsis *SOS1*, *SOS2*, and *SOS3* in alfalfa enhances proline biosynthesis and salinity tolerance (Wang et al. 2019). Additionally, increased soluble sugar levels have been linked with salinity tolerance in alfalfa (Rahman et al. 2015). A positive association has been shown between higher levels of accumulation of pinitol and sucrose in leaves and starch in roots of salt-tolerant genotype of alfalfa in response to salt stress (Bertrand et al. 2015). Alfalfa plants also synthesize osmoprotectant Pro betaine during salinity stress (Trinchant et al. 2004).

9.4.3 ROS Production, Oxidative Stress, and Antioxidant System

Excess production of ROS in plants in response to salinity stress is a secondary effect due to various factors, including water deficit induced stomatal closure, increase of leaf and plant temperature, CO_2 deficit, and photosynthesis inhibition (Gill and Tuteja 2010). Salinity stress induces accumulation of H_2O_2 (hydrogen peroxide), O_2^- (superoxide radicals), and free radicals in cellular regions, including mitochondria, chloroplast, and apoplastic space. Overproduction of ROS during salinity stress (and in response to other stresses) severely affects the balance between oxidants and antioxidants (Gill and Tuteja 2010). Salinity stress-induced oxidative stress has an additive negative effect on different crop species, including alfalfa.

Plants produce antioxidants to remove ROS that in turn helps to maintain balance between oxidants/antioxidants. Plants produce both non-enzymatic antioxidants (ascorbic acid, glutathione, vitamin E, phenolics, etc.) and enzymatic antioxidants (catalase, peroxidase, superoxide dismutase, etc.) (Ashraf 2009). These facts indicate that, in addition to non-enzymatic antioxidants, the expression status of detoxifying enzymes may determine salinity tolerance of different cultivars of alfalfa. In comparison to control, salt-treated alfalfa seedlings show higher activities for antioxidative enzymes catalase (CAT) and peroxidase (POD); and higher accumulation of antioxidant reduced glutathione (Luo et al. 2019a). A recent study comprising proteomics and metabolomics studies revealed that in response to salinity stress in alfalfa, ROS scavenging systems including CAT, POD, and proteins associated with glutathione metabolism were upregulated, which indicated that the antioxidant system played critical roles in protecting alfalfa in response to salinity stress (Li et al. 2020). Coexpression of Arabidopsis SOS pathway genes in alfalfa has been shown to enhance activities of antioxidant enzymes, CAT, POD, and superoxide dismutase (SOD), and led to enhanced salt tolerance (Wang et al. 2019).

Salt stress-induced lipid peroxidation is mediated by ROS. The level of MDA indicates the status of general peroxidation. An increase in ion leakage, malondialdehyde (MDA) level, and H_2O_2 has been observed in response to salt stress in alfalfa (Luo et al. 2019a). Under salinity, a salt-tolerant alfalfa accumulated less H_2O_2 and a lower level of MDA in comparison to the susceptible variety due to activation of the antioxidant enzymes or H_2O_2 scavengers (Rahman et al. 2015).

9.4.4 Phytohormones

Multiple phytohormones are known to play roles in response to salinity stress, including abscisic acid (ABA), auxin, ethylene, gibberellins (GA), brassinosteroids (BRs), jasmonic acid (JA), and melatonin (Kaundal et al. 2021; van Zelm et al. 2020). The role of ethylene in salinity tolerance is dependent on species. For example, ethylene is linked to salinity tolerance in Arabidopsis by retaining K^+ in roots and shoots (Yang et al. 2013). In contrast, it appears that ethylene plays a negative regulatory role in salinity tolerance in rice seedlings (Tao et al. 2015). In alfalfa, ethylene provides salinity tolerance during seed germination and seedling growth under saline stress which is dependent on putative ethylene receptor *MsETR2* (Wang et al. 2020). The abscisic acid signaling pathway has been implicated as a positive regulator of salinity tolerance in alfalfa (Li et al. 2020). Exogenous application of melatonin in alfalfa improves antioxidant capacity, reduces Na^+ accumulation that in turn enhances salinity tolerance (Cen et al. 2020).

9.4.5 Transcription Factors

Several alfalfa transcription factors have been shown to play important roles during salinity stress, which include MYB, WRKY, ethylene response factor (ERF), and Alfin1. Constitutive expression of alfalfa *MYB2L* in Arabidopsis enhanced salt tolerance ability by regulating proline biosynthesis (Song et al. 2019). Overexpression of a salinity stress-inducible transcription factor, *MsMYB4*, provided tolerance against salinity stress in Arabidopsis in an ABA-dependent manner (Dong et al. 2018). However, some MYB transcription factors also have been implicated in the negative regulation of salinity tolerance (Kim et al. 2013).

Differential expression of several WRKY transcription factors has been reported in response to salt stress (Mao et al. 2020). Overexpression of alfalfa *WRKY11* in soybean enhanced salinity tolerance at the seedling stage, suggesting that *MsWRKY11* is a positive regulator of salinity stress (Wang et al. 2018).

ERFs play critical roles in providing salinity tolerance in various plants (Chen et al. 2012a). It has been shown that the expression of alfalfa *ERF8* in tobacco plants enhanced tolerance to salinity stress (Chen et al. 2012a). Overexpression of alfalfa *ERF11* that is induced in response to NaCl and phytohormones including ethylene, enhanced salinity tolerance in Arabidopsis, suggesting that *MsERF11* may be critical during salinity stress response (Chen et al. 2012b).

Overexpression of *Alfin1*, a putative transcription factor in alfalfa, regulates the expression of *MsPRP2* and provides salinity tolerance (Winicov and Bastola 1999).

9.4.6 Role of DNA Methylation and Histone Methylation in Salinity Tolerance

The addition of a methyl group to cytosine residues of the DNA molecule is known as DNA methylation, which is a biological process that is catalyzed by a group of methyltransferase enzymes. Environmental stresses could change DNA methylation/demethylation status in plants, modulating gene expression status of various genes (Zhang et al. 2018). Salinity stress-induced DNA methylation has been shown in Arabidopsis (Jiang et al. 2014). In alfalfa seedlings, salinity stress-induced slight increase of DNA methylation has been shown in response to 20 dS m^{-1} NaCl treatment (Al-Lawati et al. 2016). The use of 5-azacytidine, a DNA demethylation agent, inhibits the growth of salt-treated alfalfa seedling, suggesting that DNA demethylation decreases salinity tolerance in alfalfa seedlings (Al-Lawati et al. 2016). A recent study indicated that activation of the *MsMYB4* transcription factor is linked to elevated levels of histone H3K4 trimethylation and H3K9 acetylation in specific regions of the promoter sequence (Dong et al. 2020). Although more detailed investigations are warranted, these findings suggest that DNA methylation and histone methylation play regulatory roles in the salinity tolerance of alfalfa.

9.4.7 Future Prospects of Genetically-Modified Salt-Tolerant Alfalfa

Approximately 450 million tons of alfalfa are produced worldwide on 30 million acres (12,140,570 ha), with the leading producers being the USA (30%), Europe (25%), and Argentina (23%) (Barros et al. 2019). In 2017, the international trade of alfalfa hay reached 8.3 million metric tons with a total value of US$ 2.3 billion (2nd Alfalfa World Congress 2018). This international congress emphasized the growing need for abiotic and biotic stress resistance breeding and the use of low-quality waters for irrigating forage crops to save freshwater for human consumption. Alfalfa was the first forage crop to be genetically modified for a low concentration of lignin to improve animal digestibility, commercialized with the trade name HarvXtra™ (Barros et al. 2019). A genetically modified glyphosate-tolerant alfalfa was deregulated in the US in 2010 and, a stacked-trait alfalfa with reduced lignin and glyphosate tolerance became available in 2015. This genetically modified alfalfa accounts for $\sim$15% of the alfalfa currently cultivated in the US and is expected to reach 50% in 10 years when Canada and Mexico are expected to deregulate GM alfalfa cultivation and import (Fernandez-Cornejo et al. 2016; Barros et al. 2019). Following on the successes of these previously desired traits incorporated into alfalfa, other of great importance should follow. Some of the desired traits would allow the crop to thrive under reduced irrigation (a must for a high-water demanding crop), irrigation with recycled and with highly saline waters (unfit for most crops), and the production of condensed tannins. Of course, not all those traits may be economically feasible, and the most important ones would have to be chosen, such as traits for drought and salinity tolerance.

9.5 Concluding Remarks

Alfalfa is considered moderately tolerant to salinity; however, salinity tolerance varies among alfalfa genotypes. Upon exposure to high salinity, alfalfa is affected by both osmotic and ionic stresses. As a forage crop, biomass yield and quality are crucial traits for alfalfa. Multiple genotypes have been identified that have a high STI for biomass yield. Recent findings suggest that branching ability under salinity may be an important trait to be considered while breeding for salinity tolerance in alfalfa (Sandhu et al. 2017; Kaundal et al. 2021).

Gene expression-, biochemical-, and physiological analyses indicate that various salt-stress signaling pathways and mechanisms known in model systems are also conserved in alfalfa. Various component traits of salinity tolerance mechanisms have been identified in alfalfa and several candidate genes underlying those traits have been recognized (Sandhu et al. 2017). Progress has been made to develop salt-tolerant transgenic alfalfa plants by expressing many key genes (native/foreign) known to play key roles in salinity tolerance in model systems (e.g., Arabidopsis). Different researchers reported tolerance levels of transgenic alfalfa at different stages of development, like during germination, at seedling stage, or at matured stage. Additionally, it is necessary to examine the salinity tolerance ability of transgenic alfalfa in real environmental conditions and over multiple generations.

The genetic dissection of salt tolerance mechanisms is challenging in alfalfa because of its self-incompatible and polyploid nature. Nevertheless, several dozen markers have been identified that are associated with salt stress tolerance by employing GWAS and other approaches that used hundreds of accessions or several advanced breeding populations of alfalfa (Yu et al. 2016; Liu and Yu 2017; Liu et al. 2019; Medina et al. 2020). Validation of various identified markers will facilitate the use of marker-assisted selection in alfalfa breeding programs.

Besides known genes, hundreds of differentially expressed genes have been identified by the transcriptomics approach in response to salinity stress in salt-tolerant alfalfa cultivars (Kaundal et al. 2021; Lei et al. 2018; Postnikova et al. 2013). Similarly, hundreds of differentially abundant proteins have been identified in salt-tolerant cultivars of alfalfa by a comparative proteomic approach (Rahman et al. 2015; Dong et al. 2018). Future functional genomics studies will expedite the characterization of the most significant genes identified in transcriptomic/proteomic studies to identify their biological roles during salinity stress. Selected candidate genes can then be employed to develop salt-tolerant alfalfa cultivars by traditional breeding and/or genetic engineering.

References

2nd Alfalfa World Congress (2018) Proceedings of the second world alfalfa congress: global iteraction for alfalfa innovation. In: Basigalup D, Spada MdC, Odorizzi A, Arolfo V (eds) Second Alfalfa World Congress. Instituto nacional de Tecnologia Agropecuaria—INTA, Cordoba, Argentina, p 2

Al-Farsi SM, Nadaf SK, Al-Sadi AM, Ullah A, Farooq M (2020a) Evaluation of indigenous Omani alfalfa landraces for morphology and forage yield under different levels of salt stress. Physiol Mol Biol Plants 26:1763–1772. https://doi.org/10.1007/s12298-020-00856-5

Al-Farsi SM, Nawaz A, Nadaf SK, Al-Sadi AM, Siddique KHM, Farooq M (2020b) Effects, tolerance mechanisms and management of salt stress in lucerne (*Medicago sativa*). Crop Pasture Sci 71:411–428. https://doi.org/10.1071/CP20033

Al-Lawati A, Al-Bahry S, Victor R, Al-Lawati AH, Yaish MW (2016) Salt stress alters DNA methylation levels in alfalfa (*Medicago* spp). Genet Mol Res 15:15018299. https://doi.org/10.4238/gmr.15018299

Apse MP, Aharon GS, Snedden WA, Blumwald E (1999) Salt tolerance conferred by overexpression of a vacuolar Na^+/H^+ antiport in Arabidopsis. Science 285:1256–1258. https://doi.org/10.1126/science.285.5431.1256

Ashraf M (2009) Biotechnological approach of improving plant salt tolerance using antioxidants as markers. Biotechnol Adv 27:84–93. https://doi.org/10.1016/j.biotechadv.2008.09.003

Bao A-K, Wang S-M, Wu G-Q, Xi J-J, Zhang J-L, Wang C-M (2009) Overexpression of the Arabidopsis H^+-PPase enhanced resistance to salt and drought stress in transgenic alfalfa (*Medicago sativa* L.). Plant Sci 176:232–240. https://doi.org/10.1016/j.plantsci.2008.10.009

Barros J, Temple S, Dixon RA (2019) Development and commercialization of reduced lignin alfalfa. Curr Opin Biotechnol 56:48–54. https://doi.org/10.1016/j.copbio.2018.09.003

Bartels D, Sunkar R (2005) Drought and salt tolerance in plants. Crit Rev Plant Sci 24:23–58. https://doi.org/10.1080/07352680590910410

Bertrand A, Dhont C, Bipfubusa M, Chalifour F-P, Drouin P, Beauchamp CJ (2015) Improving salt stress responses of the symbiosis in alfalfa using salt-tolerant cultivar and rhizobial strain. Appl Soil Ecol 87:108–117. https://doi.org/10.1016/j.apsoil.2014.11.008

Bhattarai S, Biswas D, Fu Y-B, Biligetu B (2020) Morphological, physiological, and genetic responses to salt stress in alfalfa: A Review. Agronomy 10:577. https://doi.org/10.3390/agronomy10040577

Cen H, Wang T, Liu H, Tian D, Zhang Y (2020) Melatonin application improves salt tolerance of alfalfa (*Medicago sativa* L.) by enhancing antioxidant capacity. Plants (Basel) 9. https://doi.org/10.3390/plants9020220

Chen T, Yang Q, Gruber M, Kang J, Sun Y, Ding W, Zhang T, Zhang X (2012a) Expression of an alfalfa (*Medicago sativa* L.) ethylene response factor gene *MsERF8* in tobacco plants enhances resistance to salinity. Mol Biol Rep 39:6067–6075. https://doi.org/10.1007/s11033-011-1421-y

Chen T, Yang Q, Zhang X, Ding W, Gruber M (2012b) An alfalfa (*Medicago sativa* L.) ethylene response factor gene, *MsERF11*, enhances salt tolerance in transgenic Arabidopsis. Plant Cell Rep 31:1737–1746. https://doi.org/10.1007/s00299-012-1287-z

Cornacchione MV, Suarez DL (2015) Emergence, forage production, and ion relations of alfalfa in response to saline waters. Crop Sci 55:444–457. https://doi.org/10.2135/cropsci2014.01.0062

Cornacchione MV, Suarez DL (2017) Evaluation of alfalfa (*Medicago sativa* L.) populations' response to salinity stress. Crop Sci 57:137–150. https://doi.org/10.2135/cropsci2016.05.0371

Cornacchione MV, Ferreira JF, Liu X, Suarez DL (2018) Physiological and biochemical responses of alfalfa (*Medicago sativa* L.) to salinity stress. In: Basigalup D, Spada MdC, Odorizzi A, Arolfo V (eds) Proceedings of the second world alfalfa congress: global interaction for alfalfa innovation. Instituto nacional de Tecnologia Agropecuaria—INTA Cordoba, Argentina, p 179

Díaz FJ, Grattan SR, Reyes JA, de la Roza-Delgado B, Benes SE, Jiménez C, Dorta M, Tejedor M (2018) Using saline soil and marginal quality water to produce alfalfa in arid climates. Agric Water Manag 199:11–21. https://doi.org/10.1016/j.agwat.2017.12.003

Dong W, Liu X, Li D, Gao T, Song Y (2018) Transcriptional profiling reveals that a MYB transcription factor *MsMYB4* contributes to the salinity stress response of alfalfa. PLoS One 13:e0204033. https://doi.org/10.1371/journal.pone.0204033

Dong W, Gao T, Wang Q, Chen J, Lv J, Song Y (2020) Salinity stress induces epigenetic alterations to the promoter of *MsMYB4* encoding a salt-induced MYB transcription factor. Plant Physiol Biochem 155:709–715. https://doi.org/10.1016/j.plaphy.2020.08.015

Fernandez-Cornejo J, Weshsler SJ, Milkove DL (2016) The adoption of genetically engineered alfalfa, canola, and sugarbeets in the United States Economic Information Bulletin Number 163. USDA-Economic Research Service, p 28

Ferreira JFS, Cornacchione MV, Liu X, Suarez DL (2015) Nutrient composition, forage parameters, and antioxidant capacity of alfalfa (*Medicago sativa,* L.) in response to saline irrigation water. Agriculture 5:577–597. https://doi.org/10.3390/agriculture5030577

Gaxiola RA, Li J, Undurraga S, Dang LM, Allen GJ, Alper SL, Fink GR (2001) Drought- and salt-tolerant plants result from overexpression of the AVP1 H^+-pump. Proc Natl Acad Sci USA 98:11444–11449. https://doi.org/10.1073/pnas.191389398

Gill SS, Tuteja N (2010) Reactive oxygen species and antioxidant machinery in abiotic stress tolerance in crop plants. Plant Physiol Biochem 48:909–930. https://doi.org/10.1016/j.plaphy.2010.08.016

Horie T, Hauser F, Schroeder JI (2009) HKT transporter-mediated salinity resistance mechanisms in Arabidopsis and monocot crop plants. Trends Plant Sci 14:660–668. https://doi.org/10.1016/j.tplants.2009.08.009

Hrbáčková M, Dvořák P, Takáč T, Tichá M, Luptovčiak I, Šamajová O, Ovečka M, Šamaj J (2020) Biotechnological perspectives of omics and genetic engineering methods in alfalfa. Front Plant Sci 11:592. https://doi.org/10.3389/fpls.2020.00592

Jiang C, Mithani A, Belfield EJ, Mott R, Hurst LD, Harberd NP (2014) Environmentally responsive genome-wide accumulation of de novo *Arabidopsis thaliana* mutations and epimutations. Genome Res 24:1821–1829. https://doi.org/10.1101/gr.177659.114

Jiang Z, Zhou X, Tao M, Yuan F, Liu L, Wu F, Wu X, Xiang Y, Niu Y, Liu F et al (2019) Plant cell-surface GIPC sphingolipids sense salt to trigger Ca^{2+} influx. Nature 572:341–346. https://doi.org/10.1038/s41586-019-1449-z

Kaundal A, Sandhu D, Duenas M, Ferreira JFS (2019) Expression of the *high-affinity* K^+ *transporter 1* (*PpHKT1*) gene from almond rootstock 'Nemaguard' improved salt tolerance of transgenic Arabidopsis. PLoS One 14:e0214473. https://doi.org/10.1371/journal.pone.0214473

Kaundal R, Duhan N, Acharya BR, Pudussery MV, Ferreira JF, Suarez DL, Sandhu D (2021) Transcriptional profiling of two contrasting genotypes uncovers molecular mechanisms underlying salt tolerance in alfalfa. Sci Rep 11:5210. https://doi.org/10.1038/s41598-021-84461-w

Khorasani GR, Janzen RA, McGill WB, Kennelly JJ (1997) Site and extent of mineral absorption in lactating cows fed whole-crop cereal grain silage of alfalfa silage. J Anim Sci 75:239–248

Kim JH, Nguyen NH, Jeong CY, Nguyen NT, Hong S-W, Lee H (2013) Loss of the R2R3 MYB, AtMyb73, causes hyper-induction of the *SOS1* and *SOS3* genes in response to high salinity in Arabidopsis. J Plant Physiol 170:1461–1465. https://doi.org/10.1016/j.jplph.2013.05.011

Lei Y, Xu Y, Hettenhausen C, Lu C, Shen G, Zhang C, Li J, Song J, Lin H, Wu J (2018) Comparative analysis of alfalfa (*Medicago sativa* L.) leaf transcriptomes reveals genotype-specific salt tolerance mechanisms. BMC Plant Biol 18:35. https://doi.org/10.1186/s12870-018-1250-4

Li J, Essemine J, Shang C, Zhang H, Zhu X, Yu J, Chen G, Qu M, Sun D (2020) Combined proteomics and metabolism analysis unravels prominent roles of antioxidant system in the prevention of alfalfa (*Medicago sativa* L.) against salt Stress. Int J Mol Sci 21:909. https://doi.org/10.3390/ijms21030909

Liu X-P, Yu L-X (2017) Genome-wide association mapping of loci associated with plant growth and forage production under salt stress in alfalfa (*Medicago sativa* L.). Front Plant Sci 8:853. https://doi.org/10.3389/fpls.2017.00853

Liu XP, Hawkins C, Peel MD, Yu LX (2019) Genetic loci associated with salt tolerance in advanced breeding populations of tetraploid alfalfa using genome-wide association studies. Plant Genome 12:180026. https://doi.org/10.3835/plantgenome2018.05.0026

Luo D, Zhou Q, Wu Y, Chai X, Liu W, Wang Y, Yang Q, Wang Z, Liu Z (2019a) Full-length transcript sequencing and comparative transcriptomic analysis to evaluate the contribution of osmotic and ionic stress components towards salinity tolerance in the roots of cultivated alfalfa (*Medicago sativa* L.). BMC Plant Biol 19:32. https://doi.org/10.1186/s12870-019-1630-4

Luo X, Wang B, Gao S, Zhang F, Terzaghi W, Dai M (2019b) Genome-wide association study dissects the genetic bases of salt tolerance in maize seedlings. J Integr Plant Biol 61:658–674. https://doi.org/10.1111/jipb.12797

Maas EV, Grattan SR (1999) Crop yields as affected by salinity. In: Skaggs RW, van Schilfgaarde J (eds) Agricultural drainage agronomy monograph, vol 38. ASA, CSSA, SSA Madison, WI, pp 55–108

Maas EV, Hoffman GJ (1977) Crop salt tolerance—current assessment. J Irrig Drain Div, Am Soc Civ Eng 103:115–134

Maathuis FJM, Ahmad I, Patishtan J (2014) Regulation of Na^+ fluxes in plants. Front Plant Sci 5:467. https://doi.org/10.3389/fpls.2014.00467

Mao P, Jin X, Bao Q, Mei C, Zhou Q, Min X, Liu Z (2020) WRKY transcription factors in *Medicago sativa* L.: genome-wide identification and expression analysis under abiotic stress. DNA Cell Biol 39:2212–2225. https://doi.org/10.1089/dna.2020.5726

Medina CA, Hawkins C, Liu XP, Peel M, Yu LX (2020) Genome-wide association and prediction of traits related to salt tolerance in autotetraploid alfalfa (*Medicago sativa* L.). Int J Mol Sci 21:3361. https://doi.org/10.3390/ijms21093361

Munns R, Tester M (2008) Mechanisms of salinity tolerance. Annu Rev Plant Biol 59:651–681. https://doi.org/10.1146/annurev.arplant.59.032607.092911

NRC (2000) Nutrient requirements of beef cattle. The National Academies Press, Washington, DC

Park HJ, Kim WY, Yun DJ (2016) A new insight of salt stress signaling in plant. Mol Cells 39:447–459. https://doi.org/10.14348/molcells.2016.0083

Pimentel D, Berger B, Filiberto D, Newton M, Wolfe B, Karabinakis E, Clark S, Poon E, Abbett E, Nandagopal S (2004) Water resources: agricultural and environmental issues. BioScience 54:909–918. https://doi.org/10.1641/0006-3568(2004)054%5b0909:wraaei%5d2.0.co;2

Postnikova O, Shao J, Nemchinov LG (2013) Analysis of the alfalfa root transcriptome in response to salinity stress. Plant Cell Physiol 54:1041–1055. https://doi.org/10.1093/pcp/pct056

Quintero FJ, Martinez-Atienza J, Villalta I, Jiang X, Kim WY, Ali Z, Fujii H, Mendoza I, Yun DJ, Zhu JK et al (2011) Activation of the plasma membrane Na/H antiporter Salt-Overly-Sensitive 1 (SOS1) by phosphorylation of an auto-inhibitory C-terminal domain. Proc Natl Acad Sci USA 108:2611–2616. https://doi.org/10.1073/pnas.1018921108

Radović J, Soković D, Marković J (2009) Alfalfa—most important perenial forage legume in animal husbandry Biotechnol Anim Husb 25:465–475. https://doi.org/10.2298/BAH0906465R

Rahman MA, Alam I, Kim YG, Ahn NY, Heo SH, Lee DG, Liu G, Lee BH (2015) Screening for salt-responsive proteins in two contrasting alfalfa cultivars using a comparative proteome approach. Plant Physiol Biochem 89:112–122. https://doi.org/10.1016/j.plaphy.2015.02.015

Riday H, Brummer EC (2002) Forage yield heterosis in alfalfa. Crop Sci 42:716–723. https://doi.org/10.2135/cropsci2002.7160

Sandhu D, Kaundal A (2018) Dynamics of salt tolerance: molecular perspectives. In: Gosal SS, Wani SH (eds) Biotechnologies of crop improvement. Volume 3: Genomic Approaches. Springer International Publishing, Cham, pp 25–40

Sandhu D, Cornacchione MV, Ferreira JF, Suarez DL (2017) Variable salinity responses of 12 alfalfa genotypes and comparative expression analyses of salt-response genes. Sci Rep 7:42958. https://doi.org/10.1038/srep42958

Schilling RK, Tester M, Marschner P, Plett DC, Roy SJ (2017) AVP1: one protein, many roles. Trends Plant Sci 22:154–162. https://doi.org/10.1016/j.tplants.2016.11.012

Shahid SA, Zaman M, Heng L (2018) Soil salinity: historical perspectives and a world overview of the problem. In: Guideline for salinity assessment, mitigation and adaptation using nuclear and related techniques. Springer International Publishing, Cham, pp 43–53

Singer SD, Hannoufa A, Acharya S (2018) Molecular improvement of alfalfa for enhanced productivity and adaptability in a changing environment. Plant, Cell Environ 41:1955–1971. https://doi.org/10.1111/pce.13090

Smedema LK, Shiati K (2002) Irrigation and salinity: a perspective review of the salinity hazards of irrigation development in the arid zone. Irrig Drain Syst 16:161–174. https://doi.org/10.1023/A:1016008417327

Song Y, Lv J, Qiu N, Bai Y, Yang N, Dong W (2019) The constitutive expression of alfalfa *MsMYB2L* enhances salinity and drought tolerance of *Arabidopsis thaliana*. Plant Physiol Biochem 141:300–305. https://doi.org/10.1016/j.plaphy.2019.06.007

Sun M, Qian X, Chen C, Cheng S, Jia B, Zhu Y, Sun X (2018) Ectopic expression of *GsSRK* in *Medicago sativa* reveals its involvement in plant architecture and salt stress responses. Front Plant Sci 9. https://doi.org/10.3389/fpls.2018.00226

Suyama H, Benes SE, Robinson PH, Getachew G, Grattan SR, Grieve CM (2007) Biomass yield and nutritional quality of forage species under long-term irrigation with saline-sodic drainage water: field evaluation. Anim Feed Sci Technol 135:329–345. https://doi.org/10.1016/j.anifeedsci.2006.08.010

Tao J-J, Chen H-W, Ma B, Zhang W-K, Chen S-Y, Zhang J-S (2015) The role of ethylene in plants under salinity stress. Front Plant Sci 6:1059. https://doi.org/10.3389/fpls.2015.01059

Torabi M, Halim MRA (2010) Variation of root and shoot growth and free proline accumulation in Iranian alfalfa ecotypes under salt stress. J Food Agric Environ 8:323–327. https://doi.org/10.1234/4.2010.3162

Trinchant JC, Boscari A, Spennato G, Van de Sype G, Le Rudulier D (2004) Proline betaine accumulation and metabolism in alfalfa plants under sodium chloride stress. exploring its compartmentalization in nodules. Plant Physiol 135:1583–1594. https://doi.org/10.1104/pp.103.037556

van Zelm E, Zhang Y, Testerink C (2020) Salt tolerance mechanisms of plants. Ann Rev Plant Biol 71:403–433. https://doi.org/10.1146/annurev-arplant-050718-100005

Wang J, Xu X, Ma D (2019) Establishment of an efficient alfalfa regeneration system via organogenesis and the co-expression of Arabidopsis *SOS* genes improves salt tolerance in transgenic alfalfa (*Medicago sativa* L.). Can J Plant Sci 99:348–363. https://doi.org/10.1139/cjps-2018-0105

Wang Y, Jiang L, Chen J, Tao L, An Y, Cai H, Guo C (2018) Overexpression of the alfalfa *WRKY11* gene enhances salt tolerance in soybean. PLoS One 13: e0192382. https://doi.org/10.1371/journal.pone.0192382

Wang Y, Diao P, Kong L, Yu R, Zhang M, Zuo T, Fan Y, Niu Y, Yan F, Wuriyanghan H (2020) Ethylene enhances seed germination and seedling growth under salinity by reducing oxidative stress and promoting chlorophyll content via ETR2 pathway. Front Plant Sci 11:1066. https://doi.org/10.3389/fpls.2020.01066

Winicov I, Bastola DR (1999) Transgenic overexpression of the transcription factor *Alfin1* enhances expression of the endogenous *MsPRP2* gene in alfalfa and improves salinity tolerance of the plants. Plant Physiol 120:473–480. https://doi.org/10.1104/pp.120.2.473

Yang L, Zu Y-G, Tang Z-H (2013) Ethylene improves Arabidopsis salt tolerance mainly via retaining K^+ in shoots and roots rather than decreasing tissue Na^+ content. Environ Exp Bot 86:60–69. https://doi.org/10.1016/j.envexpbot.2010.08.006

Yu LX, Liu X, Boge W, Liu XP (2016) Genome-wide association study identifies loci for salt tolerance during germination in autotetraploid alfalfa (*Medicago sativa* L.) using genotyping-by-sequencing. Front Plant Sci 7:956. https://doi.org/10.3389/fpls.2016.00956

Zhang H, Lang Z, Zhu JK (2018) Dynamics and function of DNA methylation in plants. Nat Rev Mol Cell Biol 19:489–506. https://doi.org/10.1038/s41580-018-0016-z

Zhang YM, Liu ZH, Wen ZY, Zhang HM, Yang F, Guo XL (2012) The vacuolar Na^+–H^+ antiport gene *TaNHX2* confers salt tolerance on transgenic alfalfa (*Medicago sativa*). Funct Plant Biol 39:708–716. https://doi.org/10.1071/fp12095

Zhao C, William D, Sandhu D (2020) Isolation and characterization of salt overly sensitive family genes in spinach. Physiol Plant. https://doi.org/10.1111/ppl.13125

Zhu JK (2000) Genetic analysis of plant salt tolerance using Arabidopsis. Plant Physiol 124:941–948. https://doi.org/10.1104/pp.124.3.941

Zhu JK (2002) Salt and drought stress signal transduction in plants. Annu Rev Plant Biol 53:247–273. https://doi.org/10.1146/annurev.arplant.53.091401.143329

Developing SNPs and Strategies for Genomic Analysis in Alfalfa

10

Cesar Augusto Medina and Long-Xi Yu

Abstract

Continued advances in plant breeding require innovative breeding strategies such as marker-assisted selection and genomic selection. New technologies of next generation sequencing provide efficient genotyping strategies such as genotype-by-sequencing or RADseq which can generate high density single nucleotide polymorphisms (SNPs) at the whole genome level. However, in autotetraploid alfalfa, developing markers requires analysis of allele dosage as it can affect genotyping accuracy. Similarly, different models are required for genome-wide association studies to overcome the genome complexity in autotetraploid genomes. In this chapter, we summarize the recent progress on different methodologies for effectively handling SNPs genotyping, marker development, allele dosage, association mapping, and genomic selection in alfalfa. With recent advances in sequencing technology, the cost of sequencing is reduced, and sequencing the whole genomes becomes realistic in polyploids including cultivated alfalfa. Several alfalfa genomes have been sequenced and the sequence assemblies of two cultivars of alfalfa at the chromosome level have been published last year, providing reference genomes for sequence alignment and genotype calling in tetraploid alfalfa. Toward this end, in this chapter, we also reviewed bioinformatics pipelines for SNP discovery and genotyping.

C. A. Medina
Irrigated Agriculture Research and Extension Center, Washington State University, 24,106 N Bunn Road, Prosser, WA, USA
e-mail: cesar.medinaculma@wsu.edu

L.-X. Yu (✉)
United States Department of Agriculture-Agricultural Research Service, Plant Germplasm Introduction and Testing Research, 24106 N Bunn Road, Prosser, WA, USA
e-mail: Longxi.yu@usda.gov

10.1 Introduction

Alfalfa (*Medicago sativa.* L) is one of the most valuable forages grown worldwide and is the third largest crop in the US, providing >58 Mt of hay and silage for animal feed each year. The breeding effort on improving alfalfa cultivar have been largely relayed on the phenotypic selection in field environments. Phenotypic strategy such as recurrent selection is time-consuming and costly as it gradually increases the frequency of favorable alleles and maintains genetic variability for future selection. The recurrent selection methods are most effective when integrated with genomic tools. Cultivated alfalfa is autotetraploid, outcrossing, and synthetic population with high heterozygosity. The autotetraploid

L.-X. Yu and C. Kole (eds.), *The Alfalfa Genome*, Compendium of Plant Genomes,
https://doi.org/10.1007/978-3-030-74466-3_10

alfalfa is formed by the multiplication of the chromosome sets from diploid species and often exhibits the formation of multivalents during meiosis and polysomic inheritance in the progeny (Quiros 1982). Recent advances in next generation sequencing (NGS) technologies provide platforms to generate genome-wide single nucleotide polymorphisms (SNPs). The high density SNP markers facilitate genome-wide association studies (GWAS) and genomic selection (GS) which help in the identification of DNA markers associated with important traits and use them in marker-assisted selection (MAS) to accelerate the breeding process (Medina et al. 2020). However, complex traits such as stress tolerance and yield productivity are often controlled by multiple genes. GWAS and GS use the whole genome markers for analyzing marker-trait association and selection of individuals for breeding advances based on their genetic potentials. GS does not require prior knowledge about specific loci for the trait because it is based on the association of whole genome markers with phenotypic traits to obtain the genomic estimated breeding values (GEBV) (Hawkins and Yu 2018). GEBV is obtained by training statistical models or machine learning methods. Predictive trained models are then applied to identify favorite individuals in testing populations without phenotyping.

10.2 Genotyping by Sequencing in Autotetraploids

Genotype by sequencing (GBS) is a strategy for developing massive genome-wide SNPs (Elshire et al. 2011). Briefly, genomic DNA is extracted from each individual in the population and digested with one or two methylation sensitive restriction enzymes with good coverage of the genome. In autotetraploid alfalfa, *Ape*KI is a good choice of restriction enzyme because it has an average of 1 cut site per 1.5 kb in *M. sativa*. Each digested sample is ligated to a unique barcode adapter to perform massively parallel sequencing and further demultiplexing. The samples are mixed and amplified by PCR before sequencing them using the Illumina platform. This sequencing approach produces a reduced-representation sequencing for low-cost and high throughput genotyping. During genotyping process, different types of markers can be obtained: SNPs, insertion or deletions (indels), copy number variants (CNV), and short tandem repeats (STRs). Biallelic SNPs are the most abundant and stable markers to identify genes associated with important traits (Perkel 2008). The accuracy of genotyping can be affected by bioinformatic pipelines for genotype calling, sequencing depth, and genome complexity. Multiple parameters must be adjusted to obtain high quality markers. In polyploids, multiple copies of the alleles demand dosage analysis before further analysis. Genotype calling in autotetraploids requires bioinformatics tools to distinguish among five possible alleles in biallelic SNPs including multiplex (0), simplex (1), duplex (2), triplex (3), and quadruplex (4). Furthermore, the presence of indels usually creates uninterpretable sequence reads that further reduces the number of markers. The depth of sequence reads affects the estimate of allele dosage and a higher depth (e.g. $\sim 60\times$) is required to estimate allele dosage in autotetraploid alfalfa (Gore et al. 2007; Uitdewilligen et al. 2013).

To date, most bioinformatics tools designed for diploid species are not appropriate for polyploid species such as alfalfa. When a software designed for diploid species is used in polyploid species, data must be "pseudo-diploidized" thus missing complexity information to explain the phenotypic expression. Several tools have been used for genotype calling: Universal Network-Enabled Analysis Kit (UNEAK) in the TASSEL-GBS pipeline (Glaubitz et al. 2014), GATK (McKenna et al. 2010), SAMtools (Li et al. 2009), FreeBayes (Garrison and Marth 2012), and Stacks (Rochette et al. 2019). In alfalfa, a comparison of genotype calling using UNEAK, TASSEL, and FreeBayes pipelines was performed (Yu et al. 2017a). Among them, TASSEL-GBS and UNEAK have been developed for diploids, while FreeBayes can call polyploid genotypes.

Recently, the Next Generation Sequencing Experience Platform (NGSEP) has been developed and become an important all-in-one tool for SNP genotyping with multiple functions (Duitama et al. 2014; Tello et al. 2019). NGSEP has been used in different crops including diploid rice and beans and polyploid cassava (Tello et al. 2019). NGSEP has high accuracy and efficiency for discovery and genotyping single nucleotide variants (SNVs), small and large indels, short tandem repeats (STRs), inversions, and copy number variants (CNVs) in various sequencing data sets, including GBS, whole genome sequencing (WGS) or whole exome sequencing (WES) datasets. For each potential variation site, NGSEP implements Bayesian approaches with different likelihood functions to find the most likelihood of genotypes based on sequencing quality scores. Additionally, NGSEP has different functions to generate, filter, impute and format VCF files.

Most recently, two *M. sativa* genomes at the chromosome level have been published (Chen et al. 2020; Shen et al. 2020). Chen et al. (2020) reported an allele aware *M. sativa* cv XinJiangDaYe genome with N50 of 459 kb. This genome is composted by four homologous (haplotypes) in each of 8 chromosomes and 9,789 scaffolds. Later on, Shen et al. (2020) published the chromosome-level haploid genome of *M. sativa* cv. Zhongmu No. 1 with N50 of 3.92 Mb, and 649 scaffolds (Table 10.1). Previously, sequences of its close relative *M. truncatula* A17 genome have been published and its newer version Mt5.0 is available (Pecrix et al. 2018). These three *Medicago* spp. genomes can be used as references to align alfalfa sequence reads. To date, the *M. truncatula* A17 is one of the best assembled and annotated genomes, with N50 of 18 Kb, and 32 scaffolds. In addition, the sequence assembly of mitochondrial and chloroplastic genomes is also available (Pecrix et al. 2018). Comparing these genomes, although cultivar XinJiangDaYe allows assigning markers to specific homologous chromosomes, this genome has more scaffold without chromosome assignment and shorter chromosomes compared with cultivar Zhongmu No. 1 (Table 10.1). Whereas the *M. truncatula* A17 genome provides more complete information including sequence assembly and gene annotation in the genome browser (https://medicago.toulouse.inra.fr/MtrunA17r5.0-ANR/).

Based on our previous results (Medina et al. 2020), the use of the three genomes as references

Table 10.1 Chromosome sizes in three reference genomes of *M. sativa* and *M. truncatula* for SNP discovery

Chr	XinJiangDaYe				Zhongmu	A17
	H1	H2	H3	H4		
1	82,459,472	86,910,131	79,881,340	88,815,615	102,287,719	56,706,830
2	76,462,061	74,215,936	76,375,162	76,750,018	84,774,831	51,972,579
3	93,164,418	93,375,939	96,157,931	100,399,604	104,248,068	58,931,556
4	90,245,664	93,947,428	90,228,617	90,896,203	92,728,752	64,763,011
5	81,211,777	84,165,483	80,712,490	78,626,892	109,743,485	44,819,618
6	80,303,593	89,579,199	84,649,260	64,534,737	114,337,865	42,866,092
7	88,407,277	93,528,358	91,580,142	94,657,719	95,628,712	56,236,587
8	87,242,343	84,274,390	82,440,740	81,801,543	90,714,941	49,719,271
S	9789				649	32
MT	–				–	124,033
CP	–				–	271,618

Notes H, homologous chromosomes for *M. sativa* cv. XinJiangDaYe; S, Scaffold; MT, Mitochondrial genome; CP, Chloroplastic genome; XinJiangDaYe, *M. sativa* cv. XinJiangDaYe; Zhongmu, *M. sativa* cv. Zhongmu; A17, *M. truncatula* A17 version 5.0

to alignment GBS data with bowtie2 produced different results. Using *M. truncatula* A17 as reference, we obtained 52% of reads aligned 0 times, 26% aligned exactly 1 time, and 22% aligned >1 time, with an overall alignment rate of 48%. Using *M. sativa* cv XinJiangDaYe as reference, we obtained 30% of reads aligned 0 times, 2% aligned exactly 1 time, and 68% aligned > 1 time, with an increase of overall alignment rate up to 69%. However, because most of the reads aligned are more than one time, the variant calling process was not successful. To solve this problem, we generated a haploid version of the XinJiangDaYe genome by keeping the longest chromosome among the homologous chromosomes. The alignment was improved by 26% of reads aligned 0 times, 28% aligned exactly 1 time, and 46% aligned >1 time with an overall alignment rate of 74%. Using *M. sativa* cv. Zhongmu No. 1 as reference, we obtained 9% of reads aligned 0 times, 38% aligned exactly 1 time, and 53% aligned > 1 time with an overall alignment rate of 91%. Although, the use of *M. sativa* cv XinJiangDaYe genome in haploid version and *M. sativa* cv. Zhongmu No. 1 produces similar results, the use of *M. sativa* cv. Zhongmu No. 1 is recommended for two reasons: Frist, the use of the Zhongmu No.1 produced higher values of overall alignment rates and reads aligned exactly 1 time. Second, the haploid version of XinJiangDaYe genome ignores information of homologous chromosomes.

10.3 Software for Genotype Calling in Autoteraploids

In our experience, the straightforward way to obtain the genotype calling with allele dosage is using the function `VCFConverter` and the option—`GWASPoly` of NGSEP to perform GWAS analysis with the GWASPoly package (Rosyara et al. 2016). However, there are several packages to perform genotype calling from polyploid species like polyRAD (Clark et al. 2019), superMASSA (Pereira et al. 2018), FitTetra 2.0 (Zych et al. 2019) or Updog (Gerard et al. 2018) (Table 10.2). polyRAD is a R package which implements Bayesian genotype estimation from low read depth data from several pipelines and outputs continuous or discrete numerical genotypes suitable for other analysis like GWASPoly. The advantage of polyRAD is that it is the first Bayesian genotype caller to incorporate population structure and multiple inheritance modes (Clark et al. 2019). Also, polyRAD has genotype priors which must be specified for natural or biparental populations. Recently, polyRAD has developed a pipeline for genotype calling for recent or ancient allopolyploid species with highly duplicated species using read depth distribution across a population to identify Mendelian locus (Clark et al. 2020). superMASSA performs quantitative genotype calling and dosage genotype calling implementing Bayesian network. The script VCF2SM from superMASSA takes the field read depths from the VCF file as input, and generates a VCF file with polyploid genotype calls, i.e., depicting reference and alternative allele dosages (e.g., 0/0/1/1 in GT field for a biallelic SNV ACCC with A as reference and C as alternative allele). One advantage of supperMASSA is the genotyping calling even when the ploidy of the population is unknown including genetic models into consideration, such as full-sib family and Hardy-Weinberg Equilibrium expected frequencies (Pereira et al. 2018). FitTetra and the newest version, FitTetra 2.0, is designed for Affymetrix SNP arrays and improves genotype calling in tetraploids with multiple populations and parental data support. FitTetra 2.0 is able to call a large portion of SNPs correctly with differences in performance between dyes or a nonlinear relationship between dosage and signal strength (Zych et al. 2019). Updog provides a suite of methods for genotyping polyploids by empirical Bayes approaches taking into account allele bias, overdispersion, outliers, and sequencing error. In the case of Updog, there is the need to generate two matrices to run `multidog` function, one matrix will contain depth read values and other matrix will contain reference allele observations (Gerard et al. 2018). Although in this chapter, we present a straightforward protocol for SNP genotyping in alfalfa, systematic benchmarking is required with alfalfa dataset to identify differences among other tools. Initially, we tested the allele dosage results

Table 10.2 Tools designed for allele dosage from SNP arrays or GBS data in a VCF file generated by different software

Tool	Use	Approach	Platform	References
Updog	Allele dosage NGS	Empirical Bayes genotyping procedure	R	Gerard et al. (2018)
polyRAD	Allele dosage NGS	Bayesian approach based in posterior probabilities	R	Clark et al. (2019)
fitTetra 2.0 (fitPoly)	Allele dosage array	Mixture model with arcsine transformation of ratio data	R	Voorrips et al. (2011), Zych et al. (2019)
VCF2SM (SuperMASSA)	Allele dosage array and NGS	Graphical Bayesian clustering model	Python 2.7	Serang et al. (2012), Pereira et al. (2018)
NGSEP	Variant calling and allele dosage NGS	Count of number of base calls (depth) for the 4 nucleotides.	Java	Duitama et al. (2014)
ClusterCall	Allele dosage array	Hierarchical clustering converting theta values into tetraploid genotypes	R	Schmitz Carley et al. (2017)

Notes The updated version of NGSEP v4.0 provide a customer alignment system in a function called `ReadsAligner`. We compared the alignment rates of Bowtie2 versus ReadsAligner of NGSEP v4.0 using the. Zhongmu No. 1 reference genome and obtained the follow results: Bowtie2 produced 9% of reads aligned 0 times, 38% aligned exactly 1 time and 53% aligned >1 time with an overall alignment rate of 91%, while the use of ReadsAligner produced 18% of reads aligned 0 times, 59% aligned exactly 1 time and 23% aligned >1 time with an overall alignment rate of 82%. Although the overall alignment rate was lower in ReadsAligner compared with Bowtie2, the percentage of reads aligned only 1 time was higher by `ReadsAligner`

by three tools in a dataset of 272 individuals with 6862 biallelic SNPs previously published in Medina et al. (2020). We used the VCF file to predict the allele dosage by polyRAD, updog and NGSEP identifying similarities among polyRAD and NGSEP (Fig. 10.1). However, parameter optimization must be done to obtain more accurate results.

For the generation of VCF file with NGSEP, module `MultiSampleVariantDetector` can be used to call variants over a group of read alignment samples in format `SAM`, `BAM` or `CRAM`. Different parameters should be adjusted to control false positives. For instance, ploidy can be defined using the option `-ploidy = 4`. The maximum number of alignments allowed to start at the same reference site (`–maxAlnsPerStartPos) = 100`. The minimum mapping quality to call an alignment unique (`—minMQ) = 40`. Maximum base-pair quality score (`–maxBaseQS) = 30`. Additionally, a catalog of known short tandem repeats (`catalog.str`) must be generated from the reference genome with the software of tandem repeat finder (trf409.linux64) (Benson 1999) using the supplemental script S3 from Lobaton et al. (2018). Moreover, the `catalog.str` file must be used with the option `–knownSTRs`.

The preliminary VCF file must be filtered before the imputation step using the module `VCFFilter` as follow: Minimum minor allele frequency (`–minMAF) = 0.05`, keeping only biallelic SNVs (`–s`), the minimum number of samples genotyped to keep the variant (`–m) = 70%`. This last parameter `–m` is important to reduce the number of missing values, improving imputation. For the imputation step, the function `ImputeVCF` uses the Hidden Markov Model (HMM) from the package fastPHASE (Scheet and Stephens 2006) which keeps all the VCF fields generated by NGSEP. For biparental or multi-parental populations, it is recommended to indicate the list of parents of the breeding population with option `–p`.

One of the important characteristics of the NGSEP is the generation of genotype fields. The

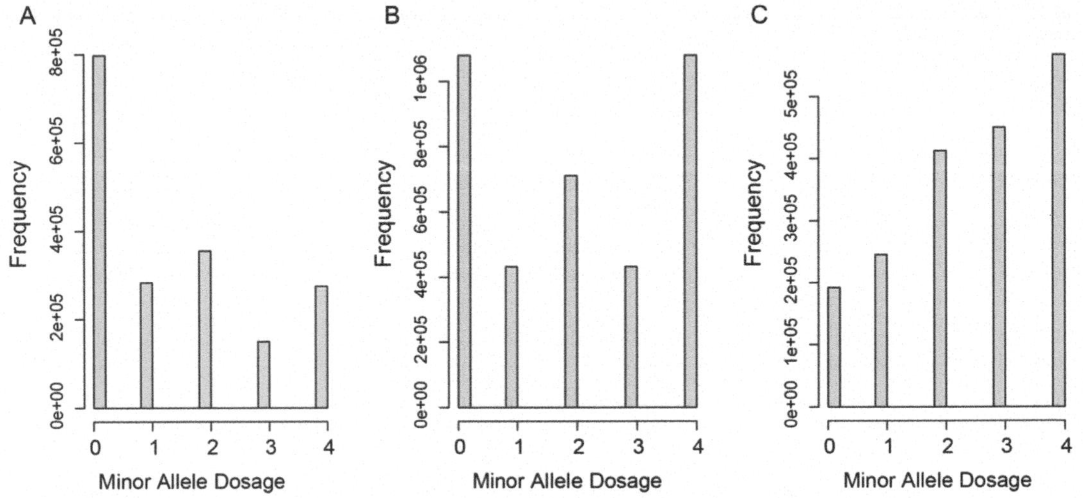

Fig. 10.1 Minor allele dosage frequency predicted by A, NGSEP, B, polyRAD and C, updog in the same dataset of 272 individuals and 6862 high-quality biallelic SNPs

VCFv4.2 file generated by NGSEP has six fields in biallelic SNPs: `GT`, `PL`, `GQ`, `DP`, `BSDP`, and `ACN`. `GT`: genotype, encoded as allele values 0 for the reference allele and 1 for the first allele listed as an alternative allele. `PL`: Phred-scaled genotype likelihoods rounded to the closest integer. `GQ`: genotype quality, encoded as a Phred score:$-10\log_{10}P$ (genotype call is wrong). `DP`: read depth for each site. `BSDP`: number of base calls (depth) for the 4 nucleotides in called SNVs sorted as `A`, `C`, `G`, `T`. The predicted copy number of each allele is taken into account the prediction of the number of copies of the region surrounding the variant. The `BSDP` field is important for genotyping polyploid species because it can count the number of reads for each nucleotide, allowing to identify allele dosage. A detailed flowchart for variant calling in VCF, subjecting to GWASPoly format and numeric format for genomic selection is presented in Fig. 10.2.

10.4 Strategies for Genome-Wide Association Studies in Autotetraploids

Genome-Wide Association Studies (GWAS) are a powerful tool for detecting genetic loci associated with phenotypic traits. GWAS is based on the linkage disequilibrium that estimates the non-random association of alleles at different loci in a given population. GWAS in polyploid species has been challenging. Until recently, Rosyara et al. (2016) developed a R package for GWAS in polyploid species (GWASPoly). In this package, biallelic SNPs were used for GWAS using structure (Q) and marker-estimated kinship (K) matrices in the linear mixed model as follows:

$$y = X\beta + ZS\tau + ZQv + Zu + \varepsilon$$

where y is the $w \times 1$ vector of phenotypic values; β is the $p \times 1$ vector of fixed-effects; X is the incidence matrix used to model environmental effects; v is the $q \times 1$ vector of effects for the subpopulations; Q is the incidence matrix for a population of size n; u is the $n \times 1$ vector of polygenic effects; Z is the $w \times n$ incidence matrix of mapping genotypes to observations; τ is the $d \times 1$ vector of SNPs effects; S is the structure incidence matrix and ε is the residuals vector $w \times 1$. GWASPoly can generate up to six different gene action models according to allele interactions, implying different degrees of dominance: general, additive, diploidized additive, diploidized general, duplex dominant (A > B & B > A), and simplex dominant (A > B & B > A).

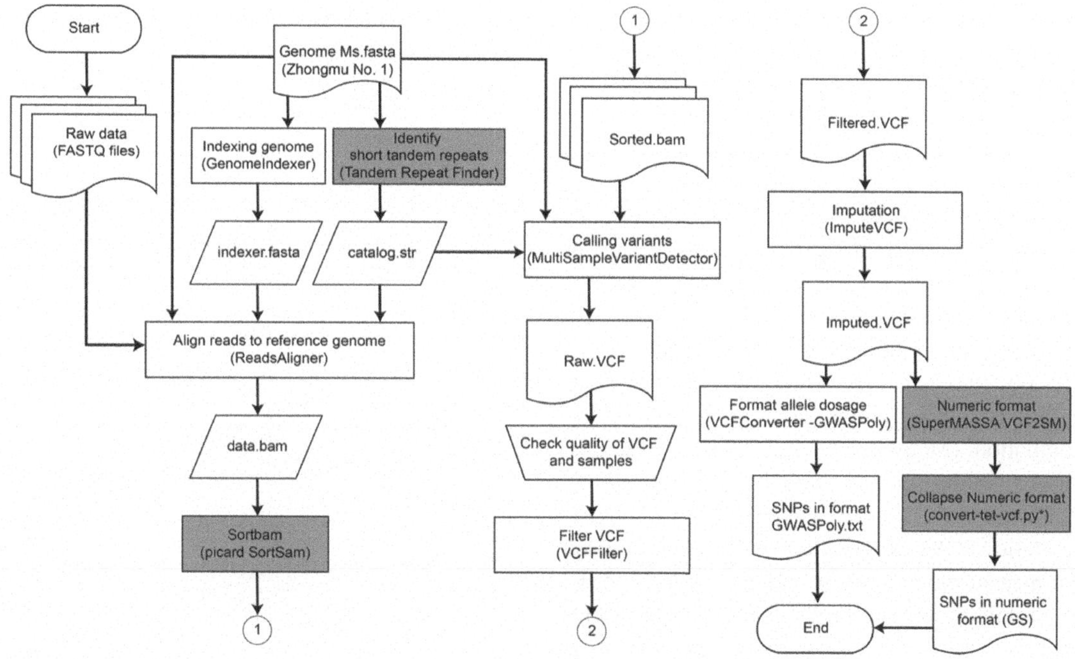

Fig. 10.2 Flow chart of bioinformatics pipelines for SNP genotyping in alfalfa using NGSEP. Start and end of the analysis are represented by oval symbol, processes are represented by rectangles. Input and output data are represented by parallelograms. Manual analysis is represented by trapezoid. Gray rectangles represent analysis by downstream software

SHEsisPlus is a software package with a web-based version for analyses of genetic association, Hardy-Weinberg equilibrium, linkage disequilibrium, and haplotype construction at multiallelic polymorphism loci, compatible with polyploid species. SHEsisPlus uses a generalization of partition-ligation-combination-subdivision expectation maximization algorithm (PLCSEM) to reduce the computational complexity and simplify phasing the haplotypes of polyploid datasets (Shen et al. 2016). However, SHEsisPlus does not take account of population structure or relatedness.

Finally, MultiGWAS is a tool for GWAS in both diploid and tetraploid species. It solved problems of replication and integration among tools because it runs four different GWAS software: GWASPoly (Rosyara et al. 2016), SHEsis (Shen et al. 2016), PLINK (Purcell et al. 2007), and TASSEL (Bourke et al. 2018). MultiGWAS choose the best gene action model to generate Venn diagrams to show marker-trait associations identified by single or multiple software (Garreta et al. 2020). MultiGWAS uses two types of statistical models: full and naïve models. The full model is the Q + K model explained above. The naïve model does not include any type of correction. One additional advantage is the use of different input of genotype file formats including GWASPoly format, genotype file matrix format, genotype field in the format 0/0/1/1 generated by the python script VCF2SM of SuperMASSA software (Pereira et al. 2018) or `BSDP` field generated by NGSEP (Duitama et al. 2014).

10.5 Genome Selection in Autotetraploids

The ability of predicting phenotypic traits based on genotype is key in plant breeding to accelerate crop improvement. However, some of the most important traits like yield are highly quantitative, with small effects by multiple genes present along all the genome, making more challenge in the discovery of QTLs and further marker development for MAS. Recently, genomic selection (GS) has become an important tool to select individuals in a population based on a whole set of genetic markers to estimate the genetic potential of individuals. In GS, thousands of genetic markers are used to train statistic or machine learning models to predict the genetic estimated breeding values (GEBV) for each individual in a population (Crossa et al. 2017). The population can be randomly partitioned as individuals into training population and testing population (Meuwissen et al. 2001). The predictive ability of the model is calculated by Pearson's correlation of GEVB based on cross-validation (CV). Regression models can be used for measuring root mean square error (RMSE) when continuous variations of phenotypic values are tested to find the best model that fits on a training dataset (Waldmann 2019).

For autotetraploids, allele dosage needs to be analyzed in the genotypic data because this information impacts genotyping accuracy, and in turn, affects the prediction of genetic values (Endelman et al. 2018; de Bem Oliveira et al. 2019). In GS analysis, VCF files with allele dosage are numerically transformed using the python scripts VCF2SM, SuperMASSA software (Pereira et al. 2018) or convert-tet-vcf.py (Hawkins 2018). The correlation between genotypic markers and phenotypic traits is tested by different models. rrBLUP and Bayesian models are the most tested in various GS studies (Annicchiarico et al. 2015; Jia et al. 2018; Rice and Lipka 2019). On the other hand, machine learning models such as support vector machine (SVM), random forest (RF) or multi-layer perceptron (MLP) have gained great popularity by their good performance on increased prediction accuracy mainly for continuous traits (Annicchiarico et al. 2015; Grinberg et al. 2020; Medina et al. 2020). Among them, machine learning methods have been widely used in computing regression because of their high accuracy and ability to deal with high-dimensional and large datasets and solving the issue of "large p small n"($p \gg n$). The basic model for genomic selection can be described as

$$y = X\beta + Z\alpha + \varepsilon$$

where y is a vector of phenotypes, β is a vector of fixed effects, α is a vector of random effects of individuals (SNP markers), X and Z are incidence matrices for β and α, respectively, and ε is a vector of residual effects. In tetraploids, the SNP matrix is the Z matrix, parametrized as 0–4 according to distinguish among five possible alleles in biallelic SNPs.

On the other hand, machine learning methods are based on regressing phenotypes on some function of SNP genotype codes $g(x)$, as

$$y = 1\mu + \begin{bmatrix} g_1(X_1) \\ g_2(X_2) \\ \vdots \\ g_n(X_n) \end{bmatrix} + \varepsilon$$

where the function $g_i(X_i)$ is an approximation of the true genetic merit of each individual, after adjusting phenotypes for environmental effects. 1 is a column of the vector of ones. The vector ε represents residuals, typically assumed $N(0, I\sigma_\varepsilon^2)$. Machine learning methods do not assume linear and additive action of markers a priori, but the type of function given by $g_i(X_i)$ determines the learning attained.

Finally, the predictive ability of the models was calculated as Pearson's correlation between GEBV and phenotypes of test populations. Our previous GS results showed that SVM was the best fitting model for biomass yield harvested in September for both 2018 and 2019, while the RF model fits the data best for yield in July 2018, May 2019, June and July 2019 based on the accuracy of Pearson's correlation and RMSE (Table 10.3) (Medina et al. 2020). In this work, machine learning methods had the best fit in Pearson's correlation and RMSE. It was likely due to the ability of these methods for identification of the top-ranking SNPs with major effects on the phenotypic variation or by capturing complex SNP–SNP interactions and nonlinear relationships, and thus increasing the genetic variance and the heritability of the trait.

10.6 Strategies for Transcriptomics in Alfalfa

Transcriptomics has been carried out in different crops and helped to understand molecular mechanisms by which environmental factors affect plant development and productivity (Postnikova et al. 2013; Shu et al. 2017; Cui et al. 2019). Differentially expressed genes in response to biotic and abiotic stresses can be used for developing markers for selecting resistance traits (Yang et al. 2011; Liu et al. 2013). Although RNA-seq has been widely used to obtain transcriptional profiles and to generate de novo genomic information, short read lengths are a major limitation for identifying full-length transcripts. To overcome this limitation, during the past few years, the use of Pacific BioSciences (PacBio) or Oxford Nanopore Technologies platforms have become popular because they dramatically increase read length. Both RNA-seq and PacBio platforms generate massive sequence reads and require a series of bioinformatics pipelines to sort out the sequence fragments and analyze transcriptomes.

Recently, we conducted a transcriptomic experiment in which the 21 alfalfa RNA samples were sequenced by PacBio and Illumina to generate the most complete and accurate transcriptome in different germplasms, tissue sources, and stress conditions. An expression matrix of transcripts per million (TPM) of a total of 888,790 isoforms in 21 treatments was obtained. A correlation matrix was generated using the R package "corrplot" to explore the profile of gene expression among treatments. They were

Table 10.3 Genomic selection (GS) results for biomass yield of alfalfa under salt stress. Eight GS models were tested using ten-fold cross-validation with Pearson's correlation values (Pearson) and root mean squared error (RMSE) by different models (Modified from Medina et al. 2020)

Dataset	Metric	rrBLUP	BayesA	BayesB	BayesC	BL	BRR	RF	SVM
Jul_18	Pearson	0.305	0.305	0.303	0.307	0.303	0.299	0.343	0.324
	RMSE	0.509	0.506	0.51	0.508	0.508	0.509	0.508	0.503
Aug_18	Pearson	0.27	0.259	0.275	0.272	0.253	0.265	0.268	0.24
	RMSE	0.409	0.411	0.407	0.408	0.408	0.408	0.414	0.414
Sep_18	Pearson	0.444	0.445	0.448	0.447	0.454	0.45	0.464	0.509
	RMSE	0.255	0.254	0.254	0.255	0.254	0.254	0.256	0.244
All_18	Pearson	0.234	0.216	0.227	0.226	0.209	0.236	0.302	0.268
	RMSE	0.377	0.38	0.376	0.379	0.375	0.377	0.37	0.371
May_19	Pearson	0.116	0.108	0.107	0.121	0.119	0.115	0.182	0.113
	RMSE	0.551	0.558	0.556	0.552	0.552	0.553	0.541	0.548
Jun_19	Pearson	0.173	0.147	0.155	0.146	0.184	0.154	0.219	0.201
	RMSE	0.477	0.481	0.478	0.478	0.474	0.478	0.467	0.469
Jul_19	Pearson	0.258	0.242	0.238	0.266	0.231	0.235	0.287	0.281
	RMSE	0.51	0.513	0.509	0.507	0.51	0.51	0.514	0.51
Sep_19	Pearson	0.249	0.231	0.257	0.24	0.247	0.236	0.276	0.301
	RMSE	0.31	0.312	0.309	0.311	0.309	0.31	0.312	0.308
All_19	Pearson	0.072	0.065	0.083	0.064	0.06	0.083	0.137	0.138
	RMSE	0.464	0.467	0.466	0.466	0.463	0.462	0.456	0.455

Notes BL, Bayesian LASSO; BRR, Bayesian ridge regression; RF, random forest; SVM, support vector machine

clustered based on the similarity of expression patterns in Wilson-DS-root, Wilson-CK-root, Saranac-SS-root, Saranac-DS-root, Saranac-CK-root, PI467895-CK-root, and PI467895-SS-root (Fig. 10.3).

A series of comprehensive pipelines have been developed for sorting out sequence reads and analyzing pan-transcriptome. The flow chart of the pipelines was illustrated in Fig. 10.4 and described as follows.

10.6.1 Sequence Quantification and Quality Control

PacBio requires high-quality RNA to ensure full-length cDNA and to capture all possible isoform variations. RNA integrity number (RIN) values must be >8. Circular Consensus Sequences (CCS) files (CCS.bam) from each library can be generated with IsoSeq Version 3 pipeline, followed by demultiplexing using Lima Isoseq resulting in individual tissue source-germplasm-treatment transcriptomes (Gordon et al. 2015). Finally, demultiplexed transcriptomes are refined with the Isoseq 3 refine software to remove polyA tails and to detect concatemers, producing full-length, non-concatemers (FLNC) reads. Lima report file can be used to generate preliminary plots of quality control, yield, and transcript length distribution (Fig. 10.5).

10.6.2 Hybrid Error Correction and Mapping

Long-reading sequence (LRS) is prone to high error rate, to solve this problem, FLNC reads can be corrected using the second-generation short read sequencing platform, such as Illumina, using a hybrid error correction strategy. LoRDEC is a software for hybrid error correction that builds de Bruijn graph using PacBio and Illumina data (Salmela and Rivals 2014). LoRDEC has the best performance with run times on the large datasets with similar accuracy and sensitivity rates to other software such as FMLRC or HALC (Fu et al. 2019). The corrected and polished

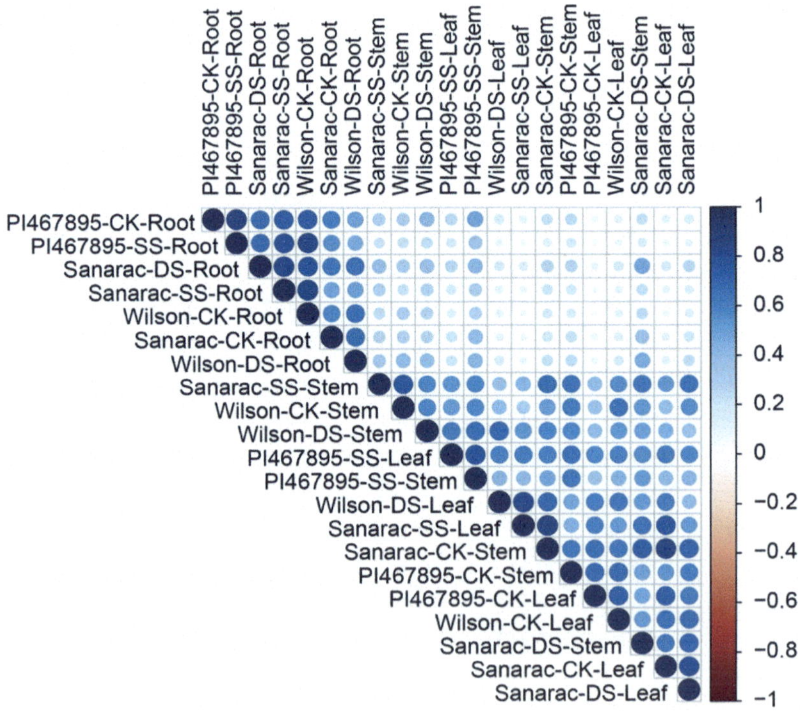

Fig. 10.3 Correlation matrix of transcripts per million ($\log_{10}$ transformed) in three tissue sources (leaf, stem and root) in three alfalfa germplasm (Saranac, Wilson, PI467895) under three conditions: drought stress (DS), salt stress (SS) and control non-stressed (CK). Blue and red colors represent low (−1) and high (1) correlations, respectively

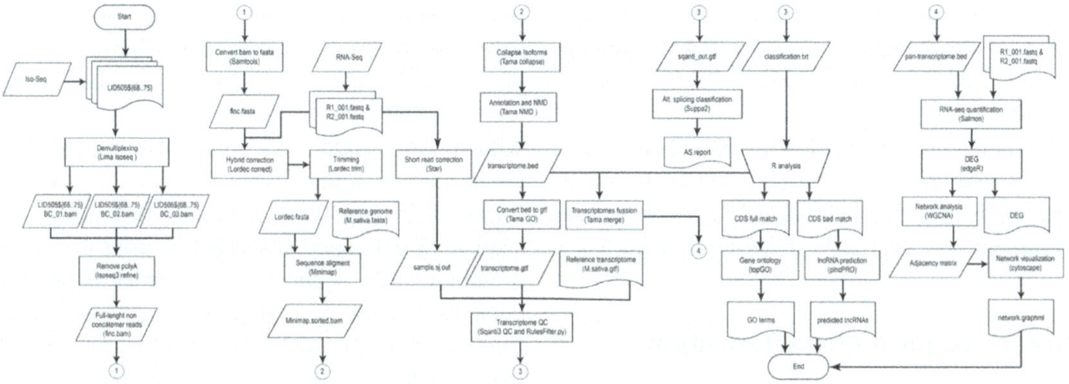

Fig. 10.4 A flow chart of bioinformatic pipelines used for analyzing Iso-Seq and RNA-Seq data. Start and end of the analysis are represented by terminator symbol, processes are represented by rectangles. Input and output data are represented by parallelograms. Manual analysis and filtering analysis by R are represented by trapezoids

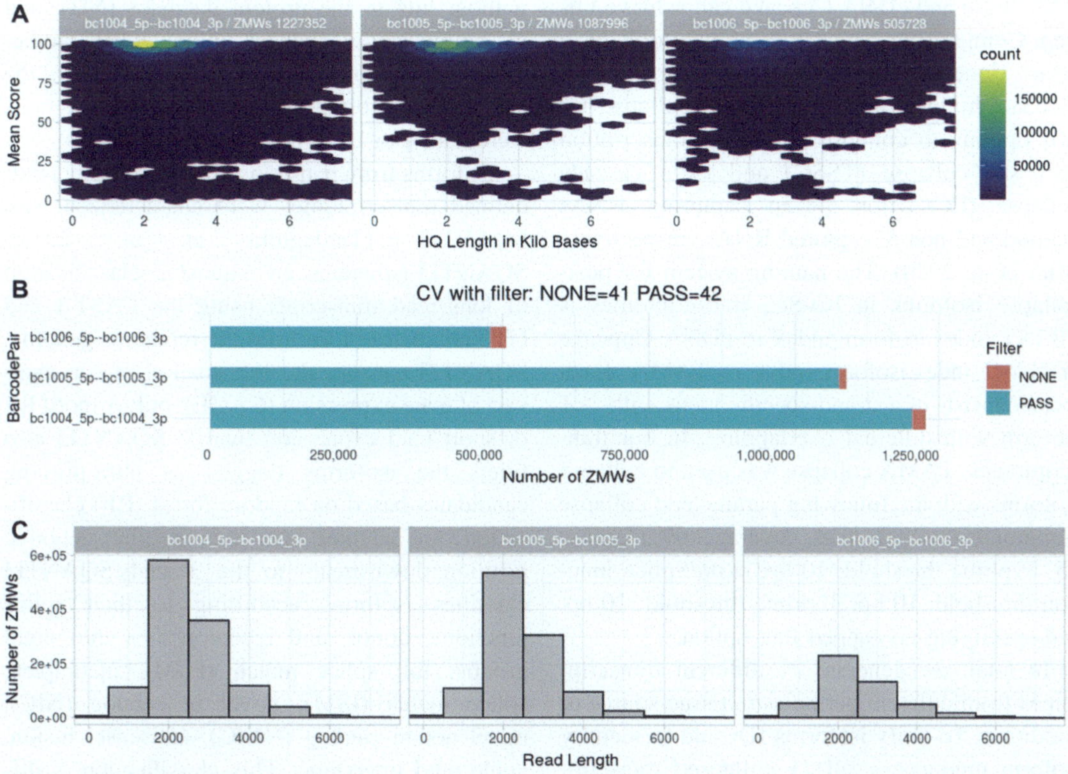

Fig. 10.5 Quality report after demultiplexing SMRTcell. A, HQ length versus barcode pair mean score (99.9% percentile) as an indicator to indicate the match of the chosen barcode pairs, where 0 is no hit and 100 is perfect match. B, Yield after the positive predictive value (PPV) of zero-mode waveguides (ZMWs). C, Histograms of read length versus number of ZMWs. Barcodes used to demultiplexing are bc_1004, bc_1005 and bc_1006

transcriptomes can be aligned to the reference genomes using the aligner ofMinimap2 (Li 2018) or GMAP (Wu and Watanabe 2005) which support spliced alignments to a reference genome. It has been reported that Minimap2 could be 40 times faster than GMAP (Li 2018). Finally, the aligned files were sorted using sort $-k$ 3, 3 $-k$ 4, 4n to obtain the sorted BAM files. In humans and mice or organisms with canonical GT/AG splice junctions, the authors recommend Minimap2with the function—splice-flank = yes. In *Medicago sativa*, a non-canonical splice junction is present (He et al. 2008), therefore, the alignment was done as—splice-flank = no.

10.6.3 Isoform Characterization

Aligned Iso-Seq data will have multiple transcripts in the same isoform also known as redundant isoforms. The next step requires collapse redundant isoforms. TAMA collapse (Kuo et al. 2020) and cDNA Cupcake can achieve this step. Comparisons between these two tools have been done in rice with similar results (Schaarschmidt et al. 2020). TAMA collapse has two options to collapse isoforms: transcription start site collapse (TSSC) and exon cascade collapse (ECC) for 5′-Cap capturing RNAs method and non 5′ captured RNAs, respectively (Kuo et al. 2020). The naming system for post-collapse isoforms in Iso-Seq transcriptomes is PB.loci_index.isoform_index in cDNA Cupcake or G.loci_index.isoform_index in TAMA. Each locus consists of a strand-specific locus with ≥ 1 isoform with different overlapping. In our transcriptomes, TAMA collapse was used to collapse isoforms with the following parameters: collapse common exons ends flags, coverage: 99, identity: 85, 5′ prime threshold: 10 bp, exon/Splice junction threshold: 10 bp, 3′ prime threshold: 10 bp, and specifying no capped flag option.

In total, we generate 21 different transcriptomes from different germplasm, tissue source or condition. To unify isoforms IDs and to identify isoform uniqueness, all 21 collapsed transcriptomes were merged using TAMA merge to generate a pan-transcriptome with 91,378 genes and 1,124,275 non-redundant isoforms. The pan-transcriptome data showed a highest number of isoforms on chromosome *4.2* (43,918) and the lowest number of isoforms were on chromosome *6.4* (9,804) (Table 10.4). Our pan-transcriptome data present a significant increment compared with 164,632 predicted isoforms during the genome sequencing (Chen et al. 2020). Finally, pan-transcriptome was annotated, and non-sense mediated mRNA decay (NMD) was predicted with TAMA GO: ORF and NMD predictions. Open reading frames were predicted in all isoforms, then they were translated to amino acid sequences. Diamond BASTP was used to search protein identity against the Uniprot100 protein database (Buchfink et al. 2015; Bateman 2019). The results of BLASTP annotation were classified into four groups: (1) open reading frames (ORFs) with complete BLASTP match (FM); (2) ORFs with incomplete BLASTP match categorized in >90, >50% or $\leq 50\%$ of amino acid sequence identity by the BLASTP; (3) ORFs without hits in the protein database (NH); and (4) amino acid sequences without ORF prediction by tama_orf_seeker.py (NO). Finally, the annotation information and NMD predictions were added to BED file.

Isoforms from pan-transcriptome or long read-defined transcriptomes were characterized with SQANTI3 (Tardaguila et al. 2018). SQANTI3 generates an extensive classification of long-read transcripts using the FASTA and GTF annotation files from the reference genome. SQANTI3 can use the information of cap analysis of gene expression (CAGE), polyA motif list or short read expression matrix. SQANTI3 also filters the isoforms flagged as intra-priming candidates based on random forest (RF) classification for adenine stretches in the genomic position downstream to the 3′ ends.SQANTI3 classifies isoforms according to their splice junctions, donor, and receptor sites into eight groups: full splice match (FSM), incomplete splice match (ISM), novel in catalog (NIC), novel not in catalog (NNIC), antisense, fusion, genic, and intergenic. This classification is different from other software like SUPPA2 (Trincado et al. 2018) or Astalavista (Foissac and Sammeth 2007) which use seven types of alternative splicing events: skipping exon (SE), alternative 5′ splice sites (A5), alternative 3′ splice site (A3), mutually exclusive exons (MX), intron retention (IR), alternative first exon (AF), and alternative last exon (AL).

Table 10.4 Gene isoforms predicted by Chen et al. (2020) (upper part) versus isoforms detected in pan-transcriptome (Unpublished) (lower part) in *M. sativa*

Chr	H1	H2	H3	H4	Total
1	5,495	5,723	5,256	5,728	22,202
2	4,661	4,403	4,644	4,658	18,366
3	5,740	5,357	5,840	5,716	22,653
4	5,554	5,752	5,696	5,696	22,698
5	4,796	4,921	4,633	4,693	19,043
6	3,087	3,468	3,454	2,366	12,375
7	4,862	5,200	5,186	5,248	20,496
8	5,337	5,356	5,273	5,095	21,061
S					5,738
[a]Total	45,270	40,180	39,982	39,200	164,632
1	41,454	41,062	39,494	41,846	163,856
2	33,549	31,129	34,302	33,533	132,513
3	38,865	36,664	38,839	38,038	152,406
4	41,115	43,918	42,073	43,159	170,265
5	32,791	29,694	29,389	32,500	124,374
6	14,548	16,454	17,937	9804	58,743
7	32,330	36,261	36,818	34,204	139,613
8	39,545	39,363	39,775	38,168	156,851
S					25,654
[b]Total	274,197	274,545	278,627	271,252	1,124,275

Notes [a]Data from genome sequences of *M. sativa* cv. XinJiangDaYe
[b]Data from Pan-transcriptome of *M. sativa* cv. Wilson, Saranac and PI467895. H, haplotype; S, Scaffold

10.6.4 lncRNAs Predictions

LncRNAs are RNAs >200 bp in length with low expression, high instability, involved in *cis*- and *trans*-acting regulation processes including plant adaptation to abiotic stress (Wang et al. 2015; Zhao et al. 2020). Software for lncRNAs prediction is based on machine learning approaches. Previous reports in *Medicago truncatula* used a coding potential calculator (Kong et al. 2007) to identify lncRNAs under osmotic, salt, and cold stress (Wang et al. 2015; Zhao et al. 2020). In our work, we used plncPRO (Singh et al. 2017), which is designed to predict lncRNAs in plants by RF after building a specific model using the mRNA and lncRNA data of *M. truncatula*.

10.6.5 Gene Ontology

GO enrichment analysis is to find GO terms that are over-represented or under-represented using annotations for a specific gene set. We selected only isoforms identified in stressed plants but not in control plants. GO terms were retrieved from Uniprot1000 DB and enrichment analysis was done with the topGO R package using the algorithm “weight01 and Kolmogorov-Smirnov test (Alexa and Rahnenfuhrer 2019).

10.6.6 Weighed Gene Co-expression Network Analysis (WGCNA)

Co-expression networks are used to discover transcriptional relationships (edges) between genes (nodes). The WGCNA R package is one of the most popular software used to generate co-expression networks (Langfelder and Horvath 2008). WGCNA uses "Guilt-by-Association" (GBA) principle. The GBA principle supposes that genes sharing similar functions are co-expressed, helping to predict new functions for known proteins and correlations between genes. We generated a co-expression network using normalized expression data from RNA-seq using leaf, stem, and root tissue sources as replicates for the same germplasm and condition. We found that 25 modules and 10,160 isoforms significantly co-expressed (Pearson correlation > 0.5 and p-value < 0.05) in PI467895-SS, Wilson-DS, Saranac-SS, and Saranac-DS. All significant modules were exported to Cytoscape to visualize transcript networks and to identify hub nodes. This approach is useful to relate the function of unknown genes and identify master genes involved in abiotic stress response. One problem with WGCNA is the generation of undirected networks, which means that the regulatory direction between two genes is unknown. The use of new tools like convolutional neural network for co-expression (CNNC) based on deep learning (Yuan and Bar-Joseph 2019) or Arboreto based on GENIE3 (Huynh-Thu et al. 2010), and gradient boosting (Moerman et al. 2019) can generate directed networks; however, they are computer demanding. In alfalfa, the generation of co-expressed networks using WCGNA have been the first approach to identify hub genes related to drought and salt stress responses.

10.7 Conclusions and Future Perspectives

With advances of new genomic technologies, SNP genotyping and marker development have become popular by providing important platforms in genomics and quantitative genetics. In alfalfa, sequencing-based approaches such as GBS and RAD-seq have been used for generating SNPs in several studies (Yu et al. 2017a, b; Yu 2017; Zhang et al. 2019; Medina et al. 2020; Lin et al. 2020). However, the generation of long reading sequences for genotyping is still under development. The use of PacBio and Oxford Nanopore Technologies (ONT) can improve genotyping efficiency, providing platforms for generating LRS. Beside SNP and insertions/deletion (INDELs), the LRS allows to discover haplotypic variations in complex genomes such as alfalfa. The PacBio has been used in alfalfa genomic and RNA sequencing and successful assembly of two alfalfa genomes (Chen et al. 2020; Shen et al. 2020). The ONT has been applied for SNP genotyping in highly duplicated genomes like allotetraploid *Brassica napus* (Malmberg et al. 2019). It is likely applicable to alfalfa genomes. However, the main constrain of ONT is the high error rate ($\sim 11\%$) which requires correction from the short read sequencing (SRS) by Illumina to polish the sequence quality (Magi et al. 2016). The improvement of the newest nanopore sequencing chemistries R10.3 or base calling software is required to reduce the error rate (Wang et al. 2020; Karst et al. 2021). R10.3 nanopore chemistry provides improved homopolymer performance and may overcome the high sequencing error rate.

The improvements in genotyping call accuracy in tetraploids have been achieved using software updog (Gerard et al. 2018) and polyRAD (Clark et al. 2019). Both the software uses the Empirical Bayes approach for genotyping polyploid individuals. However, NGSEP provides a more friendly platform for genotyping alfalfa with allele dosage, although it is necessary to find the correct threshold between sequencing depth and the number of markers generated by GBS. Targeted GBS is an upgraded approach obtaining increased numbers of marker and sequencing depth (Ott et al. 2017). This strategy selects a subset of RE fragments for amplifying a flexible genome of reduction, increasing the read depth at target sites. Additionally, precise SNP genotyping facilitates the downstream analysis. GWASPoly is a R package for GWAS and offers multiple models to detect markers associated with traits. However, a more friendly usable program is needed to facilitate beginners to work on polyploid association mapping. The use of machine learning methods such as random forest or support vector machine showed the best performance and helped in increasing the prediction of GEBV in alfalfa GS under salt stress (Medina et al. 2020). In the near future, the use of robust tools like deep learning methods, which is a subfield of machine learning, will allow to incorporate complex interactions of dominance, epistatic effects, complementary factors in GS.

References

Alexa A, Rahnenfuhrer J (2019) topGO: enrichment analysis for gene ontology. R package version 2.37.0. https://bioconductor.org/packages/release/bioc/html/topGO.html. Accessed 30 Dec 2020

Annicchiarico P, Nazzicari N, Li X, Wei Y, Pecetti L et al (2015) Accuracy of genomic selection for alfalfa biomass yield in different reference populations. BMC Genom 16:1020

Bateman A (2019) UniProt: a worldwide hub of protein knowledge. Nucleic Acids Res 47:D506–D515

Benson G (1999) Tandem repeats finder: a program to analyze DNA sequences. Nucleic Acids Res 27:573–580

Bourke PM, Voorrips RE, Visser RGFF, Maliepaard C (2018) Tools for genetic studies in experimental populations of polyploids. Front Plant Sci 9:513

Buchfink B, Xie C, Huson DH (2015) Fast and sensitive protein alignment using DIAMOND. Nat Methods 12:59–60

Chen H, Zeng Y, Yang Y, Huang L, Tang B et al (2020) Allele-aware chromosome-level genome assembly and efficient transgene-free genome editing for the autotetraploid cultivated alfalfa. Nat Commun 11:2494

Clark LV, Lipka AE, Sacks EJ (2019) polyRAD: genotype calling with uncertainty from sequencing data in polyploids and diploids. G3 Genes, Genomes, Genet g3.200913.2018

Clark LV, Mays W, Lipka AE, Sacks EJ (2020) A population-level statistic for assessing Mendelian behavior of genotyping-by-sequencing data from highly duplicated genomes. bioRxiv 1–21

Crossa J, Pérez-Rodríguez P, Cuevas J, Montesinos-López O, Jarquín D et al (2017) Genomic selection in plant breeding: methods, models, and perspectives. Trends Plant Sci 22:961–975

Cui G, Chai H, Yin H, Yang M, Hu G et al (2019) Full-length transcriptome sequencing reveals the low-temperature-tolerance mechanism of *Medicago falcata* roots. BMC Plant Biol 19:575

de Bem Oliveira I, Resende MFR, Ferrão LFV, Amadeu RR, Endelman JB et al (2019) Genomic prediction of autotetraploids; influence of relationship matrices, allele dosage, and continuous genotyping calls in phenotype prediction. G3 Genes. Genomes, Genet 9:1189–1198

Duitama J, Quintero JC, Cruz DF, Quintero C, Hubmann G et al (2014) An integrated framework for discovery and genotyping of genomic variants from high-throughput sequencing experiments. Nucleic Acids Res 42:e44

Elshire RJ, Glaubitz JC, Sun Q, Poland JA, Kawamoto K et al (2011) A robust, Simple Genotyping-by-Sequencing (GBS) approach for high diversity species. PLoS One 6:

Endelman JB, Carley CAS, Bethke PC, Coombs JJ, Clough ME et al (2018) Genetic variance partitioning and genome-wide prediction with Allele dosage information in autotetraploid potato. Genetics 209:77–87

Foissac S, Sammeth M (2007) ASTALAVISTA: dynamic and flexible analysis of alternative splicing events in custom gene datasets. Nucleic Acids Res 35:W297–W299

Fu S, Wang A, Au KF (2019) A comparative evaluation of hybrid error correction methods for error-prone long reads. Genome Biol 20:26

Garreta L, Cerón-Souza I, Palacio MRR, Reyes-Herrera PHH (2020) MultiGWAS: an integrative tool for Genome Wide Association Studies (GWAS) in tetraploid organisms. bioRxiv 1–24

Garrison E, Marth G (2012) Haplotype-based variant detection from short-read sequencing. 1207.3907

Gerard D, Ferrão LFV, Garcia AAF, Stephens M (2018) Genotyping polyploids from messy sequencing data. Genetics 210:789–807

Glaubitz JC, Casstevens TM, Lu F, Harriman J, Elshire RJ et al (2014) TASSEL-GBS: a high capacity genotyping by sequencing analysis pipeline. PLoS One 9:e90346

Gordon SP, Tseng E, Salamov A, Zhang J, Meng X et al (2015) Widespread polycistronic transcripts in fungi revealed by single-molecule mRNA sequencing. PLoS One 10:e0132628

Gore M, Bradbury P, Hogers R, Kirst M, Verstege E et al (2007) Evaluation of target preparation methods for single-feature polymorphism detection in large complex plant genomes. Crop Sci 47:S135–S148

Grinberg NF, Orhobor OI, King RD (2020) An evaluation of machine-learning for predicting phenotype: studies in yeast, rice, and wheat. Mach Learn 109:251–277

Hawkins C (2018) convert-tet-vcf. https://github.com/CharlesHawkins/convert-tet-vcf. Accessed 18 Mar 2020

Hawkins C, Yu L-X (2018) Recent progress in alfalfa (*Medicago sativa* L.) genomics and genomic selection. Crop J 6:565–575

He Z, Zou H, Wang Y, Zhu J, Yu G (2008) Maturation of the nodule-specific transcript MsHSF1c in Medicago sativa may involve interallelic trans-splicing. Genomics 92:115–121

Huynh-Thu VA, Irrthum A, Wehenkel L, Geurts P (2010) Inferring Regulatory networks from expression data using tree-based methods. PLoS One 5:

Jia C, Zhao F, Wang X, Han J, Zhao H et al (2018) Genomic Prediction for 25 agronomic and quality traits in Alfalfa (*Medicago sativa*). Front Plant Sci 9:1220

Karst SM, Ziels RM, Kirkegaard RH, Sørensen EA, McDonald D et al (2021) High-accuracy long-read amplicon sequences using unique molecular identifiers with Nanopore or PacBio sequencing. Nat Methods. https://doi.org/10.1038/s41592-020-01041-y

Kong L, Zhang Y, Ye Z-Q, Liu X-Q, Zhao S-Q et al (2007) CPC: assess the protein-coding potential of transcripts using sequence features and support vector machine. Nucleic Acids Res 35:W345–W349

Kuo RI, Cheng Y, Zhang R, Brown JWS, Smith J et al (2020) Illuminating the dark side of the human transcriptome with long read transcript sequencing. BMC Genom 21:751

Langfelder P, Horvath S (2008) WGCNA: an R package for weighted correlation network analysis. BMC Bioinformatics 9:559

Li H (2018) Minimap2: pairwise alignment for nucleotide sequences. Bioinformatics 34:3094–3100

Li H, Handsaker B, Wysoker A, Fennell T, Ruan J et al (2009) The Sequence alignment/map format and SAMtools. Bioinformatics 25:2078–2079

Lin S, Medina CA, Boge B, Hu J, Fransen S et al (2020) Identification of genetic loci associated with forage quality in response to water deficit in autotetraploid alfalfa (*Medicago sativa* L.). BMC Plant Biol 20:303

Liu Z, Chen T, Ma L, Zhao Z, Zhao PX et al (2013) Global Transcriptome sequencing using the Illumina platform and the development of EST-SSR markers in autotetraploid alfalfa. PLoS One 8:

Lobaton JD, Miller T, Gil J, Ariza D, Hoz JF et al (2018) Resequencing of common bean identifies regions of inter-gene pool introgression and provides comprehensive resources for molecular breeding. Plant Genome 11:

Magi A, Giusti B, Tattini L (2016) Characterization of MinION nanopore data for resequencing analyses. Brief Bioinform 18:940–953

Malmberg MM, Spangenberg GC, Daetwyler HD, Cogan NOI (2019) Assessment of low-coverage nanopore long read sequencing for SNP genotyping in doubled haploid canola (*Brassica napus* L.). Sci Rep 9:8688

McKenna A, Hanna M, Banks E, Sivachenko A, Cibulskis K et al (2010) The genome analysis toolkit: a MapReduce framework for analyzing next-generation DNA sequencing data. Genome Res 20:1297–1303

Medina CA, Hawkins C, Liu X-P, Peel M, Yu L-X (2020) Genome-wide association and prediction of traits related to salt tolerance in autotetraploid alfalfa (*Medicago sativa* L.). Int J Mol Sci 21:3361

Meuwissen THE, Hayes BJ, Goddard ME (2001) Prediction of total genetic value using genome-wide dense marker maps. Genetics 157:1819–1829

Moerman T, Aibar Santos S, Bravo González-Blas C, Simm J, Moreau Y et al (2019) GRNBoost2 and Arboreto: efficient and scalable inference of gene regulatory networks. Bioinformatics 35:2159–2161

Ott A, Liu S, Schnable JC, Yeh C-T Eddy, Wang K-S et al (2017) tGBS® genotyping-by-sequencing enables reliable genotyping of heterozygous loci. Nucleic Acids Res 45:e178–e178

Pecrix Y, Staton SE, Sallet E, Lelandais-Brière C, Moreau S et al (2018) Whole-genome landscape of *Medicago truncatula* symbiotic genes. Nat Plants 4:1017–1025

Pereira GS, Garcia AAF, Margarido GRA (2018) A fully automated pipeline for quantitative genotype calling from next generation sequencing data in autopolyploids. BMC Bioinform 19:398

Perkel J (2008) SNP genotyping: six technologies that keyed a revolution. Nat Methods 5:447–453

Postnikova OA, Shao J, Nemchinov LG (2013) Analysis of the Alfalfa root transcriptome in response to salinity stress. Plant Cell Physiol 54:1041–1055

Purcell S, Neale B, Todd-Brown K, Thomas L, Ferreira MAR et al (2007) PLINK: a tool set for whole-genome association and population-based linkage analyses. Am J Hum Genet 81:559–575

Quiros CF (1982) Tetrasomic segregation for multiple alleles in Alfalfa. Genetics 101:117–127

Rice B, Lipka AE (2019) Evaluation of RR-BLUP genomic selection models that incorporate peak genome-wide association study signals in maize and sorghum. Plant Genome 12:180052

Rochette NC, Rivera-Colón AG, Catchen JM (2019) Stacks 2: analytical methods for paired-end sequencing improve RADseq-based population genomics. Mol Ecol 28(21):4737–4754

Rosyara UR, De Jong WS, Douches DS, Endelman JB (2016) Software for genome-wide association studies in autopolyploids and its application to potato. Plant Genome 9:1–10

Salmela L, Rivals E (2014) LoRDEC: accurate and efficient long read error correction. Bioinformatics 30:3506–3514

Schaarschmidt S, Fischer A, Lawas LMF, Alam R, Septiningsih EM et al (2020) Utilizing PacBio Iso-Seq for novel transcript and gene discovery of abiotic stress responses in *Oryza sativa* L. Int J Mol Sci 21:8148

Scheet P, Stephens M (2006) A Fast and flexible statistical model for large-scale population genotype data: applications to inferring missing genotypes and haplotypic phase. Am J Hum Genet 78:629–644

Schmitz Carley CA, Coombs JJ, Douches DS, Bethke PC, Palta JP et al (2017) Automated tetraploid genotype calling by hierarchical clustering. Theor Appl Genet 130:717–726

Serang O, Mollinari M, Garcia AAF (2012) Efficient Exact maximum a posteriori computation for bayesian SNP genotyping in polyploids. PLoS One 7:e30906

Shen J, Li Z, Chen J, Song Z, Zhou Z et al (2016) SHEsisPlus, a toolset for genetic studies on polyploid species. Sci Rep 6:24095

Shen C, Du H, Chen Z, Lu H, Zhu F et al (2020) The Chromosome-level genome sequence of the autotetraploid alfalfa and resequencing of core germplasms provide genomic resources for Alfalfa research. Mol Plant 13:1250–1261

Shu Y, Li W, Zhao J, Zhang S, Xu H et al (2017) Transcriptome sequencing analysis of alfalfa reveals cbf genes potentially playing important roles in response to freezing stress. Genet Mol Biol 40:824–833

Singh U, Khemka N, Rajkumar MS, Garg R, Jain M (2017) PLncPRO for prediction of long non-coding RNAs (lncRNAs) in plants and its application for discovery of abiotic stress-responsive lncRNAs in rice and chickpea. Nucleic Acids Res 45:e183

Tardaguila M, de la Fuente L, Marti C, Pereira C, Pardo-Palacios FJ et al (2018) SQANTI: extensive characterization of long-read transcript sequences for quality control in full-length transcriptome identification and quantification. Genome Res 28:396–411

Tello D, Gil J, Loaiza CD, Riascos JJ, Cardozo N et al (2019) NGSEP3: accurate variant calling across species and sequencing protocols. Bioinformatics 35:4716–4723

Trincado JL, Entizne JC, Hysenaj G, Singh B, Skalic M et al (2018) SUPPA2: fast, accurate, and uncertainty-aware differential splicing analysis across multiple conditions. Genome Biol 19:40

Uitdewilligen JGAML, Wolters A-MA, D'hoop BB, Borm TJA, Visser RGF et al (2013) A next-generation sequencing method for genotyping-by-sequencing of highly heterozygous autotetraploid potato. PLoS One 8:e62355

Voorrips RE, Gort G, Vosman B (2011) Genotype calling in tetraploid species from bi-allelic marker data using mixture models. BMC Bioinform 12:172

Waldmann P (2019) On the use of the pearson correlation coefficient for model evaluation in genome-wide prediction. Front Genet 10

Wang T-Z, Liu M, Zhao M-G, Chen R, Zhang W-H (2015) Identification and characterization of long non-coding RNAs involved in osmotic and salt stress in *Medicago truncatula* using genome-wide high-throughput sequencing. BMC Plant Biol 15:131

Wang L, Qu L, Yang L, Wang Y, Zhu H (2020) NanoReviser: an error-correction tool for nanopore sequencing based on a deep learning algorithm. Front Genet 11:900

Wu TD, Watanabe CK (2005) GMAP: a genomic mapping and alignment program for mRNA and EST sequences. Bioinformatics 21:1859–1875

Yang SS, Tu ZJ, Cheung F, Xu WW, Lamb JF et al (2011) Using RNA-Seq for gene identification, polymorphism detection and transcript profiling in two alfalfa genotypes with divergent cell wall composition in stems. BMC Genom 12:199

Yu L-X (2017) Identification of single-nucleotide polymorphic loci associated with biomass yield under water deficit in Alfalfa (*Medicago sativa* L.) using genome-wide sequencing and association mapping. Front Plant Sci 8:1152

Yu L-X, Zheng P, Bhamidimarri S, Liu X-P, Main D (2017a) The impact of genotyping-by-sequencing pipelines on SNP discovery and identification of markers associated with verticillium wilt resistance in autotetraploid alfalfa (*Medicago sativa* L.). Front Plant Sci 8:89

Yu L-X, Zheng P, Zhang T, Rodringuez J, Main D (2017b) Genotyping-by-sequencing-based genome-wide association studies on Verticillium wilt resistance in autotetraploid alfalfa (*Medicago sativa* L.). Mol Plant Pathol 18:187–194

Yuan Y, Bar-Joseph Z (2019) Deep learning for inferring gene relationships from single-cell expression data. Proc Natl Acad Sci USA 116:27151–27158

Zhang F, Kang J, Long R, Yu L-X, Wang Z et al (2019) High-density linkage map construction and mapping QTL for yield and yield components in autotetraploid alfalfa using RAD-seq. BMC Plant Biol 19:165

Zhao M, Wang T, Sun T, Yu X, Tian R et al (2020) Identification of tissue-specific and cold-responsive lncRNAs in *Medicago truncatula* by high-throughput RNA sequencing. BMC Plant Biol 20:99

Zych K, Gort G, Maliepaard CA, Jansen RC, Voorrips RE (2019) FitTetra 2.0—improved genotype calling for tetraploids with multiple population and parental data support. BMC Bioinform 20:148

11 Genomic Resources for Breeding in Alfalfa: Availability, Utility, and Adoption

Maria Katherine Mejia-Guerra, Dongyan Zhao, and Moira J. Sheehan

Abstract

Alfalfa breeding still relies predominantly on recurrent phenotypic selection and breeder's experience, but the adoption of breeding strategies relying on genomics information is gaining momentum quickly as tools and knowledge become more accessible. To accelerate alfalfa improvement using genomics, the community needs to establish reliable marker sets that track parental relationships, as well as provide affordable, repeatable genotyping methods to move towards marker assisted selection and genomic selection. However, access to genomic resources is not sufficient to make genotype-based selections. It is crucial that breeders capture phenotypes widely (across desirable and undesirable individuals) and generate such data in large quantities with high quality (not addressed in this chapter). For breeders unfamiliar or new to genomic analyses, the utilization of phenotypic and genotypic data to make decisions is a knowledge gap they may not be able to bridge on their own. Therefore, a public-sector initiative has been put in place to aid breeders and hasten the adoption of new technologies, including high-throughput genotyping, to accelerate breeding and pre-breeding efforts.

M. K. Mejia-Guerra · D. Zhao · M. J. Sheehan (✉)
Breeding Insight, Cornell University, 119 CALS Surge Facility, 525 Tower Rd., Ithaca, NY 14853, USA
e-mail: moirasheehan@cornell.edu

M. K. Mejia-Guerra
e-mail: mm2842@cornell.edu

D. Zhao
e-mail: dz359@cornell.edu

11.1 Introduction

In order to use genomic analyses, breeders need to detect differences in the genomes of individuals, groups, populations, or species in a process called "genotyping". Genotyping consists of using molecular tools to determine DNA polymorphisms, and for this chapter, we will focus on two types of polymorphisms: microsatellites (or simple sequence repeats, SSRs) and single nucleotide polymorphisms (SNPs), as they are the most commonly used DNA markers in alfalfa to date. To detect markers, there are several technology platforms from which to choose, each with its own costs and limitations. For breeders just starting to think about genotyping their breeding material, many feel overwhelmed with the platform options and are understandably cautious about buying into a technology where startup costs are substantial. Aside from purchasing the original assays or arrays, the cost to use the technology, the turnaround time to data, and even the number of markers may not be ideal or feasible for the breeder's intended use.

L.-X. Yu and C. Kole (eds.), *The Alfalfa Genome*, Compendium of Plant Genomes,
https://doi.org/10.1007/978-3-030-74466-3_11

Marker assisted selection (MAS) and genomic selection (GS) rely on identifying DNA polymorphisms that are associated with trait values. Rapid, accurate, and inexpensive genotyping is needed for alfalfa breeding to adopt these genomic strategies within the breeding cycle. In the past, genotyping efforts have had relatively little impact on alfalfa breeding practices. The reasons for this are varied, but the limitations of available technologies to address the challenges of alfalfa biology, as well as a lack of experience in using data in breeding decisions, oftentimes did not justify the incurring costs. In other words, having a platform at alfalfa breeders' fingertips did not mean they had the downstream tools to use or interpret genomic data when making breeding decisions. When considering adding genotypic data to a program, the whole pipeline must be considered as a workflow that dovetails with the breeders' other activities to find the most feasible solutions.

Shifting from phenotypic selection to genomic selection has huge potential for alfalfa breeding. A cost-effective, mid-density (3000–6000) marker panel, with rapid (i.e., 3-week) turnaround time can be created as a public resource and would allow informed selection at multiple points in the breeding cycle. Early in the breeding cycle, genotyping seedlings for key disease-resistant traits and/or digestibility traits could allow culling of undesirable genotypes, reducing the number of individuals that advance and saving resources. Identifying quantitative trait loci (QTLs) or performing genome-wide association study (GWAS) experiments to identify markers, for which new MAS can be applied, will allow breeders more selection opportunities. With accurate GS models, additional key traits can be selected much earlier, and without the cost of phenotyping, allowing the cycle time to be shortened and with more desirable genotypes at each stage. Without reliable marker data, these applications are well out of reach of breeders.

In this chapter, we review (1) the history of developing genomic resources for marker discovery and mid-density genotyping in alfalfa breeding, (2) the specific barriers in alfalfa to apply molecular tools in breeding, and (3) the ongoing work of the public sector to lower or remove the barriers that alfalfa breeders face in adopting genotypic information to be used routinely to accelerate alfalfa improvement.

11.2 Transcriptomic Resources for Marker Discovery

Generating genomics resources in alfalfa (*Medicago sativa* spp.*sativa,* and related subspecies) has been slow, in contrast to resources in *Medicago truncatula*, a related model organism with a simpler genome (diploid and half the genome size) that diverged ~5.3 (estimated range from 3.7 to 7.3) million years ago. Until recently, the more readily available resources from *M. truncatula* have served as proxy genomic resources for alfalfa.

In the genomics era, the aspirational goal is to study DNA variation across each species (subspecies and related species) by the assembly of pan-genomes. Currently, the pan-genome of some plant species are being assembled (Bayer et al. 2020), but for most of the plant species, the study of DNA variation still relies on the assembly of a single reference individual. For species with genome complexity driven by high heterozygosity, variable ploidy, or large stretches of repetitive sequence, the assembly of a genome remains a difficult challenge. Because of the above, the generation of genomic resources frequently starts with the generation of transcriptome data, which only covers the portion of the DNA that encodes for genes. In alfalfa, early studies used collections of *M. truncatula* expressed sequence tags (ESTs) to identify candidate genes for some traits of interest in alfalfa breeding such as disease resistance and cold acclimation (Pennycooke et al. 2008; Yang et al. 2008).

Beyond the identification of candidate genes, transcriptome data allows one to generate gene-derived markers. For instance, markers derived from *M. truncatula* genic regions are often highly transferable to closely related species (Eujayl et al. 2004; Liu et al. 2015). SSRs derived from *M. truncatula* ESTs were proven useful to generate linkage maps, and to study population structure and diversity in alfalfa (Julier et al. 2003; Sledge et al. 2005). Liu et al. (2015), exploited a collection of 3828 EST sequences from *M. truncatula*, to test the transferability of SSR markers from ESTs with significant similarity to known genes encoding transcription factors. This study found that primers for 121 SSR markers were able to work successfully across 44 alfalfa accessions, and whose use was recommended for cultivar identification and genetic diversity studies on alfalfa germplasm (Liu et al. 2015).

As throughput and affordability increased, it was possible to generate transcriptome data directly from alfalfa and develop markers from said data. Using 12,371 ESTs directly derived from alfalfa, Wang et al. (2013) derived 28 polymorphic SSR markers that were capable of distinguishing between *M. sativa* subspecies *sativa* and the other subspecies. In a similar study, Liu et al. (2013) generated a larger set ($\sim$1649) of potential SSRs from ESTs, and from 100 randomly tested SSRs of which 82% were able to amplify sequences and $\sim$27% were polymorphic when tested in a small alfalfa panel of ten accessions. As the amount of transcriptome data increased, a group attempted to enrich the total number of available SSRs by mining alfalfa ESTs (Zhou et al. 2014). This effort resulted in the identification and testing of 750 SSRs from which 204 were polymorphic in a panel of ten alfalfa accessions.

Over time, high-throughput technologies supplanted laborious cDNA libraries and EST collections as a way to estimate levels of gene expression more broadly. The first high-throughput survey of alfalfa gene expression at a large scale relied on The Affymetrix GeneChip® Medicago Genome Array developed for *Medicago truncatula,* which was found capable of revealing differentially expressed genes (DEGs) between conditions and tissues of interest (Tesfaye et al. 2006). More recently, gene arrays have been gradually replaced by high-throughput short-read sequencing (commonly called next generation sequencing or NGS), in particular the development of RNA-seq. As surveying the transcriptome from alfalfa breeding and pre-breeding material and related species became possible, several comparative analyses using RNA-seq resulted in a better understanding of the molecular mechanisms behind alfalfa salt tolerance, tolerance to animal grazing, and fall dormancy (Wang et al. 2016; Lei et al. 2018; Liu et al. 2019).

RNA-seq data has also been used to derive alfalfa genetic markers in large numbers. Wang et al. (2014) aligned alfalfa RNA-seq data to the *Medicago truncatula* genome and identified a total of 4493 SSRs, from which 527 were tested and 372 were found to be polymorphic. As sequencing prices dropped, the use of SNPs as markers overtook SSRs as the preferred marker type. A notable example of the power of RNA-seq for SNP discovery was the study of Li et al. (2012) where RNA-seq was used to identify $\sim$0.9 million SNPs by sequencing transcriptomes of a diversity panel of 27 alfalfa accessions (Li et al. 2012). The resulting dataset was used to develop Illumina's Infinium-based array for 9277 biallelic genic SNPs to be used as a genotyping platform for alfalfa (Han et al. 2014).

11.3 Reduced Representation Libraries (RRLs)

As the cost of high-throughput sequencing continued to drop, analyzing genetic variation across a genome became a reality through the use of reduced representation libraries (RRLs), which allowed a portion of the genome to be targeted for sequencing. These RRLs are behind several low-cost genotyping strategies, such as restriction-enzyme-based genotyping-by-sequencing (GBS) (Scheben et al. 2017), and restriction site associated DNA sequencing (RAD-seq) (Peterson et al. 2012), that can yield tens of thousands of SNPs and can be applied to a large number of samples for a given species.

Before the advent of RRLs, genetic maps in *M. sativa* were developed from a combination of markers, such as amplified fragment length polymorphism (AFLP) DNA markers, randomly amplified polymorphic DNA (RAPDs), and SSRs, among others (Brummer et al. 1993; Kiss et al. 1993). These early maps were of low density, and developed only from diploid material, and while quite important as genetic resources for alfalfa at the time, lacked the desired resolution for fine mapping of alfalfa traits (Sakiroglu et al. 2012).

Since the introduction of genotyping strategies based on RRLs, it has become possible to build high-density linkage maps from diploid and tetraploid alfalfa material (Li et al. 2014b; Adhikair et al. 2018; Zhang et al. 2019). Recently, two studies were published in tetraploid material to study alfalfa yield. Zhang et al., used RAD-seq to build a dense linkage map and identify QTLs for yield and related yield traits such as plant height, and shoot diameter (Zhang et al. 2019). A similar strategy used by Adhikari et al. to study flowering time and biomass yield relied on GBS to build the linkage map and identify QTLs and molecular markers, as well as potential candidate genes associated with these traits (Adhikari et al. 2019).

GBS has also been used to understand alfalfa population diversity (Annicchiarico et al. 2016), and for genome-wide association studies (GBS-GWAS) (Biazzi et al. 2017; Sakiroglu and Brummer 2017). For genomic selection, GBS has been applied to estimate breeding values for forage yield and biomass yield (Annicchiarico et al. 2015; Li et al. 2015). The accuracy achieved in both studies (0.36 and 0.40, respectively) has great practical implications; suggesting that incorporation of genomic information can accelerate alfalfa gains per unit time compared with current non-genomic-based strategies.

11.4 Whole Genome Sequencing in Alfalfa

Whole-genome sequencing (WGS, also known as skim sequencing), offers better coverage than other discrete marker types because the sequenced fragments (reads) are in effect local haplotypes. Recently, high-throughput short-read sequencing technology has reached a price to make it affordable even to allow for large-scale marker discovery. However, WGS relies heavily on alignment to a reference genome, which wasn't available for alfalfa until recently (Li et al. 2020)

11.4.1 Genome Assembly in Diploid Alfalfa

The alfalfa community generated diploid cultivated alfalfa lines (CADL, Cultivated Alfalfa at the Diploid Level) (Bingham and McCoy 1979). The CADL lines are a desirable resource for many aspects of breeding, including easy introgression of genes from wild diploid relatives to the cultivated alfalfa. Because of the simplified genome structure, a CADL line was chosen for genome sequencing and assembly. This genome is available for download at the Medicago HapMap webpage http://www.medicagohapmap.org/downloads/CADL/. The assembly remains at contig level, but it has been reported that an improved version using PacBio long read technology is ongoing (Fajardo et al. 2016).

Recently, an assembly of the genome of *M. sativa* subsp. *caerulea* (Li et al. 2020), a diploid wild subspecies in the *Medicago sativa* complex, became available. This subspecies is thought to be the diploid progenitor of cultivated alfalfa and could be used as an alternative reference genome as has been done for other high-ploidy crops such as hexaploid bread wheat (Ling et al. 2013) and hexaploid sweetpotato (Wu et al. 2018). The assembly of *M. sativa* subsp. *caerulea* was derived from a combination of short and long reads (Illumina and Nanopore sequencing technologies) and relied on the capture of chromosome conformation coupled with paired-end sequencing (i.e., Hi–C) for the orientation of the contigs into the large pseudomolecules (Li et al. 2020). The assembled genome comprises 793.2 Mb of genomic sequence and 47,202 annotated protein-coding genes. The assembly and the gene annotation appear to be of high quality, with 97.7% of the Benchmarking Universal Single-Copy Orthologs (BUSCO (Simão et al. 2015)) genes identified.

11.4.2 Genome Assembly in Tetraploid Alfalfa

Despite the importance of alfalfa tetraploid cultivars for breeding, building a reference genome has been slowed by alfalfa's complicated autotetraploid genome architecture. The high genome complexity could be attributed to the outcrossing nature of this species and its hybridization between *M. sativa* subsp. *sativa* and *M. sativa* subsp. *falcata* to improve winter-hardiness (Riday et al. 2003). In addition, the short-read sequencing technology, predominant for the last decade, was sub-par for the task of resolving long contigs from a highly heterozygous autotetraploid in a large genome space (~3 Gb). It would take the incorporation of long reads, together with optical mapping and Hi–C, to obtain good quality and chromosome-level assemblies.

In 2020, two alfalfa reference genomes resolved to the chromosome level from two different tetraploid varieties were published. The first genome was obtained from the cultivar XinJiangDaYe (hereafter XJ), and the second was obtained from the cultivar Zhongmu-1 (hereafter ZM) (Chen et al. 2020; Shen et al. 2020). The two cultivars are widely grown in China, with the XJ cultivar being adapted to the local cold and dry weather, and the ZM cultivar being saline- and alkaline-tolerant. The transcriptome (leaf RNA-seq from plants grown in salt treatment) of the two cultivars have been compared (Lei et al. 2018), but a contrast between the two recently assembled genomes has not yet been made.

In terms of the generation of the assemblies, both genomes were obtained using a combination of sequencing technology including short (Illumina) and long reads (PacBio and Oxford Nanopore Technology). Chen et al., assembled the XJ's contigs in 32 super-scaffolds and 9789 unplaced unitigs (~419 Mb) using Hi–C paired-end reads, aiming to result in an allelic-aware genome of the 32 chromosomes (Chen et al. 2020). In contrast, Shen et al., assembled ZM's contigs into 8 pseudo-chromosomes (Shen et al. 2020), which likely resulted in a monoploid genome with collapsed haplotypes from different chromosome sets. In addition to the assembled sequence, both genomes provide annotations at the gene level, with the XJ and ZM assemblies containing 97.2 and 93.3%, respectively, of the complete set of BUSCO genes. Altogether, these data suggest that the two assemblies are near complete and of good quality.

11.4.3 Whole Genome Resequencing Across Diverse Germplasm

Until recently, WGS was not yet sufficiently economical for large-scale marker discovery. In *M. truncatula* WGS data at low coverage (~5X) allowed the creation of a HapMap (http://www.medicagohapmap.org/hapmap/about), which was of key importance to understand the population genetics of the species. The HapMap was generated from short-read sequencing of ~325 diverse *M. truncatula* germplasm, but also included several accessions from subspecies of *M. sativa*, one of which was later used for the assembly of a diploid *M. sativa* genome.

A HapMap for alfalfa is not available, but diversity studies using skim sequences are now possible for a large number of accessions. Using the assembly of the ZM genome, 137 accessions from cultivated *M. sativa* subsp. *sativa*, and 25 accessions from *M. sativa* subsp. *caerulea* were re-sequenced using short reads to characterize population migration history and genetic exchange between the subpopulations (Shen et al. 2020).

As the number of *M. sativa* subspecies' (cultivated alfalfa and related subspecies) genomes being sequenced continues to increase, it will be possible to create the alfalfa pan-genome, allowing for in-depth exploration of structural variation in the species. Cumulatively, the resources described here are expected to greatly facilitate the future improvement and application of molecular breeding strategies in alfalfa.

11.5 Barriers in Alfalfa to Exploit Genomics Information for Breeding

Compared with major row crops, e.g., maize, where genomics-enabled breeding has transformed the breeding process for decades, alfalfa breeding still relies heavily on recurrent phenotypic selection, which is slow, imprecise, and expensive. There are several barriers which hinder the incorporation of MAS and GS in alfalfa breeding.

11.5.1 Selection Cost is High

SSR markers have been the most widely used system for MAS in alfalfa (Hawkins and Yu 2018). Due to the primer hybridization issues in marker pools, only a handful of SSR markers can be genotyped in one PCR reaction, thus it is very low-throughput and can only be used for the selection of traits with known SSR markers associated. Currently, the estimated cost (DNA extraction, PCR, data interpretation, consumables, etc.) for one PCR reaction per sample is around \$20. While it is feasible for a handful of samples to be genotyped using this method, the cost of genotyping hundreds or thousands of samples can quickly become unaffordable. The fact that it is necessary to genotype many samples (usually ~ 1000) per polycross to achieve selection gain while avoiding inbreeding makes this genotyping platform infeasible as a routine method and too costly for most public alfalfa breeder budgets.

For several crops, SNP microarrays have been instrumental for genotyping pre-breeding and breeding material. One advantage of SNP microarrays is their ability to detect allelic dosage, a key requirement for genotyping tetraploid species (e.g., peanut 58 K SNP array, cotton 63 K SNP array) (You et al. 2018). In alfalfa, Illumina's Infinium-based array developed by Li et al. (2014a), was used to genotype a biparental cross between tetraploid alfalfa cultivars (i.e., Altet-4 and NECS-141) (Han et al. 2014). This study found that the array was able to detect the five SNP allelic dosages. However, as only $\sim 35\%$ (i.e., 3701) of the SNPs were found to be polymorphic and segregating in the progeny of the mapping population, the array was not cost-effective per data point for most alfalfa breeders.

SNP arrays impose strong ascertainment bias for diversity evaluation, as only a fixed set of known variants are included in the array. These variants may not be informative for a given population or crossable species and may result in fragmented, incomplete, or low-density genetic maps. In genetic diversity studies, the ascertainment bias of SNP arrays leads to false inferences about genome diversity, relatedness, and species boundaries. However, ascertainment bias can be useful in breeding programs, if the SNP array is thoughtfully designed on the entire breadth and width of the diversity utilized within the program. Unfortunately, the high costs for both the initial development and production, as well as the routine application, makes SNP arrays still prohibitive for most breeders.

GBS, which provides more coverage of the genome with more markers at a lower entry cost than SNP arrays has been successfully applied in several crops. However, the GBS cost per sample is on the order of \$30–40 (Annicchiarico et al. 2015), which hinders the implementation of this method in alfalfa breeding. Aside from the cost, GBS is a random genotyping method that has low or no ascertainment bias, meaning it is excellent for understanding genetic diversity and species boundaries, but is less well suited to biparental population genotyping and breeding applications. Breeders that are introgressing alleles or haplotypes need to follow the transmission of them in their program using stable, targeted markers. GBS cannot target the same sets of SNPs over and over in individuals or in populations due to the random nature of the platform.

11.5.2 Lack of Access to Genotyping Technologies Reliable for Handling Low-Quality DNA

To obtain reasonable gains from selection, thousands of plant samples are often needed. The process of collecting plant tissues is labor-intensive and prone to suboptimal conditions, both of which reduce the chances of getting DNA with sufficient quantity and quality required for downstream applications. Most current genotyping methods rely on high-quality genomic DNA that typically comes from low-throughput and expensive extraction methods. Conversely, affordable high-throughput DNA extraction methods usually provide poorer quality and lower quantities of genomic DNA (Anderson et al. 2018). In order to reduce the overall cost and increase sample numbers, a genotyping platform robust enough to work with low-quality DNA is needed.

11.5.3 Insufficient Bioinformatics and Analytics Tools for Data Interpretation

Bioinformatics and analytic tools have often lagged behind the development of new genotyping options. By taking advantage of the low cost of WGS for SNP discovery, a portion of those SNPs can be converted into a targeted amplicon sequencing technology marker platform. Targeted amplicon sequencing platforms allow deep sequencing of a large number (mid-density of 2000–3000 loci) of known regions in the genome (amplicons) to reveal variation, which can then be used to track traits of interest. Targeted amplicon sequencing is still relatively new, especially as applied to breeding. Some tools have been developed to tackle this type of data; however, most were developed for diploid species with relatively low genome complexity, including low heterozygosity. Processing and interpreting the genotyping results for alfalfa, an obligate outcrossing autotetraploid, remains a challenge. Moreover, breeders are experts in their area of research but their power to process large amounts of genotyping data may be limited or nonexistent.

11.6 Removal of Barriers to Applying Genomics-Enabled Selection in Alfalfa

Breeding Insight (BI), funded by The United States Department of Agriculture (USDA) Agricultural Research Service (ARS) through Cornell University, was founded to make genomics-enabled breeding a reality for small and specialty species breeding programs. Alfalfa is one of the pilot species included in this endeavor. The first step is creating a genotyping platform scaled in size, cost, and time to allow alfalfa breeders to incorporate genotyping workflows into their breeding program. BI sequenced (WGS, at low coverage) a collection of alfalfa lines that represented the crossable genetic diversity of the ARS alfalfa breeding programs. A SNP database was created for further filtering followed by the selection of a high-quality SNP set across the genome. In the meantime, BI explored genotyping platforms that could meet the alfalfa breeders' needs, i.e., low cost with DNA extraction included, high-throughput, amenable to low quality DNA, and with a less than 4-week turn-around time.

11.6.1 Selection of Alfalfa Lines that Captured Broad Genetic Diversity for US Breeding

ARS alfalfa breeders, with support from several alfalfa breeding companies, developed a strategy to choose alfalfa clones to form the diversity panel for SNP discovery. They focused on elite North American germplasm, including cultivated alfalfa cultivars from various dormancy groups and CADL plants (Table 11.1), for a total of 40 plants included in the panel. For maintenance and easy access to the research community, cuttings from the 40 plants were maintained at

Table 11.1 The diversity panel of 40 alfalfa plants used for marker development

Sample ID	Note	Provider	Additional notes
S&W dormancy 4	Elite parent	S&W seeds	Fall dormancy 4
S&W dormancy 5	Elite parent	S&W seeds	Fall dormancy 5
Legacy FD4	Elite parent	Legacy seeds	Fall dormancy 4
Legacy FD5	Elite parent	Legacy seeds	Fall dormancy 5
S&W dormancy 6	Elite parent	S&W seeds	Fall dormancy 6
S&W dormancy 7	Elite parent	S&W seeds	Fall dormancy 7
S&W dormancy 8	Elite parent	S&W seeds	Fall dormancy 8
S&W dormancy 9	Elite parent	S&W seeds	Fall dormancy 9
S&W Dormancy 9	Elite parent	S&W seeds	Fall dormancy 9
CADL-1	Cultivated alfalfa at diploid level	N. Young	S1 of sequenced plant
CADL-3	Cultivated alfalfa at diploid level	N. Young	S1 of sequenced plant
CADL-4-5	Cultivated alfalfa at diploid level	N. Young	S1 of sequenced plant
CADL-5-3	Cultivated alfalfa at diploid level	N. Young	S1 of sequenced plant
CADL-13	Cultivated alfalfa at diploid level	N. Young	S1 of sequenced plant
CADL-18	Cultivated alfalfa at diploid level	N. Young	S1 of sequenced plant
UMN3988-BIP	UMN3988	D. Samac	Biomass type
RegenSY27x	RegenSY	D. Samac	Regenerator, reference genome sequence
I-195	WAPH5	D. Samac	Highly resistant to Aphanomyces root rot
UT14-46 SP	NA	M. Peel	Tetraploid M. falcata
UT27-62	NA	M. Peel	Selection with salt tolerance
FL99	FL99	E. Rios	Fall dormancy 9
Bulldog 505	Elite parent	Ali Missaoui	Fall dormancy 5
GAMS 1403-FSH	Elite parent	Ali Missaoui	Fall dormancy 7
GAMS 1404-FSH	Elite parent	Ali Missaoui	Fall dormancy 8
GAMS 1405-FSH	Elite parent	Ali Missaoui	Fall dormancy 9
3010	Elite parent	Ali Missaoui	Fall dormancy 3
CW1010	Elite parent	Ali Missaoui	Fall dormancy 10
CUF101	From check seed	D. Samac	Fall dormancy 10 check
BIP1	Salt	M. Peel	27–62
BIP2	Salt	M. Peel	31–6
BIP3	SemiP	M. Peel	1–34

(continued)

Table 11.1 (continued)

Sample ID	Note	Provider	Additional notes
BIP4	SemiP	M. Peel	6–2
BIP5	SemiP	M. Peel	14–46
BIP6	Drought (Ut7)	M. Peel	17–43
BIP7	Drought (Ut8)	M. Peel	17–44
BIP8	Drought (Ut9)	M. Peel	18–22
BIP9	Drought (Ut10)	M. Peel	21–3
BIP10	Drought (Ut11)	M. Peel	22–30
BIP11	Drought (Ut26)	M. Peel	7–18
BIP12	Drought (Ut30)	M. Peel	13–14

three research programs: Debby Samac's group in Minnesota, Mike Peel's group in Utah, and Heathcliffe Riday's group in Wisconsin.

11.6.2 Development of the SNP Markers

BI skim sequenced the diversity panel from two biological replicates of the collection.

Sequencing was performed using the NovaSeqS4 to generate paired-end 150-bp reads, with an average 10x coverage. The recently published genome assembly of the autotetraploid alfalfa's cultivar XJ (Chen et al. 2020) was used as a reference genome. Reads were mapped using BWA against the chromosomic set of the genome assembly that included the most complete set of BUSCO genes (Chen et al. 2020). Variants were called with HaplotypeCaller using GATK version 3.8 as implemented by Sentieon, San Jose, CA, USA. For each replicate, variants were filtrated based on overall read depth, genotyping quality, retention of biallelic SNPs, and removal of indels. The high-quality SNPs from each replicate were overlapped based on position, and those in the agreement were selected for further analysis. Candidate markers, to be included in the genotyping platform, were selected by binning the genome to obtain uniform distribution and selecting preferentially SNPs in genic regions with the highest minor allele frequency.

11.6.3 Exploration of Genotyping Platforms and Vendors that Suit Breeders' Needs

A cost-effective genotyping platform is a key to the adoption of genomics-enabled breeding in alfalfa. To such end, BI has teamed up with the Excellence in Breeding (EiB) platform to provide breeders access to affordable, high-quality genotyping options, and commercial vendors with rapid turnaround times suitable for any genomic applications, including GS. BI's goal was to create an end-to-end vendor pipeline to receive plant tissue from the breeder and return genotypes back. Several requirements and criteria were used to establish the workflow (Fig. 11.1). DNA extraction must be inexpensive (<$2 per sample) and of sufficient quality to achieve good genotyping data downstream. The vendor for the targeted amplicon marker platform had to have the lowest set up cost, as well as the lowest routine genotyping cost with data returned in less than one month from tissue shipment/receipt. The workflow must eliminate the need for the breeder to be involved after shipping tissue and prior to data retrieval. After evaluating several possible vendor workflows, the EiB workflow of tissue to Intertek for DNA extraction, then direct DNA shipment to Diversity Array Technologies (DArT) for mid-density DArTag platform genotyping appeared to be the best fit for both BI and the breeding program needs (Fig. 11.1).

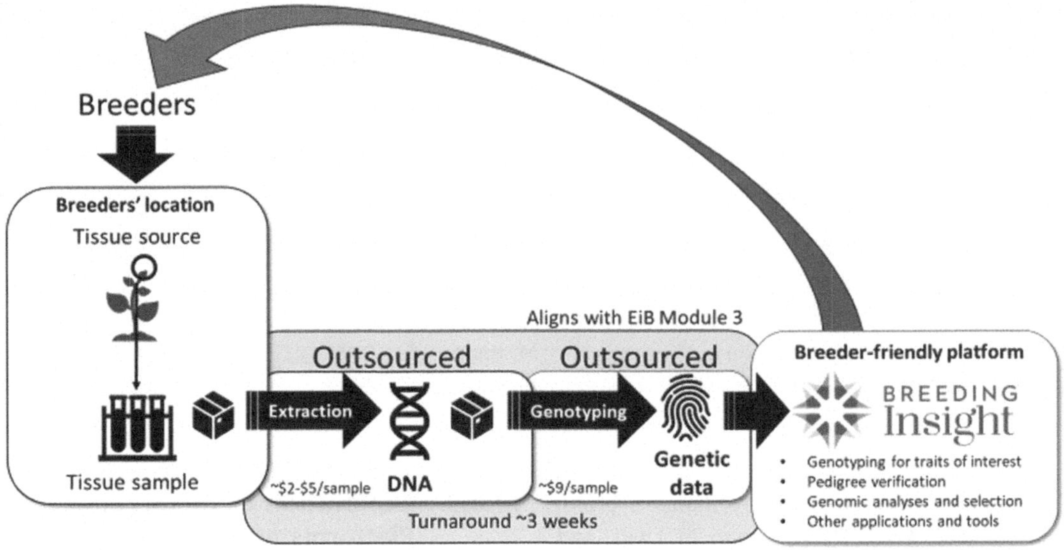

Fig. 11.1 Workflow for generation and analysis of genotyping data. Breeders will collect material from each individual program and ship tissue samples for DNA extraction. Next, DNA samples will be shipped for genotyping using the mid-density genotyping platform designed for alfalfa. Genotyping data will be sent to Breeding Insight (BI) to be properly stored and made available for data analysis in BI's breeder-friendly platform

11.6.4 Development of Bioinformatics Tools for Analyzing Genotyping Results

Without powerful bioinformatics and analytics tools, interpreting a large amount of genotyping data can be daunting or even impractical. Therefore, development of suitable bioinformatics tools has been an ongoing effort at BI. To assist alfalfa breeders' needs in a timely fashion, tool development and implementation will be carried out in two phases. For phase I, BI will use the existing tools publicly available coupled with custom scripting to generate final genotyping calls and estimate markers and trait associations based on phenotyping and genotyping results. By doing this, alfalfa breeders can quickly utilize the genotyping results to make breeding decisions in the immediate planting season. Assisting breeders on phase I is needed to facilitate the adoption of the new technologies; however, this model is not scalable as breeders' needs for speedy data analysis will increase. Because of that, BI is working to empower breeders in the adoption of using genotyping data with minimal assistance. To do so, in phase II, BI will release the resources for analysis developed in phase I in a package for the general research community. The package will be available on a breeder-friendly interface, where alfalfa breeders can run the analyses themselves.

The creation and application of a SNP panel is the first step towards molecular breeding in alfalfa. The genetic architecture of key traits such as yield, fall regrowth, and disease resistance can be elucidated and exploited for improving alfalfa cultivars and populations. Having the ability to create, store, and utilize genetic data will shorten the response time for breeders when new traits or market needs emerge. In the toolbox, alfalfa breeders have to meet sustainability and production security for U.S. alfalfa growers, genomic data will be a powerful new tool.

Acknowledgments We thank Debby Samac, Mike Peel, Heathcliffe Riday, Ali Missaoui, Esteban Rios, Nevin Young, S&W Seeds, and Legacy Seeds for their contributions to the alfalfa diversity panel, and overall to the

ARS alfalfa research community for the valuable discussion during the process of marker discovery and the design of the alfalfa DArTag array. We also acknowledge Kirsten Richardson and Vanessa Greenlee for the proofreading of this manuscript.

References

Adhikair L, Orville ML, Jonathan M, Missaoui AM (2018) Dissecting key adaptation traits in the polyploid perennial medicago sativa using gbs-snp mapping. Front Plant Sci 9:934

Adhikari L, Makaju SO, Missaoui AM (2019) QTL Mapping of flowering time and biomass yield in tetraploid alfalfa (*Medicago sativa* L). BMC Plant Biol 19(1):359

Anderson CB, Franzmayr BK, Hong SW, Larking AC, Van Stijn TC, Roger Moraga RT, Faville MJ, Griffiths AG (2018) Protocol: a versatile, inexpensive, high-throughput plant genomic DNA extraction method suitable for genotyping-by-sequencing. Plant Methods 14:75

Annicchiarico P, Nazzicari N, Li X, Wei Y, Pecetti L, Brummer EC (2015) Accuracy of genomic selection for alfalfa biomass yield in different reference populations. BMC Genom 16:1020

Annicchiarico P, Nazzicari N, Ananta A, Carelli M, Wei Y, Brummer EC (2016) Assessment of cultivar distinctness in alfalfa: a comparison of genotyping-by-sequencing, simple-sequence repeat marker, and morphophysiological observations. Plant Genome 9(2):1–12

Bayer PE, Golicz AA, Scheben A, Batley J, Edwards D (2020) Plant pan-genomes are the new reference. Nat Plants 6(8):914–920

Biazzi E, Nazzicari N, Pecetti L, Brummer EC, Palmonari A, Tava A, Annicchiarico P (2017) Genome-wide association mapping and genomic selection for alfalfa (*Medicago sativa*) forage quality traits. PLoS One 12(1):

Bingham ET, McCoy TJ (1979) Cultivated alfalfa at the diploid level: origin, reproductive stability, and yield of seed and forage. Crop Sci 19(1):97–100

Brummer EC, Bouton JH, Kochert G (1993) Development of an RFLP map in diploid alfalfa. Theor Appl Genet 86(2–3):329–332

Chen H, Zeng Y, Yang Y, Huang L, Tang B, Zhang H, Hao F, Liu W, Li Y, Liu Y, Zhang X, Zhang R, Zhang Y, Li Y, Wang K, He H, Wang Z, Fan G, Yang H, Bao A, Shang Z, Chen J, Wang W, Qiu Q (2020) Allele-aware chromosome-level genome assembly and efficient transgene-free genome editing for the autotetraploid cultivated alfalfa. Nat Commun 11(1):2494

Eujayl I, Sledge MK, Wang L, May GD, Chekhovskiy K, Zwonitzer JC, Mian MA (2004) *Medicago truncatula* EST-SSRs reveal cross-species genetic markers for *Medicago* spp. Theor Appl Genet 108(3):414–422

Fajardo DA, Ramaraj T, Devitt N, Tang HB, Cameron CT, Brummer EC, Town CD, Udvardi MK, Monteros MJ, Farmer AD, Miller JR, Young ND, Mudge J (2016) Sequencing and genome assembly of cultivated alfalfa at the diploid level (CADL) *Medicago sativa*. In: Proceedings of plant & animal genome conference XXIV, San Diego, CA

Han Y, Khu D, Xuehui L, Farmer A, Martinez JM, Brummer EC, Monteros MJ (2014) High density array for SNP genotyping and mapping in tetraploid alfalfa. In: Sokolović D, Huyghe C, Radović J (eds) Quantitative traits breeding for multifunctional grasslands and turf. Springer, Dordrech, Netherlands, pp 255–259

Hawkins C, Yu L-X (2018) Recent progress in alfalfa (*Medicago sativa* L.) genomics and genomic selection. Crop J 6(6):565–575

Julier B, Flajoulot S, Barre P, Cardinet G, Santoni S, Huguet T, Huyghe C (2003) Construction of two genetic linkage maps in cultivated tetraploid alfalfa (*Medicago sativa*) using microsatellite and AFLP markers. BMC Plant Biol 3:9

Kiss GB, Csanádi G, Kálmán K, Kaló P, Okrész L (1993) Construction of a basic genetic map for alfalfa using RFLP, RAPD, isozyme and morphological markers. Mol Gen Genet 238(1–2):129–137

Lei Y, Xu Y, Hettenhausen C, Lu C, Shen G, Zhang C, Li J, Song J, Lin H, Wu J (2018) Comparative analysis of alfalfa (*Medicago sativa* L.) leaf transcriptomes reveals genotype-specific salt tolerance mechanisms. BMC Plant Biol 18(1):35

Li X, Acharya A, Farmer AD, Crow JA, Bharti AK, Kramer RS, Wei Y, Han Y, Gou J, May GD, Monteros MJ, Brummer EC (2012) Prevalence of single nucleotide polymorphism among 27 diverse alfalfa genotypes as assessed by transcriptome sequencing. BMC Genom 13:568

Li X, Han Y, Wei Y, Acharya A, Farmer AD, Ho J, Monteros MJ, Brummer EC (2014a) Development of an alfalfa SNP array and its use to evaluate patterns of population structure and linkage disequilibrium. PLoS ONE 9(1):e84329

Li X, Wei Y, Acharya A, Jiang Q, Kang J, Brummer EC (2014b) A saturated genetic linkage map of autotetraploid alfalfa (*Medicago sativa* L.) developed using genotyping-by-sequencing is highly syntenous with the *Medicago truncatula* genome. G3 4(10):1971–1979

Li X, Wei Y, Acharya A, Hansen JL, Crawford JL, Viands DR, Michaud R, Claessens A, Brummer EC (2015) Genomic prediction of biomass yield in two selection cycles of a tetraploid alfalfa breeding population. Plant Genome 8(2): eplantgenome2014.12.0090

Li A, Liu A, Du X, Chen JY, Yin M, Hu HY, Shrestha N, Wu SD, Wang HQ, Dou QW, Liu ZP, Liu JQ, Yang YZ, Ren GP (2020) A chromosome-scale genome assembly of a diploid alfalfa, the progenitor of autotetraploid alfalfa. Hort Res 7(1):19

Ling HQ, Zhao S, Liu D, Wang J, Sun H, Zhang C, Fan H, Li D, Dong L, Tao Y, Gao C, Wu H, Li Y, Cui Y, Guo X, Zheng S, Wang B, Yu K, Liang Q, Yang W, Lou X, Chen J, Feng M, Jian J, Zhang X, Luo G, Jiang Y, Liu J, Wang Z, Sha Y, Zhang B, Wu H, Tang D, Shen Q, Xue P, Zou S, Wang X, Liu X, Wang F, Yang Y, An X, Dong Z, Zhang K, Zhang X, Luo MC, Dvorak J, Tong Y, Wang J, Yang H, Li Z, Wang D, Zhang A, Wang J (2013) Draft genome of the wheat a-genome progenitor *Triticum irartu*. Nature 496(7443):87–90

Liu Z, Chen T, Ma L, Zhao Z, Zhao PX, Nan Z, Wang Y (2013) Global transcriptome sequencing using the illumina platform and the development of EST-SSR markers in autotetraploid alfalfa. PLoS One 8(12):

Liu W, Jia X, Liu Z, Zhang Z, Wang Y, Liu Z, Xie W (2015) Development and characterization of transcription factor gene-derived microsatellite (TFGM) markers in *Medicago truncatula* and their transferability in leguminous and non-leguminous species. Molecules 20(5):8759–8771

Liu ZY, Baoyin T, Li XL, Wang ZL (2019) How fall dormancy benefits alfalfa winter-survival? Physiologic and transcriptomic analyses of dormancy process. BMC Plant Biol 19(1):205–213

Pennycooke JC, Cheng H, Stockinger EJ (2008) Comparative genomic sequence and expression analyses of *Medicago truncatula* and alfalfa subspecies falcatacold-acclimation-specific genes. Plant Physiol 146(3):1242–1254

Peterson BK, Weber JN, Kay EH, Fisher HS, Hoekstra HE (2012) Double digest RADseq: an inexpensive method for de novo SNP discovery and genotyping in model and non-model species. PLoS One 7(5):

Riday H, Brummer EC, Campbell TA, Luth D, and Cazcarro P (2003) Comparisons of genetic and morphological distance with heterosis between *Medicago sativa* subsp. *sativa* and subsp. *falcata*. Euphytica 131(1):37–45

Sakiroglu M, Brummer EC (2017) Identification of loci controlling forage yield and nutritive value in diploid alfalfa using GBS-GWAS. Theor Appl Genet 130 (2):261–268

Sakiroglu M, Sherman-Broyles S, Story A, Moore KJ, Doyle JJ, Charles Brummer E (2012) Patterns of linkage disequilibrium and association mapping in diploid alfalfa (*M. sativa* L.). Theor Appl Genet 125 (3):577–590

Scheben A, Batley J, Edwards D (2017) Genotyping-by-sequencing approaches to characterize crop genomes: choosing the right tool for the right application. Plant Biotechnol J 15(2):149–161

Shen C, Du H, Chen Z, Lu H, Zhu F, Chen H, Meng X, Liu Q, Liu P, Zheng L, Li X, Dong J, Liang C, Wang T (2020) The chromosome-level genome sequence of the autotetraploid alfalfa and resequencing of core germplasms provide genomic resources for alfalfa research. Mol Plant 13(9):1250–1261

Simão FA, Waterhouse RM, Ioannidis P, Kriventseva EV, Zdobnov EM (2015) BUSCO: assessing genome assembly and annotation completeness with single-copy orthologs. Bioinformatics 31(19):3210–3212

Sledge MK, Ray IM, Jiang G (2005) An expressed sequence tag SSR map of tetraploid alfalfa (*Medicago sativa* L.). Theor Appl Genet 111(5):980–992

Tesfaye M, Silverstein KAT, Bucciarelli B, Samac DA, Vance CP (2006) The Affymetrix Medicago GeneChip® array is applicable for transcript analysis of alfalfa (*Medicago sativa*). Funct Plant Biol 33(8):783–788

Wang J, Zhao Y, Ray I, Song M (2016) Transcriptome responses in alfalfa associated with tolerance to intensive animal grazing. Sci Rep 6:19438

Wang Z, Yan H, Fu X, Li X, Gao H (2013) Development of simple sequence repeat markers and diversity analysis in alfalfa (*Medicago sativa* L.). Mol Biol Rep 40(4):3291–3298

Wang Z, Yu G, Shi B, Wang X, Qiang H, Gao H (2014) Development and characterization of simple sequence repeat (SSR) markers based on RNA-sequencing of *Medicago sativa* and *in silico* mapping onto the *M. truncatula* genome. PLoS One 9(3):e92029

Wu S, Lau KH, Cao Q, Hamilton JP, Sun H, Zhou C, Eserman L, Gemenet DC, Olukolu BA, Wang H, Crisovan E, Godden GT, Jiao C, Wang X, Kitavi M, Manrique-Carpintero N, Vaillancourt B, Wiegert-Rininger K, Yang X, Bao K, Schaff J, Kreuze J, Gruneberg W, Khan A, Ghislain M, Ma D, Jiang J, Mwanga ROM, Leebens-Mack J, Coin LJM, Yencho GC, Buell CR, Fei Z (2018) Genome sequences of two diploid wild relatives of cultivated sweetpotato reveal targets for genetic improvement. Nat Commun 9(1):4580

Yang S, Gao M, Xu C, Gao J, Deshpande S, Lin S, Roe BA, Zhu H (2008) Alfalfa benefits from *Medicago truncatula*: The RCT1 gene from *M. truncatula*confers broad-spectrum resistance to anthracnose in alfalfa. Proc Natl Acad Sci USA 105(34):12164–12169

You Q, Yang X, Peng Z, Xu L, Wang J (2018) Development and applications of a high throughput genotyping tool for polyploid crops: single nucleotide polymorphism (SNP) array. Front Plant Sci 9:104

Zhang F, Kang J, Long R, Yu LX, Wang Z, Zhao Z, Zhang T, Yang Q (2019) High-density linkage map construction and mapping QTL for yield and yield components in autotetraploid alfalfa using RAD-Seq. BMC Plant Biol 19(1):165

Zhou Q, Chen T, Wang Y, Liu Z (2014) The development of 204 novel EST-SSRs and their use for genetic diversity analyses in cultivated alfalfa. Biochem Syst Ecol 57:227–230

Zhou P, Silverstein KA, Ramaraj T, Guhlin J, Denny R, Liu J, Farmer AD, Steele KP, Stupar RM, Miller JR, Tiffin P, Mudge J, Young ND (2017) Exploring structural variation and gene family architecture with *de novo* assemblies of 15 Medicago genomes. BMC Genom 18(1):261

Genomic Selection for Higher Yield and Quality in Alfalfa

12

P. Annicchiarico, N. Nazzicari, and L. Pecetti

Abstract

Genomic selection (GS) has high potential interest for improving alfalfa biomass yield and forage quality, to alleviate challenges for phenotypic selection (PS) represented by low narrow-sense heritability, long selection cycles, high evaluation costs, and multi-trait selection. This report discusses various factors that may affect the prediction ability and the cost-efficient exploitation of GS in breeding programs, considering as well specific aspects relative to genotyping-by-sequencing (GBS)-generated markers. We provided an original comparison of six statistical models for GS and four SNP calling procedures for GBS data (based on *M. truncatula* or *M. sativa* genomes, the dDocent-mock reference genome, and the UNEAK pipeline) in terms of predictive ability for biomass yield, leaf protein content, and stem NDF digestibility. Current GBS costs and other considerations support the application of GS to predict additive genetic variation effects (as allowed for by phenotyping half-sib progenies of genotyped parent plants) of plants belonging to relatively broad-based reference populations, following a preliminary stage of stratified mass selection. We outlined a procedure for comparing GS versus PS in terms of selection efficiency according to predicted genetic gains per unit time and same selection cost, which suggested predictive accuracy around 0.15 as a threshold value for considering GS more cost-efficient than PS for biomass yield. A similar threshold may apply to alfalfa forage quality traits selected concurrently with crop yield. Pioneer genomic selection studies for biomass yield or forage quality traits of alfalfa and other perennial forages are generally encouraging for GS implementation. However, information on GS prediction accuracy is still lacking or extremely limited for biomass yield in environments featuring different prevailing stresses (e.g., drought, cold, salinity) or specific crop managements (e.g., severe grazing, intercropping). Crucial research issues for alfalfa GS optimization are represented by cost-efficient allele dosage estimation, quality of cross-population predictions (which may affect GS strategies and the definition of genetic bases by breeding programs), the value of parsimonious GS models incorporated into new genotyping tools (e.g., RAD capture ones), and most of all, the comparison

P. Annicchiarico (✉) · N. Nazzicari · L. Pecetti
Centre for Animal Production and Aquaculture, Council for Agricultural Research and Economics, Lodi, Italy
e-mail: paolo.annicchiarico@crea.gov.it

N. Nazzicari
e-mail: nelson.nazzicari@crea.gov.it

L. Pecetti
e-mail: luciano.pecetti@crea.gov.it

L.-X. Yu and C. Kole (eds.), *The Alfalfa Genome*, Compendium of Plant Genomes,
https://doi.org/10.1007/978-3-030-74466-3_12

of GS versus PS in terms of actual genetic gains per unit time achieved with similar selection costs.

12.1 Introduction

Alfalfa (alias lucerne; *Medicago sativa* L. subsp. *sativa*, $2n = 4x = 32$) is the most-grown perennial forage legume in temperate countries (Annicchiarico et al. 2015a). Its agricultural importance is two-fold, as it provides remarkable environmental benefits (Julier et al. 2017; Fernandez et al. 2019) while contributing to the farm and regional self-sufficiency of feed proteins under an increasing scarcity of this resource (Pilorgé and Muel 2016). However, the economic sustainability of alfalfa, which depends largely on its biomass production over the crop cycle, is threatened by rates of genetic yield gain that are well below those of major grain crops and lower than those of other major perennial legumes (Annicchiarico et al. 2015a). For example, yield gains in the U.S. derived essentially from improved tolerance to biotic stresses, and as such, were manifest only under stress conditions (Lamb et al. 2006). Alfalfa yield gains are hindered by several factors that also apply to other perennial forages, such as small breeding investment, long selection cycles, and impossibility to select real hybrids or pure lines and to capitalize on harvest index for crop yield improvement. In addition, alfalfa biomass yield is characterized by large non-additive genetic variance due to complementary alleles in the repulsion phase at different loci and intra-locus allelic interactions (Bingham et al. 1994; Woodfield and Bingham 1995), which results in narrow-sense heritability (h^2) in the range 0.21–0.30 (Riday and Brummer 2005; Annicchiarico 2015; Acharya et al. 2020). Finally, crop yield gains may be limited by large genotype × environment interaction (GEI) even within relatively small target regions whose environments differ in water availability for the crop or other factors (Annicchiarico and Piano 2005; Pembleton et al. 2010; Annicchiarico et al. 2011; Hakl et al. 2019), unless exploiting specific-adaptation effects by breeding specifically for distinct areas and/or growing conditions within a target region (Annicchiarico 2021).

Other economically important alfalfa traits are forage nutritive value and seed yield, which displayed very low (Annicchiarico et al. 2015a; Lamb et al. 2006) and moderate (Holland and Bingham 1994) rates of genetic gain, respectively. Selecting alfalfa cultivars for traits associated with higher forage quality proved feasible (Hall et al. 2000) and is expected to receive increasing attention, especially for forage targeted to intensive dairy systems. As a matter of fact, forage quality encompasses several component traits, such as protein concentration, proportion of undegradable protein, forage intake by animals as indicated by low values of neutral detergent fiber (NDF), digestibility of the NDF fraction, and other indicators of forage digestibility such as low acid detergent fiber (ADF) and low acid detergent lignin (ADL) (Marten et al. 1988; Fahey and Hussein 1999; Oba and Allen 1999). Major forage quality traits, which are under polygenic control (Biazzi et al. 2017), exhibited h^2 values in the range 0.44–0.64 in Hill and Barnes (1977) and 0.18–0.39 in Guines et al. (2002). The leaf-to-stem ratio may be used as a synthetic positive indicator of forage quality, owing to its close relationship with digestibility and intake of forage (Kephart et al. 1990). The h^2 of this trait ranged from fairly modest (0.38) in Guines et al. (2002) to high (0.75) in Annicchiarico (2015). The relative size of GEI for alfalfa quality traits is definitely lower than that for alfalfa biomass yield (Julier and Huyghe 1997; Sheaffer et al. 1998; Biazzi et al. 2017). Findings from various reports suggest that the simultaneous selection for biomass yield and quality is not hindered substantially by negative correlations (Fonseca et al. 1999; Julier et al. 2000; Annicchiarico 2015).

Various studies based on large marker numbers (>500) revealed many quantitative trait loci (QTLs) associated with biomass yield, each featuring a relatively small effect (Ray et al. 2015; Annicchiarico et al. 2015b; Yu et al. 2017). The reports by Robins et al. (2007) and Yu et al. (2017) highlighted the occurrence of large marker × environment interaction for biomass yield, which reflected the large GEI frequently observed for this trait. Several QTLs emerged as well for key traits contributing to forage quality in genome-wide association studies (GWAS) (Biazzi et al. 2017; Lin et al. 2020). Possible challenges for marker-based forage quality may arise from modest colocation of markers linked to (i) different quality traits (Biazzi et al. 2017), (ii) same trait in leaves and stems (Biazzi et al. 2017), and (iii) same trait across environments with contrasting drought stress levels (Lin et al. 2020).

Improving quantitative traits by marker-assisted selection (MAS) based on the identification and introgression of QTLs may be highly complex and rather inefficient for traits controlled by many loci of small effect. In contrast, using genome-wide markers to build a genomic selection (GS) model can enable many small-effect loci to be incorporated into prediction equations that exploit the linkage disequilibrium of markers with causative loci without explicitly identifying such loci (Meuwissen et al. 2001; Heffner et al. 2009; Viana et al. 2016), thereby offering a more valuable alternative to phenotypic selection (PS). Hereafter, PS mostly designates field-based selection, which may encompass the evaluation of non-replicated individual plants, or replicated clones, half-sib progenies or progenies of selfed genotypes. The GS model is developed by the joint analysis of phenotyping and genotyping data of a germplasm sample (training population) that represents well the target genetic base (breeding population). Phenotyping data are usually represented by best linear unbiased prediction (BLUP) values, by which genotype main effects are shrunk as a function of the deviation from unity of the broad-sense heritability on a genotype mean basis (DeLacy et al. 1996). Following its successful validation on independent material of the same genetic base, the GS model is applied to large germplasm sets sorted out of the genetic base, in order to largely reduce or even eliminate the PS effort (Hawkins and Yu 2018). Simulation and empirical studies, performed essentially on cereal crops, proved that GS is superior to MAS to predict breeding values for complex traits such as crop yield (Bernardo and Yu 2007; Heffner et al. 2011).

GS (also referred to as genome-wide selection) was first applied to dairy cattle improvement (Hayes et al. 2009a), where it represented a breakthrough mainly because of the much shorter selection cycles that it allowed for (Wiggans et al. 2017). Its application to plant breeding, which may be particularly convenient for crops with long selection cycles such as alfalfa, has been delayed by its requirement for high marker numbers until recently, when next generation sequencing techniques allowed to genotype large germplasm sets by thousands of single nucleotide polymorphism (SNP) markers at a relatively low cost. In particular, the development of genotyping methods that neglect sequence discovery and explore SNP polymorphism in DNA fragments cut by a restriction enzyme, such as genotyping-by-sequencing (GBS; Elshire et al. 2011), has given high impulse to crop GS, because of their lower costs relative to SNP array platforms (Elbasyoni et al. 2018). However, the exploitation of GBS requires more biostatistical work than array data and decisions on various technical aspects, as discussed in Sect. 12.3. Other genotyping techniques, e.g., a restriction-site associated DNA (RAD) capture tool (Ali et al. 2016) or a low-cost array that incorporate large sets of SNP markers with trait-prediction value selected out of a larger set of GBS-generated markers, may become of interest in the future, to decrease the genotyping costs and/or simplify the exploitation of genotyping data for ordinary application of GS.

The actual value of GS in a breeding program does not depend only on the model predictive ability but also on savings in time and evaluation costs. Therefore, it ought to be assessed in terms of genetic gain per unit time and unit cost relative to PS (Heffner et al. 2010; Rajsic et al. 2016; Annicchiarico et al. 2017a). Another potential advantage of GS is the greater opportunity to select simultaneously for several traits, which is especially important for perennial forages because of high phenotyping costs and the need to select at least 8–10 parent genotypes for a synthetic variety. For example, selecting 10 genotypes for four traits at the modest selection rate of 20% for each trait requires a working population of $[10 \times (1 / 0.20)^4] = 6{,}250$ individuals, a number that is hardly workable for PS (particularly when involving a time- and resource-consuming trait such as biomass yield over several harvests). Such a number is getting within reach for GS, particularly in the perspective of decreasing genotyping costs. Most importantly, GS-based multi-trait selection implies a negligible increase of evaluation costs relative to one-trait selection (as genotyping costs are constant no matter how many selected traits its data are used for), unlike PS-based selection.

Crop GS has been the target of several review papers, e.g., Heffner et al. (2009), Lorenz et al. (2011), Crossa et al. (2011; 2017), Daetwyler et al. (2013) and Lin et al. (2014), which can provide a thorough overview of methods, challenges, and opportunities of this technique. The objective of this paper is to focus on issues that can be of major interest for GBS-based GS aimed to improve the biomass yield and forage quality of alfalfa.

12.2 Genomic Selection in Alfalfa Breeding Programs

12.2.1 Incorporation in Selection Schemes

The inclusion of GS in breeding programs is graphically exemplified in Fig. 12.1 to highlight differences between an outbred crop bred as a synthetic variety, such as alfalfa, and a hypothetical inbred crop bred as a pure line. In both crops, a breeding population is created from recombination of elite, genetically-contrasting material; a representative germplasm sample (training population) is sorted out for phenotyping and genotyping aimed to develop a genome-enabled prediction model for one or a few quantitative traits; and the GS model, upon validation, is applied to a large number of newly genotyped individuals sorted out of the breeding population. For inbred crops, training and breeding populations include inbred lines, and GS aims to narrow down dramatically the number of lines that undergo a final stage of PS. For alfalfa, the training population is represented by individual plants that are genotyped and then phenotyped based on results of their half-sib progenies as in Fig. 12.1, their S_1 progenies, or as cloned parents; and the final application of GS to the breeding population usually aims to identify parents of a new synthetic variety, without a step of PS of candidate parents.

Phenotyping half-sib progenies instead of cloned parents is recommended for alfalfa GS model development because (Annicchiarico et al. 2015a): (i) half-sib progenies allow to model additive genetic variation effects, namely, the effects that can be exploited for synthetic variety breeding; (ii) the availability of a moderate amount of progeny seed per parent facilitates the phenotyping in different environments and/or under dense planting, and the long-term conservation of parental germplasm. The latter advantage applies as well to S_1 progeny phenotyping (which offers ideal opportunities for parent conservation). The superiority of half-sib progeny-based selection that emerged in a recent comparison of nine alfalfa selection schemes encompassing the evaluation of replicated clones and half-sib, S_1 or S_2 families without or with within-family selection (Annicchiarico and Pecetti 2021) reinforced the interest in constructing and exploiting genome-enabled models that predict half-sib progeny-based breeding values of candidate parents. The selection cycle may be repeated various times according to a recurrent selection scheme (Li and Brummer 2012), but this requires selection of a larger parent set than that in Fig. 12.1 to avoid a quick decrease of genetic variation and genetic gain across cycles. As a matter of fact, the selection of various cultivars from subsequent cycles of recurrent selection may be hindered by the difficulty of achieving sufficient distinctness of candidate varieties for variety registration, a problem often encountered even for genetically unrelated material (Gilliland et al. 2020). GS may be concurrently defined for, and applied to, different breeding populations, as discussed below in relation to the origin and genetic variation of breeding populations.

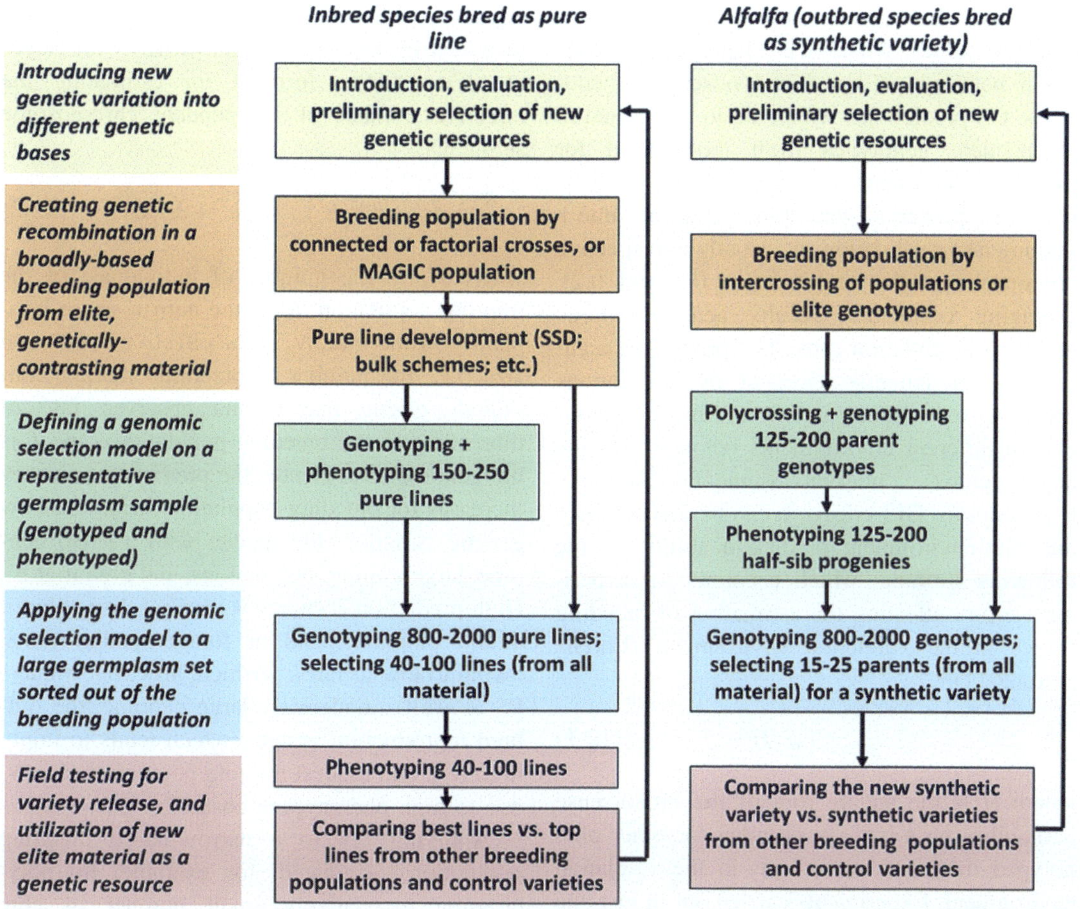

Fig. 12.1 Example of breeding schemes integrating genomic selection for pure line selection of an inbred crop and synthetic variety selection of an outbred crop (such as alfalfa)

GS may be applied directly to the breeding population or to a subset of elite candidate parents selected from the breeding population by a preliminary stage of stratified mass selection. Such a stage, envisaged of short-duration (16 months), proved to be a cost-efficient strategy for alfalfa biomass improvement prior to half-sib progeny-based selection under field conditions, suggesting that it may be useful also for GS at current phenotyping and genotyping costs (Annicchiarico and Pecetti 2021). For a new breeding population, mass selection on a large genotype sample could be performed concurrently with phenotyping and genotyping of a smaller genotype sample aimed at GS model construction to ensure the timely application of GS to material issued by mass selection.

In some cases, breeders may be interested in genomic predictions based on allele frequencies of populations or families rather than allele values of individuals, e.g., when predicting the value of accessions in germplasm collections or F_2 families issued by paired crosses (Guo et al. 2018).

12.2.2 Factors Affecting the Prediction Accuracy of Genomic Selection, and Definition of Breeding Populations

Prediction accuracy (r_{Ac}) and predictive ability (r_{Ab}) of a GS model are closely linked concepts, albeit not quite synonymous. The former is the correlation between predicted and observed breeding values, whereas the latter is the correlation between predicted and observed phenotypes. GS predictions, which are to be applied to independent genotypes from those used for model construction, are assessed by a cross-validation procedure that splits genotypes into a training and a validation set, usually in respective proportions of 0.9 and 0.1 (or 0.8 and 0.2), averaging correlation results across a large number of validation runs. The prediction accuracy is conveniently assessed by correlations involving model construction and model validation in different environments belonging to the target region (i.e., inter-environment predictions), to account for GEI effects. It can be derived from the inter-environment predictive ability by the following formula, which accounts for experiment errors affecting the estimation of breeding values in the validation environment (Lorenz et al. 2011):

$$r_{Ac} = r_{Ab}/H \qquad (12.1)$$

where H is the square root of the broad-sense heritability on a half-sib progeny (or other phenotyped material) mean basis in the validation environment. Experiments carried out in just one site of a target region, as it frequently occurs for perennial crops because of budget constraints, can be used for assessing intra-environment predictions based on cross validations in terms of predictive ability (possibly referred to as prediction accuracy, but with r_{Ac} values coinciding with r_{Ab} values in this case). GS predictions tend to be overestimated in this case because of no account of GEI effects within the target region. However, GEI effects for alfalfa biomass yield across different crop cycles in the same site or neighboring sites were quite modest according to the lack of cultivar × crop cycle interaction (Annicchiarico 1992) and the high repeatability of location-specific cultivar responses across different crop cycles (Annicchiarico 2021), because of the buffering effect of a multi-year crop cycle on GEI due to the year factor.

The prediction accuracy (r_{Ac}) of a GS model is expected to vary according to the following equation obtained by combining the formula provided by Daetwyler et al. (2010) with Hayes et al.'s (2009b) formula for estimating the effective number of independent chromosome segments:

$$r_{Ac} = \sqrt{\left(\left(N_p h^2\right) / \left(N_p h^2 + 2N_e L\right)\right)} \qquad (12.2)$$

where N_p is the number of individuals of the training population, h^2 is the narrow-sense heritability of the trait, N_e is the effective population size (i.e., the number of randomly mating individuals giving rise to the observed rate of inbreeding in the breeding population), and L is the genome size. Hence, the prediction accuracy increases for breeding populations with a narrow genetic variation, the species with a small genome, large training populations, and high trait h^2. High prediction accuracy is hardly achievable for alfalfa biomass yield or forage quality traits, owing to low to fair h^2 (which, however, hinders PS as well), moderately large genome, the outbred reproduction system (which tends to imply high N_e), and modest training population size N_p due to high phenotyping costs.

Applying GS to a narrow-based breeding population, originated, for example, by polycrossing a relatively small number of elite genotypes, can increase the GS prediction accuracy for the target population by decreasing N_e in formula (12.2). However, the inference space of the resulting GS model is expected to be particularly narrow and with low predictive value for other narrow-based breeding populations, thereby requiring extensive phenotyping and genotyping efforts to develop GS models for several different breeding populations. Additional possible drawbacks of narrow-based populations are greater exposure to inbreeding depression and lower opportunity for pooling and recombining favorable alleles. Breeding populations with moderately large genetic variation, such as those arising from intercrossing a large number of elite individuals or a few genetically distant elite populations or from a large number of paired crosses between genetically contrasting individuals, may be preferred in the absence of large budgets for phenotyping work, because GS predictions, albeit of lower accuracy, may apply to a large portion of a breeder's genetic base. This option is supported by the moderate cross-population predictive ability observed across two genetically unrelated, broad-based alfalfa breeding populations, one relative to semi-dormant Po Valley germplasm (encompassing genotypes from 18 landraces or varieties), and the other relative to Mediterranean germplasm with low fall dormancy (created by repeated intercrossing of one variety and two landraces with contrasting geographic origin and adaptation pattern). In particular, the decrease of predictive ability exhibited by top-performing models passing from intra-population to cross-population prediction was in the range 25–30% when using a subset of SNP markers that the two populations had in common, and 38–45% when considering all available markers for separate intra-population predictions (Annicchiarico et al. 2015b). One reason contributing to the only moderate loss of predictive ability could be the much greater extent of within-population variation compared with among-population variation (i.e., the modest population differentiation), which was observed not only across cultivars but also across breeding pools that were sharply contrasting for geographic origin (Annicchiarico et al. 2017b) and even across populations of different subspecies of the *Medicago sativa* complex (Muller et al. 2005). Modest population differentiation would imply the copresence of many useful alleles in contrasting populations.

Another strategy of GS is phenotyping simultaneously various training populations that represent as many breeding populations to develop various population-specific GS models and one generic (across-population) GS model, and then making decisions on the basis of their predictive accuracies. Substantial phenotyping effort may be required, however, to obtain a sufficiently large training sample for each population. In a GS study on biomass yield of perennial ryegrass (*Lolium perenne* L.), the generic GS model (which could profit from larger training population size) exhibited predictive ability comparable to that of best-performing population-specific GS models (Faville et al. 2018).

Obviously, predictions decrease if the genetic relatedness between the training and the breeding population is not maximized (as occurs when the former is not fully representative of the latter). In the presence of a definite genetic structure in the breeding population, optimizing the training population by genotyping samples that reflect the population structure, as devised in Isidro et al. (2015), can improve the GS prediction accuracy. It is also possible to account for population structure in the GS prediction model. However, various studies reported negligible genetic structure even for alfalfa breeding populations with moderately large intra-population diversity, probably because of the modest population differentiation in this species. In particular, negligible genetic structure emerged for 124 genotypes of the Po Valley breeding population and 154 genotypes of the Mediterranean population that were described in Annicchiarico et al. (2015b), as well as for genotypes of the breeding population evaluated by Li et al. (2015) (which derived from intercrossing 100 individuals of three cultivars).

Large training population size, which can increase GS predictive ability by raising N_p in formula (12.2), is usually prevented in alfalfa by high phenotyping costs, particularly for biomass yield evaluation. However, GS predictions of practical value for biomass could be obtained for a broadly-based breeding population of alfalfa using a training population of just 124 genotypes (Annicchiarico et al. 2015b). In a simulation study, Viana et al. (2016) reported a modest increase in prediction accuracy when the training population of an open-pollinated crop exceeded the threshold of 200 individuals, for the scenario of low trait h^2 and 5,000 SNP markers available. Budget constraints that may hinder the achievement of this training population threshold are likely to be overcome in the future by developing reliable high-throughput tools for biomass yield prediction based on remote or proximal sensing (Pittman et al. 2015; Noland et al. 2018), thereby avoiding or limiting the work load of harvesting and weighing several hundred plots across several harvests.

The optimal number of markers for GS depends on the length of the linkage disequilibrium as determined essentially by N_e and the breeding population structure (being much smaller, for example, for the hypothetical case of a biparental progeny relative to a landrace population). The response of the GS prediction accuracy to marker number is curvilinear, however, and decreasing values beyond a plateau may occur due to inaccurate estimation of marker value, as highlighted by Ben Hassen et al. (2018) in a rice diversity panel genotyped by over 43,500 SNP markers. Some procedures were proposed to eliminate redundant markers (Ben Hassen et al. 2018; Ramstein et al. 2016).

12.2.3 Statistical Models for Genomic Selection

There are many statistical models potentially adoptable for GS (Heffner et al. 2009; Lorenz et al. 2011; Heslot et al. 2012; Lin et al. 2014; Wang et al. 2018). A popular model is Ridge regression best linear unbiased prediction (RR-BLUP) (Meuwissen et al. 2001), which assumes a linear mixed additive model where each marker is assigned an effect as a solution of the equation

$$Y = \mu + Gu + \varepsilon \qquad (12.3)$$

where Y is the vector of observed phenotypes, μ is the mean of Y, G is the genotype matrix (e.g., {0,1,2}), $u \sim N(0, I\sigma^2_u)$ is the vector of marker effects, and $\varepsilon \sim N(0, I\sigma^2_e)$ is the vector of residuals. This model assumes that the effects of all loci have a common variance, making it suitable for traits influenced by a large number of minor genes. The Genomic best linear unbiased prediction (G-BLUP) model (Clark and van der Werf 2013) can be mathematically equivalent to RR-BLUP under certain conditions (Habier et al. 2007), but exploits genomic relationships between individuals to estimate breeding values, has reduced dimensions, and does not require thousands of iterations for its construction (Wang et al. 2018).

Bayesian models assume relatively few markers with large effects, allowing markers to have different effects and variances (Meuwissen et al. 2001; Habier et al. 2011). These models assign prior densities to marker effects, thereby inducing different types of shrinkage. Both Bayes A and Bayes B assume marker effects to follow a t distribution, but Bayes B imposes a stronger shrinkage because it assumes, in addition, that most loci have no effect on the trait and are, therefore, excluded from the prediction model (Meuwissen et al. 2001). There are various other proposed models, such as Bayes Cπ (Habier et al. 2011) and Bayesian Lasso (Park and Casella 2008). The shrinkage produced by Bayes Cπ is calculated from the posterior distribution based on the experimental data, whereas that imposed by Bayesian Lasso is stronger for regression coefficients of marker effects with low value.

There is a set of semi-parametric or non-parametric models that are characterized by flexibility and no assumption of linear relationships. This makes them potentially capable of capturing a portion of non-additive genetic effects, which are, however, of dubious interest for synthetic variety breeding (unless selecting quite narrow-based cultivars). Reproducing kernel Hilbert space (RKHS) regression methods are semi-parametric and may incorporate different genetic models (infinitive or not, additive or not) (Gianola and van Kaam 2008). Support vector regression (Schölkopf and Smola 2002) is a machine-learning, non-parametric method based on the computation of a linear regression function in a high dimensional feature space in which the input data are mapped by a linear (SVR-lin) or gaussian (SVR-gau) kernel function. Random Forest (RF) and Artificial neural networks are other major non-parametric models.

A number of empirical comparisons of statistical models indicated that there is no model that is superior to the others for every trait, as expected from the fact that they assume different genetic architectures. Several models ought to be empirically assessed for each target trait, although best-performing models may show quite comparable predictive ability. Bayesian models are likely to be more convenient in the presence of fewer QTLs, whereas RR-BLUP and G-BLUP are expected to perform well only for highly polygenic traits, such as crop yield (Heffner et al. 2009; Lin et al. 2014). Heslot et al. (2012) reported the good performance of Bayesian Lasso, Bayes B, and RF for different traits, but also recommended to assess RR-BLUP. Model comparisons relative to alfalfa indicated (i) somewhat greater merit of SVR-lin and RR-BLUP over three Bayesian models, and low value of RF, for biomass yield (Annicchiarico et al. 2015b), and (ii) some advantage of RR-BLUP, Bayes B and Bayesian Lasso over SVR-lin, for forage quality traits (Biazzi et al. 2017).

In Table 12.1, we report the results of an original model comparison based on phenotyping data of biomass yield in Annicchiarico et al. (2015b) and crude protein content of leaves and NDF digestibility (NDFD) after 24 h of stems in Biazzi et al. (2017), and GBS-generated SNP markers issued by four SNP calling procedures. In agreement with many other studies, our results did not show large differences between models for prediction ability. They highlighted (i) a slight advantage of RR-BLUP for predicting a highly polygenic trait such as biomass yield, and (ii) a slight advantage of some Bayesian model for forage quality traits, such as Bayes B for leaf protein or G-BLUP (solved according to a Bayesian approach) for stem NDFD, for which the level of polygenic control is less complex than biomass based on results in Annicchiarico et al. (2015b) and Biazzi et al. (2017). A second GS model comparison for two forage quality traits relative to the entire plant aerial biomass, which was based only on SNP markers issued by the UNEAK pipeline (which were much fewer than those issued by other SNP calling procedures: see Table 12.2), confirmed the modest difference in predictive ability among models while highlighting a good performance of Bayes Cπ for both traits (Fig. 12.2).

Imputing population structure (previously assessed by various possible methods such as principal component analysis, non-metric multidimensional scaling or cluster analysis) can be very useful for GWAS. Its imputation as a fixed factor in GS models is possible (Guo et al. 2014) but often leads to no substantial increase of prediction accuracy, owing to various possible reasons (the genetic relationships may partly be accounted for by the adopted GS models; partitioning the genomic variation into within-group and among-group components may result in modest within-group prediction accuracy due to smaller population size and reduced genetic diversity, and a less parsimonious overall model; etc.) (Lyra et al. 2018; Lehermeier et al. 2015). An alternative multivariate approach proposed by Lehermeier et al. (2015), which accounts for among-population heterogeneity of marker effects, may be more useful in some cases.

Table 12.1 Prediction ability of six genomic selection models for three alfalfa traits, averaged across results of four SNP calling procedures

Model	Biomass yield	Leaf crude protein	Stem NDF digestibility
RR-BLUP	0.363	0.333	0.335
G-BLUP	0.352	0.326	0.348
Bayes B	0.351	0.340	0.337
Bayes Cπ	0.353	0.321	0.338
Bayesian Lasso	0.339	0.322	0.340
RKHS	0.360	0.329	0.344

Predictions based on 10 repetitions of 10-fold cross validations, assessed for SNP calling procedures based on *M. truncatula* or *M. sativa* genomes, UNEAK or the dDocent-mock reference genome. Genotype missing data thresholds are 10, 20 or 30% for UNEAK (with selection of the top-performing model), and 3% for the other SNP calling procedures

Table 12.2 Number of polymorphic markers, and prediction ability for each of three alfalfa traits averaged across results of six statistical models of genomic selection, for four SNP calling procedures

		Prediction ability		
SNP calling procedure	Marker number	Biomass yield	Leaf crude protein	Stem NDF digestibility
UNEAK	4,264–7,731	0.329	0.366	0.304
dDocent-mock genome	84,208	0.375	0.326	0.348
M. truncatula genome	45,513	0.375	0.290	0.363
M. sativa genome	53,169	0.334	0.311	0.347

Predictions based on 10 repetitions of 10-fold cross validations, assessed for the six statistical models listed in Table 12.1. Genotype missing data thresholds are 10, 20 or 30% for UNEAK (with selection of the top-performing model), and 3% for the other SNP calling procedures

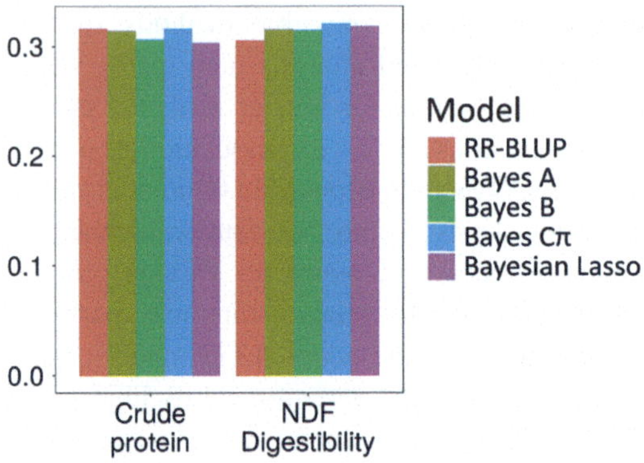

Fig. 12.2 Prediction ability of Ridge Regression BLUP and four Bayesian models for two alfalfa forage quality traits, based on 11,450 polymorphic SNP markers produced by the UNEAK pipeline (allowing for a genotype missing data threshold of 30%)

Phenotyping experiments performed in environments representative of different target regions could be used to develop region-specific genome-enabled predictions by separate analyses for each environment or by developing a GS model that takes account of GEI effects (Crossa et al. 2017). The latter, which may adopt different modeling options (Jarquín et al. 2014; Cuevas et al. 2017), usually displays better environment-specific predictions than models based only on data of the single environments, especially in the presence of genetic correlation between environments for the response of phenotyped material. Genetic gains expected from GS for the specific regions (which are proportional to the GS predictive accuracy) may be compared with genetic gains expected for GS aimed to breed for wide adaptation across several regions. More generally, GS is subjected to the same decisions on adaptation targets as PS, and multi-environment phenotyping data analyzed by state-of-the-art procedures (DeLacy et al. 1996; Annicchiarico 2002) can concur to identify subregions for specific breeding (Annicchiarico 1992; Annicchiarico et al. 2011) and to compare wide-adaptation versus specific-adaptation strategies (Annicchiarico 2021).

The simultaneous genome-enabled selection for two or more traits can be carried out in a similar manner as performed for more than one phenotypically selected trait, namely, by means of a genomic selection index (Dekkers 2007; Matias et al. 2019). Simultaneously modeling multiple quantitative traits (Jia and Jannink 2012) may result in greater predictive accuracy than GS modeling for the single traits, particularly in the presence of genetic correlation between predicted traits and when predictions for a given trait of unknown individuals may profit from the available phenotypic information for a correlated trait (Wang et al. 2018). Sometimes, forage yield and one forage quality trait may conveniently be combined into a single, meaningful trait for GS, such as the production per unit area of crude protein, digestible dry matter, or Milk Feed Units.

12.2.4 Genomic Selection Predictive Ability for Biomass Yield and Forage Quality

Half-sib progeny breeding values for biomass yield have been the object of genomic predictions in some pioneer studies for alfalfa and another major outbred perennial forage such as perennial ryegrass. In alfalfa, the intra-environment predictive ability of GS based on at least 10,000 polymorphic SNP markers achieved 0.35 for 154 genotypes of a Mediterranean training population featuring low fall dormancy, and 0.32 for 124 genotypes of a semi-dormant Po Valley training population, both evaluated under moisture-favorable conditions (Annicchiarico et al. 2015b). However, increasing drought stress in managed environments tended to imply progressively lower GS predictive ability for genotypes belonging to the Mediterranean population (up to 0.11 under intense stress; Annicchiarico et al., unpublished data). In perennial ryegrass, the predictive ability achieved 0.36 for a multi-population training set including over 500 genotypes phenotyped in a few environments of New Zealand and genotyped by over 44,900 SNP markers (Faville et al. 2018). The predictive ability for clonally-evaluated alfalfa genotypes was 0.25 for 322 genotypes evaluated in China genotyped by over 44,750 SNP markers (Jia et al. 2018), and 0.51 for mean yield of 190 genotypes across three sites of the U.S.A. or Canada genotyped by about 10,000 SNP markers (Li et al. 2015). In the latter study, a model constructed from data of two or three sites exhibited up to 0.39 prediction accuracy for material evaluated in a subsequent selection stage.

Biazzi et al. (2017) assessed genome-enabled predictions for half-sib progeny-based breeding values of 154 alfalfa genotypes in a Mediterranean training population that were relative to leaf-to-stem ratio and leaf and stem contents of crude protein, NDF, ADL, and NDFD after 24 h across three growing environments, using 11,450 polymorphic SNP markers. They reported predictive ability values that were moderate (0.30–0.40) for stem NDFD and leaf protein content, modest for leaf ADL and NDFD (0.18–0.20), and low (<0.13) for the other traits. Obviously, predictions were affected by the extent of genetic variation and GEI for the target trait as jointly accounted for by trait broad-sense heritability (H^2). For example, NDFD exhibited much greater H^2 for stems than leaves (0.57 versus 0.22). Given the practical importance of trait values relative to the entire plant, we used the phenotyping data of leaf and stem dry matter and quality collected in Biazzi et al. (2017) to compute crude protein and NDFD values for the overall plant and used these phenotyping data along with the available genotyping data to assess GS predictive ability values for the two traits according to five GS models. The results indicated moderate predictive ability (>0.30) for both traits (Fig. 12.2). In the study on clonally-evaluated alfalfa genotypes by Jia et al. (2018), the prediction ability of NDFD after 30 and 48 h achieved 0.20 and 0.25, respectively, whereas predictions were in the range 0.18–0.20 for NDF and ADF and in the range 0.05–0.10 for forage digestibility, ADL, and crude protein content. On the whole, these studies provided consistent evidence for the moderate GS predictive ability of NDFD, which is comforting in view of the crucial impact of this trait on cattle dry matter intake and milk yield (Oba and Allen 1999).

Genome-enabled predictions reported for alfalfa so far were based on SNP calling procedures that pooled the three possible heterozygote classes (Aaaa, AAaa, and AAAa) into a unique class marked as a diploid heterozygote (Aa), while the two tetraploid homozygotes (AAAA and aaaa) were marked as diploid homozygotes (AA and aa). This limitation, which prevented to use allele dosage effects for predictions, was due to an insufficient number of reads for estimating allele dosage classes. Allele dosage information is expected to be available for predictions in the near future, by greater sequencing effort allowed for by decreasing genotyping costs. SNP calling and allele dosage issues are discussed in the following section in relation to GBS data.

12.3 Technical Issues Relative to Genotyping-by-Sequencing Data for Genomic Selection

12.3.1 Restriction Enzyme

Methylation-sensitive restriction enzymes are preferable to insensitive ones because they tend to skip genomic regions that include repetitive or non-informative DNA (which is highly methylated) (Pootakham et al. 2016). The enzyme *Apek*I proposed in Elshire et al.'s (2011) protocol proved convenient for GBS in alfalfa and other legume species, especially when combined with KAPA *Taq* polymerase (Annicchiarico et al. 2017a), and has been used for major GS studies to date for alfalfa (Annicchiarico et al. 2015b; Li et al. 2015; Biazzi et al. 2017; Jia et al. 2018) and other perennial forages (Faville et al. 2018; Guo et al. 2018). Two-enzyme systems proved useful for crops with larger genomes (Poland et al. 2012) but may have potential also for alfalfa (Bernadette Julier, personal communication, 2020).

12.3.2 SNP Calling and Missing Data Imputation

Technical advances in GBS, such as the passage from 100 bp single-end to 150 bp double-end raw reads issued from sequencing, and the development of more performing procedures for discovery and calling of SNP markers, have increased the number and the quality of polymorphic markers available for GS. Many GS studies have relied on SNP calling of GBS data by the UNEAK pipeline (Lu et al. 2013). Another option has been the SNP calling of markers that aligned to the *M. truncatula* genome (Tang et al. 2014), e.g., by using the bwa and FreeBayes tools of the dDocent pipeline, thereby facilitating the possible identification of QTLs by GWAS, albeit at the cost of less markers available for GS. For example, about 74% of the 11,450 GBS-generated SNP markers issued by the UNEAK pipeline in Biazzi et al. (2017) aligned to *M. truncatula* chromosomes. Novel opportunities have arisen from the development of the dDocent de novo pipeline (Puritz et al. 2014) or the GBS-SNP-CROP pipeline (Melo et al. 2016) that produce a mock reference genome based on the linkage disequilibrium of SNP markers observed in a small subset of test genotypes, and an haplotype-based FreeBayes pipeline usable to distinguish nulliplex/simplex/duplex/triplex/quadriplex genotypes (Yu et al. 2017). The SNP calling of markers aligned to the alfalfa genome has recently become possible, after the publication of the first sequenced genome by Chen et al. (2020).

Different SNP calling procedures can produce different marker numbers. In an alfalfa study, Yu et al. (2017) compared the pipelines TASSEL-GBS and UNEAK and the haplotype-based FreeBayes pipeline that provides allele dosage information. TASSEL-GBS, relative to UNEAK, issued a higher number of markers but lower heterozygosity of markers, whereas the FreeBayes pipeline ranked last for marker number. The pipeline choice also affected the results of a GWAS, which revealed partly different loci associated with *Verticillium* wilt resistance.

Table 12.2 summarizes the results of an unprecedented comparison of four SNP calling procedures for GBS-generated markers, namely, UNEAK, dDocent-mock reference genome, and markers aligned to the *M. truncatula* genome in Tang et al. (2014) or the *M. sativa* genome in Chen et al. (2020). The procedures were compared for the prediction ability of GS models that they generated for biomass yield and leaf protein content and stem NDFD of 154 alfalfa genotypes of a Mediterranean training population. GS-based predictions of these traits in earlier studies (Annicchiarico et al. 2015b; Biazzi et al. 2017) were based on SNP markers issued by the

UNEAK pipeline. We envisaged different thresholds of genotype missing data, namely, 10, 20 or 30%, for the UNEAK pipeline, while setting a 3% missing data threshold for the other procedures, because of the much smaller marker number generated by UNEAK due to its highly conservative method of SNP polymorphism assessment. Predictions for each SNP calling procedure were averaged across the six statistical models listed in Table 12.1, considering the best configuration for the UNEAK pipeline, which related to 30% genotype missing data (providing greater marker number) in nearly all cases. The mock reference genome pipeline, which was tailored to the data set and produced more indeterminate SNP polymorphism information (e.g., by pooling data of repetitive DNA genome fractions), provided definitely greater marker number than SNP calling procedures based on sequenced genomes (Table 12.2). The *M. truncatula* genome-based pipeline generated GS models with high predictive ability for biomass yield and stem NDFD, whereas models based on SNP data from the mock genome-based pipeline tended to perform fairly well for all traits. Results for the UNEAK procedure were inconsistent, as its generated models tended to display lower predictive ability than those issued from other procedures for biomass yield and stem NDFD, but were top-performing for leaf protein content (Table 12.2). While awaiting confirmation from more thorough studies, these results confirm that greater marker number does not necessarily imply better predictive ability of the generated GS models, because of the danger of less accurate estimation of the effects of redundant markers (Ben Hassen et al. 2017). SNP calling based on the sequenced *M. sativa* genome, which could exploit only one sequenced genotype so far (Chen et al. 2020), did not imply a marked disadvantage of generated GS models for trait predictive ability (Table 12.2) and offers the advantage of much greater usefulness when data are also used for GWAS studies. The future availability of an improved sequenced alfalfa genome may increase the number and quality of markers issued by this SNP calling procedure.

The imputation of genotype missing data can be important for GBS because of its possibly large number of missing data (due to variation between sequencing runs, intrinsically noisy data, and overlapping of reads from different loci). The imputation can be difficult for an outcrossing, tetraploid species such as alfalfa, particularly in the absence of a sequenced genome. K-nearest neighbors and Random Forest imputation showed less imputation errors than singular value decomposition, with the first method preferable for its greater computational efficiency (Nazzicari et al. 2016). However, a wider range of imputation methods may be envisaged (e.g. Beagle) after the availability of the sequenced genome.

12.3.3 Estimation of Allele Dosage

As anticipated, alfalfa genotypes have been treated as diploids in GS studies by distinguishing one heterozygote class (pooling the actual Aaaa, AAaa, and AAAa classes) and two homozygote classes. It was necessary to remove heterozygous loci with less than four aligned reads, and homozygous loci with less than 11 reads, to keep the probability to falsely call a AAAa heterozygote as a AAAA homozygote (or a Aaaa heterozygote as a aaaa homozygote) below 5% (Nazzicari et al. 2016). While requiring a 11x coverage for this SNP calling target, a coverage of 48–60x would be needed for a statistically sound estimation of allele dosage (Uitdewilligen et al. 2013). An alternative to the SNP calling into five classes of allele dosage (AAAA, AAAa, AAaa, Aaaa, aaaa) is represented by the SNP calling according to the observed A/a allele ratio, i.e., the allowance of continuous variation for allele dosage regardless of a definite statistical threshold, which requires, however, decisions on a minimum threshold of reads for ratio attribution, e.g., $\geq$ 40x (Oliveira et al. 2019). This approach, which is computationally simpler and avoids problems associated with misclassification of genotypic classes, produced genomic selection models about as accurate as those based on estimated genotype classes in a blueberry study (Oliveira et al. 2019).

Another approach that allows for allele class estimation on the ground of a minimum threshold of reads as low as ≥ 25x is based on a probabilistic graphical model implemented in SuperMASSA software (Pereira et al. 2018).

No assessment of the potential advantage for genome-enabled predictions of allele dosage imputation is available to date for alfalfa. A study by Lara et al. (2019) on the perennial autotetraploid forage grass *Panicum maximum* reported a remarkable advantage of allele dosage imputation based on Pereira et al.'s (2018) probabilistic model over the diploid model pooling the heterozygote classes, with increases of predictive ability of about 50% for leaf dry matter, 42% for crude protein content, and 18% for in vitro digestibility. The advantage of allelic dosage imputation was minimal for predicting agronomic traits of interspecific hybrids of the tropical grass *Urochloa* spp. (Matias et al. 2019), possibly because of the segmental allotetraploid (partly autotetraploid and partly allotetraploid) genome of this material.

12.4 Efficiency of Genomic versus Phenotypic Selection

The value of GS for a breeding program ought to be assessed in relation to opportunities offered by the adopted PS procedure, through comparisons based on predicted and/or actual genetic gains per unit time achieved with similar evaluation costs. Actually, alfalfa PS may be carried out according to several possible breeding schemes, including the selection of parent plants for synthetic varieties based on individual plants (mass selection), clonal evaluation, or evaluation of half-sib progenies or progenies produced by one or two selfing stages without or with additional within-family selection (Rumbaugh et al. 1988; Posselt 2010). While the actual breeding scheme for PS may vary among breeding programs, half-sib progeny-based selection of cloned parents (preceded by a preliminary stage of stratified mass selection) may be considered as a valuable reference scheme for alfalfa in view of its popularity (Rumbaugh et al. 1988) and its high genetic gain per unit time or unit of annual direct costs that emerged for biomass yield in a recent comparison of nine breeding methods (Annicchiarico and Pecetti 2021). Also, this scheme allows for a straightforward comparison with GS models constructed from half-sib progeny phenotyping data.

A comparison of GS versus half-sib progeny-based PS for selection efficiency may be based on the predicted genetic gain per year produced by GS (Δ_G) versus that issued by PS (Δ_P)

$$\begin{aligned}\Delta_G/\Delta_P &= (i_G r_{Ac}/t_G)\,/\,(i_P h/t_P)\\ &= (i_G/i_P)\,(r_{Ac} t_P/h t_G)\end{aligned} \qquad (12.4)$$

where i and t are the standardized selection differential and the selection cycle duration in years, respectively, for one cycle of GS (i_G and t_G) or PS (i_P and t_P); r_{Ac} is the inter-environment predictive accuracy of GS estimated as described in Sect. 12.2; and h is the square root of the narrow-sense heritability. The intra-environment predictive ability r_{Ab} may substitute for r_{Ac}, in the absence of repetition of phenotyping experiments in different environments of the target region. Half-sib progeny-based PS inclusive of progeny seed generation (one year), progeny testing (three years) and final intermating of selected genotypes (one year) would imply $t_P = 5$, whereas $t_G = 1$. The i_G/i_P ratio can be relevant for taking account of possible differences in genotype evaluation cost between GS and PS, in order to perform a fair comparison by assuming same selection costs per year of the two selection strategies. This assumption is verified if

$$(N_G C_G/t_G) = (N_P C_P/t_P) \qquad (12.5)$$

where N_G and N_P are numbers of evaluated genotypes, and C_G and C_P are costs per evaluated genotype, respectively, for GS and PS over one selection cycle. The equality in formula (12.5) is satisfied when the ratio of N_G to N_P is

$$N_G/N_P = (C_P t_G)\,/\,(C_G/t_P) \qquad (12.6)$$

A recent estimate of direct costs over a five-year PS cycle in Italy (where half-sib progeny-based selection was performed in a three-year selection experiment with two replicates) was € 170 per test genotype (Annicchiarico and Pecetti 2021), whereas € 68 is the estimated cost per genotype of GS (inclusive of DNA extraction, outsourced GBS at the fee of € 49 per sample plus 20% VAT, and final intermating of selected genotypes). Based on these values, and assuming $t_P = 5$ and $t_G = 1$, then the ratio of N_G to N_P in formula (12.6) is 0.5 (meaning that the number of evaluated genotypes per yearly selection cycle by GS has to be 50% of that evaluated by PS over a five-year cycle). This ratio implies a two-fold greater selected fraction per cycle for GS (α_G) relative to PS (α_P), to keep constant the number of selected parent plants per cycle of the two selection strategies. For example, hypothesizing $\alpha_P = 0.1$ (hence $i_P = 1.75$) implies $\alpha_G = 0.2$ (hence $i_G = 1.40$), and $i_G / i_P = 0.80$; whereas $\alpha_P = 0.05$ implies $\alpha_G = 0.1$ and $i_G / i_P = 0.85$. For alfalfa biomass yield, the h value reportedly lies in the range 0.46–0.55 according to h^2 values in earlier studies (Riday and Brummer 2005; Annicchiarico 2015; Acharya et al. 2020), suggesting to set $h = 0.50$ in the absence of specific values for the target breeding population. Upon these assumptions, and envisaging $i_G / i_P = 0.80$, $\Delta_G > \Delta_P$ if r_{Ac} (or r_{Ab}) > 0.125 according to formula (12.4). This threshold is just indicative, as it may be lower for PS based on three or more experiment replicates or lower α_P, and higher when taking account of phenotyping and other costs involved in GS model construction. Predictive ability around 0.15 could probably be considered as a safe indicator of greater efficiency of GS over PS for biomass yield in most situations. Such a threshold was definitely overcome by prediction accuracy values reported by Li et al. (2015). Alfalfa biomass yield data from a two-replicate evaluation experiment for a Po Valley training population indicated $r_{Ab} = 0.32$ (Annicchiarico et al. 2015b) and $h^2 = 0.21$ (Annicchiarico 2015), which, along with the other assumptions above, would imply $\Delta_G/\Delta_P = 2.78$ according to the formula (12.4) (i.e., nearly three-fold greater predicted efficiency for GS over PS). For biomass yield of perennial ryegrass, Faville et al. (2018) reported two-fold greater predicted efficiency for GS relative to half-sib family-based PS.

The use of GS for one or more forage quality traits is likely to be envisaged when GS is concurrently used for biomass yield, thereby lowering the overall cost of GS relative to PS (as the GS cost per genotype is substantially invariant to the number of selected traits, unlike the cost of PS). Neglecting differences in direct costs between GS and PS for quality traits, and considering that the h value for quality traits may be in the range 0.45–0.75 according to Hill and Barnes (1977) and Guines et al. (2002), r_{Ac} or $r_{Ab} > 0.15$ would indicate greater predicted gains for GS relative to PS even when $h = 0.75$, when setting $t_P = 5$. This is definitely the case for two key traits such as NDF digestibility and crude protein content in Fig. 12.2, whose r_{Ab} values exceeded 0.30 (thereby suggesting at least two-fold greater selection efficiency for GS relative to PS).

Thorough comparisons of GS versus PS in terms of actual genetic gains, which would be of crucial importance to assess the real value of GS for breeding programs, are not available yet for biomass yield or forage quality of alfalfa or other major perennial forages. Preliminary results by Brummer et al. (2019) for alfalfa biomass yield indicated that a GS model was successful for divergent selection of higher- and lower-yielding synthetic populations, but produced distinctly lower yield gains than PS. A possible reason contributing to lower observed efficiency of GS relative to expectations based on predictive accuracy values in Li et al. (2015) could be the fact that the GS model was constructed from phenotypic data of individual genotypes rather than their half-sib progenies and, as such, predicted also non-additive genetic effects that could not be fixed by selection of synthetic varieties.

12.5 Conclusions

Genomic selection is likely to become increasingly important for the improvement of biomass yield and forage quality of alfalfa and other major perennial forages. However, its ordinary adoption by breeding programs requires technical optimization in various respects along with modifications to selection schemes (e.g., in the definition and construction of breeding populations), as well as convincing supporting evidence for greater cost-efficiency relative to PS. Important contributions for alfalfa GS may be provided by future research work aimed to assess:

(i) the advantage for trait predictions of presence versus absence of allele dosage and of different procedures for allele dosage estimation, in relation to possibly greater genotyping costs required for achieving higher read number thresholds;
(ii) the value of parsimonious GS models incorporated into new genotyping tools (e.g., RAD capture ones), as determined by reduction of genotyping costs along with likely losses of GS predictive accuracy caused by less markers available (possibly counterbalanced in part by less missing data);
(iii) intra-population and cross-population predictions for breeding populations featuring different origin/genetic distance and level of intra-population diversity, to understand whether cost-efficient GS can be achieved by very few breeding populations with large genetic variation or many narrow-based breeding populations.

The cooperation among alfalfa breeding programs of different countries can be strategic for sharing costs (especially phenotyping ones) and technical challenges associated with the implementation of GS, as well as for developing robust GS models. Such a work can be fundamental also to widen our knowledge on GS prediction accuracy for biomass yield in environments featuring different prevailing stresses (e.g., drought, cold, salinity) or specific crop management (e.g., severe grazing, intercropping), and for other key traits such as forage quality ones and seed yield. As highlighted from the reviewed information, our current knowledge on these aspects is still extremely limited.

A research area of crucial importance will be the comparison of GS versus PS (applied on same genetic base) in terms of actual genetic gains per unit of time and similar costs. This can be obtained through proof-of-concept experiments or, ideally, by full-scale selection efforts possibly focusing on different traits and/or target environments. Such selection efforts are also expected to deliver the first marketed GS-produced alfalfa varieties.

Acknowledgements The experimental data currently object of original analyses were produced by the FP7-ArimNet project 'Resilient, water- and energy-efficient forage and feed crops for Mediterranean agricultural systems (REFORMA)' funded by the Italian Ministry of Agriculture, Food and Forestry Policy, the project 'High quality alfalfa for the dairy chain (Qual&Medica)' funded by Fondazione Cassa di Risparmio di Bologna and Regione Emilia-Romagna, and genotyping work funded by the Samuel Roberts Noble Foundation.

References

Acharya JP, Lopez Y, Gouveia BT, Oliveira IB, Resende MFR Jr., Muñoz PR, Rios EF (2020) Breeding alfalfa (*Medicago sativa* L.) adapted to subtropical agroecosystems. Agronomy 10:742

Ali OA, O'Rourke SM, Amish SJ, Meek MH, Luikart G, Jeffres C, Miller MR (2016) RAD Capture (Rapture): flexible and efficient sequence-based genotyping. Genetics 202:389–400

Annicchiarico P (1992) Cultivar adaptation and recommendation from alfalfa trials in Northern Italy. J Genet Breed 46:269–278

Annicchiarico P (2002) Genotype × environment interactions: challenges and opportunities for plant breeding and cultivar recommendations. FAO Plant Production and Protection Paper No. 174, Food and Agricultural Organization, Rome

Annicchiarico P (2015) Alfalfa forage yield and leaf/stem ratio: narrow-sense heritability, genetic correlation, and parent selection procedures. Euphytica 205:409–420

Annicchiarico P (2021) Breeding gain from exploitation of regional adaptation: an alfalfa case study. Crop Sci (accepted paper; https://doi.org/10.1002/csc2.20423)

Annicchiarico P, Pecetti L (2021) Comparison among nine alfalfa breeding schemes based on actual biomass yield gains. Crop Sci (accepted paper; https://doi.org/10.1002/csc2.20464)

Annicchiarico P, Piano E (2005) Use of artificial environments to reproduce and exploit genotype × location interaction for lucerne in northern Italy. Theor Appl Genet 110:219–227

Annicchiarico P, Pecetti L, Abdelguerfi A, Bouizgaren A, Carroni AM, Hayek T, Bouzina M, Mezni M (2011)

Adaptation of landrace and variety germplasm and selection strategies for lucerne in the Mediterranean basin. Field Crops Res 120:283–291
Annicchiarico P, Barrett B, Brummer EC, Julier B, Marshall AH (2015a) Achievements and challenges in improving temperate perennial forage legumes. Crit Rev Plant Sci 34:327–380
Annicchiarico P, Nazzicari N, Li X, Wei Y, Pecetti L, Brummer EC (2015b) Accuracy of genomic selection for alfalfa biomass yield in different reference populations. BMC Genom 16:1020
Annicchiarico P, Nazzicari N, Wei Y, Pecetti L, Brummer EC (2017a) Genotyping-by-sequencing and its exploitation for forage and cool-season grain legume breeding. Front Plant Sci 8:679
Annicchiarico P, Wei Y, Brummer EC (2017b) Genetic structure of putative heterotic populations of alfalfa. Plant Breed 136:671–678
Ben Hassen M, Cao TV, Bartholomé J, Orasen G, Colombi C, Rakotomalala J, Razafinimpiasa L, Bertone C, Biselli C, Volante A, Desiderio F, Jacquin L, Valè G, Ahmadi N (2018) Rice diversity panel provides accurate genomic predictions for complex traits in the progenies of biparental crosses involving members of the panel. Theor Appl Genet 131:417–435
Bernardo R, Yu J (2007) Prospects for genomewide selection for quantitative traits in maize. Crop Sci 47:1082–1090
Biazzi E, Nazzicari N, Pecetti L, Brummer EC, Palmonari A, Tava A, Annicchiarico P (2017) Genome-wide association mapping and genomic selection for alfalfa (*Medicago sativa*) forage quality traits. PLoS ONE 12:e0169234
Bingham ET, Groose RW, Woodfield DR, Kidwell KK (1994) Complementary gene interactions in alfalfa are greater in autotetraploids than diploids. Crop Sci 34:823–829
Brummer EC, Li X, Wei Y, Hanson JL, Viands DR (2019) The imperative of improving yield of perennial forage crops: will genomic selection help? Grassl Sci Eur 24:370–372
Chen H, Zeng Y, Yang Y, Huang L, Tang B, Zhang H, Hao F, Liu W, Li Y, Liu Y, Zhang X, Zhang R, Zhang Y, Li Y, Wang K, He H, Wang Z, Fan G, Yang H, Bao A, Shang Z, Chen J, Wang W, Qiu Q (2020) Allele-aware chromosome-level genome assembly and efficient transgene-free genome editing for the autotetraploid cultivated alfalfa. Nature Comm 11:2494
Clark SA, van der Werf J (2013) Genomic best linear unbiased prediction (gBLUP) for the estimation of genomic breeding values. In: Gondro C, van der Werf J, Hayes B (eds) Methods in molecular biology. Springer, New York, pp 221–230
Crossa J, Pérez P, de los Campos G, Mahuku G, Dreisigacker S, Magorokosho C (2011) Genomic selection and prediction in plant breeding, J Crop Improv 25:239–261
Crossa J, Pérez-Rodríguez P, Cuevas J, Montesinos-López O, Jarquín D, de los Campos G, Burgueño J, González-Camacho JM, Pérez-Elizalde S, Beyene Y, Dreisigacker S, Singh R, Zhang X, Gowda M, Roorkiwal M, Rutkoski J, Varshney RK (2017) Genomic selection in plant breeding: methods, models, and perspectives. Trends Plant Sci 22:961–975
Cuevas J, Crossa J, Montesinos-López OA, Burgueño J, Pérez-Rodríguez P, de los Campos G (2017) Bayesian genomic prediction with genotype × environment interaction kernel models. G3 (Bethesda) 7:41–53
Daetwyler HD, Pong-Wong R, Villanueva B, Woolliams JA (2010) The impact of genetic architecture on genome-wide evaluation methods. Genetics 185: 1021–1031
Daetwyler HD, Calus MPL, Pong-Wong R, de los Campos G, Hickey JM (2013) Genomic prediction in animals and plants: simulation of data, validation, reporting, and benchmarking. Genetics 193:347–365
Dekkers JC (2007) Prediction of response to marker-assisted and genomic selection using selection index theory. J Anim Breed Genet 124:331–341
DeLacy IH, Basford KE, Cooper M, Bull IK, McLaren CG (1996) Analysis of multi-environment trials–an historical perspective. In: Cooper M, Hammer GL (eds) Plant adaptation and crop improvement. CABI, Wallingford, UK, pp 39–124
Elbasyoni IS, Lorenz AJ, Guttieri M, Frels K, Baenziger PS, Poland J, Akhunov E (2018) A comparison between genotyping-by-sequencing and array-based scoring of SNPs for genomic prediction accuracy in winter wheat. Plant Sci 270:123–130
Elshire RJ, Glaubitz JC, Sun Q, Poland JA, Kawamoto K, Buckler ES, Mitchell SE (2011) A robust, simple genotyping-by-sequencing (GBS) approach for high diversity species. PLoS ONE 6:e19379
Fahey GC Jr, Hussein HS (1999) Forty years of forage quality research: accomplishments and impact from an animal nutrition perspective. Crop Sci 39:4–12
Faville MJ, Ganesh S, Cao M, Jahufer MZZ, Bilton TP, Easton HS, Ryan DL, Trethewey JAK, Rolston MP, Griffiths AG, Moraga R, Flay C, Schmidt J, Tan R, Barrett BA (2018) Predictive ability of genomic selection models in a multi-population perennial ryegrass training set using genotyping-by-sequencing. Theor Appl Genet 131:703–720
Fernandez AL, Sheaffer CC, Tautges NE, Putnam DH, Hunter MC (2019) Alfalfa, wildlife, and the environment, 2nd edn. National Alfalfa and Forage Alliance, St. Paul, MN
Fonseca C, Viands D, Hansen J, Pell A (1999) Associations among forage quality traits, vigor, and disease resistance in alfalfa. Crop Sci 39:1271–1276
Gianola D, van Kaam JB (2008) Reproducing kernel hilbert spaces regression methods for genomic assisted prediction of quantitative traits. Genetics 178:2289–2303
Gilliland TJ, Annicchiarico P, Julier B, Ghesquière M (2020) A proposal for enhanced EU herbage VCU and DUS testing procedures. Grass Forage Sci 75:227–241

Guines F, Julier B, Ecalle C, Huyghe C (2002) Genetic control of quality traits of lucerne (*Medicago sativa* L.). Aust J Agric Res 53:401–407
Guo Z, Tucker DM, Basten CJ, Gandhi H, Ersoz E, Guo B, Xu Z, Wang D, Gay G (2014) The impact of population structure on genomic prediction in stratified populations. Theor Appl Genet 127:749–762
Guo X, Cericola F, Fè D, Pedersen MG, Lenk I, Jensen CS, Jensen J, Janss LL (2018) Genomic prediction in tetraploid ryegrass using allele frequencies based on genotyping by sequencing. Front Plant Sci 9:1165
Habier D, Fernando RL, Dekkers JCM (2007) The impact of genetic relationship information on genome-assisted breeding values. Genetics 177:2389–2397
Habier D, Fernando RL, Kizilkaya K, Garrick DJ (2011) Extension of the Bayesian alphabet for genomic selection. BMC Bioinformatics 12:186
Hakl J, Mofidian SMA, Kozová Z, Fuksa P, Jaromír Š (2019) Estimation of lucerne yield stability for enabling effective cultivar selection under rainfed conditions. Grass Forage Sci 74:687–695
Hall MH, Smiles WS, Dickerson RA (2000) Morphological development of alfalfa cultivars selected for higher quality. Agron J 92:1077–1080
Hawkins C, Yu LX (2018) Recent progress in alfalfa (*Medicago sativa* L.) genomics and genomic selection. Crop J 6:565–575
Hayes BJ, Bowman PJ, Chamberlain AJ, Goddard ME (2009a) Invited review: genomic selection in dairy cattle: progress and challenges. J Dairy Sci 92:433–443
Hayes BJ, Visscher PM, Goddard ME (2009b) Increased accuracy of artificial selection by using the realized relationship matrix. Genet Res 91:47–60
Heffner EL, Sorrells ME, Jannink J-L (2009) Genomic selection for crop improvement. Crop Sci 49:1–12
Heffner EL, Lorenz AJ, Jannink J-L, Sorrells ME (2010) Plant breeding with genomic selection: gain per unit time and cost. Crop Sci 50:1681–1690
Heffner EL, Jannink J-L, Iwata H, Souza E, Sorrells ME (2011) Genomic selection accuracy for grain quality traits in biparental wheat populations. Crop Sci 51:2597–2606
Heslot N, Yang H-P, Sorrells ME, Jannink J-L (2012) Genomic selection in plant breeding: a comparison of models. Crop Sci 52:146–160
Hill RR, Barnes RF (1977) Genetic variability for chemical composition of alfalfa. II. Yield and traits associated with digestibility. Crop Sci 17:948–952
Holland JB, Bingham ET (1994) Genetic improvement for yield and fertility of alfalfa cultivars representing different eras of breeding. Crop Sci 34:953–957
Isidro J, Jannink J-L, Akdemir D, Poland J, Heslot N, Sorrells ME (2015) Training set optimization under population structure in genomic selection. Theor Appl Genet 128:145–158
Jia Y, Jannink J-L (2012) Multiple-trait genomic selection methods increase genetic value prediction accuracy. Genetics 192:1513–1522
Jia C, Wu X, Chen M, Wang Y, Liu X, Gong P, Xu Q, Wang X, Gao H, Wang Z (2018) Identification of genetic loci associated with crude protein and mineral concentrations in alfalfa (*Medicago sativa*) using association mapping. BMC Plant Biol 17:97
Jarquín D, Crossa J, Lacaze X, Du Cheyron P, Daucourt J, Lorgeou J, Piraux F, Guerreiro L, Pérez P, Calus M, Burgueño J, de los Campos G (2014) A reaction norm model for genomic selection using high-dimensional genomic and environmental data. Theor Appl Genet 127:595–607
Julier B, Huyghe C (1997) Effect of growth and cultivar on alfalfa digestibility in a multi-site trial. Agronomie 17:481–489
Julier B, Huyghe C, Ecalle C (2000) Within- and among-cultivar genetic variation in alfalfa: forage quality, morphology, and yield. Crop Sci 40:365–369
Julier B, Gastal F, Louarn G, Badenhausser I, Annicchiarico P, Crocq G, Le Chatelier D, Guillemot E, Emile JC (2017) Alfalfa (lucerne) in European cropping systems. In: Murphy-Bokern D, Stoddard F, Watson C (eds) Legumes in Cropping Systems. CAB International, Wallingford, UK, pp 168–191
Kephart KD, Buxton DR, Hill RR (1990) Digestibility and cell-wall components of alfalfa following selection for divergent herbage lignin concentration. Crop Sci 30:207–212
Lamb JFS, Sheaffer CC, Rhodes LH, Sulc RM, Undersander DJ, Brummer EC (2006) Five decades of alfalfa cultivar improvement: impact on forage yield, persistence, and nutritive value. Crop Sci 46:902–909
Lara LAC, Santos MF, Jank J, Chiari L, Vilela M, Amadeu RR, dos Santos JPR, Pereira G, Zeng Z-B, Garcia AAF (2019) Genomic selection with allele dosage in *Panicum maximum* Jacq. G3 (Bethesda) 9:2463–2475
Lehermeier C, Schön CC, de Los Campos G (2015) Assessment of genetic heterogeneity in structured plant populations using multivariate whole-genome regression models. Genetics 201:323–337
Li X, Brummer EC (2012) Applied genetics and genomics in alfalfa breeding. Agronomy 2:40–61
Li X, Wei Y, Acharya A, Hansen JL, Crawford JL, Viands DR, Michaud R, Claessens A, Brummer EC (2015) Genomic prediction of biomass yield in two selection cycles of a tetraploid alfalfa breeding population. Plant Genome 8:2
Lin Z, Hayes BJ, Daetwyler HD (2014) Genomic selection in crops, trees and forages: a review. Crop Pasture Sci 65:1177–1191
Lin S, Medina CA, Boge B, Hu J, Fransen S, Norberg S, Yu LX (2020) Identification of genetic loci associated with forage quality in response to water deficit in autotetraploid alfalfa (*Medicago sativa* L.). BMC Plant Biol 20:303
Lorenz AJ, Chao S, Asoro FG, Heffner EL, Hayashi T, Iwata H, Smith KP, Sorrells ME, Jannink J-L (2011) Genomic selection in plant breeding: knowledge and prospects. Adv Agron 110:77–123

Lu F, Lipka AE, Glaubitz J, Elshire R, Cherney JH, Casler MD, Buckler ES, Costich DE (2013) Switchgrass genomic diversity, ploidy, and evolution: novel insights from a network-based SNP discovery protocol. PLoS Genet 9:e1003215

Lyra DH, Granato ÍSC, Morais PPP, Alves FC, dos Santos ARM, Yu X, G T, Yu J, Fritsche-Neto R (2018) Controlling population structure in the genomic prediction of tropical maize hybrids. Mol Breed 38:126

Marten GC, Buxton DR, Barnes RF (1988) Feeding value (forage quality). In: Hanson AA, Barnes DK, Hill RR Jr (eds) Alfalfa and Alfalfa improvement. ASA, CSSA, SSSA, Madison, WI, pp 463–491

Matias FI, Alves FC, Meireles KGX, Barrios SCL, do Valle CB, Endelman JB, Fritsche-Neto R (2019) On the accuracy of genomic prediction models considering multi-trait and allele dosage in *Urochloa* spp. interspecific tetraploid hybrids. Mol Breed 39:100

Melo ATO, Bartaula R, Hale I (2016) GBS-SNP-CROP: a reference-optional pipeline for SNP discovery and plant germplasm characterization using variable length, paired-end genotyping-by-sequencing data. BMC Bioinf 17:29

Meuwissen THE, Hayes BJ, Goddard ME (2001) Prediction of total genetic value using genome-wide dense marker maps. Genetics 157:1819–1829

Muller MH, Poncet C, Prosperi JM, Santoni S, Ronfort J (2005) Domestication history in the *Medicago sativa* species complex: inferences from nuclear sequence polymorphism. Mol Ecol 15:1589–1602

Nazzicari N, Biscarini F, Cozzi P, Brummer EC, Annicchiarico P (2016) Marker imputation efficiency for genotyping-by-sequencing data in rice (*Oryza sativa*) and alfalfa (*Medicago sativa*). Mol Breed 36:69

Noland RL, Wells MS, Coulter JA, Tiede T, Baker JM, Martinson KL, Sheaffer CC (2018) Estimating alfalfa yield and nutritive value using remote sensing and air temperature. Field Crops Res 222:189–196

Oba M, Allen MS (1999) Evaluation of the importance of the digestibility of neutral detergent fiber from forage: effects on dry matter intake and milk yield of dairy cows. J Dairy Sci 82:589–596

Oliveira IB, Resende MF, Ferrão LFV, Amadeu RR, Endelman JB, Kirst M, Coelho ASG, Munoz PR (2019) Genomic prediction of autotetraploids; influence of relationship matrices, allele dosage, and continuous genotyping calls in phenotype prediction. G3 (Bethesda) 9:1189–1198

Park T, Casella G (2008) The bayesian lasso. J Am Stat Assoc 103:681–686

Pembleton KG, Smith RS, Rawnsley RP, Donaghy DJ, Humphries AW (2010) Genotype by environment interactions of lucerne (*Medicago sativa* L.) in a cool temperate climate. Crop Pasture Sci 61:493–502

Pereira GS, Garcia AAF, Margarido GRA (2018) A fully automated pipeline for quantitative genotype calling from next generation sequencing data in autopolyploids. BMC Bioinf 19:398

Pilorgé E, Muel F (2016) What vegetable oils and proteins for 2030? Would the protein fraction be the future of oil and protein crops? OCL 23:D402

Pittman JJ, Arnall DB, Interrante SM, Moffet CA, Butler TJ (2015) Estimation of biomass and canopy height in bermudagrass, alfalfa, and wheat using ultrasonic, laser, and spectral sensors. Sensors 15: 2920–2943

Poland JA, Brown PJ, Sorrells ME, Jannink J-L (2012) Development of high-density genetic maps for barley and wheat using a novel two-enzyme genotyping-by-sequencing approach. PLoS ONE 7:

Pootakham W, Sonthirod C, Naktang C, Jomchai N, Sangsrakru D, Tangphatsornruang S (2016) Effects of methylation-sensitive enzymes on the enrichment of genic SNPs and the degree of genome complexity reduction in a two-enzyme genotyping-by-sequencing (GBS) approach: a case study in oil palm (*Elaeis guineensis*). Mol Breed 36:154

Posselt UK (2010) Breeding methods in cross-pollinated species. In: Boller B, Posselt UK, Veronesi F (eds) Fodder crops and amenity grasses. Handbook of Plant Breeding, Springer, New York, pp 39–87

Puritz JB, Hollenbeck CM, Gold JR (2014) dDocent: a RADseq, variant-calling pipeline designed for population genomics of non-model organisms. PeerJ 2: e431

Rajsic P, Weersink A, Navabi A, Pauls KP (2016) Economics of genomic selection: the role of prediction accuracy and relative genotyping costs. Euphytica 210:259–276

Ramstein GP, Evans J, Kaeppler SM, Mitchell RB, Vogel KP, Buell CR, Casler MD (2016) Accuracy of genomic prediction in switchgrass (*Panicum virgatum* L.) improved by accounting for linkage disequilibrium. G3 (Bethesda) 6:1049–1062

Ray IM, Han YEL, Meenach CD, Santantonio N, Sledge MK, Pierce CA, Sterling TM, Kersey RK, Bhandari HS, Monteros MJ (2015) Identification of quantitative trait loci for alfalfa forage biomass productivity during drought stress. Crop Sci 55:2012–2033

Riday H, Brummer EC (2005) Heterosis in a broad range of alfalfa germplasm. Crop Sci 45:8–17

Robins JG, Luth D, Campbell TA, Bauchan GR, He C, Viands DR, Hansen JL, Brummer EC (2007) Genetic mapping of biomass production in tetraploid alfalfa. Crop Sci 47:1–10

Rumbaugh MD, Caddel JL, Rowe DE (1988) Breeding and quantitative genetics. In: Hanson AA, Barnes DK, Hill RR (eds) Alfalfa and alfalfa improvement. CSSA, SSSA Publishers, Madison, WI, ASA, pp 777–808

Schölkopf B, Smola AJ (2002) Learning with kernels: support vector machines, regularization, optimization, and beyond. MIT Press, Cambridge, MA

Sheaffer CC, Cash D, Ehlke NJ, Henning JC, Jewett JG, Johnson KD, Peterson MA, Smith M, Hansen JL, Viands DR (1998) Entry × environment interactions for alfalfa forage quality. Agron J 90:774–780

Tang H, Krishnakumar V, Bidwell S, Rosen B, Chan A, Zhou S, Gentzbittel L, Childs KL, Yandell M, Gundlach H, Mayer KF, Schwartz DC, Town CD (2014) An improved genome release (version Mt4.0) for the model legume *Medicago truncatula*. BMC Genomics 15:312
Uitdewilligen JGAML, Wolters AMA, D'hoop BB, Borm TJA, Visser RGF, van Eck HJ (2013) A next-generation sequencing method for genotyping-by-sequencing of highly heterozygous autotetraploid potato. PLoS ONE 8:e62355
Viana JMS, Piepho H-P, Fonseca F (2016) Quantitative genetics theory for genomic selection and efficiency of genotypic value prediction in open-pollinated populations. Sci Agric 74:41–50
Wang X, Xu Y, Hu Z, Xu C (2018) Genomic selection methods for crop improvement: current status and prospects. Crop J 6:330–340
Wiggans GR, Cole JB, Hubbard SM, Sonstegard TS (2017) Genomic selection in dairy cattle: the USDA experience. Annu Rev Anim Biosci 5:309–327
Woodfield DR, Bingham ET (1995) Improvement in two-allele autotetraploid populations of alfalfa explained by accumulation of favorable alleles. Crop Sci 35:988–994
Yu LX, Zheng P, Bhamidimarri S, Liu XP, Main D (2017) The impact of genotyping-by-sequencing pipelines on SNP discovery and identification of markers associated with verticillium wilt resistance in autotetraploid alfalfa (*Medicago sativa* L.). Front Plant Sci 8:89

Identification and Characterization of Disease Resistance Genes in Alfalfa and *Medicago truncatula* for Breeding Improved Cultivars

13

Deborah A. Samac, Long-Xi Yu, and Ali M. Missaoui

Abstract

Alfalfa plants are susceptible to numerous diseases caused by fungi, oomycetes, bacteria, and nematode pests that significantly reduce forage yield, quality, and productive stand life. Disease resistance is the most common and effective means of reducing damage from pathogens. Using phenotypic recurrent selection, plant breeders have successfully developed alfalfa cultivars with high levels of resistance to major yield-limiting diseases. This chapter summarizes the cases in which the mechanisms of disease resistance and inheritance of resistance have been investigated. DNA markers associated with resistance have been identified for several diseases. However, alfalfa lags behind other major crops in the identification of disease resistance genes and the use of molecular markers in cultivar improvement. A better understanding of disease resistance and the use of marker-assisted selection of superior germplasm should accelerate the breeding and development of more productive cultivars.

D. A. Samac (✉)
USDA-ARS-Plant Science Research Unit, St. Paul, MN 55108, USA
e-mail: debby.samac@usda.gov

L.-X. Yu
USDA-ARS-Plant Germplasm Introduction and Testing Research, Prosser, WA 99350, USA
e-mail: Longxi.yu@usda.gov

A. M. Missaoui
Department of Crop and Soil Sciences, Institute of Plant Breeding Genetics and Genomics, The University of Georgia, Athens, GA 30602, USA
e-mail: cssamm@uga.edu

13.1 Introduction

Plant breeders have been very successful in developing multi-disease-resistant alfalfa cultivars through recurrent phenotypic selection. For specific major yield-limiting diseases, high resistance with greater than 50% resistant plants has been achieved in most modern cultivars (Elgin et al. 1988). However, developing such resistant cultivars requires screening thousands of seedlings, a time-consuming and laborious process that may limit progress in developing other desirable traits such as improved forage quality, higher plant biomass yield, or tolerance to abiotic stresses. In contrast to other major crops, few DNA markers associated with disease resistance have been identified in alfalfa and the use of markers in alfalfa improvement is limited at this time. The lack of an alfalfa reference genome sequence and inexpensive marker methodologies has hampered progress in the molecular breeding of alfalfa for disease resistance. Identification of markers associated with resistance to major diseases would accelerate the development of more resistant cultivars and help clarify host–pathogen

L.-X. Yu and C. Kole (eds.), *The Alfalfa Genome*, Compendium of Plant Genomes,
https://doi.org/10.1007/978-3-030-74466-3_13

relationships. For example, pathogen races have been identified for only three alfalfa pathogens, *Colletotrichum trifolii*, causing anthracnose of alfalfa; *Aphanomyces euteiches*, the causal agent of Aphanomyces root rot; and *Peronospora trifolii,* the causal agent of downy mildew. However, in the case of *A. euteiches*, the number of races attacking alfalfa is unclear and breeders must utilize a wide range of strains when selecting resistant plants, without knowledge of the specific pathogenicity of each strain. It is possible that races are present in populations of other pathogens but have not been recognized due to the difficulty in developing differential cultivars with single resistance genes. Identification of resistance genes or markers associated with resistance genes would facilitate tracking of genes within alfalfa breeding lines and stacking of multiple resistance genes. This chapter will summarize the current knowledge on resistance to major diseases in alfalfa and the related annual species *Medicago truncatula* (barrel medic) and progress toward the identification of specific resistance genes and mechanisms of resistance.

13.2 Mechanisms of Disease Resistance

Plants have developed multiple layers of defenses for protection from pathogenic microbes. Physical barriers such as a waxy cuticle layer on leaves and stems prevent the adherence and penetration of many potential pathogens. Additionally, there is growing evidence that microbial communities on plant surfaces and within plants (endophytes) help to repel pathogens through competition for nutrients and/or production of antimicrobial compounds (Schlatter et al. 2017; Köhl et al. 2019; White et al. 2019). Plants also continually produce an arsenal of compounds to inhibit penetration and colonization by pathogens (Lacerda et al. 2014). Additionally, plants have developed a network of interacting factors that contribute to immunity (Andersen et al. 2018). The first layer of immunity is the recognition of pathogens by receptors in the plasma membrane. These receptors typically recognize conserved microbial structures such as bacterial flagella, fungal carbohydrates, and viral nucleic acids known as pathogen-/microbe-associated molecular patterns (PAMPs/MAMPs). These receptors trigger defenses before cellular invasion while another class of receptors perceives damage by pathogenic enzymes. Successful pathogens dampen or delay these basal defense responses. The second layer of immunity is the recognition of pathogen proteins (effectors) or detecting modification of an effector's target protein by receptors known as resistance (R) genes, usually found in the cytoplasm. Because these proteins have conserved motifs, nucleotide-binding domains, and leucine-rich repeats, they are often called NLR genes. This molecular interaction typically occurs with a specific cultivar and specific pathogen strain, pathotype, or race. Recognition triggers rapid localized plant cell death at the site of penetration known as a hypersensitive response (HR) that halts further colonization by the pathogen through the production of reactive oxygen species, cell wall modification, and/or production of antimicrobial proteins and compounds. In legumes, products of the phenylpropanoid pathway termed phytoalexins are a major class of such general defensive products. In alfalfa, the major phytoalexin is medicarpin, which has been shown to inhibit the growth of fungi (Blount et al. 1992), oomycetes (Vaziri et al. 1981), and nematodes (Baldridge et al. 1998). NLR genes are often clustered in the genome, which facilitates their rapid evolution but complicates their identification through genetic and genomic studies (Ameline-Torregrosa et al. 2008b). The autotetraploid outcrossing nature of alfalfa adds further difficulty to mapping and identifying resistance genes.

13.3 Identifying and Improving Disease Resistance in Alfalfa Germplasm

Early in the cultivation of alfalfa in North America, diseases were recognized as limiting yield and having a severe impact on winter survival and stand life. Some of the first breeding of alfalfa aimed at improving resistance to bacterial wilt caused by the gram-positive bacterium *Clavibacter insidiosus* (Barnes et al. 1971). Breeding methods for improving disease resistance in alfalfa have changed little over time with reliance on poly-cross methods and recurrent phenotypic selection. During the 1970s and 1980s, there was a concerted effort by plant pathologists and alfalfa geneticists to evaluate germplasm in core collections of alfalfa and annual medics to identify accessions with resistance to major diseases. These efforts were greatly aided by the development and use of standardized tests by the alfalfa community to evaluate germplasm and to characterize cultivars for disease resistance. The tests developed and curated by the North American Alfalfa Improvement Conference (NAAIC) describe methods for reliably achieving plant infection and standard scoring methods to characterize plant responses and check cultivars to use as resistant and susceptible controls (www.naaic.org). Data on responses to 13 diseases was entered into the USDA Germplasm Resources Information Network (GRIN) database. In general, two to three cycles of selection and intermating resulted in an alfalfa population with sufficient levels of resistance to show improved field performance against a particular disease. Currently, NAAIC has developed 22 standard tests for diseases and nematodes, as well as standard tests for major insect pests and a range of agronomic traits. Although resistance to many diseases has been identified, and in many cases transferred to commercial cultivars, less is known regarding the mechanisms of resistance and resistance genes (Table 13.1). Genetic mapping of resistance has been done for several diseases to identify quantitative trait loci (QTL), but so far no R gene from alfalfa has been cloned to demonstrate activity. DNA markers associated with resistance were identified for a few diseases, downy mildew (Obert et al. 2000), Phytophthora root rot (Musial et al. 2005), Stagonospora crown and root rot (Musial et al. 2007), common leaf spot (Wang et al. 2012), and Verticillium wilt (Yu et al. 2017a), but to our knowledge, they have not been used in breeding programs.

Many of the same diseases and pests that attack alfalfa also attack annual medic species (Tivoli et al. 2006). Native to the Mediterranean Basin, some annual medics are cultivated as forage crops or cover crops. In order to develop more molecular tools for legume crops, *M. truncatula* was selected as a model species due to its small genome size, rapid cycling time, and wealth of genetic and genomic resources (de Bruijn 2020). Resistance genes and QTL associated with resistance to several diseases have been characterized in *M. truncatula* (Table 13.2) with the expectation that this information can improve disease resistance in alfalfa, grain legumes, and annual medics. Interestingly, the race-specific resistance identified in alfalfa for *C. trifolii* and *A. euteiches* has not been found in *M. truncatula*, which appears to have non-race-specific resistance toward these pathogens (Djébali et al. 2009; Pilet-Nayel et al. 2009; Yang et al. 2008). As detailed below, the *R* gene in *M. truncatula* for resistance to *C. trifolii*, *RCT1*, was cloned and shown to function in alfalfa (Yang et al. 2008), indicating that all of the downstream genes required for effective resistance are present in alfalfa.

The interaction of pathogens with alfalfa and characterization of defenses and resistance genes have been investigated for a few alfalfa diseases. Below we provide summaries of cases in which identification of resistant germplasm, host reactions to infection, and inheritance and/or mapping of resistance genes has been carried out.

13.4 Anthracnose

Of the alfalfa–pathogen interactions, anthracnose is the best understood. It is a major disease of alfalfa in the United States, Argentina, Australia, and Europe. The disease is caused primarily by *Colletotrichum trifolii* in the United States and *C. destructivum* in Canada (Graham et al. 1976). Additionally, *C. dematium* f. *truncatum* causes mild symptoms, and *C. linicola* causes anthracnose on alfalfa in Serbia (Vasic et al. 2014) and China (Wang et al. 2018). The disease is favored by warm and humid conditions and is most severe in late summer, often leading to summer decline in stands (Irwin and Armour 2015). In the field, symptoms occur on stems as sunken straw-colored, oval to diamond-shaped lesions with a dark border in which acervuli producing orange-colored conidia erupt with dark hair-like setae. Lesions can coalesce and girdle stems under conducive conditions resulting in straw-colored dead stems with a classic "shepherd's crook" curled apex. The fungus spreads through stems into crown tissues where it causes crown rot, distinguished by blue-black discoloration, and into the taproot causing root rot. Crown rot is more common in the western United States where it can kill plants leading to significant stand decline.

Resistance to anthracnose is highly heritable and can be increased by recurrent selection (Devine et al. 1971). Breeding led to the development of the resistant cultivars Arc and SaranacAR, which differentiate two distinct pathotypes of the fungus named race 1 and race 2. Elegant genetic analysis of resistance by Elgin and Ostazeski (1985) showed that resistance to race 1 in Arc is conditioned by the single dominant gene *An1* with tetrasomic inheritance, while resistance to race 1 and race 2 in SaranacAR is conditioned by the single dominant gene *An2*. The two genes act independently and are not linked. *An1* was found infrequently in SaranacAR but was masked by the presence of *An2*. A gene for resistance to only race 2 has not been identified. Later, Elgin and O'Neill (1988) showed that *An1* in Arc and SaranacAR are the same gene. Induced resistance can occur (O'Neill et al. 1989), so simultaneous or sequential inoculations of multiple races should be avoided in breeding programs. Cultivars with resistance to anthracnose were found to have a significant yield advantage (Elgin et al. 1981). Genetics of resistance were found to be different in alfalfa germplasm from Australia selected for anthracnose resistance. Single tetrasomic gene models were not supported using Australian *C. trifolii* isolates and germplasm. In some plants, resistance was simply inherited and incompletely dominant and in other plants, resistance was quantitatively inherited and incompletely recessive (Mackie and Irwin 1998; Mackie et al. 2003; Irwin et al. 2006). Interestingly, in the Australian germplasm, distinct genes appear to control resistance to epidermal infection and infection by injection of spores into stems (Mackie and Irwin 1998) while in Arc and SaranacAR, plants resistant to epidermal infection are also resistant to stem injection (Ostazeski and Elgin 1982).

Race 1 is the predominant pathotype in the U. S., with race 2 reported in the Mid-Atlantic states and more recently in Wisconsin (Samac et al. 2014). A potential race 3 was identified in Oklahoma (Allen et al. 1982), which was later identified as most likely *C. destructivum* (O'Neill 1996). Race 4 was identified in Ohio (Ariss and Rhodes 2007) and in Australia (Mackie et al. 2003). Most recently, a highly virulent pathotype causing disease on both Arc and SaranacAR was found in Wisconsin and Minnesota that was identified as a new race, race 5 (Rodgers et al. 2019). Thus, a continuation of pathogen surveillance and resistance breeding is needed for combating the ongoing emergence of new yield-reducing pathotypes.

Resistance to anthracnose in alfalfa is characterized by a strong post-penetration hypersensitive response. In both resistant and susceptible plants, the early events of infection are indistinguishable. In the first 72 h after inoculation, spores germinate and form melanized appressoria at the junctions of epidermal cell walls (Mould et al. 1991a). The plant responds to the penetration peg touching the cuticle by forming papillae and thickening the cell walls in contact with appressoria. The resistant interaction is distinguished by vacuolar fragmentation and aggregation of cytoplasm around the penetration peg in the epidermal cell penetrated by the fungus. Production of phenolic compounds in the cytoplasm and deposition of lignin and suberin in the cell wall likely prevent diffusion of fungal enzymes from the penetrated cell, protecting neighboring cells (Mould et al. 1991b). Resistance is due to sequestering fungal infection within the necrotic epidermal cell, although host responses are not restricted to a single cell. Production of phytoalexins may kill the sequestered hyphae to complete the resistant response. Saunders and O'Neill (2004) found that upregulation of genes in the phenylpropanoid pathway was associated with increases in the phytoalexins medicarpin and sativan in resistant plants. In contrast, in susceptible plants, the cytoplasm of the penetrated epidermal cell undergoes autolysis and the hyphae grow within cortical cells. No biotrophic phase occurs in alfalfa, although a short biotrophic phase occurs in other hosts infected with *Colletotrichum* species. The sequence of infection events is similar in *M. truncatula* inoculated with *C. trifolii* (Torregrosa et al. 2004), although no race specificity has been documented (Ameline-Torregrosa et al. 2008a; Yang et al. 2008). In the resistant accession Jemalong 6, a hypersensitive response occurs with fungal penetration that is accompanied by the production of reactive oxygen species and autofluorescent phenolic compounds. There is also a strong upregulation of genes involved in plant defense, cell wall modification, production of phytoalexins, and oxylipin signaling. (Torregrosa et al. 2004; Samac et al. 2011). As in alfalfa, resistance was found to be a dominant trait (Torregrosa et al. 2004).

Mapping of QTL for resistance to anthracnose has been done in alfalfa. Using plants from the Australian alfalfa cultivar Sequel, mapping populations analyzed with RAPD and AFLP markers identified QTL for epidermal cell resistance to race 1 on the top of LG4 (Irwin et al. 2006). From the cultivar Trifecta, QTL for dominant incomplete resistance to race 1 and race 4 were identified on LG8, with QTL for incomplete recessive resistance to race 2 on LG4 (Mackie et al. 2007).

In *M. truncatula*, resistance to race 1 and race 2 was mapped using recombinant inbred lines with Jemalong A17 as the resistant parent (Ameline-Torregrosa et al. 2008a). A major QTL for resistance to both races explaining approximately 40% of phenotypic variation was identified on the top of LG4 with a minor QTL from the susceptible parent on LG6. The major QTL contained a cluster of NLR genes. Although separate loci for resistance to the two races were not distinguished, there was genetic evidence for separate genes for resistance to race 1 and race 2. Additional genetic determinants were postulated to act as modulators of resistance. In separate experiments, an F2 mapping population was developed using Jemalong A17 as the resistant parent in which resistance to race 1 fit a model for a single dominant gene, which was named *RCT1* (Yang et al. 2007). The F2 population was used for high-resolution genetic and physical mapping of *RCT1* primarily using SNPs (Yang et al. 2008). Notably, susceptible plants were used for mapping, which was more informative for identifying recombination events and avoided errors in disease phenotyping. A marker tightly linked to *RCT1* was used to identify a BAC clone containing the QTL for resistance. Sequencing of the 200 kb interval between linked markers identified five NLR genes (Yang et al. 2008). Of those, three had complete open reading frames and two were truncated. The three full-length genes were cloned and expressed in susceptible alfalfa plants under their own promoters. Expression of one candidate gene conferred resistance to races 1, 2, and 4 (Yang et al. 2008). The gene is a TIR-NBS-LRR type of resistance gene that consists of five exons. Alternative splicing results in retention of intron 4, which is common in this type of resistance gene. Both the regular and alternative transcripts are required for resistance (Tang et al. 2013). Although a homolog of *RCT1* was cloned from the race 1-resistant alfalfa cultivar Arc, its relationship to *An1* and role in disease resistance are not clear. *RCT1* may be ancestral, predating the divergence of alfalfa and *M. truncatula*, or may have more recently acquired novel specificity for resistance to multiple races (Yang et al. 2008). Possibly, variants from a single gene, such as those derived by alternative splicing, increase the spectrum of pathogen detection by R genes (van Wersch and Li 2019). Nonetheless, identification of resistance genes in alfalfa germplasm conferring broad-spectrum resistance to *C. trifolii* would be highly desirable for genetic improvement of alfalfa.

13.5 Aphanomyces Root Rot

Aphanomyces root rot, caused by *Aphanomyces euteiches*, was first described as a serious disease of pea (*Pisum sativum*) (Jones and Drechsler 1925). It was subsequently found to also cause root rot of several forage legumes including alfalfa, red clover (*Trifolium pratense*), common vetch (*Vicia sativa*), and white clover (*T. repens*); grain legumes, bean (*Phaseolus vulgaris*), faba bean (*Vicia faba*), lentil (*Lens culinaris*), and chickpea (*Cicer arietinum)* (Delwiche et al. 1987; Grau et al. 1991; Moussart et al. 2008); and to cause disease on several legumes native to North American prairies (Malvick et al. 2009). It was recognized as a serious and widespread disease of alfalfa when cultivars with resistance to Phytophthora root rot showed signs of seedling root rot (Delwiche et al. 1987). Isolates of *A. euteiches* from the United States appear to form subpopulations that differ in pathogenicity depending on plant species (Grau et al. 1991). For example, isolates from alfalfa are more aggressive on alfalfa and pea than on bean, while bean isolates are highly pathogenic on bean but not on pea and alfalfa. Much less host preference is found in *A. euteiches* isolates from Europe and disease on alfalfa was reported to not occur in the field (Moussart et al. 2008). The pathogen shows a high amount of genetic and pathogenic diversity (Malvick and Percich 1998; Malvick et al. 2009) suggesting that although it is homothallic in culture, outcrossing may occur at a low frequency leading to genetic variability (Grünwald and Hoheisel 2006). Isolates that infect alfalfa have been further divided into two pathotypes (races) based on the reaction of differential cultivars (Malvick and Grau 2001).

Disease symptoms of Aphanomyces root rot on alfalfa seedlings are virtually indistinguishable from those caused by *Phytophthora medicaginis*. Seedlings are stunted with chlorotic cotyledons that become necrotic as the disease progresses. Infected roots and hypocotyls first appear gray and water-soaked, then turn yellow–brown. In contrast to seedlings infected by *P. medicaginis*, seedlings infected by *A. euteiches* often remain upright for some time after infection. In established plants, the pathogen causes a sublethal rot of fibrous and lateral roots in wet soil conditions, reducing nodulation, and leading to yield loss and reduced stand life (Malvick and Grau 2016). *P. medicaginis* and *A. euteiches* can infect alfalfa plants simultaneously (Vandemark et al. 2010). Interestingly, infection by *A. euteiches* inhibits colonization by *P. medicaginis* suggesting that *A. euteiches* may stimulate plant defenses that inhibit *P. medicaginis* or produce compounds that restrict colonization by *P. medicaginis*. Surveys in several states reported that *A. euteiches* is more prevalent than *P. medicaginis* (Vincelli et al. 1994; Munkvold and Carlton 1995). However, because the surveys relied on baiting the pathogens from soil using seedlings, results may have been confounded by the difference in plant colonization by the two pathogens. Specific and sensitive quantitative PCR assays may help to resolve questions on the relative abundance of *P. medicaginis* and *A. euteiches* in soil (Gangneux et al. 2014; Bithell et al. 2020).

Low levels of resistance to Aphanomyces root rot were identified in alfalfa cultivars (Holub and Grau 1990b; Grau 1992) and improved cultivars were developed rapidly with resistance to Aphanomyces root rot. Heritability of resistance to Aphanomyces root rot is high (Holub and Grau 1990b; Vandemark et al. 2004) suggesting that it is conditioned by a small number of genes. Field studies showed that under wet soil conditions, cultivars with dual resistance to Aphanomyces root rot and Phytophthora root rot had improved forage yields (Holub and Grau 1990a;

Wiersma et al. 1995; Vincelli et al. 2000). Nonetheless, infection of dual resistant cultivars by *A. euteiches* in some locations led to the identification of a second pathotype, race 2 (Munkvold et al. 2001; Malvick and Grau 2001). The discovery of race 2 isolates in alfalfa fields that had never been planted with race 1-resistant cultivars and in fields with no history of alfalfa cultivation indicates that both races are endemic and that race 2 did not emerge in response to selection pressure from deploying race 1-resistant cultivars (Malvick et al. 2009). Surveys in several states found that race 1 and race 2 isolates are widespread (Malvick and Grau 2001; Malvick et al. 2009). Surveys in Minnesota of 53 alfalfa fields found that race 2 isolates were more prevalent with 45% of soil samples positive for race 1 and 11% positive for race 2 (Samac et al. 2017). Similarly, a survey of 40 New York alfalfa fields found 27% positive for race 1 and 50% positive for race 2 (Samac et al. 2017). However, the assay is dependent on differential cultivars, WAPH-1 with resistance to race 1 and WAPH-5 with resistance to race 1 and race 2. Thus, the presence of a race 2 strain in the soil sample can mask the presence of a race 1 strain and can make race 2 strains appear more common.

A persistent question is whether there are more than two races of *A. euteiches* infecting alfalfa in North America. In approximately 7% of the Minnesota soils surveyed for Aphanomyces root rot, there was poor seedling emergence of all check cultivars, which was at least partially explained by the presence of highly aggressive strains of *Pythium* species and *Fusarium* species causing seed rot and seedling damping-off (Berg et al. 2017). However, in 16% of soil samples, there was low survival of WAPH-5 seedlings, suggesting the possible presence of pathotypes that could overcome the race 1 and race 2 resistance. A total of 60 strains were isolated from five of such soil samples and used in the bioassay with the differential cultivars WAPH-1 and WAPH-5. All strains were identified as either race 1 or race 2 in this bioassay (Samac et al. 2017). However, the assay is limited by the availability of only the two differential cultivars. It is possible that the race 2 resistance locus is comprised of a cluster of resistance genes conditioning resistance to multiple pathotypes.

To help clarify the resistance gene–race relationships, mapping of resistance to race 1 and race 2 strains was carried out in an F1 mapping population developed from crossing a plant susceptible to both races with a plant highly resistant to both races selected from the cultivar 53V52 (Samac et al. 2018). Similar to the interaction with *P. medicaginis*, zoospores of *A. euteiches* are attracted to the root hair zone of alfalfa roots, with no difference in attraction by resistant or susceptible plants (Fig. 13.1). However, highly resistant plants show no disease symptoms after inoculation. Inspection of cross-sections through inoculated areas of resistant plants found that infections progress no farther than the initial epidermal cell that is penetrated by a germinating zoospore (Fig. 13.1). In contrast, in susceptible plants, the pathogen rapidly colonizes the cortical cells, grows in the intercellular spaces, and causes massive cellular degradation. Resistance is also associated with browning of the infected cell and a few neighboring cells, which fluoresce under UV light, indicating the presence of phenolic compounds. Transcript profiling found strong upregulation of genes in the phenylpropanoid pathway, jasmonic acid synthesis, receptor kinases, transcription factors, and defense response genes such as those encoding chitinase, glucanase, and peroxidases (Samac, unpublished). No oospores formed within resistant plants while numerous oospores formed within the roots of susceptible plants (Fig. 13.1). The rapid HR and restriction of pathogen growth in resistant plants suggest that resistance to race 1 and race 2 is conditioned by NLR resistance genes, most likely separate genes for each race interaction. For mapping, an F1 population was selected that showed segregation to race 1 of 4.6:1 (resistant: susceptible) and to race 2 of 1:3 (resistant: susceptible). Individual F1 plants were maintained in the greenhouse, and vegetative cuttings were made for disease resistance phenotyping of adventitious roots inoculated with zoospores. A total of 373 plants were used for genotyping-by-sequencing (Samac et al. 2018).

Highly significant SNP markers for resistance to race 1 were identified at the top of chromosome 1, approximately 38 kb from a cluster of NBL genes in the Cultivated Alfalfa at the Diploid Level (CADL) genome sequence (Samac et al. 2018). For race 2 resistance, highly significant markers were identified at the top of chromosome 2 using two different *A. euteiches* strains. No putative NBL genes were identified near the markers in either the *M. truncatula* or CADL genome sequences, which is not unexpected since both *M. truncatula* and CADL are susceptible to race 2 strains.

Resistance to Aphanomyces root rot was identified in *M. truncatula* using diverse pathogen strains. Vandemark and Grünwald (2004) screened 30 accessions with a race 2 strain of *A. euteiches* and found seven accessions that showed resistance similar to that of WAPH-5 and two as susceptible as the alfalfa susceptible check cultivar Saranac. The remaining accessions had intermediate responses. Using a broad host range isolate from pea, Moussart et al. (2008) screened 34 accessions and identified resistant to moderately susceptible accessions. Subsequently, dominant monogenic resistance was identified in the accession DZA045.5 (Pilet-Nayel et al. 2009). The pathogen could not be re-isolated from resistant plants, suggesting that resistance is mediated by a hypersensitive response. Genetic mapping of resistance identified a QTL named *AER1* explaining 88% of the total phenotypic variation at the top of chromosome 3 near a cluster of NBL genes (Pilet-Nayel et al. 2009). The same locus was identified using a second pea isolate as well as race 1 and race 2 alfalfa strains, in addition to minor alleles on chromosomes 1 and 8 (Hamon et al. 2010). Partial resistance with reduced colonization of the pathogen was found in Jemalong A17. Resistance was associated with dark brown root symptoms, reduced H_2O_2 production, higher catalase activity, increased soluble phenolics as well as peroxidase activity in epidermal cells, and lignin deposition in the pericycle cells (Djébali et al. 2009, 2011). The resistance QTL, *prAe1*, explains 34% of the total resistance phenotype and maps to the top of chromosome 3 (Djébali et al. 2009). Additional mapping discovered that the *AER1* and *prAe1* loci overlap in the same genomic region. Most recently, QTL mapping and genome-wide association studies (GWAS) in *M. truncatula* found about thirty loci associated with partial resistance to *A. euteiches*. The two most significant and independent loci identified three candidate genes inside the 440 kbp *AER1/prAe1* genomic region, among which an F-box protein-coding gene was associated with resistance (Bonhomme et al. 2014). Partial or quantitative disease resistance is of particular interest as it may be more durable and protect against a wider range of strains than resistance conditioned by NBL genes.

13.6 Phytophthora Root Rot

Phytophthora root rot of alfalfa was first described in California as the cause of poor seedling establishment in wet soils (Erwin 1954). Subsequent surveys found that the pathogen has worldwide distribution (Frosheiser and Barnes 1973) and prior to the development of resistant cultivars, the disease caused significant damage to seedlings and adult plants. The disease remains a production issue in locations with poor soil drainage as a component of "wet-soil syndrome" (Alva et al. 1985). The most severe losses occur during the seeding year in which the pathogen can cause pre- and post-emergence damping-off (Lueschen et al. 1976; Irwin et al. 1979). Symptoms on roots of young plants include yellow to brown girdling lesions at the junction of lateral roots or at the root tip. If the soil dries sufficiently, plants may produce lateral roots above the rotted area but these plants only develop shallow root systems and will be less productive. Severely affected plants show yellow to reddish-brown discoloration of foliage, stunting, and stand thinning leading to weed invasion (Frosheiser and Barnes 1973; Lueschen et al. 1976).

The pathogen was originally described as *Phytophthora megasperma* f. sp. *medicaginis*; however, multilocus sequence analysis of mitochondrial genes separates *P. megasperma* and *P. medicaginis* into two separate clades, although the two species are morphologically similar (Martin et al. 2012). Isolates classified as *P. megasperma* cause disease on alfalfa but are less aggressive than isolates of *P. medicaginis* (Hansen and Maxwell 1991). *P. medicaginis* also causes root rot on chickpea (*Cicer arietinum*). Most annual medics appear susceptible to *P. medicaginis* (De Haan et al. 2002), although resistance in two accessions of *M. truncatula* was reported (D'Souza 2009).

The disease cycle initiates with the germination of oospores, a long-lived resting structure that is the product of sexual reproduction. The hypha of a germinating oospore may infect a plant directly or produce a sporangium with numerous asexual zoospores, which are motile in water by means of flagella. The zoospores of *P. medicaginis* are attracted to all parts of alfalfa roots except the extreme root cap (Miller and Maxwell 1984). Within 2 min of contact with roots, the zoospores encyst, become attached to the root and lose flagella. Both susceptible and resistant plants are equally attractive. Encysted zoospores germinate 60–90 min after inoculation and directly penetrate roots. Growth in roots is rapid and after two hours, hyphae penetrate from epidermal cells into the outer cortex. At 12 h after inoculation, differences between resistant and susceptible plants are observed. In resistant plants, the cells in contact with hyphae plasmolyze, and growth is limited to intercellular spaces in the epidermis and cortex (Miller and Maxwell 1984). Colonization is more extensive in susceptible plants with intercellular growth penetrating to the stele. Marks and Mitchell (1971) characterized the response by resistant plants as a hypersensitive response, observing granulation and discoloration of cells in contact with hyphae, likely due to the production of phenolic compounds. They also observed that the stele diameter in resistant plants was significantly larger than in susceptible plants. Production of phytoalexins, particularly medicarpin, was associated with resistance and medicarpin inhibited the growth of *P. medicaginis*, suggesting that it has an important role in resistance to Phytophthora root rot (Vaziri et al. 1981).

Resistance to Phytophthora root rot was initially developed by selecting healthy plants from a field disease nursery (Froshelser 1980). The percentage of resistant plants in most unselected populations varied from 0 to 16%. However, this could be rapidly increased through recurrent selection. Starting from the cultivar Vernal with 10% resistant plants, two cycles of selection increase resistance to 50%, and a third cycle increased resistance to 63% (Frosheiser and Barnes 1973). Disease-resistant germplasm showed significantly enhanced forage yield and persistence (Frosheiser and Barnes 1973; Lueschen et al. 1976; Gray et al. 1988). Seedling disease screens are now the most common means of selecting resistant plants since resistance has been introduced into parental germplasm with desired agronomic characteristics, and very high percentages of resistant plants are reported in recent cultivar releases. Methods for measuring pathogen DNA concentration in inoculated roots have been developed (Vandemark and Barker 2003), but breeders still rely on phenotypic screening using disease symptoms to select resistant plants. Interestingly, resistance to Phytophthora root rot has been very stable with the widespread long-term deployment of resistant cultivars, although evidence of pathotypes with virulence against specific plant genotypes was identified (Marks and Mitchell 1971; Nygaard and Grau 1989; Liew and Irwin 1997). Putative "failure" of resistant cultivars was found to be caused by other soilborne pathogens such as *Aphanomyces euteiches* and *Pythium* species (Nygaard and Grau 1989; Munkvold and Carlton 1995). Irwin et al. (1995) suggest that resistance stability is due to the heterogeneity of cultivars with different resistance mechanisms and different resistance genes such that there is insufficient selection pressure on new pathotypes to increase in frequency.

Genetic analysis of resistance supports multiple resistance mechanisms in alfalfa germplasm in addition to the association of stele diameter and medicarpin production with resistance. Lu et al. (1973) determined inheritance of resistance in plants from the cultivars Vernal and Lahontan which had been shown to respond to pathogen infection with a hypersensitive response and to have a larger stele diameter than susceptible plants (Marks and Mitchell 1971). Based on the rapid progress to increase resistance in several genetic backgrounds, they hypothesized that resistance is simply inherited. Segregation ratios supported that susceptibility in these plant materials is conditioned by a single gene, *Pm*, with incomplete dominance, inferring that resistance is a recessive trait. A wide range of disease susceptibility genes have been identified in plants (Van Schie and Takken 2014). All plant genes that facilitate infection and support colonization of a pathogen can be considered susceptibility genes. Therefore, mutation or loss of a susceptibility gene can limit the ability of a pathogen to cause disease. Resistance from loss or alteration of a susceptibility gene is generally recessive. In contrast to the results of Lu et al. (1973), several studies found resistance to be a dominant trait. In diploid and tetraploid *M. sativa* plants, resistance was conditioned by two incompletely dominant complementary genes, *Pm1* and *Pm2* (Irwin et al. 1981a, b). In plants from a diploid *M. falcata*, resistance was inherited as a dominant allele at either of two independent segregating loci, *Pm3* and *Pm4* (Havey and Maxwell 1987). Two additional distinct dominant resistance genes, *Pm5* and *Pm6,* were identified in tetraploid plants by Havey et al. (1987). Thus, resistance appears to be complex and dependent on genetic background. Because a hypersensitive response appears to be part of disease resistance in at least some germplasm, it seems likely that an NBL gene(s) will be associated with Phytophthora root rot resistance.

A limited amount of genetic mapping for Phytophthora root rot resistance was done using plants from a nondormant cultivar (Musial et al. 2005). A genetic linkage map was constructed with a backcross population of 120 plants segregating for resistance to *P. medicaginis*. Three QTL were identified on non-homologous linkage groups that explained 6–15% of the variance, indicating the presence of at least three genes. Interestingly, two QTL increased susceptibility. Additional mapping studies in different genetic backgrounds that include investigation of morphological and biochemical defense responses of plants may help to clarify the basis of resistance to Phytophthora root rot in alfalfa.

13.7 Rust

Alfalfa rust is caused by the fungal pathogen *Uromyces striatus*, a member of the Pucciniaceae (Schoch et al. 2020). Rust fungi are one of the largest and most diverse clusters within the basidiomycetes (Aime et al. 2006). They are prevalent worldwide affecting an extensive range of crop species, including most food and fodder legumes (Sillero et al. 2006; Dean et al. 2012). Rust fungi are obligate biotrophs with the ability to infect and reproduce on an wide range of legume species. An investigation of the host range of a monouredinial isolate of *Uromyces striatus* from alfalfa on a collection of 345 alfalfa-related species from 27 genera showed that 141 species from 11 genera were susceptible to the isolate, highlighting the scope of the host range of this pathogen and its importance in the epidemiology of the fungus (Skinner and Stuteville 1995). Inoculation of 18 *Melilotus* species with an isolate of *U. striatus* from alfalfa showed that 10 became heavily rusted, in addition to all the *M. alba* plants included as susceptible controls, suggesting they are all potential hosts (Stuteville 2002). These findings provide ample support that *U. striatus* has a wide range of hosts which makes the possibility of invasion of new territories and host species serious concerns.

Environmental factors like temperature, humidity, and light are key conditions for rust infection and spread in alfalfa. In cold Northern latitudes, rust outbreaks usually occur after mid-July because the fungus does not overwinter in those regions. However, if humidity is high in summer, rust spores are blown from the south where the spores overwinter on alfalfa or other hosts and when the conditions become favorable they infect alfalfa fields. Aggressive growth of *U. striatus* was observed at temperatures between 21 to 29 °C and high humidity conditions (https://corn.agronomy.wisc.edu/Management/

pdfs/IPMManual_2_Alfalfa.pdf). Besides the growing environment, the rust infection rate in alfalfa may also fluctuate with the leaf surface and host cellular properties. The duration of leaf wetness and temperature influence the efficiency of early pathogen invasion, the incubation period, and the rate of pustule appearance on alfalfa leaves (Webb and Nutter 1997). Rust-infected alfalfa plants display masses of red-brown spores from circular pustules on leaves, petioles, and sometimes stems, where the spores can be rubbed off easily. Uredinia were described as mostly hypophyllous and the urediniospores are one-celled, 20–25 × 18–22 μm, and have a globular or ellipsoid shape. Walls were 1.0 to 1.2 μm thick, echinulate, and with predominantly four pores (Stuteville 2002). Rust infection causes leaf wilting and premature defoliation, impeding plant growth and development, and leading to substantial yield losses (Webb and Nutter 1997). Infection of newly established alfalfa seedlings in late summer results in stand weakening and reduction of winterhardiness, making the plants more susceptible to winterkill and therefore compromising stand persistence. Also, rust-infected hay may cause allergic reactions to horses in addition to biomass quality deterioration and lower digestibility and energy intake.

Alfalfa rust can be controlled by agronomic practices or using resistant cultivars. Controlling *U. striatus* alternate hosts might be a strategy to reduce the pathogen load in the field. Rust damage is lower when the alfalfa fields are harvested frequently, but more mature plants or those grown for seed production are at risk of defoliation by rust. The effective practice to minimize rust damage is timely clipping and harvesting infected fields early. The application of commercial fungicides such as Headline® SC at the rate of 6 to 9 oz/acre is an effective way to control alfalfa rust and other fungal diseases. Glyphosate, the active ingredient in Roundup herbicide, was also suggested as another alternative for controlling *U. striatus* in Roundup Ready alfalfa (Samac and Foster-Hartnett 2012). However, planting alfalfa cultivars with durable resistance remains the most effective and sustainable way to protect alfalfa against rust.

Breeding for broad-spectrum resistance to rust in alfalfa requires understanding the host response to the pathogen as well as the specific biology of rust resistance. Alfalfa rust is probably one of the least characterized diseases. Even though there is an established standard test for rust resistance in alfalfa using MSA-CW3An3 as a resistant check and Saranac and Moapa 69 as susceptible checks (Stuteville 1991), there are no recommended cultivars with rust resistance in the NAFA alfalfa variety ratings. Races of *Uromyces striatus* have not been characterized either.

Most research on alfalfa rust interactions was predominantly conducted in *Medicago truncatula* using histological and functional genomic studies to unravel the host–*U. striatus* interaction and to identify potentially important genes. Investigation of resistance to the alfalfa rust in a collection of *M. truncatula* accessions exhibiting a wide range of resistance reactions showed that resistance to *U. striatus* was due to the ability of some accessions to restrict haustorium formation by aborting fungal colonies in early stages (Rubiales and Moral 2004). Characterization of a collection of 113 mostly European accessions of *M. truncatula* revealed that stomatal surface characteristics did not interfere with the ability of *U. striatus* germ tubes to infect. But after penetration, the resistant ecotypes reacted with various degrees of cell death in different stages of haustorial establishment, some accessions showed hypersensitive reactions by developing necrotic lesions, and one accession exhibited a pre-haustorial type of defense without hypersensitive response (Kemen et al. 2005).

During infection, obligate biotrophs typically deliver fungal effectors into the host cell. In the case of the rust fungi *Uromyces fabae* and *U. striatus*, the first confirmed fungal protein localized within the host cytoplasm was RTP1p. Analysis of 28 RTP1 homologs in these rust fungi showed that members of the RTP family are ubiquitous and most likely specific to the order Pucciniales. The RTPp effectors were similar in structure to cysteine protease inhibitors and inhibited proteolytic activity in *Pichia pastoris* cultures, suggesting that the RTP1p homologs have an inhibitory activity, probably associated with effector function during biotrophic interactions with the host (Pretsch et al. 2013).

Depending upon the nature of the plant–pathogen interactions, plants generally display two types of disease resistance, host resistance and nonhost resistance (NHR), with host resistance being mostly cultivar specific (Mysore and Ryu 2004). The reaction of *M. truncatula* to one specific (*U. striatus*) and two non-specific rusts (*U. viciae-fabae* and *U. lupinicolus*) showed that similar pre- and post-haustorial formation mechanisms of resistance appear to be working in *M. truncatula* against the two types of rust fungi, an indication that NHR to rust fungi in this species involves a combination of specific and non-specific responses (Vaz Patto and Rubiales 2014). Infection with *U. striatus* has also been shown to modify the phytochemical profile of alfalfa leading to the synthesis of compounds with varying degrees of inhibitory activity against fungal diseases and protective efficacy of over 80% against *Puccinia striiformis* (Li et al. 2018). Analysis of the expression of transcription factors (TF) in resistant and susceptible accessions of *M. truncatula* at the onset of infection with *U. striatus* identified thirteen putative TF differentially expressed between resistant and susceptible genotypes that are known to be important in cellular defense, suggesting that resistance could be mediated both by genes that are constitutively expressed and by genes that are activated/repressed when plants are challenged with the pathogen (Madrid et al. 2010).

Applied genetic and genomic studies of rust resistance in alfalfa are scarce. Recent linkage analysis of rust severity (RS) phenotype using SNP loci derived from genotyping-by-sequencing of a pseudotestcross F1 population from a cross of one resistant (CW1010) and one susceptible (3010) alfalfa cultivar identified five QTL in the CW1010 genetic map and three QTL in 3010. The most significant QTL (Us-RustR1) explained 13% ($R^2 = 0.13$) of the variance. The cumulative effect of the five QTL detected in CW1010 explained 38% of the total variation, while the three QTL from the susceptible genotype 3010 accounted for 21% of total phenotypic variation. The number of QTL suggests that rust resistance in alfalfa is polygenic and most likely incomplete (Adhikari and Missaoui 2019). Incomplete resistance to rust caused by various species of *Uromyces* has been documented in other cool-season legumes such as pea (*Pisum sativum*), faba bean (*Vicia faba*), and chickpea (*Cicer arietinum*). Sources of incomplete or partial resistance to rust were also described in lentil (*Lens culinaris*) (Rubiales et al. 2013). Some of the rust QTL sequences identified in this study were similar to *M. truncatula* genes and proteins reported for their response to rust or other biotic and abiotic stresses, such as the eukaryotic aspartyl protease family proteins involved in the defense against the fungal pathogen *Botrytis* in Arabidopsis (Li et al. 2016), leucine zipper transcription factor homologs, PGK, and FHA that are known for their response to disease stresses in plants (Alves et al. 2013; Sekhwal et al. 2015; Feng et al. 2016).

The overlaps and similarities in response pathways highlighted in the various studies of *U. striatus* interaction with alfalfa and *M. truncatula* provide evidence for the involvement of multiple molecular pathways in alfalfa for resistance to rust. Alfalfa germplasm screening to identify sources of resistance combined with further genetic and expression analysis studies will contribute to the development of the necessary conventional breeding approaches and molecular tools to use in developing rust-resistant alfalfa cultivars.

13.8 Verticillium Wilt

The fungus *Verticillium alfalfa* (previously *V. albo-atrum;* Inderbitzin et al. 2011) causing a severe wilt disease on alfalfa was introduced in North America in the 1970s, likely on infested seed (Graham et al. 1977). Molecular studies with a small sample of strains collected in the 1970s and 1980s from a limited geographical range concluded that the population is genetically homogeneous and highly clonal (Griffen et al. 1997; Mahuku and Platt 2002; Qin et al. 2006). The pathogen enters the plant through the roots and rapidly colonizes the xylem tissue throughout the plant and is spread from plant to plant by forage harvest (Pennypacker and Leath 1993). Susceptible plants show foliar symptoms of wilting and leaf chlorosis followed by stunting and stand decline (Graham et al. 1977). Up to 50% yield losses have been observed (Peaden et al. 1985), and cultivars with at least 60% resistant plants are needed for maintaining yields under disease pressure (Busch et al. 1985). In resistant plants, several mechanisms limit growth and spread of the fungus within the plant (Pennypacker and Leath 1993). However, some plants tolerate the fungus and support pathogen growth throughout the plant without showing disease symptoms, although yield is affected significantly (Pennypacker et al. 1990). The lack of symptoms may be the result of insensitivity to a fungal toxin or other effector (Fradin and Thomma 2006). Relying on foliar symptoms alone in cultivar development risks introducing tolerance into cultivars. Thus, the development of DNA markers was seen as a high priority for alfalfa improvement.

The genetics of plant resistance to Verticillium wilt (VW) was investigated in several alfalfa cultivars. Viands (1985) compared the inheritance of VW resistance in alfalfa plants selected from the European cultivars, Maris Kabul and Vertus, and suggested that a major dominant gene may control Verticillium wilt resistance in Maris Kabul, while more complex additive effects were found in Vertus. Conventional selection for Verticillium wilt resistance in alfalfa requires evaluating host responses to the pathogen and is time-consuming. Developing molecular markers associated with Verticillium wilt resistance and the use of marker-assisted selection (MAS) would greatly accelerate breeding programs.

To identify loci associated with resistance to Verticillium wilt, a genome-wide association study was initiated in two alfalfa populations segregating for Verticillium wilt resistance developed by Forage Genetics International (FGI) and S & W Seed Company. Genotyping-by-sequencing was used for developing single nucleotide polymorphism markers for marker-trait association. Markers associated with Verticillium wilt resistance were found in the FGI population. The most significant markers were located on chromosome 8, which contributed major effects on the phenotypic variation observed and represented novel loci associated with Verticillium wilt resistance (Yu et al. 2017b). Markers associated with Verticillium wilt resistance were also identified in the S & W population, and the most significant markers were located on chromosome 6 (Yu et al. 2017a). Multiple loci in different genetic regions associated with Verticillium wilt resistance in different populations highlighted the complexity of the genetic basis of resistance and indicate that quantitative resistance exists in alfalfa. To identify quantitative trait loci (QTL) for Verticillium wilt resistance in alfalfa, a full-sib population segregating for resistance was used. High-density linkage maps for both resistant and susceptible parents were constructed using single-dose alleles of SNP markers generated by genotyping-by-sequencing. Five QTL associated with Verticillium wilt resistance were identified and they were on four linkage groups (4D, 6B, 6D, and 8C). Of those, three QTL (qVW-6D-1, qVW-6D-2, and qVW-8C) had a higher logarithm of odds. Two putative candidates of nucleotide-binding site leucine-rich repeat disease resistance genes were identified in the QTL intervals of qVW-6D-2 and qVW-8C, respectively (Yu et al. 2020). These results agreed with genome-wide association studies, in which similar resistance loci were identified.

Studies on genetic mapping of Verticillium wilt resistance in *M. truncatula* have been carried out in three RIL populations (Ben et al. 2013b). Three QTL were identified. Major QTL were located on linkage group 7 for the parent line A17, and additional QTL were located on linkage groups 2 and 6 for parent DZA45.5. The analysis found that only a small number of major

QTL are involved in resistance to Verticillium wilt in *M. truncatula*, suggesting simple genetic control. In another study (Negahi et al. 2014), recombinant inbred lines (RILs) derived from the cross of A17 and F83005.5 were used and infected with a potato isolate of *V. albo-atrum*, LPP0323. Four QTL for resistance to LPP0323 were identified on LGs 1, 2, 6, and 8. The phenotypic variance explained by each QTL ranged from 4 to 21%. Additive gene effects showed that favorable alleles for resistance all came from the resistant parent. The four QTL are distinct from those described for an alfalfa *V. albo-atrum* isolate described by Ben et al. (2013b), confirming the existence of several resistance mechanisms in this species. Recently, genome-wide association studies (GWAS) were performed and identified several QTL associated with Verticillium wilt resistance on chromosome 1, 7, and 8 (Mazurier et al. 2016). Both phenotypic and genetic analyses suggested that different resistance mechanisms exist in *M. truncatula* populations for Verticillium wilt.

13.9 Utilization of Disease Resistance Genes in Alfalfa Improvement

Recurrent phenotypic selection is an effective means to develop multi-pest-resistant alfalfa cultivars. However, it is slow, labor-intensive, and inefficient. It requires a large amount of greenhouse space for seedling diseases or season-long field evaluations for diseases of adult plants, and is influenced by environmental conditions, inoculation techniques, and human error in scoring symptoms. Fully characterizing elite parents for disease resistance by inoculation involves laborious test crosses and progeny testing, which may not be practical in commercial operations. DNA markers that are tightly linked to or integral to the function of specific R genes, so-called functional markers, are in routine use in cultivar development in other major crops for characterizing parental germplasm and accelerating cultivar development. Lack of knowledge of R genes in alfalfa may be one impediment to increasing yield and other desirable agronomic traits because undesirable genes or alleles may be inherited with R genes when using recurrent selection. Current efforts to develop reference alfalfa genome sequences in diverse genetic backgrounds and to develop inexpensive universal marker platforms should soon enable researchers to identify markers linked to resistance genes for major diseases. Markers would be of considerable value in cases where resistance is additive, requiring multiple genes with small individual effects. Markers are also valuable for defining resistance loci for candidate gene identification conditioning disease resistance. Cloning and verification of R genes is not required for marker development or use but would lead to a better understanding of alfalfa defense mechanisms. At present, the costs of commercializing a genetically modified alfalfa cultivar make utilization of cloned genes such as *RCT1* from *M. truncatula* impractical. However, markers could be used to uncover alfalfa homologs of *RCT1* for broad non-race-specific resistance. Genetic modification could be warranted if a gene from a secondary or tertiary gene pool (other *Medicago* species) confers a large agronomic benefit. We expect that the ongoing genomics research in alfalfa will accelerate breeding programs in general and breeding for disease resistance in particular. The availability of complete genome sequences offers great opportunities for rapid and efficient development of molecular markers tightly linked to R genes.

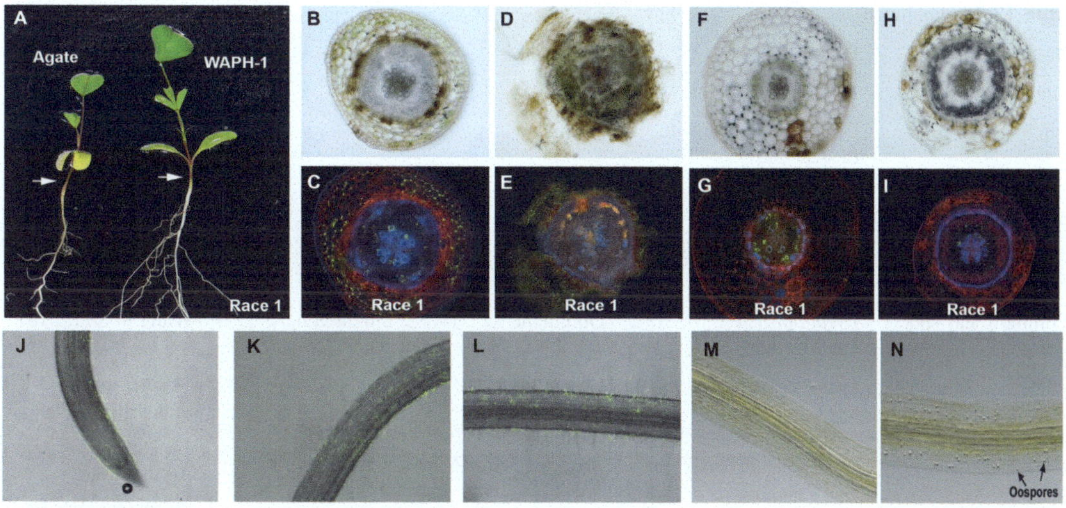

Fig. 13.1 Resistance to Aphanomyces root rot in alfalfa is associated with a hypersensitive response. A. Symptoms on the susceptible cultivar Agate and resistant cultivar WAPH-1. Arrow points to the location where sections were made. B, C. Cross section of Agate hypocotyl at 7 days after inoculation. D, E. Cross section of Agate hypocotyl at 14 days after inoculation. F, G. Cross section of WAPH-1 hypocotyl at 7 days after inoculation. H, I. Cross section of WAPH-1 hypocotyl at 14 days after inoculation. J. Root tip with germinating zoospores. K. Elongation zone of root with germinating zoospores. L. Root hair zone with germinating zoospores. M. Root of WAPH-1 14 days after inoculation without oospores. N. Root of Agate 14 days after inoculation with oospores. Panels C, E, G, I, J, K, L stained with wheat germ agglutinin-FITC to visualize *A. euteiches* (Djébali et al. 2009). Viewed under UV-illumination, the pathogen fluoresces green and phenolic compounds fluoresce red. Green staining in the vascular cylinder in panel G suggests the presence of N-acetyl-glucosamine, not the pathogen

Table 13.1 Disease resistance genes identified in alfalfa

Disease	Pathogen	Inheritance	Gene(s)	References
Bacterial diseases				
Bacterial wilt	*Clavibacter insidiosus*	Dominant	*BW1*	Viands and Barnes (1980)
		Additive	*BW2, BW3*	Viands and Barnes (1980)
Oomycete diseases				
Downy mildew	*Peronospora trifoliorum*	Incomplete dominance	*Dm*	Pedersen and Barnes (1965)
		Dominant	*PtR1, PtR2, PtR3, PtR4, PtR5*	Skinner and Stuteville (1985a)
		Additive		Skinner and Stuteville (1985b)
Phytophthora root rot	*Phytophthora medicaginis*	Recessive	*Pm*	Lu et al. 1973)
		Dominant	*Pm1, Pm2*	Irwin et al. (1981a, b)
		Dominant	*Pm3, Pm4*	Havey and Maxwell (1987)
		Dominant	*Pm5, Pm6*	Havey et al. (1987)

(continued)

Table 13.1 (continued)

Disease	Pathogen	Inheritance	Gene(s)	References
Fungal diseases				
Anthracnose	*Colletotrichum trifolii* race 1	Dominant	*An1*	Elgin and Ostazeski (1985)
	Colletotrichum trifolii race 1 and 2	Dominant	*An2*	Elgin and Ostazeski (1985)
	Colletotrichum trifolii race 1	Dominant, recessive		Mackie and Irwin (1998)
Common leaf spot	*Pseudopeziza medicaginis*	Dominant	Single gene	Wang et al. (2012)
Rust	*Uromyces striatus*	Dominant, temperature sensitive		Skinner and Stuteville (1989)
		Polygenic		Adhikari and Missaoui (2019)
Sclerotinia Crown and Stem Rot	*Sclerotinia trifoliorum*	Polygenic		Halimi and Rowe (1998)
Fusarium wilt	*Fusarium oxysporum* f. sp. *medicaginis*	Dominant	*FW1*	Hijano et al. (1983)
		Incomplete dominance, additive	*FW2*	Hijano et al. (1983)
Stagospora leaf spot and root rot	*Stagonospora meliloti*	Dominant	Single gene	Irwin et al. (2004)
		Polygenic, additive		Musial et al. (2007)
Verticillium wilt	*Verticillium alfalfae*	Dominant		Viands (1985)
		Polygenic, additive		Viands (1985)
Nematodes				
Alfalfa stem nematode	*Ditylenchus dipsaci*	Dominant	*Sn*	Grundbacher and Stanford (1962)
Root knot nematode	*Meloidogyne hapla*	Dominant	*Rk1*	Goplen and Stanford (1960)
	M. javanica	Dominant	*Rk2*	Goplen and Stanford (1960)
Root lesion nematode	*Pratylenchus penetrans*	Polygenic, additive		Thies et al. (1994)
Viruses				
Alfalfa mosaic	*Alfalfa mosaic virus*	Recessive	*am1*	Crill et al. (1971)
		Recessive, temperature sensitive		Iwai et al. (1992)

Table 13.2 Disease resistance genes or QTL identified in *Medicago truncatula*

Disease	Pathogen	Inheritance	Gene(s) or QTL designations	References
Bacterial diseases				
Bacterial wilt	*Ralstonia solanacearum*	QTL	*MtQRRS1*	Vailleau et al. (2007); Ben et al. (2013a)
Aphanomyces root rot	*Aphanomyces euteiches*	Dominant	*AER1*	Pilet-Nayel et al. (2009)
		Recessive	*prAe1*	Djébali et al. (2009)
Phytophthora root rot	*Phytophthora medicaginis*	Dominant		D'Souza (2009)
Fungal diseases				
Anthracnose	*Colletotrichum trifolii*	Dominant	*RCT1*	Ameline-Torregrosa et al. (2008a), Yang et al. (2007, 2008)
Ascochyta blight	*Didymella pinodes*	QTL		Madrid et al. (2014)
Powdery mildew	*Erysiphe pisi*	QTL	*Epp1, Epa1, Epa2*	Ameline-Torregrosa et al. (2008b)
Root canker	*Rhizoctonia solani*	QTL		Anderson et al. (2013)
Spring black stem and leaf spot	*Phoma medicaginis*	QTL	*rnpm1, rnpm2*	Kamphuis et al. (2008)
Verticillium wilt	*Verticillium alfalfae*	QTL	Mt*Va1*, Mt*Va2*, Mt*Va3*	Ben et al. (2013b)
Nematodes				
Root knot nematode	*Meloidogyne incognita*	Dominant		Dhandaydham et al. (2008)
Viruses				
Subterranean clover mottle	*Subterranean clover mottle virus*	Dominant	*RSCMoV1*	Saqib et al. (2009)

References

Adhikari L, Missaoui AM (2019) Quantitative trait loci mapping of leaf rust resistance in tetraploid alfalfa. Physiol Mol Plant Pathol 106:238–245. https://doi.org/10.1016/j.pmpp.2019.02.006

Aime MC, Matheny PB, Henk DA, Frieders EM, Nilsson RH et al (2006) An overview of the higher level classification of Pucciniomycotina based on combined analyses of nuclear large and small subunit rDNA sequences. Mycologia 98:896–905. https://doi.org/10.3852/mycologia.98.6.896

Allen S, Barnes G, Caddel J (1982) A new race of *Colletotrichum trifolii* on alfalfa in Oklahoma. Plant Dis 66:922–924

Alva AK, Lanyon LE, Leath KT (1985) Excess soil water and Phytophthora root rot stresses of Phytophthora root rot sensitive and resistant alfalfa cultivars. Agron J 77:437–442. https://doi.org/10.2134/agronj1985.00021962007700030019x

Alves MS, Dadalto SP, Gonçalves AB, De Souza GB, Barros VA et al (2013) Plant bZIP transcription factors responsive to pathogens: a review. Intl J Mol Sci 14:7815–7828. https://doi.org/10.3390/ijms14047815

Ameline-Torregrosa C, Cazaux M, Danesh D, Chardon F, Cannon S et al (2008a) Genetic dissection of resistance to anthracnose and powdery mildew in *Medicago truncatula*. Mol Plant-Microbe Interact 21:61–69. https://doi.org/10.1094/MPMI-21-1-0061

Ameline-Torregrosa C, Wang B-B, O'Bleness MS, Deshpande S, Zhu H et al (2008b) Identification and characterization of nucleotide-binding site-leucine-rich repeat genes in the model plant *Medicago truncatula*. Plant Physiol 146:5–21. https://doi.org/10.1104/pp.107.104588

Andersen EJ, Ali S, Byamukama E, Yen Y, Nepal MP (2018) Disease resistance mechanisms in plants.

Genes (Basel) 9:339. https://doi.org/10.3390/genes9070339

Anderson JP, Lichtenzveig J, Oliver RP, Singh KB (2013) *Medicago truncatula* as a model host for studying legume infecting *Rhizoctonia solani* and identification of a locus affecting resistance to root canker. Plant Pathol 62:908–921. https://doi.org/10.1111/j.1365-3059.2012.02694.x

Ariss JJ, Rhodes LH (2007) A new race of *Colletotrichum trifolii* identified on alfalfa in Ohio. Plant Dis 91:1362. https://doi.org/10.1094/PDIS-12-19-2742-PDN

Baldridge GD, O'Neill NR, Samac DA (1998) Alfalfa (*Medicago sativa* L.) resistance to the root-lesion nematode, *Pratylenchus penetrans*: defense-response gene mRNA and isoflavonoid phytoalexin levels in roots. Plant Mol Biol 38:999–1010. https://doi.org/10.1023/A:1006182908528

Barnes DK, Hanson CH, Frosheiser FI, Elling LJ (1971) Recurrent selection for bacterial wilt resistance in alfalfa. Crop Sci 11:545–546. https://doi.org/10.2135/cropsci1971.0011183X001100040025x

Ben C, Debellé F, Berges H, Bellec A, Jardinaud M-F et al (2013a) MtQRRS1, an R-locus required for *Medicago truncatula* quantitative resistance to *Ralstonia solanacearum*. New Phytol 199:758–772. https://doi.org/10.1111/nph.12299

Ben C, Toueni M, Montanari S, Tardin M-C, Fervel M et al (2013b) Natural diversity in the model legume *Medicago truncatula* allows identifying distinct genetic mechanisms conferring partial resistance to Verticillium wilt. J Exp Bot 64:317–332. https://doi.org/10.1093/jxb/ers337

Berg LE, Miller SS, Dornbusch MR, Samac DA (2017) Seed rot and damping-off of alfalfa in Minnesota caused by *Pythium* and *Fusarium* species. Plant Dis 101:1860–1867. https://doi.org/10.1094/PDIS-02-17-0185-RE

Bithell SL, Moore K, Herdina MA, Harden S et al (2020) Phytophthora root rot of chickpea: inoculum concentration and seasonally dependent success for qPCR based predictions of disease and yield loss. Australas Plant Pathol 50:91–103. https://doi.org/10.1007/s13313-020-00752-2

Blount JW, Dixon RA, Paiva NL (1992) Stress responses in alfalfa (*Medicago sativa* L.) XVI. Antifungal activity of medicarpin and its biosynthetic precursors; implications for the genetic manipulation of stress metabolites. Physiol Mol Plant Pathol 41:333–349. https://doi.org/10.1016/0885-5765(92)90020-V

Bonhomme M, André O, Badis Y, Ronfort J, Burgarella C et al (2014) High-density genome-wide association mapping implicates an F-box encoding gene in *Medicago truncatula* resistance to *Aphanomyces euteiches*. New Phytol 201:1328–1342. https://doi.org/10.1111/nph.12611

Busch LV, Christie BR, Smith EA, Boland G (1985) Testing alfalfa cultivars for resistance to an alfalfa strain of *Verticillium albo-atrum*. Can J Plant Pathol 7:203–205. https://doi.org/10.1080/07060668509501504

Crill P, Hanson E, Hagedorn D (1971) Resistance and tolerance to alfalfa mosaic virus in alfalfa. Phytopathology 61:369–371. https://doi.org/10.1094/phyto-61-369

D'Souza K (2009) Pathosystem development, characterisation and genetic dissection of the soil pathogen *Phytophthora medicaginis* and the model legme *Medicago truncatula*: a view to application of disease resistance in susceptible legume species. PhD Dissertation, Murdoch University

de Bruijn F (ed) (2020) The model legume. Medicago truncatula. Wiley

De Haan RL, Sheaffer CC, Samac DA, Moynihan JM, Barnes DK (2002) Evaluation of annual *Medicago* for Upper Midwest agroecosystems. J Agron Crop Sci 188:417–425. https://doi.org/10.1046/j.1439-037X.2002.00591.x

Dean R, Van Kan JA, Pretorius ZA, Hammond-Kosack KE, Di Pietro A et al (2012) The Top 10 fungal pathogens in molecular plant pathology. Mol Plant Pathol 13:414–430. https://doi.org/10.1111/j.1364-3703.2011.00783.x

Delwiche PA, Grau CR, Holub EB, Perry J (1987) Characterization of *Aphanomyces euteiches* isolates recovered from alfalfa in Wisconsin. Plant Dis 71:155–161. https://doi.org/10.1094/PD-71-0155

Devine TE, Hanson CH, Ostazeski SA, Campbell TA (1971) Selection for resistance to anthracnose (*Colletotrichum trifolii*) in four alfalfa populations. Crop Sci 11:854–855. https://doi.org/10.2135/cropsci1971.0011183x001100060026x

Dhandaydham M, Charles L, Zhu H, Starr JL, Huguet T et al (2008) Characterization of root-knot nematode resistance in *Medicago truncatula*. J Nematol 40:46–54

Djébali N, Jauneau A, Ameline-Torregrosa C, Chardon F, Jaulneau V et al (2009) Partial resistance of *Medicago truncatula* to *Aphanomyces euteiches* is associated with protection of the root stele and is controlled by a major QTL rich in proteasome-related genes. Mol Plant-Microbe Interact 22:1043–1055. https://doi.org/10.1094/MPMI-22-9-1043

Djébali N, Mhadhbi H, Lafitte C, Dumas B, Esquerré-Tugayé M-T et al (2011) Hydrogen peroxide scavenging mechanisms are components of *Medicago truncatula* partial resistance to *Aphanomyces euteiches*. Eur J Plant Pathol 131:559–571. https://doi.org/10.1007/s10658-011-9831-1

Elgin J, O'Neill N (1988) Comparison of genes controlling race 1 anthracnose resistance in Arc and Saranac AR alfalfa. Crop Sci 28:657–659. https://doi.org/10.2135/cropsci1988.0011183X002800040020x

Elgin J, Ostazeski S (1985) Inheritance of resistance to race 1 and race 2 anthracnose in Arc and Saranac AR alfalfa. Crop Sci 25:861–865. https://doi.org/10.2135/cropsci1985.0011183X002500050032x

Elgin J, Welty R, Gilchrist D (1988) Breeding for disease and nematode resistance. In: Hanson A, Barnes D,

Hill RJ (eds) Alfalfa and Alfalfa improvement. ASA-CSSA-SSSA, Madison, WI, pp 827–858
Elgin JH, Barnes DK, Busbice TH, Buss GR, Clark NA et al (1981) Anthracnose resistance increases alfalfa yields. Crop Sci 21:457–460. https://doi.org/10.2135/cropsci1981.0011183X002100030026x
Erwin DC (1954) Root rot of alfalfa caused by *Phytophthora cryptogea*. Phytopathology 44:700–704
Feng B, Ma S, Chen S, Zhu N, Zhang S et al (2016) PARylation of the forkhead-associated domain protein DAWDLE regulates plant immunity. EMBO Rep 17:1799–1813. https://doi.org/10.15252/embr.201642486
Fradin EF, Thomma B (2006) Physiology and molecular aspects of Verticillium wilt diseases caused by *V dahliae* and *V albo-atrum*. Mol Plant Pathol 7:71–86. https://doi.org/10.1111/j.1364-3703.2006.00323.x
Frosheiser FI, Barnes DK (1973) Field and greenhouse selection for Phytophthora root rot resistance in alfalfa. Crop Sci 13:735–738. https://doi.org/10.2135/cropsci1973.0011183x001300060044x
Frosheiser FI (1980) Conquering Phytophthora root rot with resistant alfalfa cultivars. Plant Dis 64:909–912. https://doi.org/10.1094/PD-64-909
Gangneux C, Cannesan M-A, Bressan M, Castel L, Moussart A et al (2014) A sensitive assay for rapid detection and quantification of *Aphanomyces euteiches* in soil. Phytopathology 104:1138–1147. https://doi.org/10.1094/PHYTO-09-13-0265-R
Goplen BP, Stanford EH (1960) Autotetraploidy and linkage in alfalfa- a study of resistance to two species of root-knot nematodes. Agron J 52:337–342. https://doi.org/10.2134/agronj1960.00021962005200060011x
Graham J, Devine T, Hanson C (1976) Occurrence and interaction of three species of *Colletotrichum* on alfalfa in Mid-Atlantic United States. Phytopathology 66:538–541
Graham J, Peaden RN, Evans DW (1977) Verticillium wilt of alfalfa found in the United States. Plant Dis Report 61:337–340
Grau CR (1992) Registration of WAPH-1 alfalfa germplasm with resistance to Aphanomyces root rot. Crop Sci 32:287–288. https://doi.org/10.2135/cropsci1992.0011183X003200010074x
Grau CR, Muehlchen A, Tofte J, Smith R (1991) Variability in virulence of *Aphanomyces euteiches*. Plant Dis 75:1153–1156. https://doi.org/10.1094/PD-75-1153
Gray F, Bohl W, Wofford D (1988) Relationships of increasing levels of resistance of Phytophthora root rot in alfalfa to stand longevity and yield. Plant Dis 72:1064–1067. https://doi.org/10.1094/PD-72-1064
Griffen AM, Bainbridge BW, Heale JB (1997) Ribosomal, mitochondrial and amplified DNA polymorphisms in *Verticillium albo-atrum* pathogenic to hops, lucerne and other plants. Mycol Res 101:1085–1091. https://doi.org/10.1017/S0953756297003742
Grundbacher FJ, Stanford EH (1962) Genetic factors conditioning resistance in alfalfa to the stem nematode. Crop Sci 2:211–217. https://doi.org/10.2135/cropsci1962.0011183x000200030011x
Grünwald NJ, Hoheisel GA (2006) Hierarchical analysis of diversity, selfing, and genetic differentiation in populations of the oomycete *Aphanomyces euteiches*. Phytopathology 96:1134–1141. https://doi.org/10.1094/PHYTO-96-1134
Halimi ES, Rowe DE (1998) Stem-tip inoculation of alfalfa with *Sclerotinia trifoliorum*: II. Performance of S1 and F1 progenies from resistant and susceptible selections. Crop Sci 38:23–26. https://doi.org/10.2135/cropsci1998.0011183X003800010004x
Hamon C, Baranger A, Miteul H, Lecointe R, Le Goff I et al (2010) A complex genetic network involving a broad-spectrum locus and strain-specific loci controls resistance to different pathotypes of *Aphanomyces euteiches* in *Medicago truncatula*. Theor Appl Genet 120:955–970. https://doi.org/10.1007/s00122-009-1224-x
Hansen EM, Maxwell DP (1991) Species of the *Phytophthora megasperma* complex. Mycologia 83:376–381. https://doi.org/10.2307/3759999
Havey MJ, Maxwell DP (1987) Inheritance of Phytophthora root rot resistance in two diploid alfalfa species. Crop Sci 27:225–228. https://doi.org/10.2135/cropsci1987.0011183x002700020019x
Havey MJ, Maxwell DP, Irwin JAG (1987) Independent inheritance of genes conditioning resistance to phytophthora root rot from diploid and tetraploid alfalfa. Crop Sci 27:873–879. https://doi.org/10.2135/cropsci1987.0011183x002700050007x
Hijano EH, Barnes DK, Frosheiser FI (1983) Inheritance of resistance to Fusarium wilt in alfalfa. Crop Sci 23:31–34. https://doi.org/10.2135/cropsci1983.0011183X002300010010x
Holub EB, Grau CR (1990a) Productivity and survival of alfalfa ramets in soil naturally infested with *Aphanomyces euteiches* and *Phytophthora megasperma*. Can J Plant Pathol 12:83–91. https://doi.org/10.1080/07060669009501046
Holub EB, Grau CR (1990b) Specificity of resistance to *Aphanomyces euteiches* in seedling alfalfa. Plant Dis 74:164–168. https://doi.org/10.1094/PD-74-0164
Inderbitzin P, Bostock RM, Davis RM, Usami T, Platt H et al (2011) Phylogenetics and taxonomy of the fungal vascular wilt pathogen *Verticillium*, with the descriptions of five new species. PLoS ONE 6:e28341. https://doi.org/10.1371/journal.pone.0028341
Irwin J, Armour D (2015) Anthracnose. In: Samac D, Rhodes L, Lamp W (eds) Compendium of alfalfa diseases and pests, 3rd edn. APS Press, St. Paul, pp 27–30
Irwin JA, Cahill DM, Drenth A (1995) *Phytophthora* in Australia. Aust J Agric Res 46:1311–1337. https://doi.org/10.1071/AR9951311
Irwin JA, Miller S, Maxwell DP (1979) Alfalfa seedling resistance to *Phytophthora megasperma*.

Phytopathology 69:1051–1055. https://doi.org/10.1094/Phyto-69-1051

Irwin JAG, Aitken KS, Mackie JM, Musial JM (2006) Genetic improvement of lucerne for anthracnose (*Colletotrichum trifolii*) resistance. Australas Plant Pathol 35:573–579. https://doi.org/10.1071/AP06059

Irwin JAG, Mackie JM, Marney TS, Musial JM, Roberts S (2004) Incidence of *Stagonospora meliloti* and *Acrocalymma medicaginis* in lucerne crowns and roots in eastern Australia, their comparative aggressiveness to lucerne and inheritance of reaction to *S. meliloti* in lucerne. Australas Plant Pathol 33:61–67. https://doi.org/10.1071/AP03083

Irwin JAG, Maxwell DP, Bingham ET (1981a) Inheritance of resistance to *Phytophthora megasperma* in diploid alfalfa. Crop Sci 21:271–276. https://doi.org/10.2135/cropsci1981.0011183x002100020017x

Irwin JAG, Maxwell DP, Bingham ET (1981b) Inheritance of resistance to *Phytophthora megasperma* in tetraploid alfalfa. Crop Sci 21:277–283. https://doi.org/10.2135/cropsci1981.0011183x002100020018x

Iwai H, Horikawa Y, Figueiredo G, Hiruki C (1992) A recessive resistance gene of alfalfa to alfalfa mosaic virus and impact of high ambient temperature on virus resistance. Plant Pathol 41:69–75. https://doi.org/10.1111/j.1365-3059.1992.tb02318.x

Jones F, Drechsler C (1925) Root rot of peas in the United States caused by *Aphanomyces euteiches* (N. SP.). J Agric Res 30:293–325

Kamphuis LG, Lichtenzveig J, Oliver RP, Ellwood SR (2008) Two alternative recessive quantitative trait loci influence resistance to spring black stem and leaf spot in *Medicago truncatula*. BMC Plant Biol 8:30. https://doi.org/10.1186/1471-2229-8-30

Kemen E, Hahn M, Mendgen K, Struck C (2005) Different resistance mechanisms of *Medicago truncatula* ecotypes against the rust fungus *Uromyces striatus*. Phytopathology 95:153–157. https://doi.org/10.1094/phyto-95-0153

Köhl J, Kolnaar R, Ravensberg WJ (2019) Mode of action of microbial biological control agents against plant diseases: relevance beyond efficacy. Front Plant Sci 10:845. https://doi.org/10.3389/fpls.2019.00845

Lacerda AF, Vasconcelos ÃAR, Pelegrini PB, Grossi de Sa MF (2014) Antifungal defensins and their role in plant defense. Front Microbiol 5:116. https://doi.org/10.3389/fmicb.2014.00116

Li Y, Kabbage M, Liu W, Dickman MB (2016) Aspartyl protease-mediated cleavage of BAG6 is necessary for autophagy and fungal resistance in plants. Plant Cell 28:233–247. https://doi.org/10.1105/tpc.15.00626

Li Y, Li G, Yu H, Jiao X, Gao K (2018) Antifungal activities of isoflavonoids from *Uromyces striatus* infected alfalfa. Chem Biodivers 15:e1800407. https://doi.org/10.1002/cbdv.201800407

Liew ECY, Irwin JAG (1997) Differential disease reactions on lucerne genotypes inoculated with *Phytophthora medicaginis* isolates from lucerne and chickpea. Aust J Agric Res 48:545–551. https://doi.org/10.1071/A96143

Lu NS, Barnes DK, Frosheiser FI (1973) Inheritance of Phytophthora root rot resistance in alfalfa. Crop Sci 13:714–717. https://doi.org/10.2135/cropsci1973.0011183x001300060037x

Lueschen WE, Barnes DK, Rabas DL, Frosheiser FI, Smith DM (1976) Field performance of alfalfa cultivars resistant and susceptible to Phytophthora root rot. Agron J 68:281–285. https://doi.org/10.2134/agronj1976.00021962006800020019x

Mackie JM, Irwin JAG (1998) Genetics and race variability of the lucerne-*Colletotrichum trifolii* pathosystem in Australia. Aust J Agric Res 49:713–722. https://doi.org/10.1071/A97133

Mackie JM, Musial JM, Armour DJ, Phan HTT, Ellwood SE et al (2007) Identification of QTL for reaction to three races of *Colletotrichum trifolii* and further analysis of inheritance of resistance in autotetraploid lucerne. Theor Appl Genet 114:1417–1426. https://doi.org/10.1007/s00122-007-0527-z

Mackie JM, Musial JM, O'Neill NR, Irwin JAG (2003) Pathogenic specialisation within *Colletotrichum trifolii* in Australia, and lucerne cultivar reactions to all known Australian pathotypes. Aust J Agric Res 54:829–836. https://doi.org/10.1071/AR03079

Madrid E, Barilli E, Gil J, Huguet T, Gentzbittel L et al (2014) Detection of partial resistance quantitative trait loci against *Didymella pinodes* in *Medicago truncatula*. Mol Breed 33:589–599. https://doi.org/10.1007/s11032-013-9976-z

Madrid E, Gil J, Rubiales D, Krajinski F, Schlereth A, Millán T (2010) Transcription factor profiling leading to the identification of putative transcription factors involved in the *Medicago truncatula-Uromyces striatus* interaction. Theor Appl Genet 121:1311–1321. https://doi.org/10.1007/s00122-010-1390-x

Mahuku GS, Platt HW (2002) Molecular evidence that *Verticillium ablo-atrum* Grp 2 isolates are distinct from *V. albo-atrum* Grp 1 and *V. tricorpus*. Mol Plant Pathol 3:71–79. https://doi.org/10.1046/j.1464-6722.2001.00096.x

Malvick DK, Grau CR (2001) Characteristics and frequency of *Aphanomyces euteiches* races 1 and 2 associated with alfalfa in the Midwestern United States. Plant Dis 85:740–744. https://doi.org/10.1094/PDIS.2001.85.7.740

Malvick DK, Grau CR (2016) Aphanomyces root rot. In: Samac DA, Rhodes LH, Lamp W (eds) Compendium of alfalfa diseases and pests, 3rd edn. APS Press, St Paul, pp 12–15

Malvick DK, Grünwald NJ, Dyer AT (2009) Population structure, races, and host range of *Aphanomyces euteiches* from alfalfa production fields in the central USA. Eur J Plant Pathol 123:171–182. https://doi.org/10.1007/s10658-008-9354-6

Malvick DK, Percich JA (1998) Genotypic and pathogenic diversity among pea-infecting strains of *Aphanomyces euteiches* from the Central and Western United States. Phytopathology 88:915–921. https://doi.org/10.1094/PHYTO.1998.88.9.915

Marks G, Mitchell J (1971) Factors involved with the reaction of alfalfa to root rot caused by *Phytophthora megasperma*. Phytopathology 61:510–514

Martin FN, Abad ZG, Balci Y, Ivors K (2012) Identification and detection of *Phytophthora* : Reviewing our progress, identifying our needs. Plant Dis 96:1080–1103. https://doi.org/10.1094/PDIS-12-11-1036-FE

Mazurier M, Sbeiti AL, Benameur S, Tardin M-C, Gras MC (2016) Genome wide analysis of quantitative disease resistance against Verticillium wilt in *Medicago truncatula*. Conference paper: 12th International Verticillium Symposium in Ljubljana

Miller SA, Maxwell DP (1984) Light microscope observations of susceptible, host resistant, and nonhost resistant interactions of alfalfa with *Phytophthora megasperma*. Can J Bot 62:109–116. https://doi.org/10.1139/b84-019

Mould M, Boland G, Robb J (1991a) Ultrastructure of the *Colletotrichum trifolii-Medicago sativa* pathosystem. I Pre-Penetration events. Physiol Mol Plant Pathol 38:179–194. https://doi.org/10.1016/S0885-5765(05)80123-7

Mould M, Boland G, Robb J (1991b) Ultrastructure of the *Colletotrichum trifolii-Medicago sativa* pathosystem. II. Post-Penetration Events. Physiol Mol Plant Pathol 38:195–210. https://doi.org/10.1016/S0885-5765(05)80124-9

Moussart A, Even MN, Tivoli B (2008) Reaction of genotypes from several species of grain and forage legumes to infection with a French pea isolate of the oomycete *Aphanomyces euteiches*. Eur J Plant Pathol 122:321–333. https://doi.org/10.1007/s10658-008-9297-y

Munkvold GP, Carlton W (1995) Prevalence and distribution of *Aphanomyces euteiches* and *Phytophthora medicaginis* in Iowa alfalfa fields. Plant Dis 79:1251–1253. https://doi.org/10.1094/PD-79-1251

Munkvold GP, Carlton WM, Brummer EC, Meyer JR, Undersander DJ et al (2001) Virulence of *Aphanomyces euteiches* isolates from Iowa and Wisconsin and benefits of resistance to *A. euteiches* in alfalfa cultivars. Plant Dis 85:328–333. https://doi.org/10.1094/PDIS.2001.85.3.328

Musial JM, Aitken KS, Mackie JM, Irwin JAG (2005) A genetic linkage map in autotetraploid lucerne adapted to northern Australia, and use of the map to identify DNA markers linked to resistance to *Phytophthora medicaginis*. Aust J Agric Res 56:333–344. https://doi.org/10.1071/AR04317

Musial JM, Mackie JM, Armour DJ, Phan HTT, Ellwood SE et al (2007) Identification of QTL for resistance and susceptibility to *Stagonospora meliloti* in autotetraploid lucerne. Theor Appl Genet 114:1427–1435. https://doi.org/10.1007/s00122-007-0528-y

Mysore KS, Ryu CM (2004) Nonhost resistance: how much do we know? Trends Plant Sci 9:97–104. https://doi.org/10.1016/j.tplants.2003.12.005

Negahi A, Ben C, Gentzbittel L, Maury P, Nabipour AR et al (2014) Quantitative trait loci associated with resistance to a potato isolate of *Verticillium albo-atrum* in *Medicago truncatula*. Plant Pathol 63:308–315. https://doi.org/10.1111/ppa.12100

Nygaard SL, Grau CR (1989) *Phytophthora megasperma* virulence to alfalfa measured using single-isolate zoospore suspensions. Can J Plant Pathol 11:101–108. https://doi.org/10.1080/07060668909501122

O'Neill N (1996) Pathogenic variability and host resistance in the *Colletotrichum trifolii/Medicago sativa* pathosystem. Plant Dis 80:450–457. https://doi.org/10.1094/PD-80-0450

O'Neill N, Elgin J, Baker C (1989) Characterization of induced resistance to anthracnose in alfalfa by races, isolates, and species of *Colletotrichum*. Phytopathology 79:750–756. https://doi.org/10.1094/Phyto-79-750

Obert DE, Skinner DZ, Stuteville DL (2000) Association of AFLP markers with downy mildew resistance in autotetraploid alfalfa. Mol Breed 6:287–294. https://doi.org/10.1023/A:1009672008702

Ostazeski SA, Elgin Jr JH (1982) Use of hypodermic inoculations of alfalfa for identifying host reactions and races of *Colletotrichum trifolii*. Crop Sci 22:545–546. https://doi.org/10.2135/cropsci1982.0011183X002200030026x

Peaden RN, Gilbert RG, Christen AA (1985) Control of *Verticillium albo-atrum* on alfalfa. Can J Plant Pathol 7:511–514. https://doi.org/10.1080/07060668509501506

Pedersen MW, Barnes DK (1965) Inheritance of downy mildew resistance in alfalfa. Crop Sci 5:4–5. https://doi.org/10.2135/cropsci1965.0011183x000500010002x

Pennypacker B, Leath KT (1993) Anatomical response of resistant alfalfa infected with *Verticillium albo-atrum*. Phytopathology 83:80–85. https://doi.org/10.1094/Phyto-83-80

Pennypacker B, Leath KT, Hill RJ (1990) Growth and physiological response of resistant alfalfa clones infected with *Verticillium albo-atrum*. Phytopathology 80:1247–1253. https://doi.org/ https://doi.org/10.1094/Phyto-80-1247

Pilet-Nayel M-L, Prospéri J-M, Hamon C, Lesné A, Lecointe R et al (2009) *AER1*, a major gene conferring resistance to *Aphanomyces euteiches* in *Medicago truncatula*. Phytopathology 99:203–208. https://doi.org/10.1094/PHYTO-99-2-0203

Pretsch K, Kemen A, Kemen E, Geiger M, Mendgen K et al (2013) The rust transferred proteins-a new family of effector proteins exhibiting protease inhibitor function. Mol Plant Pathol 14:96–107. https://doi.org/10.1111/j.1364-3703.2012.00832.x

Qin Q-M, Vallad GE, Wu BM, Subbarao KV (2006) Phylogenetic analyses of phytopathogenic isolates of *Verticillium* spp. Phytopathology 96:582–592. https://doi.org/10.1094/PHYTO-96-0582

Rodgers C, McCaslin M, Witte D, Ho, JC-H, Caeron JN (2019) Anthracnose resistant alfalfa plants. United States Patent Application US 2019/0045734A1

Rubiales D, Moral A (2004) Prehaustorial resistance against alfalfa rust (*Uromyces striatus*) in *Medicago truncatula*. Eur J Plant Pathol 110:239–243. https://doi.org/10.1023/B:EJPP.0000019792.19573.64

Rubiales D, Rojas-Molina MM, Sillero JC (2013) Identification of pre- and posthaustorial resistance to rust (*Uromyces viciae-fabae*) in lentil (*Lens culinaris*) germplasm. Plant Breed 132:676–680. https://doi.org/10.1111/pbr.12096

Samac DA, Allen S, Witte D, Miller D, Peterson J (2014) First report of race 2 of *Colletotrichum trifolii* causing anthracnose on alfalfa (*Medicago sativa*) in Wisconsin. Plant Dis 98:843. https://doi.org/10.1094/PDIS-08-13-0808-PDN

Samac DA, Bucciarelli B, Dornbusch MR, Miller SS, Yu L-X (2017) Identification of markers associated with race-specific resistance to Aphanomyces root rot in alfalfa. Phytopathology 107:S5.1. DOI: https://doi.org/10.1094/PHYTO-107-12-S5.1

Samac DA, Dornbusch MR, Bucciarelli B, Miller SS, Yu L-X (2018) Genetic mapping of resistance to Aphanomyces root rot in alfalfa. In: Proceedings of the North American Alfalfa improvement conference. Logan, UT

Samac DA, Foster-Hartnett D (2012) Effect of glyphosate application on foliar diseases in glyphosate-tolerant alfalfa. Plant Dis 96:1104–1110. https://doi.org/10.1094/pdis-08-11-0715-re

Samac DA, Peñuela S, Schnurr JA, Hunt EN, Foster-Hartnett D et al (2011) Expression of coordinately regulated defence response genes and analysis of their role in disease resistance in *Medicago truncatula*. Mol Plant Pathol 12:786–798. https://doi.org/10.1111/j.1364-3703.2011.00712.x

Saqib M, Ellwood SR, Jones RAC, Jones MGK (2009) Resistance to subterranean clover mottle virus in *Medicago truncatula* and genetic mapping of a resistance locus. Crop Pasture Sci 60:480–489. https://doi.org/10.1071/CP08373

Saunders J, O'Neill N (2004) The characterization of defense responses to fungal infection in alfalfa. Biocontrol 49:715–728. https://doi.org/10.1007/s10526-004-5281-4

Schlatter D, Kinkel L, Thomashow L, Weller D, Paulitz T (2017) Disease suppressive soils: new insights from the soil microbiome. Phytopathology 107:1284–1297. https://doi.org/10.1094/PHYTO-03-17-0111-RVW

Schoch CL, Ciufo S, Domrachev M, Hotton CL, Kannan S et al (2020) NCBI Taxonomy: a comprehensive update on curation, resources and tools. Database (oxford) 2020. https://doi.org/10.1093/database/baaa062

Sekhwal MK, Li P, Lam I, Wang X, Cloutier S et al (2015) Disease resistance gene analogs (RGAs) in plants. Int J Mol Sci 16:19248–19290. https://doi.org/10.3390/ijms160819248

Sillero JC, Fondevilla S, Davidson J, Patto MCV, Warkentin TD et al (2006) Screening techniques and sources of resistance to rusts and mildews in grain legumes. Euphytica 147:255–272. https://doi.org/10.1007/s10681-006-6544-1

Skinner DZ, Stuteville DL (1985a) Genetics of host-parasite interactions between alfalfa and *Peronospora trifoliorum*. Phytopathology 75:119–121. https://doi.org/10.1094/Phyto-75-119

Skinner DZ, Stuteville DL (1985b) Quantitatively inherited reactions of alfalfa to *Peronospora trifoliorum*. Phytopathology 75:717–721. https://doi.org/10.1094/Phyto-75-717

Skinner DZ, Stuteville DL (1989) Influence of temperature on expression of resistance to rust in diploid alfalfa. Crop Sci 29:675–677. https://doi.org/10.2135/cropsci1989.0011183X002900030027x

Skinner DZ, Stuteville DL (1995) Host range expansion of the alfalfa rust pathogen. Plant Dis 79:456–460. https://doi.org/10.1094/PD-79-0456

Stuteville DL (1991) Standard Tests to Characterize Alfalfa Cultivars. Rust D-9. https://www.naaic.org/stdtests/rust.pdf

Stuteville DL (2002) First report of *Uromyces striatus* on white sweetclover in Kansas. Plant Dis 86:1404. https://doi.org/10.1094/pdis.2002.86.12.1404a

Tang F, Yang S, Gao M, Zhu H (2013) Alternative splicing is required for *RCT1*-mediated disease resistance in *Medicago truncatula*. Plant Mol Biol 82:367–374. https://doi.org/10.1007/s11103-013-0068-6

Thies JA, Basigalup D, Barnes DK (1994) Inheritance of resistance to *Pratylenchus penetrans* in alfalfa. J Nematol 26:452–459

Tivoli B, Baranger A, Sivasithamparam K, Barbetti MJ (2006) Annual *Medicago*: from a model crop challenged by a spectrum of necrotrophic pathogens to a model plant to explore the nature of disease resistance. Ann Bot 98:1117–1128. https://doi.org/10.1093/aob/mcl132

Torregrosa C, Cluzet S, Fournier J, Huguet T, Gamas P et al (2004) Cytological, genetic, and molecular analysis to characterize compatible and incompatible interactions between *Medicago truncatula* and *Colletotrichum trifolii*. Mol Plant-Microbe Interact 17:909–920

Vailleau F, Sartorel E, Jardinaud M-F, Chardon F, Genin S et al (2007) Characterization of the interaction between the bacterial wilt pathogen *Ralstonia solanacearum* and the model legume plant *Medicago truncatula*. Mol Plant-Microbe Interact 20:159–167. https://doi.org/10.1094/MPMI-20-2-0159

Van Schie CCN, Takken FLW (2014) Susceptibility genes 101: how to be a good host. Annu Rev Phytopathol 52:551–581. https://doi.org/10.1146/annurev-phyto-102313-045854

van Wersch S, Li X (2019) Stronger when together: clustering of plant NLR disease resistance genes. Trends Plant Sci 24:688–699. https://doi.org/10.1016/j.tplants.2019.05.005

Vandemark GJ, Ariss JJ, Hughes TJ (2010) Real-time PCR suggests that *Aphanomyces euteiches* is associated with reduced amounts of *Phytophthora medicaginis* in alfalfa that is co-inoculated with both

pathogens. J Phytopathol 158:117–124. https://doi.org/10.1111/j.1439-0434.2009.01583.x
Vandemark GJ, Barker BM (2003) Quantifying *Phytophthora medicaginis* in susceptible and resistant alfalfa with a real-time fluorescent PCR assay. J Phytopathol 151:577–583. https://doi.org/10.1046/j.0931-1785.2003.00768.x
Vandemark GJ, Barker BM, Hughes TJ (2004) Heritability of resistance to *Aphanomyces euteiches* races 1 and 2 in alfalfa. Euphytica 136:45–50. https://doi.org/10.1023/B:EUPH.0000019518.62425.ba
Vandemark GJ, Grünwald NJ (2004) Reaction of *Medicago truncatula* to *Aphanomyces euteiches* Race 2. Arch Phytopathol Plant Prot 37:59–67. https://doi.org/10.1080/03235400410001662148
Vasic T, Bulajic A, Krnjaja V, Zivkovic S, Andjelkovic B (2014) First report of anthracnose on alfalfa caused by *Colletotrichum linicola* in Serbia. Plant Dis 98:1276. https://doi.org/10.1094/PDIS-03-14-0258-PDN
Vaz Patto MC, Rubiales D (2014) Unveiling common responses of *Medicago truncatula* to appropriate and inappropriate rust species. Front Plant Sci 5:618. https://doi.org/10.3389/fpls.2014.00618
Vaziri A, Keen NT, Erwin DC (1981) Correlation of medicarpin production with resistance to *Phytophthora megasperma* f. sp. *medicaginis* in alfalfa seedlings. Phytopathology 71:1235–1238. https://doi.org/ https://doi.org/10.1094/Phyto-71-1235
Viands DR (1985) Comparison of 'Maris Kabul' with 'Vertus' alfalfa for resistance to Verticillium wilt. Crop Sci 25:1096–1100. https://doi.org/10.2135/cropsci1985.0011183X002500060047x
Viands DR, Barnes DK (1980) Inheritance of resistance to bacterial wilt in two alfalfa gene pools: qualitative analysis. Crop Sci 20:48–54. https://doi.org/10.2135/cropsci1980.0011183X002000010012x
Vincelli P, Henning J, Hendrick T, Brown J, Osborne LJ et al (2000) Improved seedling health, yield, and stand persistence with alfalfa resistant to Aphanomyces root rot. Agron J 92:1071–1076. https://doi.org/10.2134/agronj2000.9261071x
Vincelli P, Nesmith W, Eshenaur B (1994) Incidence of *Aphanomyces euteiches* and *Phytophthora medicaginis* in Kentucky alfalfa fields. Plant Dis 78:645–647. https://doi.org/10.1094/PD-78-0645
Wang Y, Bi B, Yuan QH, Li XL, Gao JM (2012) Association of AFLP and SCAR markers with common leafspot resistance in autotetraploid alfalfa (*Medicago sativa*). Genet Mol Res 11:606–616. https://doi.org/10.4238/2012.March.14.4
Wang Y, Ma J, Yuam Q, Li XL, Miao LH (2018) First report of *Colletotrichum linicola* causing anthracnose on alfalfa in China. Plant Dis 102:1039. https://doi.org/10.1094/PDIS-09-17-1376-PDN
Webb DH, Nutter FW (1997) Effects of leaf wetness duration and temperature on infection efficiency, latent period, and rate of pustule appearance of rust in alfalfa. Phytopathology 87:946–950. https://doi.org/10.1094/PHYTO.1997.87.9.946
White JF, Kingsley KL, Zhang Q, Verma R, Obi N et al (2019) Review: endophytic microbes and their potential applications in crop management. Pest Manag Sci 75:2558–2565. https://doi.org/10.1002/ps.5527
Wiersma DW, Grau CR, Undersander DJ (1995) Alfalfa cultivar performance with differing levels of resistance to *Phytophthora* and *Aphanomyces* root rots. J Prod Agric 8:259–264. https://doi.org/10.2134/jpa1995.0259
Yang S, Gao M, Deshpande S, Lin S, Roe BA et al (2007) Genetic and physical localization of an anthracnose resistance gene in *Medicago truncatula*. Theor Appl Genet 116:45–52. https://doi.org/10.1007/s00122-007-0645-7
Yang S, Gao M, Xu C, Gao J, Deshpande S et al (2008) Alfalfa benefits from *Medicago truncatula*: the *RCT1* gene from *M. truncatula* confers broad-spectrum resistance to anthracnose in alfalfa. Proc Natl Acad Sci U S A 105:12164–12169. https://doi.org/10.1073/pnas.0802518105
Yu L-X, Zhang F, Culma CM, Lin S, Niu Y et al (2020) Construction of high-density linkage maps and identification of quantitative trait loci associated with Verticillium wilt resistance in autotetraploid alfalfa (*Medicago sativa* L.). Plant Dis 104:1439–1444. https://doi.org/10.1094/PDIS-08-19-1718-RE
Yu L-X, Zheng P, Bhamidimarri S, Liu X-P, Main D (2017a) The impact of genotyping-by-sequencing pipelines on SNP discovery and identification of markers associated with Verticillium wilt resistance in autotetraploid alfalfa (*Medicago sativa* L). Front Plant Sci 8:89. https://doi.org/10.3389/fpls.2017.00089
Yu L-X, Zheng P, Zhang T, Rodringuez J, Main D (2017b) Genotyping-by-sequencing-based genome-wide association studies on Verticillium wilt resistance in autotetraploid alfalfa (*Medicago sativa* L). Mol Plant Pathol 18:187–194. https://doi.org/10.1111/mpp.12389

Genetic and Genomic Assessments for Improving Drought Resilience in Alfalfa

14

Long-Xi Yu, Cesar Augusto Medina, and Michael Peel

Abstract

Alfalfa (*Medicago sativa* L.) production is challenged by adverse environmental factors. Developing alfalfa varieties with resistance to these stresses is imperative for sustainable alfalfa production worldwide. Abiotic stresses such as drought and high salinity affect alfalfa production resulting in severe economic losses. Conventional breeding procedures to develop alfalfa cultivars are time-consuming and costly. The use of molecular tools such as marker-assisted selection (MAS) and genomics-assisted breeding (GAB) can accelerate the breeding process. Recent advances in next-generation sequencing have provided a new strategy to generate cost-effective, high-density, genome-wide single nucleotide polymorphisms (SNPs). In conjunction with genome-wide association studies (GWAS) and/or genomic selection (GS), more powerful platforms can be developed to improve gains in alfalfa breeding. Given that alfalfa cultivars are genetically broad-based synthetic populations, it has been a challenge understanding genetic mechanisms by which environmental factors affect alfalfa plant growth, development, and production. Genomic approaches such as genotyping-by-sequencing (GBS), association mapping (AM), GAB, and GS can be applied to overcome the bottleneck of traditional breeding. Studies have been performed in recent years describing the generation of large marker sets in alfalfa and performing GAB for enhanced drought resistance. In addition, transcriptomics and gene regulation network in response to drought stress have been investigated in alfalfa. In this chapter, we provide the background of traditional breeding for drought tolerance and review recent progress in the utilization of genomic approaches toward improving drought resilience in alfalfa.

L.-X. Yu (✉)
United States Department of Agriculture-Agricultural Research Service, Plant Germplasm Introduction and Testing Research, 24106 N Bunn Road, Prosser, WA, USA
e-mail: longxi.yu@usda.gov

C. A. Medina
Irrigated Agriculture and Extension Center, Washington State University, 24106 N Bunn Road, Prosser, WA, USA
e-mail: cesar.medinaculma@wsu.edu

M. Peel
United States Department of Agriculture-Agricultural Research Service, Forage & Range Research Lab, Logan, UT, USA
e-mail: mike.peel@usda.gov

14.1 Introduction

Alfalfa is the third largest crop after corn and soybean in US Agriculture. The estimated value of alfalfa hay in US is $8.8 billion annually.

L.-X. Yu and C. Kole (eds.), *The Alfalfa Genome*, Compendium of Plant Genomes,
https://doi.org/10.1007/978-3-030-74466-3_14

Alfalfa meal and cubes are exported to other countries with a value of $49.4 million to the United States economy (USDA 2017). Alfalfa is also called the "Queen of the Forages" due to its high biomass production and high nutritional quality as animal feed. Alfalfa typically contains between 18 and 22% crude protein and is also an excellent source of vitamins and minerals. In addition to the traditional uses of alfalfa as an animal feed, alfalfa is consumed by humans in the form of alfalfa sprouts and as health food products. Moreover, alfalfa has the potential to be used as a bio-fuel crop and as a bioresource for the production of industrial enzymes such as lignin peroxidase, alpha-amylase, cellulase, and phytase. Alfalfa is also an important source of biological nitrogen fixation. The average rate of nitrogen fixation of alfalfa is about 200 kg of nitrogen per acre per year, thus reducing the need to apply expensive nitrogen fertilizers.

Changing trends in multipurpose uses increase the demand for alfalfa. However, the production of alfalfa is challenged by adverse environmental factors. Abiotic stresses such as drought and high salinity affecting alfalfa production cause severe economic losses. Water use is high for alfalfa production, so increased crop water use efficiency (WUE) and drought tolerance are key factors for sustainable production of alfalfa under water-limited conditions.

Compared to other crop species, little is known about mechanisms by which genetic factors contribute to drought tolerance in alfalfa. Research on the molecular biology of drought responses in alfalfa was initiated in the 1990s (Luo et al. 1991, 1992; Laberge et al. 1993). Later on, microarrays were used for the analysis of the alfalfa transcriptome to identify genes responsive to dehydration (Chen et al. 2008). Recently, proteomics and metabolite profiling were performed to identify proteins and soluble metabolites that respond to drought in leaves and nodules of alfalfa (Aranjuelo et al. 2011). RNA-seq has been performed using Chilean and Wisfal accessions with contrasting WUE. Over 40,000 single nucleotide polymorphisms (SNPs) have been identified (Han et al. 2011; Kang et al. 2011; Li et al. 2012). Genome regions have been identified by different groups that are associated with forage yield in mesic environments (Li et al. 2010), flowering date and plant height (Herrmann et al. 2010), and biomass under drought condition (Ray et al. 2015; Zhang et al. 2015; Yu 2017). Traditional and molecular breeding efforts to improve drought tolerance in alfalfa have been undertaken (McKersie et al. 1996; Zhang et al. 2005; Vasconcelos et al. 2008; Suárez et al. 2009; Li et al. 2010). Modern alfalfa germplasm has been used for genetic mapping (Segovia-Lerma et al. 2003; Ariss and Vandemark 2007) and field studies related to forage quality and WUE (Lenssen et al. 1991; Ray et al. 1998, 2004; Segovia-Lerma et al. 2003). The WUE was measured by carbon isotope discrimination and revealed that *M. sativa* ssp. *falcata* has higher WUE compared with other germplasm, although its yield was relatively low (Ray et al. 1998, 2004; Segovia-Lerma et al. 2003; Santantonio et al. 2019). Backcross populations were developed using the *M. sativa* ssp. *falcata* variety Wisfal (high WUE) and Chilean germplasm (low WUE) as parents and a genetic map was constructed in these populations (Sledge et al. 2005). Most recently, quantitative trait loci (QTL) associated with biomass and WUE under drought stress have been identified using conventional QTL mapping (Ray et al. 2015; Santantonio et al. 2019) and genome-wide association studies (GWAS) (Zhang et al. 2015; Yu 2017).

14.2 History of Breeding Drought-Tolerant Alfalfa in the Semi-Arid Western USA

Breeding drought-tolerant alfalfa for semi-arid regions of the western USA can in large part be traced back to the introduction of falcata or yellow-flowered alfalfa (*Medicago sativa* subsp. falcata Arcang) by N.E. Hansen to South Dakota in 1906 (Oakley and Garver 1917; Rumbaugh 1979). Hansen made collections throughout northern Europe and Asia recognizing falcata for drought tolerance in addition to tolerance to cold winters, with high nutritive value and persistence. The earliest breeding work for alfalfa, in general, was from 1903 to 1915, during that time "Grimm", "Baltic", "Cossack", and "Ladak" were developed (Boe et al. 2020). Ladak, one of these early ones was selected in 1914 by the introduction from the Ladakh Province in northern India (Garver 1946). Ladak was the most productive, vigorous, and persistent among 10 cultivars evaluated in hay and grazing systems at three semi-arid locations in North Dakota during the early 1950s. It was superior to Grimm, "Nomad" (Oregon in 1941), Semipalatinsk, "Rhizoma " (developed by University of British Columbia in 1950), and "Ranger" (developed by USDA and Nebraska Agricultural Experiment Station in 1942). It has also contributed to the development of "Rambler" (Heinrichs and Bolton 1958), "Travios" (Rumbaugh et al. 1964), "Roamer" (Heinrichs 1967), and "Drylander" (Heinrichs 1971). The stability and performance of this material have withstood the test of time and demonstrated by Ladak which is used as a check in dryland alfalfa tests and is often among top performing cultivars in dryland trials conducted in Northern Utah in the early 2000s (Peel M, Unpublished data).

Since the early 1970s when breeding for severe dry conditions emphasis has been on developing yellow-flowered alfalfa (*M. falcata*) cultivars in the USA and Canada. This began with "Anik", released in Canada in 1971 (Pankiw and Siemens, 1976) followed three decades later by three more cultivars Yellowhead (McLeod et al. 2009), Don (Peel et al. 2009), and Sholty (tested as SDSU Experimental Population SD201) and two improved germplasms, IAMF101 and IAMF102 (Riday and Brummer 2007). The drought tolerance of these materials is measured in their ability to survive harsh conditions, when compared to typical *M. sativa* types of alfalfa, yield production of this yellow-flowered alfalfa is often much less (Boe et al. 2020). In addition, these are all synthetic cultivars which are developed by compositing (planting together) two or more alfalfa genotypes with bulk seed harvested and planted in successive generations. A large amount of variations within these cultivars require the modern technologies described below to improve genetic gains.

14.3 The Use of Genomic Tools for Breeding Drought-Resistant Alfalfa

Conventional breeding procedures for developing new alfalfa cultivars are time-consuming and costly. Marker-assisted selection (MAS) can accelerate the breeding process and reduce the cost of labor. Developing and identifying DNA markers associated with drought tolerance is a first step toward using MAS to develop resilience to stressors in alfalfa. Incorporating MAS into breeding programs can accelerate progress in developing resistant varieties. MAS is an indirect process by which plants are selected based on the detection of specific alleles closely linked with traits of interest. Although MAS has been widely adapted to the commercial development of several important crops, including corn, wheat, and soybeans, MAS is rarely employed for the development of improved alfalfa varieties. Enabling technologies that must be developed before alfalfa breeders can employ MAS include identifying genetic loci from alfalfa that are robustly associated with important traits such as abiotic tolerance. However, the genetic approach in this species is slow because of its autotetraploid allogamous nature and the synthetic structure of the varieties (Julier et al. 2003). Quantitative traits such as resistance to abiotic stress are most likely under the control of multiple genes and interact with environmental factors. Identification of resistance loci that contribute to phenotypic variation in such complex traits is a primary challenge in plant breeding and population genetics. The use of the integrated framework that merges a QTL mapping approach called "genome-wide association studies (GWAS)" with a high-throughput sequencing methodology called "next-generation sequencing (NGS)" is a powerful strategy to develop high-density genome-wide single nucleotide polymorphisms (SNPs). This framework provides a statistical basis for analyzing marker-trait association in germplasm panels or breeding populations. The use of genomics approaches provides a new strategy to improve alfalfa varieties with enhanced resistance to abiotic stressors. Genomics approaches such GWAS, NGS, and comparative genomics can be used to discover genetic markers or genes that influence traits of interest. Markers closely linked with resistance can then be used for MAS and genomic selection (GS) in breeding programs to accelerate the breeding process and increase genetic gain toward improving alfalfa yield potential under stress conditions.

14.4 Marker Development and Marker-Based Selection

MAS is the process of DNA test for the selection of individuals with the trait of interest. It helps to identify parents for crossing to generate progeny and breeding populations. Significant SNPs closely linked to the trait loci can be used as DNA markers after validation. High-throughput platforms such as Kompetitive allele-specific PCR (KASP) (www.lgcgenomics.com) or TaqMan (www.thermofisher.com) assays can be used to test the co-segregation of marker loci with respective phenotypic traits. Multiplex primer combinations will be used for evaluating the resistance locus or candidate gene, and all markers will be scored in a given genotype. Single markers with two-character states will be tested for significant phenotypic differences between genotype groups by the t-test for each trait, and Mann–Whitney U test for chip quality. Marker combinations will be analyzed using analysis of variance (ANOVA) for each trait, and the Kruskal–Wallis test for assay quality. Once validated, the marker can be used for marker-assisted selection in breeding programs for improving alfalfa with enhanced resistance to stress.

Backcrossing has been widely used to incorporate one or a few genes into an adapted or elite variety in plant breeding. The use of DNA markers in backcrossing greatly increases the efficiency of selection. The validated marker is used for screening parents with resistance to drought and salt stress. The selected parents are crossed with an elite variety with high yield and quality to generate F1 progeny. Then F1 plants are crossed back with the elite variety to generate BC1 populations followed by marker-assisted recurrent selection. The selection continues until obtaining new varieties with improved abiotic resistance while maintaining high yield and quality (Fig. 14.1). Marker-assisted backcrossing (MAB) is used in combination with or to replace screening for the target genetic loci. This may be particularly useful for traits that have laborious or time-consuming phenotypic screening procedures, such as yield and quality traits. It can also be used to select recessive alleles, which is difficult in the conventional breeding.

Genomic selection is another marker-based selection strategy that allows breeders to access a broad range of variation for improving quantitative traits. Rather than detecting QTL per se, the purpose of genomic selection is to develop a predictive model that can identify individuals with the highest breeding values in a population. This new approach uses high-throughput genotyping to generate high densities of DNA markers genome-wide in large plant breeding populations. In this regard, genomic selection requires the availability of both phenotypic and genotypic data for a large reference population (also called the training population). This data set is then used to estimate model parameters such that the total genetic variation for a trait is explained by the markers analyzed. Consequently, it is essential that sufficient molecular markers be identified to provide good genome coverage. Once the model is established, it must be tested for its accuracy. This can be done by using the majority of the training population to create a prediction model, which is then used to predict a genomic estimated breeding value (GEBV) for each of the remaining individuals in the training population based only on their genotype data. Once validated, the model can then be applied to a breeding population to calculate GEBVs of individuals based only on their genotypic information. Such GEBVs represent

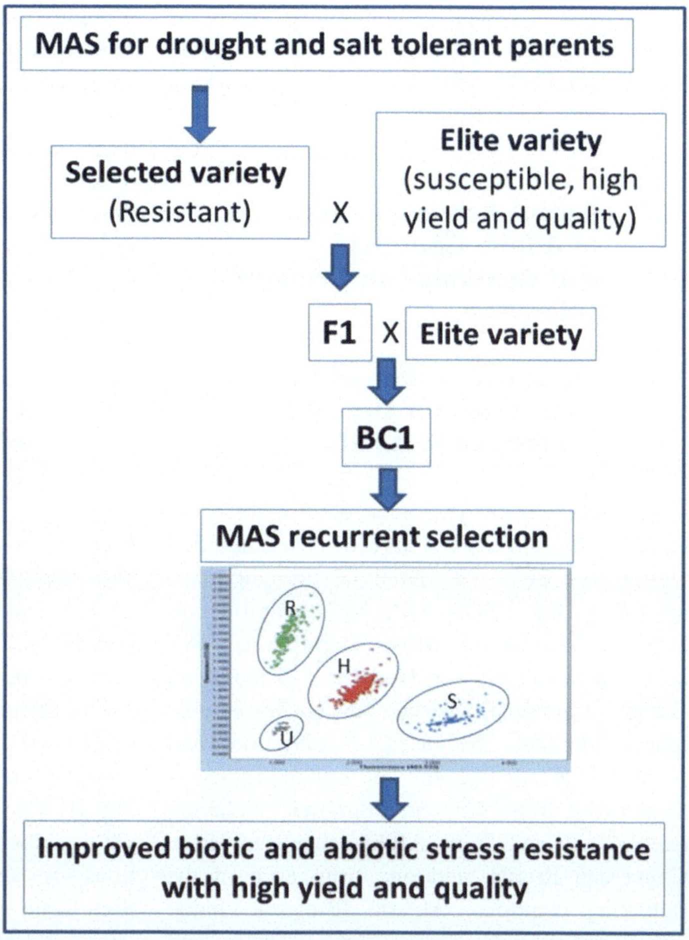

Fig. 14.1 Strategy for marker-assisted selection (MAS) for abiotic resistance in alfalfa breeding programs. R = resistance allele, S = susceptible allele, H = heterozygous, U = undetermined

the overall predicted value of an individual as a potential target. Given that selection is based on predictions, precision phenotyping is a critical part of genomic selection to ensure the accuracy of statistical models (Heffner et al. 2009; Crossa et al. 2010). There are many models to choose from, and several recent studies have compared popular models. Heslot et al. (2012) compared several statistic models for prediction accuracy, based on the correlation between observed phenotypic values and predicted breeding values. They include Bayesian and non-Bayesian ridge regression and lasso models, weighted Bayesian shrinkage, Bayes Cπ, empirical Bayes, reproducing kernel Hilbert space, random forest regression, support vector regression, and a hidden-layer artificial neural network trained to predict breeding values as well as linear combinations of multiple models. It has been suggested that the Bayesian lasso and the weighted Bayesian shrinkage regression models had the highest prediction accuracy in genomic selection. Additional models for the prediction of breeding values have been reviewed (See Hawkins and Yu 2018, for review). After genomic selection cycles, the selected individuals/lines can be advanced for variety testing and used to develop synthetic varieties. In addition, as new germplasm is incorporated into the breeding population, new phenotypic and marker data can be collected to update selection models. Alfalfa families with high GEBVs are selected through

GS cycles for advanced generations to create the synthetic varieties. After testing and evaluation, the new cultivars with improved resistance to abiotic stress are released.

14.5 Recent Achievements of Alfalfa Genetics and Genomics on Drought Tolerance

14.5.1 Characterization of Genetic Resources for Drought Tolerance in Alfalfa

To identify genetic resources of drought resistance in alfalfa, Zhang et al. (2017) selected 200 alfalfa accessions with potential drought tolerance from the USDA-ARS NPGS alfalfa collection (Fig. 14.2). Most germplasm was collected in Canada and Northern US including British Columbia, Saskatchewan, Manitoba, Idaho, Montana, Nebraska, Washington, and North and South Dakota. The remaining accessions were from different countries, including twelve collected from Afghanistan, two from China and Russia, and one from each of the following countries, Algeria, Bulgaria, India, Lebanon, Germany, Spain, Turkey, Oman, and Yemen. A drought-tolerant cultivar, "Wilson" was used as the drought tolerance check. The genetic diversity and agronomic and physiological traits associated with drought tolerance were characterized in greenhouse and field. They found that drought stress significantly decreased biomass yield by 61.9% compared with well-watered control. A positive correlation was found between relative water content (RWC) and drought susceptible index (DSI), while negative correlations were obtained between RWC and leaf senescence (LS) and between RWC and canopy temperature (CT). Alfalfa accessions with high RWC showed relative lower yield reduction, cooler CT, and delayed LS. Biomass yield under drought negatively correlated with total protein (TP) and relative feed value (RFV), suggesting that the efforts to improve yield under drought tend to negatively affect forage quality. Therefore, maintaining forage quality should be considered for breeding alfalfa with drought resistance and enhanced WUE. The study identified a group of alfalfa accessions with higher drought resistance scores (Fig. 14.2b). Among those, 34 accessions had high scores than the resistant check "Wilson". They can be used as genetic resources to develop alfalfa varieties with improved drought resistance.

14.5.2 Conventional QTL Mapping of Drought Resistance Traits

QTL associated with drought tolerance have been identified in alfalfa (Table 14.1). Ray et al. (2015) used two backcross (BC1) mapping populations (n = 253) derived from a cross between *M. sativa subsp. sativa* and *M. sativa subsp. falcata* for mapping quantitative trait loci (QTL) associated with forage productivity during drought stress. Half-sib families derived from 206 BC1 individuals were evaluated for forage yield under drought conditions in multiple locations in New Mexico and Oklahoma, USA. Significant effects were detected in each population and environment. Interval mapping identified 10 and 15 QTL associated with increased or reduced forage yield during drought. The average phenotypic variance of each QTL explained 3–6% biomass yield and the intend of these effects were considerably consistent over environments. Later on, the same group performed genetic mapping of WUE and carbon and nitrogen metabolism in two populations of alfalfa consisted of 29 and 96 half-sib families (Santantonio et al. 2019). QTL associated with shoot biomass (SB), carbon isotope crimination (CID), and C and N contents during drought stress were identified. Significant positive correlations were detected between SB and CID and between C and N content. The alignment of QTL positions into the *M. truncatula* Gaertn genome sequence encompassed multiple genes with functions of abiotic stress responses.

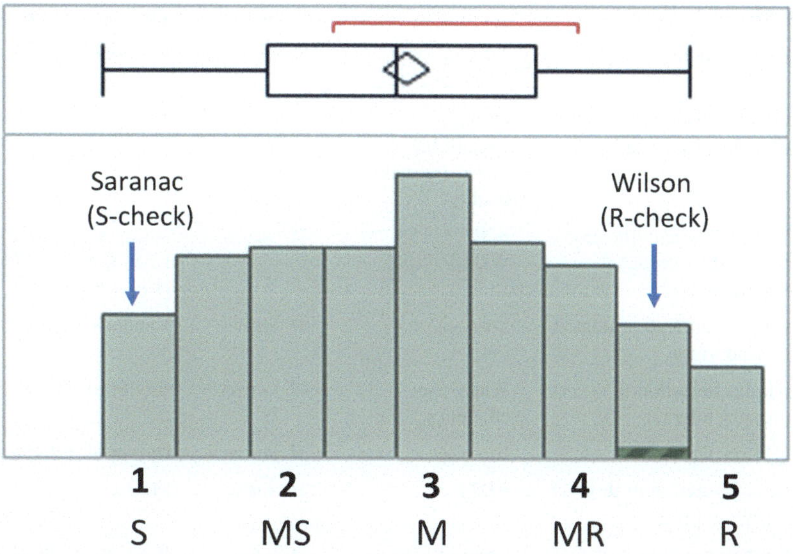

Fig. 14.2 Two hundred alfalfa accessions were evaluated for drought tolerance. Alfalfa plants were scored using 1–5 scares with 1 = susceptible (left) and 5 = resistant (right). The distribution of drought scores among accessions including resistant check "Wilson" and susceptible check "Saranac". Of those, 34 resistant lines above Wilson have been identified

14.5.3 Genome-Wide Association Studies on Drought Tolerance

Genome-wide associations with genotyping-by-sequencing were employed in alfalfa for the identification of markers associated with drought tolerance (Table 14.1). Zhang et al. (2015) used a diversity panel of alfalfa accessions comprised of 198 cultivars and landraces selected from the USDA-ARS National Plant Germplasm System alfalfa collection and analyzed two important traits of drought resistance index (DRI) and relative leaf water content (RWC). Marker-trait association identified nineteen and fifteen loci associated with DRI and RWC, respectively. Alignments of target sequences flanking to the resistance loci against the reference genome of *M. truncatula* revealed multiple chromosomal locations. Markers associated with DRI are located on all chromosomes while markers associated with RWC are located on chromosomes 1, 2, 3, 4, 5, 6, and 7. Co-localizations of significant markers between DRI and RWC were found on chromosomes 3, 5, and 7. Most loci associated with DRI in this work agree with the reported QTL associated with biomass under drought in alfalfa (Ray et al. 2015). The same association panel was evaluated in the field trials to identify marker loci associated with biomass yield under drought (Yu 2017). A total of 28 markers at 22 genetic loci were associated with yield under water deficit, whereas only three markers were associated with the same trait under well-watered condition. Comparisons of marker-trait associations between water deficit and well-watered conditions showed non-similarity. Most of the markers were identical across harvest periods within the treatment, although different levels of significance were found among the three harvests. The loci associated with biomass yield under water deficit located throughout all chromosomes in the alfalfa genome agreed with previous reports (Ray et al. 2015; Zhang et al. 2015). Characterizations of drought stress responses at the transcriptional level were investigated in two contrasting alfalfa varieties, Longdong and Algonquin (Table 14.1) (Quan et al. 2016). The drought-tolerant Longdong with smaller leaf size and lower stomata density showed less water loss than the drought-sensitive Algonquin. Transcriptional expressions

Table 14.1 Recent achievements in genetics and genomics of drought tolerance in alfalfa

Achievement	Trait or gene addressed	Species	References
Genetic resources for drought resistance	Drought susceptible index, biomass, quality	*Medicago sativa*	Zhang et al. (2018)
MicroRNA identification involved in drought tolerance	miRNA 156, WD40	*M. sativa*	Aung et al. (2015), Arshad et al. (2015, 2018), Li et al. (2017), Hannoufa et al. (2018), Feyissa et al. (2020), Zhao et al. (2020)
QTL identification related to seed vigor	Seed vigor	*M. sativa*	Vandecasteele et al. (2011)
Alfalfa Selection of drought tolerant	Water use efficiency	*M. sativa*	Ray et al. (1998, 2004), Condon et al. (2004)
QTL in Drought	Biomass	*M. sativa*	Ray et al. (2015)
Overexpression of drought tolerance gene	*EDT1*	*M. sativa*	Zheng et al. (2017)
Gene expression in response to drought	Gene expression	*M. sativa*	Kang et al. (2011), Kang and Udvardi (2012)
Improve drought tolerance by cytokinin regulation in Rhizobium	IAA responsive	*M. sativa*	Xu et al. (2012), Defez et al. (2017)
Genetic mapping in drought	WSE, Carbon, Nitrogen	*M. sativa*	Santantonio et al. (2019)
Identification and expression of MYB transcription factors	MYB transcription factor	*M. sativa*	Zhou et al. (2019)
Enhancing drought and salt tolerance by expression of MYB	*MsMYB2L*	*A. thaliana*	Song et al. (2019)
Transcriptome of drought responses	Differentially expressed genes	*M. sativa*	Quan et al. (2016) Medina et al. (2020)
Improving drought tolerance by genetic engineering	Wax gene *WXP1*	*M. sativa*	Zhang et al. (2005), Jiang et al. (2009)
Forage quality in response to drought	Multiple quality factors	*M. sativa*	Lin et al. (2020)
GWAS on drought resistance	Biomass, DRI, RWC	*M. sativa*	Zhang et al. (2015), Yu (2017)

of eight drought-responsive genes showed that Longdong exhibited significantly higher transcripts of drought-responsive genes in leaf and root under drought. Under drought stress, the expression of drought-responsive gene, *MtP5CS*, in the root of Longdong was significantly increased and was higher than that of Algonquin.

14.5.4 Gene Expression in Response to Drought

Gene expressions in response to drought were compared between two contrasting varieties: Wisfal (drought-resistant) and Chilean (drought-sensitive) (Kang et al. 2011). Genes encoding transcription factors and other regulatory proteins and genes involved in the biosynthesis of osmolytes and flavonoids were differentially regulated between the two varieties. Genes involved in isoflavonoid biosynthesis were highly expressed in Wisfal during drought. In contrast, isoflavonoid biosynthesis genes were repressed in roots of both varieties during drought, although repression was generally slower and less pronounced in Wisfal. It has been suggested that Isoflavonoids are a family of phenolic compounds largely restricted to legumes that function in nodulation and plant defense.

Reactive oxygen species (ROS) production and scavenging in plants under drought stress have been studied intensively in recent years. Kang and Udvardi (2012) reported a global analysis of gene expression for the major ROS generating and scavenging proteins in alfalfa under drought stress. They compared transcripts between drought-tolerant and drought-sensitive alfalfa varieties and found that no qualitative differences in ROS gene regulation between them (Kang and Udvardi 2012). However, tissue-specific patterns of gene expression of ROS-scavenging gene families in response to drought were observed, including ascorbate peroxidase, monodehydroascorbate reductase, and peroxiredoxin.

Overexpressions of exogenous genes into alfalfa to enhance drought tolerance have been achieved (Table 14.1). Zhang et al. (2005) reported that an AP2 domain-containing putative transcription factor gene designated *WXP1* from *M. truncatula* was able to activate wax production and confer drought tolerance in alfalfa. Overexpression of the *WXP1* induced wax-related genes and increased cuticular wax loading on leaves of transgenic alfalfa (Zhang et al. 2005). Transgenic alfalfa plants showed reduced water loss and chlorophyll leaching and enhanced drought tolerance with delayed leaf wilting and quicker recovery after re-watering. Zheng et al. (2017) transformed the Arabidopsis Enhanced Drought Tolerance 1 (*AtEDT1*) gene into alfalfa and conferred drought tolerance. Drought stress treatment resulted in higher survival rates and biomass and reduced water loss in the transgenic plants compared to the wild-type plants.

14.5.5 MicroRNA Regulation of Drought Responses

MicroRNAs (miRNAs) are small non-coding RNAs that regulate gene expression at the post-transcriptional level. The role of microRNAs on enhancing tolerance to abiotic stresses in plants has been reviewed (Hannoufa et al. 2018). Among a large number of MicroRNA discovered, microRNA156 (miR156) is one of the most conservative microRNAs in plants and was obtained extensive studies in alfalfa (Table 14.1). Arshad et al. (2017) reported that the overexpression of miR156 in alfalfa plants enhanced drought tolerance. A recent study on the miR156 overexpression in alfalfa resulted in under-expression of a WD40 family gene (*WD40-2*) under drought stress. Transgenic alfalfa with reduced expression of this gene showed enhanced tolerance to drought (Arshad et al. 2018). Most recently, a transcriptional analysis in alfalfa plants with silenced *SPL13* target gene showed tissue- and genotype-specific gene expression in transgenic alfalfa with overexpressed miR156 (Feyissa et al. 2020). Based on this analysis, a model of drought tolerance mechanism of miR156 overexpression regulated *SPL13* in a tissue-specificity has been proposed. In this model, drought-induced miR156 leads to *SPL13* gene silencing. Reduced expression of *SPL13* driven by miR156 and enhanced expression of *WD40–1/DFR*, in turn, improve drought tolerance in alfalfa.

14.5.6 Genomic Selection for Drought Tolerance

Genomic selection has been used in breeding programs for agronomic traits in alfalfa. Li et al. (2015) performed genomic selection for dryland yield in an alfalfa population composed of commercial cultivars. The population was tested for two cycles of genomic selection and achieved the highest prediction accuracy of 0.66, whereas 0.32–0.35 accuracies were obtained for biomass yield in two alfalfa composite populations consisted of a total of 278 individuals (Annicchiarico et al. 2015). Most recently, we reported genome-wide association and prediction of salt tolerance in alfalfa (Medina et al. 2020), where eight GS models were used for predicting estimated breeding values for salt tolerance in advance breeding populations. The highest prediction accuracy of 0.79 was achieved for the biomass yield under salt stress by using a support vector machine and random forest.

14.5.7 Transcriptomes in Response to Drought Stress

Transcriptomic analysis has been carried out in crops and helped in understanding molecular mechanisms by which environmental factors affect plant development. Although NGS has been widely used to obtain transcriptional profiles and to generate de novo genomic information, short read lengths is a major limitation for identifying isoforms with high quality. To overcome this limitation, during the past few years, the use of Pacific Biosciences (PacBio) or Oxford Nanopore Technologies (ONT) platforms has become popular because they dramatically increase read length avoiding bias during the alternative splice isoforms detection (Rhoads and Au 2015; Dumschott et al. 2020). Recently, two autotetraploid alfalfa genomes have been sequenced using PacBio and de novo genome assembly has been reported (Chen et al. 2020; Shen et al. 2020). These genome sequences provide important references for genomic assessments of various aspects of this species.

The PacBio Isoform Sequencing (Iso-Seq) method generates full-length cDNA sequences with 5' and 3' UTRs in one single read in order to predict and validate gene models and detect alternative splice events (ASE) with high accuracy. Additionally, Iso-seq generates information about transcriptome complexity including long non-coding RNAs (lncRNAs) or nonsense-mediated decay (NMD) of mRNA (Rhoads and Au 2015; Abdel-Ghany et al. 2016). lncRNAs are non-coding RNAs greater than 200 bp in length with low expression and high instability (Kapranov et al. 2007; Rai et al. 2019). lncRNAs are involved in cis- and trans-acting gene regulation processes and can be transcribed from introns, exons, intergenic, and overlapping regions (Rai et al. 2019). Transcriptomic and bioinformatic analyses have been used to predict lncRNAs in *Medicago truncatula*, highlighting their importance in plant adaptation to abiotic stress (Wang et al. 2015; Zhao et al. 2020). NMD of mRNAs is involved in the post-transcriptional fine-tuning of gene expression. The NMD process targets mRNAs containing premature translation termination codons and elicits their rapid degradation (Guan et al. 2006). In plants, NMD of mRNAs is involved in response to biotic (Rayson et al. 2012) and abiotic stresses (Kalyna et al. 2012; Gloggnitzer et al. 2014). Forty percent and 36% of isoforms were predicted as candidates for NMD using Iso-seq in maize and sorghum, respectively (Wang et al. 2018). Additionally, coupling Iso-seq reads with RNA-seq data can be used to generate weighted gene co-expression networks (Langfelder and Horvath, 2008) of highly co-expressed isoforms and hub genes.

In alfalfa, RNA-seq was used to develop a gene expression atlas (O'Rourke et al. 2015) and to identify differentially expressed genes under different conditions including defoliation (Cheng et al. 2018), salt stress (Postnikova et al. 2013; Dong et al. 2018), freezing stress (Shu et al. 2017), and bacterial pathogen infection (Nemchinov et al. 2017). Most recently, Duan et al. (2020) identified hub genes and modules closely related to floral pigmentation in two alfalfa cultivars (Duan et al. 2020). Full-length transcriptomes were obtained from the root tips of "Zhongmu No. 1" under osmotic and ionic stresses (Luo et al. 2019). However, only parts of genes and isoforms were captured in the study as a single tissue source of a variety was used. No report has yet been made for identifying lncRNA in autotetraploid alfalfa. A robust transcriptome in alfalfa using more tissue sources and varieties with extreme responses to specific stresses is required to generate a completed transcriptomic atlas that can be used in alfalfa genomics-assisted breeding programs.

Most recently, our group used two alfalfa varieties of Wilson and Saranac with contrast resistances to drought for unraveling the alfalfa response to drought through full-length transcriptome analysis (unpublished). RNA samples were extracted from three tissue sources (leaf, stem, and root) of plants subjected to drought and control non-stress treatments. These samples were sequenced using PacBio SMRT technology and the Illumina HiSeq platform to obtain comparative transcriptional profiles in response to drought and salt stresses. A total of 20 and 710 million sequence reads were obtained from PacBio and Illumina sequencing, respectively. The long reads from PacBio were corrected using a hybrid approach by comparing them with short reads from Illumina to improve the quality of transcriptomes by LoRDEC (Salmela and Rivals 2014). After correction, a total of 91,378 high-quality unique transcripts and 1,124,275 isoforms were obtained among all treatments. Transcriptomic analysis identified the transcriptional and post-transcriptional differences among leaves, stems, and roots of Wilson and Saranac under drought stress and non-stressed control. Transcript densities were analyzed by aligning transcript sequences to the tetraploid alfalfa reference genome (Chen et al. 2020). The transcript density of allele-aware chromosome-level was distributed on eight chromosomes with 32 haplotypes (Fig. 14.3). A lower gene density was identified in centromeric regions of chromosomes. Deferentially expressed isoforms in the resistant variety Wilson and non-resistant variety Saranac by drought stress (DS) and control (CK) treatments were classified using the SUPPA2 pipeline. The variations of splicing events were found in different tissue sources, stress conditions, and varieties (Fig. 14.4).

The software of SUPPA2 was used for analyzing differential gene expression and 265 differentially expressed isoforms were identified in Wilson under drought stress (Fig. 14.4a, b). Among them, 138 isoforms were upregulated and 127 were downregulated by drought. In Saranac, 249 isoforms were differentially expressed under drought with 127 upregulated and 122 downregulated (Fig. 14.4c, d). Post-transcriptional modifications including alternative splicing events (ASE), fusion genes, and nonsense-mediated mRNA decay (NMD) events were found. Among those, more than half of ASEs were intron retention (IR), followed by alternative 3′ splice-site (A3), alternative 5′ splice-site (A5), alternative first exon (AF), alternative last exon (AL), and mutually exclusive exons (MX) (Fig. 14.4). The annotation of differentially expressed isoforms suggested their functions in stress responses. For instance, isoform G81783 was annotated as ethylene-responsive transcription factor 1 (ERF1) that plays a regulatory role in plant responses to abiotic stresses. *MsERF8* of *M. sativa* was reported to increase tolerance to salinity in transgenic tobacco plants (Chen et al. 2012). Similarly, a wheat ethylene-responsive factor 1 (*TaERF1*) increased multiple stress tolerance (Xu et al. 2007). In Arabidopsis, positive regulation of ERF1 increased tolerance to drought and salt stress by binding to different cis-elements like the DRE element or GCC box (Cheng et al. 2013). Six isoforms (G20067, G25787, G30379, G32331, G65565, and G87127) were annotated as F-box proteins which act as substrate receptors in the SCF complex involved in drought stress responses(Cho et al. 2017). F-box domain proteins have been characterized as key regulators in protein degradation in *M. truncatula* (Song et al. 2015). Isoform G16956.1 was annotated as TIFY transcription factor. TIFY genes (*ClTIFY1* and *ClTIFY2*) were highly induced by drought and salt treatments (Yang et al. 2019). Isoform G80228.6 was annotated as BAX inhibitor 1 (*BI-1*) that involved in anti-cell death in animals and plants. In Arabidopsis, *BI-1* encoded protective protein located in the endoplasmic reticulum regulated the activation of plant cell death (Watanabe and Lam 2008). *BI-1* of Arabidopsis conferred drought tolerance in sugarcane (Ramiro et al. 2016).

Recently, alfalfa nodulated by Rhizobium had overproduction of IAA under drought condition

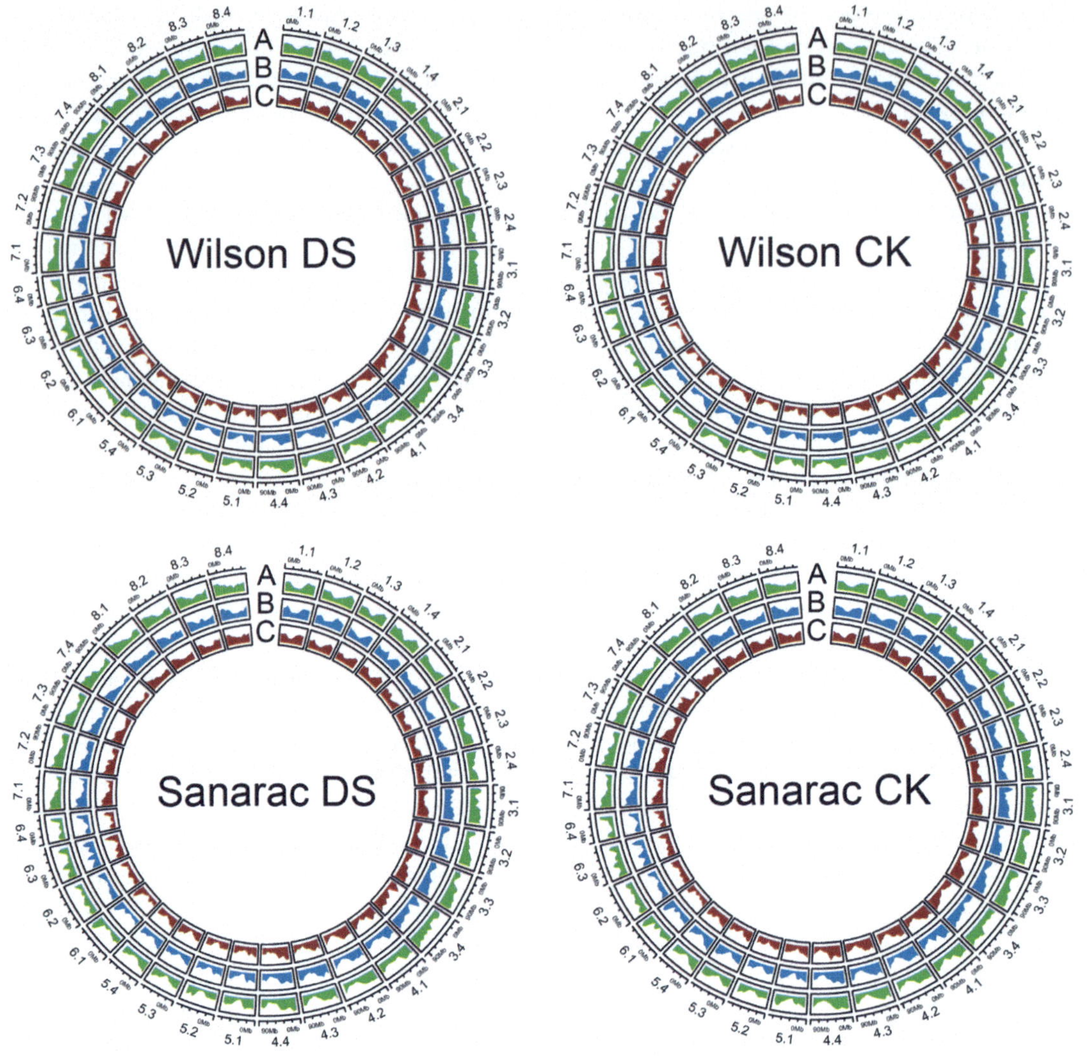

Fig. 14.3 Transcript density in allele aware chromosome-level with 8 chromosomes and 32 haplotypes (4 haplotypes in each chromosome) in tetraploid alfalfa based on the alignment to the reference genome (Chen et al. 2020). Two varieties: Wilson and Saranac and three tissue types: (A) leaf, (B) stem, and (C) root were used with control non-stressed (CK) or drought stress (DS) treatments

(Defez et al. 2017), where alfalfa plants nodulated by the *Ensifermeliloti* wild-type Ms-1021 was compared with the Ms-RD64 derivative strains for IAA-overproduction. They found that Ms-RD64 plants showed significantly weaker damage by drought, with a lower biomass reduction, and a higher RuBisCO protein level compared to the Ms-1021 plants. The production of low-molecular-weight osmolytes, such as proline and pinitol, was greater in the stressed Ms-RD64 plants. Additionally, thylakoid membrane proteins and non-photochemical quenching increased after the stress in these plants. The reduced leaf wilting of MS-R64 plants observed was correlated to the significant down-regulation of the *MtAA03* gene involved in the ABA biosynthesis. Moreover, the MS-R64 plants accumulated 40% less ethylene compared to the Ms-1021 plants by drought. It was suggested that the overproduction of rhizobial IAA in root

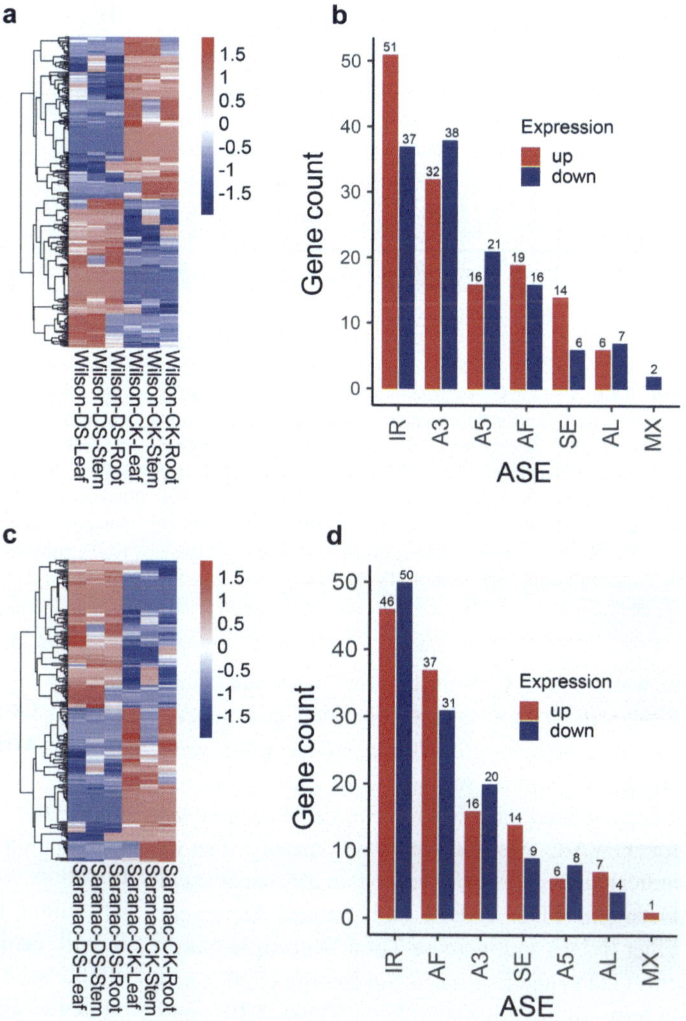

Fig. 14.4 Differential expressed genes (DEG) according to percent spliced in (PSI) values: <−0.5 and >0.5 and p-value > 0.05 generated by SUPPA2. (**a**) Heatmap of Wilson-CK and Wilson-DS in leaf, stem, and root. (**b**) alternative splice events (ASE) classification of DEG in Wilson. (**c**) Heatmap of Saranac-CK and Saranac-DS in leaf, stem, and root. (**d**) ASE classification of DEG in Saranac. Control non-stressed (CK), salt stress (SS), drought stress (DS), Exon skipping (SE), alternative 5′ splice-site (A5), alternative 3′ splice-site (A3), intron retention (RI), alternative first exon (AF), alternative last exon (AL), and mutually exclusive exons (MX)

nodules triggered the elevation of endogenous IAA levels and resulting in alfalfa plants with better drought tolerance in nodulated alfalfa plants.

14.5.8 Genetic Factors that Affect Forage Quality Under Drought

Alfalfa forage quality is affected by environmental factors, such as drought and soil salinity. Lin et al. (2020) evaluated 26 forage quality traits in a panel of 198 alfalfa accessions of the core collection for drought in the field under three irrigation regimes: well-watered, mild, and severe water deficits. Genetic architectures of phenotypic plasticities for forage quality traits were investigated. More than one hundred significant markers associated with forage quality under different water treatments were identified using genome-wide association studies with genotyping-by-sequencing. Among them, 34 markers were associated with multiple traits within the irrigation regime. Most of the associated markers were independent of water deficit, suggesting genetic controls for forage quality traits are independent of the stress treatment. Twenty-four loci associated with forage quality were annotated to functional genes that may play roles in cell development or in response

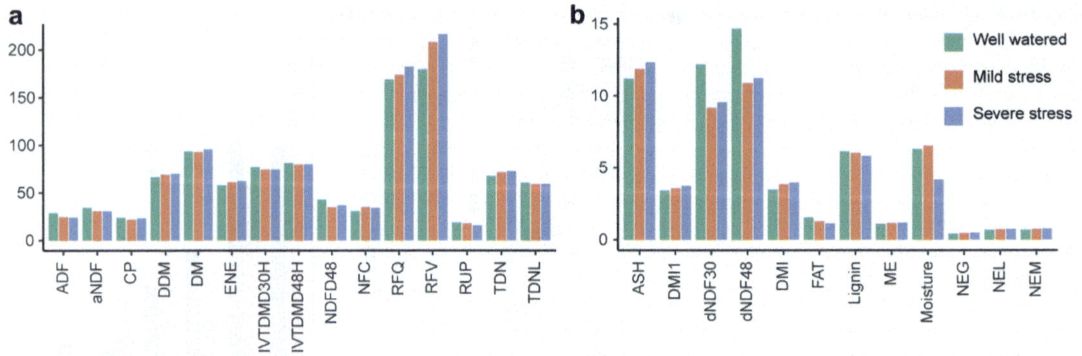

Fig. 14.5 Responses of forage quality factors to well-watered, mild, and severe water stresses in alfalfa (**a** and **b**). ADF, acid detergent fiber; a NDF, neutral detergent fiber analyzed with amylase; CP, crude protein; DM, dry matter; DDM, digestible dry matter; DMI, dry matter intake using NDF; DMI1, dry matter intake using NDF and NDFD; ENE, estimated net energy; IVDDM30, 30-h in vitro digestible dry matter; IVDDM48, 48-h in vitro digestible dry matter; ME, metabolizable energy; NEM, net energy for maintenance; NEG, net energy for gain; NDF30, 30-h digestible NDF; dNDF48, 48-h digestible NDF; NDFD48, 48-h NDFD; NEL, net energy for lactation; NFC, nonfibrous carbohydrates; RFQ, relative forage quality index; RFV, relative feed value index; RUP, rumen undegradable protein; TDN, total digestible nutrients; TDNL, total digestible nutrients for legume

to water stress. Genetic markers associated with forage quality traits were identified and genetic regions responsive for the respective traits were compared. Similar regions were found between energy-related traits when mean values were used for GWAS. Measurements of quality factors indicated that drought tended to decrease Acid Detergent Fiber (ADF) and Neutral Detergent Fiber (NDF) and increase Total Digestible Nutrients (TDN) and Estimated Net Energy (ENE), and in turn, improve Relative Feed Value (RFV) and Relative Forage Quality (RFQ) in alfalfa (Fig. 14.5). It has been reported that mild drought stress may be beneficial for forage quality as drought-stressed alfalfa will accelerate its shift to reproductive growth (Cassida 2012). Water deficit promotes a reduction in vegetative growth and promotes early maturity. Once a stem has flowered, there will be little additional growth in the stem. Stem internode growth is suppressed, resulting in a greater leaf: stem ratio. In the short term of drought stress, the greater proportion of leaves in the drought-stressed forage improves feed quality and digestibility. However, if drought stress has been too severe, and for an extended period, the stress effect becomes permanent and the plant may not recover.

14.6 Concluding Remarks and Prospects

Improved methods of selection based on molecular markers can be used in MAS for the identification of resistance sources for desired agronomic traits such as abiotic stress tolerance. It will facilitate breeding progress and provide alfalfa seed companies with improved breeding strategies for use in the production of improved commercial varieties of alfalfa. This will increase the yield potential of alfalfa to meet the demand for alfalfa as both animal feed and potential feedstock for bioenergy. Additionally, due to its nitrogen fixation ability, stress-tolerant alfalfa would be an ideal rotation crop in cropping systems to increase profitability and sustainability. The current procedure for breeding resistant alfalfa cultivar requires 8–10 years. The genomics-assisted selection will provide unambiguous identification of resistant plants in a timely fusion. The economic impact of the application of genomic tools in breeding drought-tolerant alfalfa will benefit animal feed supplies and sustainable bioenergy in the arid and semi-arid regions of the U.S. Alfalfa varieties with improved resistance to drought should increase biomass production of water-stressed soils and meet the demand of additional feedstock for bioenergy production and increase sustainability, and reduce costs to consumers.

Although MAS has been widely adapted to the commercial development of several important crop species, including corn and soybeans, MAS is rarely employed for the development of improved alfalfa varieties. Genetic progress in this species is slow because of its autotetraploid allogamous nature and the synthetic structure of the varieties (Julier et al. 2003). Enabling technologies that must be developed before alfalfa breeders can employ MAS include identifying genetic marker loci from alfalfa that are robustly associated with important traits such as abiotic tolerance. Besides, quantitative traits such as abiotic stress resistance are most likely under the control of multiple genes and interact with environmental factors. Identification of resistance loci that contribute to variation in such complex traits is a primary challenge in plant breeding and population genetics. Genotyping accuracy in tetraploid species requires a critical improvement, and there is little point in building models for polyploid inheritance without accurate genotype data. To date, several tools for analyzing allele dosage and association mapping in autotetraploid species have been developed and more reference genomes in tetraploid alfalfa are available. The transition of diploid to tetraploid reference genomes would not only help plant biologists to understand the genetic evolution of diploid to tetraploid alfalfa, but also assist breeders in designing new strategies for genetic improvement of alfalfa with enhanced resistance to environmental stress such as drought and high salinity.

References

Abdel-Ghany SE, Hamilton M, Jacobi JL, Ngam P, Devitt N, Schilkey F, Ben-Hur A, Reddy ASN (2016) A survey of the sorghum transcriptome using single-molecule long reads. Nat Commun. https://doi.org/10.1038/ncomms11706

Annicchiarico P, Nazzicari N, Li X, Wei Y, Pecetti L, Brummer EC (2015) Accuracy of genomic selection for alfalfa biomass yield in different reference populations. BMC Genomics 16(1):1020. https://doi.org/10.1186/s12864-015-2212-y

Aranjuelo I, Molero G, Erice G, Avice JC, Nogués S (2011) Plant physiology and proteomics reveals the leaf response to drought in alfalfa (*Medicago sativa* L). J Exp Bot 62:111–123. https://doi.org/10.1093/jxb/erq249

Ariss JJ, Vandemark GJ (2007) Assessment of genetic diversity among nondormant and semidormant alfalfa populations using sequence-related amplified polymorphisms. Crop Sci 47:2274–2284. https://doi.org/10.2135/cropsci2006.12.0782

Arshad M, Feyissa BA, Amyot L, Aung B, Hannoufa A (2017) MicroRNA156 improves drought stress tolerance in alfalfa (*Medicago sativa)* by silencing SPL13. Plant Sci 258:122–136. https://doi.org/10.1016/j.plantsci.2017.01.018

Arshad M, Gruber M, Hannoufa A (2018) Transcriptome analysis of microRNA156 overexpression alfalfa roots under drought stress. Sci Rep 8:9363. https://doi.org/10.1038/s41598-018-27088-8

Boe A, Kephart KD, Berdahl JH, Peel MD, Brummer EC, Xu L, Wu Y(2020) Breeding alfalfa for semiarid regions in the Northern Great Plains: history and additional genetic evaluations of novel germplasm. Agronomy 10:1686. https://doi.org/10.3390/agronomy10111686

Breseghello F, Sorrells ME (2006) Association mapping of kernel size and milling quality in wheat (*Triticum aestivum* L.) cultivars. Genetics 172:1165–1177. https://doi.org/10.1534/genetics.105.044586

Cassida K (2012) Managing alfalfa during drought. Michigan State University Extension. https://www.canr.msu.edu/news/managing_alfalfa_during_drought

Chen D, Liang MX, DeWald D, Weimer B, Peel MD et al (2008) Identification of dehydration responsive genes from two non-nodulated alfalfa cultivars using Medicago truncatula microarrays. Acta Physiol Plant 30:183–199. https://doi.org/10.1007/s11738-007-0107-5

Chen H, Zeng Y, Yang Y, Huang L, Tang B et al (2020) Allele-aware chromosome-level genome assembly and efficient transgene-free genome editing for the autotetraploid cultivated alfalfa. Nat Commun. https://doi.org/10.1038/s41467-020-16338-x

Chen T, Yang Q, Gruber M, Kang J, Sun Y, Ding W, Zhang T, Zhang X (2012) Expression of an alfalfa (*Medicago sativa* L.) ethylene response factor gene MsERF8 in tobacco plants enhances resistance to salinity. Mol Biol Rep. https://doi.org/10.1007/s11033-011-1421-y

Cheng MC, Liao PM, Kuo WW, Lin TP (2013) The arabidopsis ETHYLENE RESPONSE FACTOR1 Regulates abiotic stress-responsive gene expression by binding to different cis-acting elements in response to different stress signals. Plant Physiol. https://doi.org/10.1104/pp.113.221911

Cheng Q, Bai S, Ge G, Li P, Liu L, Zhang C, Jia Y (2018) Study on differentially expressed genes related to defoliation traits in two alfalfa varieties based on RNA-Seq. BMC Genomics. https://doi.org/10.1186/s12864-018-5180-1

Cho SK, Ryu MY, KimJH HJS, Oh TR, Kim WT, Yang SW (2017) RING E3 ligases: Key regulatory elements are involved in abiotic stress responses in plants. BMB Rep 50(8):393–400. https://doi.org/10.5483/BMBRep.2017.50.8.128

Crossa J, de Los CG, Pérez P, Gianola D, Burgueño J et al (2010) Prediction of genetic values of quantitative traits in plant breeding using pedigree and molecular markers. Genetics 186:713–724. https://doi.org/10.1534/genetics.110.118521

Condon AG (2004) Breeding for high water-use efficiency. J Exp Bot 55:2447–2460

de Vasconcelos ES, Barioni Junior W, Cruz CD, Ferreira RP, Rassini JB, Vilela D, (2008) Alfalfa genotype selection for adaptability and stability of dry matter production. Acta Sci Agron 30:339–343. https://doi.org/10.4025/actasciagron.v30i3.3511

Defez R, Andreozzi A, Dickinson M, Charlton A, Tadini L et al (2017) Improved drought stress response in alfalfa plants nodulated by an IAA Overproducing Rhizobium Strain. Front Microbiol 8:2466. https://doi.org/10.3389/fmicb.2017.02466

Dong W, Liu X, Li D, Gao T, Song Y (2018) Transcriptional profiling reveals that a MYB transcription factor MsMYB4 contributes to the salinity stress response of alfalfa. PLoS One. https://doi.org/10.1371/journal.pone.0204033

Duan HR, Wang LR, Cui GX, ZhouXH DXR, Yang HS (2020) Identification of the regulatory networks and hub genes controlling alfalfa floral pigmentation variation using RNA-sequencing analysis. BMC Plant Biol. https://doi.org/10.1186/s12870-020-2322-9

Dumschott K, Schmidt MHW, Chawla HS, Snowdon R, Usadel B (2020) Oxford Nanopore sequencing: new opportunities for plant genomics? J Exp Bot. https://doi.org/10.1093/jxb/eraa263

Elshire RJ, Glaubitz JC, Sun Q, Poland JA, Kawam K et al. (2011) A robust, simple Genotyping-by-Sequencing (GBS) approach for high diversity species. PLoS One 6(5):e19379. https://doi.org/10.1371/journal.pone.0019379

Feyissa BA, Renaud J, Nasrollahi V, Kohalmi SE, Hannoufa A (2020) Transcriptome-IPMS analysis reveals a tissue-dependent miR156/SPL13 regulatory mechanism in alfalfa drought tolerance. BMC Genomics 21:721. https://doi.org/10.1186/s12864-020-07118-4

Garreta L, Cerón-Souza I, Palacio MR, Reyes-Herrera PH (2020) MultiGWAS: An integrative tool for Genome Wide Association Studies (GWAS) in tetraploid organisms. https://doi.org/10.1101/2020.08.16.252791

Garrison E, Marth G (2012) Haplotype-based variant detection from short-read sequencing—Free bayes—Variant Calling—Longranger. arXiv:1207.3907

Garver S (1946) Alfalfa in South Dakota: twenty-one years research at the Redfield Station. South Dakota Agricultural Experiment Station Bulletin 383

Gloggnitzer J, Akimcheva S, Srinivasan A, Kusenda B, Riehs N, Stampfl H, Bautor J, Dekrout B, Jonak C, Jiménez-Gómez JM, Parker JE, Riha K (2014) Nonsense-mediated mRNA decay modulates immune receptor levels to regulate plant antibacterial defense. Cell Host Microbe. https://doi.org/10.1016/j.chom.2014.08.010

Guan Q, Zheng W, Tang S, Liu X, Zinkel RA, Tsui KW, Yandell BS, Culbertson MR (2006) Impact of nonsense-mediated mRNA decay on the global expression profile of budding yeast. PLoS Genet. https://doi.org/10.1371/journal.pgen.0020203

Han Y, Kang Y, Torres-Jerez I, Cheung F, Town CD et al (2011) Genome-wide SNP discovery in tetraploid alfalfa using 454 sequencing and high-resolution melting analysis. BMC Genomics 12:350. https://doi.org/10.1186/1471-2164-12-350

Hannoufa A, Matthews C, Feyissa BA, Gruber MY, Arshad M (2018) Progress toward deep sequencing-based discovery of stress-related micrornain plants and available bioinformatics tools. In: Cánovas F, Lüttge U, Leuschner C, Risueño MC (eds) Progress in Botany, vol 81. Springer, Cham. https://doi.org/10.1007/124_2018_25

Hawkins C, Yu LX (2018) Recent progress in alfalfa (Medicago sativa L) genomics and genomic selection. Crop J https://doi.org/10.1016/j.cj.2018.01.006

Heffner EL, Sorrells ME, Jannink JL (2009) Genomic selection for crop improvement. Crop Sci 49:1–12. https://doi.org/10.2135/cropsci2008.08.0512

Heinrichs DH (1967) Roamer alfalfa. Can J Plant Sci 47:220–221

Heinrichs DH (1971) Drylander alfalfa . Can J Plant Sci 51:430–432

Heinrichs DH, Bolton JL (1958) Rambler alfalfa. Publication 1030. Canada Department of Agriculture. Ottawa, Ontario

Herrmann D, Barre P, Santoni S, Julier B (2010) Association of a CONSTANS-LIKE gene to flowering and height in autotetraploid alfalfa. Theor Appl Genet 121:865–876. https://doi.org/10.1007/s00122-010-1356-z

Heslot N, Yang HP, Sorrells ME, Jannink, JL (2012) Genomic selection in plant breeding: a comparison of models. Crop Sci 52:146– 160. https://doi.org/10.2135/cropsci2011.06.0297

Jiang Q, Zhang J-Y, Guo X, et al (2009) Physiological characterization of transgenic alfalfa (Medicago sativa) plants for improved drought tolerance. Int J Plant Sci 170:969–978

Julier B, Flajoulot S, Barre P, Cardinet G, Santoni S et al (2003) Construction of two genetic linkage maps in cultivated tetraploid alfalfa (*Medicago sativa*) using microsatellite and AFLP markers. BMC Plant Biol 3:9. https://doi.org/10.1186/1471-2229-3-9

Kalyna M, Simpson CG, Syed NH, Lewandowska D, Marquez Y et al (2012) Alternative splicing and nonsense-mediated decay modulate expression of important regulatory genes in Arabidopsis. Nucleic Acids Res. https://doi.org/10.1093/nar/gkr932

Kang H, Han Y, Torres-Jerez I, Sinharoy S, Tang Y, Monteros MJ, Udvardi MK (2011) System responses to long-term drought and re-watering of two contrasting alfalfa varieties. Plant J 68:871–889. https://doi.org/10.1111/j.1365-313X.2011.04738.x

Kang Y, Udvardi M (2012) Global regulation of reactive oxygen species scavenging genes in alfalfa root and shoot under gradual drought stress and recovery. Plant Signal Behav 7(5):539–543. https://doi.org/10.4161/psb.19780

Kapranov P, Cheng J, Dike S, Nix DA, Duttagupta R et al (2007) RNA maps reveal new RNA classes and a possible function for pervasive transcription. Science. https://doi.org/10.1126/science.1138341

Laberge S, Castonguay Y, Vezina LP (1993) New cold and drought- regulated gene from *Medicago sativa*. Plant Physiol 101(4):1411–1412. https://doi.org/10.1104/pp.101.4.1411

Langfelder P, Horvath S (2008) WGCNA: An R package for weighted correlation network analysis. BMC Bioinformatics. https://doi.org/10.1186/1471-2105-9-559

Lenssen AW, Sorensen EL, Posler GL, Harbers LH (1991) Basic alfalfa germplasms differ in nutritive content of forage. Crop Sci 31:293–296. https://doi.org/10.2135/cropsci1991.0011183x003100020012x

Li GD, Nie ZN, Boschma SP, Dear BS, Lodge GM, Hayes RC, Clark B, Hughes SJ, Humphries AW (2010) Persistence and productivity of *Medicago sativa* subspecies sativa, caerulea, falcata and varia accessions at three intermittently dry sites in south-eastern Australia. Crop Pasture Sci 61:645–658. https://doi.org/10.1071/CP09360

Li H, Handsaker B, Wysoker A, Fennell T, Ruan J, Homer N, Marth G, Abecasis G, DurbinR, (2009) The sequence alignment/map format and SAMtools. Bioinformatics. https://doi.org/10.1093/bioinformatics/btp352

Li X, Acharya A, Farmer AD, Crow JA, Bharti AK, Kramer R, Wei Y, Han Y, Gou J, May GD, Monteros MJ, Brummer EC (2012) Prevalence of single nucleotide polymorphism among 27 diverse alfalfa genotypes as assessed by transcriptome sequencing. BMC Genomics 13:568. https://doi.org/10.1186/1471-2164-13-568

Li X, Wei Y, Acharya A, Hansen JL, Crawford JL, Viands DR, Michaud R, Claessens A, Brummer EC (2015) Genomic prediction of biomass yield in two selection cycles of a tetraploid alfalfa breeding population. Plant Genome 8(2). https://doi.org/10.3835/plantgenome2014.12.0090

Lin S, Medina CA, Boge B, Hu J, Fransen S, Norberg S, Yu L-X (2020) Identification of genetic loci associated with forage quality in response to water deficit in autotetraploid alfalfa (*Medicago sativa* L) BMC Plant Biol 20:303. https://doi.org/10.1186/s12870-020-02520-2

Liu X-P, Hawkins C, Peel M, Yu L-X (2018) Genetic loci associated with salt tolerance in advanced breeding populations of tetraploid alfalfa (*Medicago sativa* L) using genome-wide association studies. Plant Genome. https://doi.org/10.3835/plantgenome2018.05.0026

Liu X-P, Yu L-X. (2017) Genome-wide association mapping of loci associated with plant growth and forage production under salt stress in alfalfa (*Medicago sativa* L). Front Plant Sci. https://doi.org/10.3389/fpls.2017.00853

Li Y, Wan L, Bi S, et al (2017) Identification of drought-responsive MicroRNAs from roots and leaves of Alfalfa by high-throughput sequencing. Genes (Basel) 8:119

Luo D, Zhou Q, Wu Y, Chai X, Liu W, Wang Y, Yang Q, Wang Z, Liu Z (2019) Full-length transcript sequencing and comparative transcriptomic analysis to evaluate the contribution of osmotic and ionic stress components towards salinity tolerance in the roots of cultivated alfalfa (*Medicago sativa* L). BMC Plant Biol. https://doi.org/10.1186/s12870-019-1630-4

Luo M, Liu JH, Mohapatra S, Hill RD, Mohapatra SS (1992) Characterization of a gene family encoding abscisic acid- and environmental stress-inducible proteins of alfalfa. J Biol Chem 267:15367–15374. https://doi.org/10.1016/S0021-9258(19)49543-4

Luo M, Lin L, Hill RD, Mohapatra SS (1991) Primary structure of an environmental stress and absicic acid inducible alfalfa protein. Plant Mol Biol 17(6):1267–1269. https://doi.org/10.1007/BF00028745

McKersie BD, Bowley SR, Harjanto E, Leprince O (1996) Water deficit tolerance and field performance of transgenic alfalfa overexpressing superoxide dismutase. Plant Physiol 111:1177–1181. https://doi.org/10.1101/gr.107524.110

McLeod JG, Muri R, Jefferson PG, Bittman S, McCartney D (2009) Yellowhead alfalfa. Can J Plant Sci 89:653–655

Medina CA, Hawkins C, Liu X-P, Peel M, Yu L-X (2020) Genomic-wide association and predication of traits related to salt tolerance in autotetraploid alfalfa (*Medicago sativa* L). Intl J Mol Sci 21(9):1–25. https://doi.org/10.3390/ijms21093361

Nemchinov LG, Shao J, Lee MN, Postnikova OA, Samac DA (2017) Resistant and susceptible responses in alfalfa (*Medicago sativa*) to bacterial stem blight caused by Pseudomonas syringae pv. syringae. PLoS One. https://doi.org/10.1371/journal.pone.0189781

O'Rourke JA, Fu F, Bucciarelli B, Yang SS, Samac DA, et al. (2015) The Medicago sativa gene index 1.2: a web-accessible gene expression atlas for investigating expression differences between Medicago sativa subspecies. BMC Genomics. https://doi.org/10.1186/s12864-015-1718-7

Oakley RW, Garver S (1917) *Medicago falcata*, a yellow-flowered alfalfa. USDA Bulletin No 428

Pankiw P, Siemens B (1976) Anik alfalfa. Canadian J. Plant Sci 56:203–205

Peel MD, Asay KH, Waldron BL, Jensen KB, Robins JG, Mott IW (2009) 'Don', a diploid falcata alfalfa for Western U.S. rangelands. J Plant Regist 3:115–118

Postnikova OA, Shao J, Nemchinov LG (2013) Analysis of the alfalfa root transcriptome in response to salinity stress. Plant Cell Physiol. https://doi.org/10.1093/pcp/pct056

Quan W, Liu X, Wang H, Chan Z (2016) Comparative physiological and transcriptional analyses of two contrasting drought tolerant alfalfa varieties. Front Plant Sci 6:1256. https://doi.org/10.3389/fpls.2015.01256

Rai MI, Alam M, Lightfoot DA, Gurha P, Afzal AJ (2019) Classification and experimental identification of plant long non-coding RNAs. Genomics. https://doi.org/10.1016/j.ygeno.2018.04.014

Ramiro DA, Melotto-Passarin DM, Barbosa M de A, et al (2016) Expression of Arabidopsis Bax Inhibitor-1 in transgenic sugarcane confers drought tolerance. Plant Biotechnol J 14:1826–1837

Ray IM, Segovia-Lerma A, Murray LW (2004) Diallel analysis of carbon isotope discrimination and its association with forage yield among nine historically recognized alfalfa germplasms. Crop Sci 44:1970–1975. https://doi.org/10.2135/cropsci2004.1970

Ray IM, Han Y, Lei E, Meenach LE, Santantonio N, Sledge MK et al (2015) Identification of QTL for alfalfa forage biomass productivity during drought. Crop Sci 55:2012–2033. https://doi.org/10.2135/cropsci2014.12.0840

Ray IM, Townsend MS, Henning JA (1998) Variation for yield, water-use efficiency, and canopy morphology among nine alfalfa germplasms. Crop Sci. https://doi.org/10.2135/cropsci1998.0011183X003800050038x

Rayson S, Arciga-Reyes L, Wootton L, de Torres ZM, Truman W, Graham N, Grant M, Davies B (2012) A role for nonsense-mediated mrna decay in plants: pathogen responses are induced in arabidopsis thaliana NMD mutants. PLoS ONE. https://doi.org/10.1371/journal.pone.0031917

Rhoads A, Au KF (2015) PacBio sequencing and its applications. Genomics, Proteomics and Bioinformatics. https://doi.org/10.1016/j.gpb.2015.08.002

Riday H, Brummer EC (2007) Registration of two improved yellow-flowered alfalfa germplasms. J Plant Regist 1:131–133

Rosyara UR, De Jong WS, Douches DS, Endelman JB (2016) Software for genome-wide association studies in autopolyploids and its application to potato. The Plant Genome 9(2):1–10. https://doi.org/10.3835/plantgenome2015.08.0073

Rumbaugh MD (1979) NE Hansen's contributions to alfalfa breeding in North America. Bulletin 665. Agricultural Experiment Station. South Dakota State University. Brookings, SD 57007

Rumbaugh MD, Colburn JD, Semeniuk G (1964) Registration of Travois alfalfa. Crop Sci 4:117

Salmela L, Rivals E (2014) LoRDEC: accurate and efficient long read error correction. Bioinformatics. https://doi.org/10.1093/bioinformatics/btu538

Santantonio N, Pierce CA, Steiner RL, Ray IM (2019) Genetic mapping of water use efficiency and carbon and nitrogen metabolism in drought-stressed alfalfa. Crop Sci 59(1):92–106. https://doi.org/10.2135/cropsci2018.05.0307

Segovia-Lerma A, Cantrell RG, Conway JM, Ray IM (2003) AFLP based assessment of genetic diversity among nine alfalfa germplasms using bulk DNA templates. Genome 46:51–58. https://doi.org/10.1139/g02-100

Shen C, Du H, Chen Z, Lu H, Zhu F et al (2020) The chromosome-level genome sequence of the autotetraploid alfalfa and resequencing of core germplasms provide genomic resources for alfalfa research. Mol Plant 13(9):1250–1261. DOI: https://doi.org/10.1016/j.molp.2020.07.003

Shu Y, Li W, Zhao J, Zhang S, Xu H, Liu Y, Guo C (2017) Transcriptome sequencing analysis of alfalfa reveals cbf genes potentially playing important roles in response to freezing stress. GenetMol Biol. https://doi.org/10.1590/1678-4685-gmb-2017-0053

Sledge MK, Ray IM, Jiang G (2005) An expressed sequence tag SSR map of tetraploid alfalfa (*Medicago sativa* L). Theor Appl Genet 111:980–992. https://doi.org/10.1007/s00122-005-0038-8

Song JB, Wang YX, Li HB, et al (2015) The F-box family genes as key elements in response to salt, heavy mental, and drought stresses in Medicago truncatula. Funct Integr Genomics 15:495–507

Song Y, Lv J, Qiu N, et al (2019) The constitutive expression of alfalfa MsMYB2L enhances salinity and drought tolerance of Arabidopsis thaliana. Plant Physiol Biochem 141:300–305

Suárez R, Calderón C, Iturriaga G (2009) Enhanced tolerance to multiple abiotic stresses in transgenic alfalfa accumulating trehalose. Crop Sci 49:1791–1799. https://doi.org/10.2135/cropsci2008.09.0573

Uitdewilligen J, Wolters AMA, D'hoop BB, Borm TJA, Visser RGF, van Eck HJ, (2013) A next-generation sequencing method for genotyping-by-sequencing of highly heterozygous autotetraploid potato. PLoS One. https://doi.org/10.1371/journal.pone.0062355

USDA (2017) Census of Agriculture 2017. USDA-National Agricultural Statistics Service. https://www.nass.usda.gov/Publications/AgCensus/2017/index.php#full_report

Vandecasteele C, Teulat-Merah B, Morère-Le Paven M-C, et al (2011) Quantitative trait loci analysis reveals a correlation between the ratio of sucrose/raffinose family oligosaccharides and seed vigour in Medicago truncatula. Plant Cell Environ 34:1473–1487

Voorrips RE, Gort G, Vosman B (2011) Genotype calling in tetraploid species from bi-allelic marker data using mixture models. BMC Bioinformatics 12:172. https://doi.org/10.1186/1471-2105-12-172

Wang B, Regulski M, Tseng E, Olson A, Goodwin S, Richard McCombie W, Ware D (2018) A comparative transcriptional landscape of maize and sorghum obtained by single-molecule sequencing. Genome Res. https://doi.org/10.1101/gr.227462.117

Wang TZ, Liu M, Zhao MG, Chen R, Zhang WH (2015) Identification and characterization of long non-coding RNAs involved in osmotic and salt stress in Medicago truncatula using genome-wide high-throughput sequencing. BMC Plant Biol. https://doi.org/10.1186/s12870-015-0530-5

Watanabe N, Lam E (2008) BAX inhibitor-1 modulates endoplasmic reticulum stress-mediated programmed cell death in arabidopsis. J Biol Chem. https://doi.org/10.1074/jbc.M706659200

Xu ZS, Xia LQ, Chen M, Cheng XG, Zhang RY et al (2007) Isolation and molecular characterization of the *Triticum aestivum* L ethylene-responsive factor 1 (TaERF1) that increases multiple stress tolerance. Plant Mol Biol https://doi.org/https://doi.org/10.1007/s11103-007-9237-9

Xu J, Li X-L, Luo L (2012) Effects of engineered sinorhizobium meliloti on cytokinin synthesis and tolerance of alfalfa to extreme drought stress. Appl Environ Microbiol 78:8056–8061

Yang Y, Ahammed GJ, Wan C, et al (2019) Comprehensive analysis of TIFY transcription factors and their expression profiles under jasmonic acid and abiotic stresses in watermelon. Int J Genomics 2019:1–13

Yu L-X (2017) Identification of single-nucleotide polymorphic markers associated with biomass yield under water deficit in alfalfa (*Medicago sativa* L.) using genome-wide sequencing and association mapping. Front Plant Sci https://doi.org/10.3389/fpls.2017.01152

Yu L-X, Liu X, Boge W, Liu X-P (2016) Genome-wide association study identifies loci for salt tolerance during germination in autotetraploid alfalfa (*Medicago sativa* L) using genotyping-by-sequencing. Front Plant Sci 7:956 https://doi.org/10.3389/fpls.2016.00956 https://doi.org/https://doi.org/10.3389/fpls.2016.00956

YYu L-X, Lorenz A, Rutkoski J, Singh RP, Bhavani S, Huerta-Espino J, Sorrells ME, (2011) Association mapping and gene-gene interaction for stem rust resistance in spring wheat germplasm. Theor Appl Genet 123:1257–1268. https://doi.org/10.1007/s00122-011-1664-y

Yu L-X, Zheng P, Bhamidimarri S, Liu X-P, Main D (2017) The impact of genotyping-by-sequencing pipelines on snp discovery and identification of markers associated with verticillium wilt resistance in autotetraploid alfalfa (*Medicago sativa* L). Front Plant Sci 8:89 https://doi.org/https://doi.org/10.3389/fpls.2017.00089

Yu L-X, Zheng P, Zhang T, Rodringuez J, Main D (2016) Genotyping-by-sequencing based genome-wide association studies on Verticillium wilt resistance in autotetraploid alfalfa (*Medicago sativa* L). Mol Plant Pathol 18(2):187–194. https://doi.org/10.1111/mpp.12389

Zhang JY, Broeckling CD, Blancaflor EB, Sledge MK, Sumner LW, Wang ZY (2005) Overexpression of WXP1, a putative Medicago truncatula AP2 domain-containing transcription factor gene, increases cuticular wax accumulation and enhances drought tolerance in transgenic alfalfa (*Medicago sativa*). Plant J 42:689–707. https://doi.org/10.1111/j.1365-313X.2005.02405.x

Zhang T, Yu L-X, McCord P, Miller D, Bhamidimarri S, et al. (2014) Identification of molecular markers associated with verticillium wilt resistance in alfalfa (*Medicago sativa*L.) using high-resolution melting. PLoS One 9(12):e115953 https://doi.org/https://doi.org/10.1371/journal.pone.0115953

Zhang T, Yu L-X, Zheng P, Li Y, Rivera M, Main D, Greene S (2015) Identification of loci associated with drought resistance traits in heterozygous autotetraploid alfalfa (*Medicago sativa* L.) using genome-wide association studies with genotyping by sequencing. PLoS One 10(9):e0138931 https://doi.org/https://doi.org/10.1371/journal.pone.0138931

Zhang T, Kesoju S, Greene SL, et al (2018) Genetic diversity and phenotypic variation for drought resistance in alfalfa (Medicago sativa L.) germplasm collected for drought tolerance. Genet Resour Crop Evol 65:471–484

Zhao M, Wang T, Sun T, Yu X, Tian R, Zhang WH (2020) Identification of tissue-specific and cold-responsive lncRNAs in *Medicago truncatula* by high-throughput RNA sequencing. *BMC* Plant Biol. https://doi.org/10.1186/s12870-020-2301-1

Zhou Q, Jia C, Ma W, et al (2019) MYB transcription factors in alfalfa (Medicago sativa): genome-wide identification and expression analysis under abiotic stresses. Peer J 7:e7714

Zheng G, Fan C, Di S, Wang X, Xiang C, Pang Y (2017) Over-Expression of Arabidopsis EDT1 Gene Confers Drought Tolerance in Alfalfa (*Medicago sativa* L.). Front Plant Sci. https://doi.org/https://doi.org/10.3389/fpls.2017.02125

Self-incompatibility, Inbreeding Depression, and Potential to Develop Inbred Lines in Alfalfa

15

Atit Parajuli, Long-Xi Yu, Michael Peel, Deven See, Steve Wagner, Steve Norberg, and Zhiwu Zhang

Abstract

Alfalfa (*Medicago sativa* L.) is a perennial and outcrossing legume crop predominantly grown for hay, silage, or pasture. Intensive selection has made a dramatic improvement on fitness traits, including winter survival and disease resistance. However, there is a minimal improvement on other economically important traits, such as hay yield, which is still similar to 30 years ago. Intensive phenotyping on this type of trait is prohibitive to apply high selection pressure to identify any superior outcross individuals. Severe inbreeding depression inhibited the development of inbred lines to accumulate favorite alleles and use heterosis. This review highlights the outcomes of inbreeding depression and the causes, including unmasking deleterious alleles and triggering self-incompatibility. We tracked the research efforts that unveil the genetic bases underlying deleterious alleles and self-incompatibility. The magnitudes of inbreeding depression were compared with the rate of heterozygous halved time in diploid and tetraploid. To fill the gap between the controversy and existing hypotheses, we speculated a numeric dominant model of inheritance to align the gap. The numeric dominant model is similar to the Mendelian dominance model defining a genotype to exhibit a dominant phenotype if there is a dominant allele (alphabet dominant). The difference is that the numeric dominance model defines a genotype to exhibit a dominant phenotype with the number of dominant alleles equal to or less than the recessive alleles. The review is completed by the

A. Parajuli · Z. Zhang (✉)
Department of Crop and Soil Sciences, Washington State University, Pullman, WA 99164, USA
e-mail: Zhiwu.Zhang@WSU.Edu

A. Parajuli
e-mail: atit.parajuli@wsu.edu

L.-X. Yu
Plant Germplasm Introduction and Testing Research, USDA-ARS, Prosser, WA 99350, USA
e-mail: longxi.yu@usda.gov

M. Peel
Forage & Range Research Lab, USDA-ARS, Logan, UT 84322, USA
e-mail: mike.peel@usda.gov

D. See
Wheat Health, Genetics, and Quality Research Unit, USDA-ARS and Washington State University, Pullman, WA 99164, USA
e-mail: deven_see@wsu.edu

S. Wagner
Alforex Seeds, N4505 CTH M, West Salem, WI 54669, USA
e-mail: steve.wagner@corteva.com

S. Norberg
Regional Forage Specialist and Irrigated Cropping Systems, Franklin County Extension Office, Washington State University, 404 West Clark Street, Pasco, WA 99301, USA
e-mail: s.norberg@wsu.edu

L.-X. Yu and C. Kole (eds.), *The Alfalfa Genome*, Compendium of Plant Genomes,
https://doi.org/10.1007/978-3-030-74466-3_15

discussion on the development of pseudo-inbred and a hypothesis to identify deleterious alleles using bulked segregation analysis and consequently purge deleterious alleles using marker-assisted selection toward the success of the development of real inbred lines in alfalfa.

15.1 Introduction

Alfalfa (*Medicago sativa* L.) is a perennial and highly outcrossing legume forage crop grown predominantly for hay, silage, and pasture. Alfalfa exist in two ploidy levels (diploid, $2n = 2x = 16$ and tetraploid, $2n = 4x = 32$). Cultivated alfalfa is tetraploid with a base chromosome number of $x = 8$ and a genome size of ~3.15 Gb (Chen et al. 2020; Shen et al. 2020). It is the most important and widely cultivated forage crop around the world due to its nutritious forage. It also plays a significant role in improving soil fertility as it forms a symbiotic association with soil bacterium *Sinorhizobium meliloti* for atmospheric nitrogen fixation, thereby improving nitrogen content in the soil for following crops. The perennial nature of the crop helps prevent soil erosion through deep root systems.

However, genetic improvements in terms of forage yield have been limited in alfalfa (Riday and Brummer 2002; Lamb et al. 2006; USDA-NASS 2018). Studies (Holland and Bingham 1994; Hill et al. 1988; Wiersma et al. 1997) suggested no significant yield improvement (<1%) in alfalfa over the last half century (Fig. 15.1). In comparison, the yield increment in maize is substantial with 2% per year since the extensive adoption of single cross hybrid in the 1960s (Duvick 1992). In alfalfa, genomic complexity, high inbreeding depression, and self-incompatibility complicate breeding procedure for higher production. Synthetic varieties from multiple crosses fix favorable alleles and utilize intra-locus allelic interaction for higher production (Hill 1987). However, genetic equilibrium upon intercrossing of available germplasm (Barnes et al. 1977) may affect future improvement process (Holland and Bingham 1994). So, future improvement requires maintaining genetic diversity to prevent a genetic bottleneck.

According to Lamb et al. (2006), the stagnant yield in alfalfa is due to the focus of the breeding program on the improvement of non-yield traits rather than breeding for yield. The current alfalfa breeding program utilizes recurrent phenotypic selection with multiple crosses between selected parents, to accumulate desirable alleles at high frequency (Li and Brummer 2012), for producing synthetic variety. However, the low heritability of traits, size of the breeding population that can be evaluated in the field, and limited resources of breeding programs to evaluate genotype x environment interaction (Li and Brummer 2012) seriously undermine the phenotypic selection process. Additionally, genomic complexity, perennial nature that requires multiple harvests, inability to exploit heterosis in commercial cultivars (Tucak et al. 2012), and inefficient

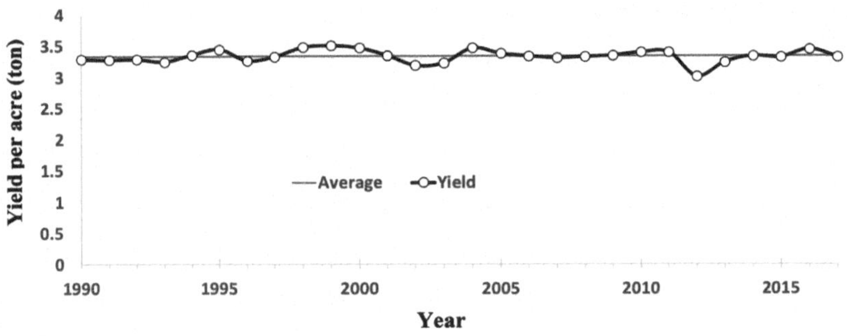

Fig. 15.1 The trend of alfalfa hay yield from 1990–2017 in the US. The plot is based on the data from USDA-NASS (2018). The blue dots and lines indicate the yield at each year and the red line indicates the average during the period

selection method utilizing additive genetic variance (Casler and Brummer 2008; Tucak et al. 2009; Kumar 2011) also contribute to slow progress in yield improvement in alfalfa.

Bhandari et al. (2007) demonstrated the possibility of the development of high-yielding hybrids with performance equal to or exceeding the performance of elite cultivars, by crossing high-yielding alfalfa accessions. The results indicated genetic improvement in alfalfa through the exploitation of both additive as well as non-additive gene effects (Bhandari et al. 2007; Tucak et al. 2012). According to (Dudley and Busbice 1969; Rowe and Hill 1981), approximately two-thirds of the variance for yield in alfalfa was non-additive. These results indicate the potential of alfalfa to exploit heterosis for higher yield. However, high heterozygosity of the plants and severe inbreeding depression upon selfing prevent the development of pure inbred lines (Julier et al. 2003). As a result, synthetic varieties are seen as the most feasible means of genetic gain in alfalfa cultivars (Hill 1987).

Prior studies on alfalfa have shown the existence of heterosis (Bhandari et al. 2007; Al Lawati et al. 2011). Heterosis is the superior performance of progeny in relation to their parents. It forms the basis of genetic gain in maize and other crops. According to Falconer and MacKay (1996), expression of heterosis in population crosses requires directional dominance at loci controlling the trait of interest and differing allele frequencies at those loci in population or lines to be crossed. In alfalfa, a major barrier to hybrid production is to accumulate the homozygous alleles for the trait of interest due to high inbreeding depression and self-incompatibility that prevents the development of pure inbred lines for hybrid development. The problems associated with creating inbred lines and their use for the synthesis of new populations have not been adequately studied in alfalfa (Galiolla et al. 2017).

Previous reviews on inbreeding depression of alfalfa pointed out the loss of heterozygosity (Jones and Bingham 1995), multiple allelic interaction (Busbice and Wilsie 1966), and complementary gene action (Bingham et al. 1994) as major reasons for inbreeding depression. Viands et al. (1988) pointed out the importance of loss of higher order allelic interaction in alfalfa self-incompatibility study, but they were unable to explain the genetic mechanism behind self-incompatibility. Also, alfalfa experiences ambiguous situations of severe inbreeding depression similar to diploid species (Busbice and Wilsie 1966; Dewey 1966, 1969) that is not clearly explained. In order to explain this ambiguity, we speculated a numeric dominant model of inheritance in this review. Further, we also hypothesize developing pure inbred lines through bulked segregation analysis(BSA) followed by marker-assisted selection(MAS) to identify and purge deleterious alleles.

15.2 Inbreeding Depression

Mating between individuals that are related by ancestry is termed as Inbreeding. Generally, in cross-pollinated species, inbreeding causes loss of vigor and decline in fitness, known as Inbreeding Depression (Darwin 1867). Darwin (1867) found considerable effects of inbreeding with shorter plants, late flowering, less weight, and fewer seeds than the outcrossing species and a significant reduction (41%) in seed production and decline (13%) in height (Frankham et al. 2002). The effect of inbreeding is prevalent in the inbreeding species through decreased fitness and vigor. Inbreeding causes loss of heterozygosity, thereby decreasing the mean value of traits associated with fitness and consequently leading to inbreeding depression (Lynch and Walsh 1998). Another hypothesis argues that most mutations are deleterious and generally recessive (Davenport 1908; McKay 2001), and increased homozygosity due to continuous inbreeding extends the possibility of unmasking these harmful deleterious alleles (Charlesworth and Charlesworth 1999) consequently resulting in inbreeding depression (Keller and Waller 2002).

To generalize, deleterious recessive alleles in a cross-pollinating species remain hidden by their dominant counterparts and are not expressed phenotypically. However, upon subsequent inbreeding, these deleterious alleles progressively fix to their homozygous form and express phenotypically resulting in individuals having reduced vigor and fitness. However, the molecular basis underlying inbreeding depression as well as genes or gene pathways associated with inbreeding depression in crops are unknown (Paige 2010). Also, the number of loci responsible for causing inbreeding depression of fitness and its components are unknown as well (Frankham et al. 2002).

Alfalfa is a highly allogamous crop. Studies over different geographical areas show 89% natural crossing of alfalfa plants under field condition (Tysdal et. al. 1942; Knowles 1943; Bolton 1948; Pederson 1953). Most alfalfa plants are found to be partial to complete self-incompatible (Viands et al. 1988) with a highly heterozygous population. Selfing these highly heterozygous individuals causes severe inbreeding depression due to the expression of deleterious recessive alleles in homozygous form. The higher sensitivity of alfalfa to inbreeding is shown by Williams 1931; Tysdal et al. 1942; Busbice et al. 1972. Kirk (1927) was the first to report a progressive decrease of vigor in alfalfa due to inbreeding. Wilsie (1958) demonstrated loss of self-fertility from 80–90% and vigor from 20–30%, upon one generation of selfing. Posler et al. (1972) also found out decline in fertility with the advance generation of inbreeding rapidly with selfing than sib mating. Similarly, Ray and Bingham (1992) observed inbreeding depression under selfing first and second generation to be 49 and 26%, respectively. High inbreeding depression upon successive selfing alfalfa plants, which makes it difficult to develop pure inbred lines for hybrid production, is the main reason behind its lack of genetic gain.

15.2.1 Inbreeding Depression Due to Deleterious Alleles

Allard (1960) demonstrated that populations exhibiting inbreeding depression carry a large number of deleterious alleles in the recessive form that are hidden at the heterozygous loci. This does not imply all recessive alleles to be deleterious and dominant beneficial. Natural selection simply selects against the dominant deleterious alleles; however, the deleterious recessive alleles survive in their recessive condition, masked by their dominant counterpart in respective loci and passed from generation to generation in their recessive state. It is theorized that inbreeding depression is the result of increased homozygosity at loci affecting the trait of interest. As loci become more homozygous, deleterious recessive alleles uncovered from the effects of more favorable dominant alleles (Jones and Bingham 1995), resulting in inbreeding depression.

According to Burton et al. (1978), three conditions are required for inbreeding depression to occur: (1) Presence of favorable and less favorable alleles at loci affecting the trait of interest in a population, (2) Dominance effect must be present among these alleles, and (3) Inbreeding must result in loss of heterozygosity at these loci. The synthetic nature of alfalfa plants has harbored a high amount of genetic load (Jones and Bingham 1995) which are masked by their favorable dominant alleles. Genetic load is a measure of the number of recessive deleterious alleles in a population (Ness and Knight 2004). Upon inbreeding, loss of heterozygosity results in unmasking of these harmful unfavorable alleles in their homozygous recessive form, resulting in inbreeding depression. Desirable alleles with additive and dominant effect contributing to heterosis and population improvement rarely have large positive effects, rather act in a cumulative manner; however, recessive deleterious alleles contributing to inbreeding depression have shown to have large effects on plant fitness and vigor (Genter 1973).

The concept of decreased heterozygosity during inbreeding as a sole cause does not completely explain inbreeding depression in polyploids, especially alfalfa. As an autotetraploid species, inbreeding depression in alfalfa should be at a slower rate than in diploids as heterozygosity is halved by 3.80 generations of selfing in autotetraploids than 1 generation of selfing in diploids (Dewey 1966; Jones and

Bingham 1995), ultimately requiring an increased number of generations of selfing for similar levels of homozygosity as in diploids. However, alfalfa exhibits greater inbreeding depression, similar to those expected in diploids (Busbice and Wilsie 1966; Dewey 1966, 1969). In addition to increasing homozygosity of recessive alleles at heterozygous loci, Busbice and Wilsie (1966) assumed the loss of multiple allelic interaction within a locus to be responsible for severe inbreeding depression in alfalfa.

Fixing of genes to their homozygous form decreases allelic interaction within a locus due to subsequent inbreeding, resulting in high inbreeding depression. Similarly, Bingham et al. (1994), proposed the idea of complementary gene action in autotetraploid alfalfa as a reason for high inbreeding depression. According to them, autotetraploid alfalfa has greater complementary gene interaction than diploids, and progressively selfing these autotetraploid causes rapid loss of gene interaction due to a decrease of heterozygosity. This rapid loss of complementary gene interaction due to inbreeding explains the severe inbreeding depression in alfalfa. Even though it is difficult to point out the main reason behind high inbreeding depression in autotetraploid alfalfa, the study shows the change in allelic frequency from heterozygosity to homozygosity as the major cause of inbreeding depression.

We speculated a numeric dominant model of inheritance to explain the possible interaction between dominant and recessive alleles (Fig. 15.1). Under the conventional alphabet dominant model, a genotype containing one dominant allele exhibits a dominant phenotype. In contrast, a genotype exhibits a dominant phenotype only if the genotype contains the same or more dominant allele than the recessive allele. For diploid species, there is no difference between the alphabet dominant model and the numeric dominant model. However, the two models create different phenotypes among the same genotypes in polyploid species, such as autotetraploid alfalfa. Under the alphabet dominant model, selfing a heterozygous with the genotype of AAaa only generates 1/16 recessive homozygous (aaaa) exhibiting inbreeding depression, far less than the proportion (¼) of recessive homozygous of diploid species such as maize. Under the numeric dominant model, both genotypes of aaaa and Aaaa exhibit recessive phenotype and occupy the same proportion of recessive phenotype in diploid (¼). This may explain why the autotetraploid alfalfa exhibits the same level of inbreeding depression as the diploid species.

15.2.2 Inbreeding Depression Due to Self-Incompatibility

Similar to Inbreeding depression, self-incompatibility (SI) is also responsible for reduced seed set (decreased fitness) in alfalfa however, the mechanism related to the lower fitness due to self-incompatibility and inbreeding depression differs significantly. Self-incompatibility is defined as 'the inability of a fertile hermaphrodite seed plant to produce zygotes after self-pollination' (Nettancourt 2001). It is a genetically determined physiological barrier preceding fertilization, that allows cells of the pistils to discriminate between 'self' and 'non-self' pollen grains or pollen tubes, that interfere with pollen germination at the surface of the stigma or pollen tube growth within the pistil (Nasrallah 2017), precluding fertilization and seed production. Inbreeding depression, on the other hand, is the reduced survival and fertility of off-springs of related individuals due to inbreeding (Charlesworth and Wils 2009). It is caused by the unmasking of deleterious recessive alleles to their homozygous forms.

Since deleterious recessive alleles are generally present in the population at a lower frequency, masked by their dominant counterpart at heterozygous loci, progressive inbreeding unmasks these deleterious alleles into their homozygous form leading to reduced fitness and vigor of the subsequent off-springs. Brink and Cooper (1938) showed that the Alfalfa flower exhibits partial self-incompatibility. Different studies (Cooper 1935; Brink and Cooper 1936; Cooper et al. 1937; Cooper and Brink 1940) have indicated the differential pollen tube growth of self and foreign-pollen, along with partial incompatibility and embryo abortion following self-fertilization as compared to embryo development from cross-fertilization. Bauchan et al. (1990) observed pollen–stigma interaction, pollen tube–style interaction, and pollen tube–ovule interactions within the locule in self-incompatible plants from the population of two pest resistance varieties W10 AC3 and BMP8 AC3.

15.3 Genetic Basis of Self-Incompatibility

Most flowering plants with perfect flowers (male reproductive organ and female reproductive organ on the same flower) have the tendency to self-fertilize, which in the long run have deleterious effects on population survival. The prevalence of self-incompatibility in a species restricts the occurrence of inbreeding, which ultimately improves genetic variability. Mather (1943) describes self-incompatibility as 'the failure, following mating or pollination, of a male and female gamete to achieve fertilization, where each of them is capable of uniting with other gametes of the breeding group after similar mating or pollination'. So, self-incompatibility involves the rejection of related pollen by the pistil from the same flower before or at the time of fertilization. In general, self-incompatibility in plants is the consequence of the interaction between pollen–stigma, pollen tube–style, and pollen tube–ovule. The self-incompatible pistil recognizes genetically related (self) and genetically unrelated (non-self) pollen, thereby rejecting the selfed pollen either on the surface of the stigma or during pollen tube growth in the stylar region.

The self-incompatibility reaction at the surface of the stigma is called early acting SI systems, whereas the self-incompatibility reaction inside the stylar region is called late acting SI systems. According to Rea and Nasrallah (2008), recognition of self-pollen is based on allele-specific interactions (direct or indirect) of the pistil and pollen expressed products of *S-locus* genes, that trigger a cellular response in pistil or pollen, terminating with inhibition of pollen tube development. Studies till date have identified three molecular mechanisms (Rea and Nasrallah 2008) of self-incompatibility based on molecular analysis of plant families; *Brassicaceae, Solanaceae, Rosaceae, Scrophulariaceae,* and B*Papaveraceae*, in which it was identified that the recognition and response phases of self-incompatibility depend upon the site of pollen inhibition (i.e., whether it occurs early or late during the pollen tube's journey through the pistil, which in turn is determined by characteristics of the stigma surface) (Rea and Nasrallah 2008).

The first mechanism, the inhibition of pollen at the surface of the stigma, also known as early acting Self-incompatibility system, occurring in the Brassicaceae family, is highly localized involving the contact zone between pollen grain and stigmatic epidermal cells. Prevention of pollen germination and pollen tube growth at the surface of the stigma occurs due to the action of cell surface-localized receptors and ligands encoded by two *S-locus* genes; *SRK (S-locus Receptor Kinase)* that encodes single-pass transmembrane serine/threonine kinase at the plasma membrane of the stigmatic epidermal cell (Takasaki et al. 2000) and *S-locus Cysteine-Rich protein (SCR)* (Schopfer et al. 1999) encoding small peptide and functioning as ligand receptor for *SRK* receptor in the pollen coat. Interaction between *SRK-SCR* upon contact of pollen grain on the stigmatic surface results in specific recognition of selfed pollen (Shimosato et al. 2007), resulting in an incompatibility reaction. Therefore, in *Brassicaceae*, incompatibility reaction is the result of the interaction of *SCR* and *SRK*, that activates receptor's kinase in signaling epidermal layer of stigma into recognition and rejection of selfed pollen (Rea and Nasrallah 2008).

The second mechanism, the late acting SI systems, involves the death of a pollen tube inside the stylar region which is also known as programmed cell death. In the case of programmed cell death, observed in *Papaveraceae* family, the germination of the pollen tube at the surface of stigma is accompanied by the death of the pollen tube in the stylar region (Geitmann 1999). In this system of SI, only the female determinant is reported that is responsible for the secretion of the *S protein*, which results in the influx of calcium inside the pollen tube altering the normal cytosolic tip focused calcium gradient that is responsible for maintaining the growth of pollen tip (Franklin-Tong et al. 2002). The calcium influx triggers events like actin depolymerization and disruption of the actin cytoskeleton (Snowman et al. 2002; Thomas et al. 2006) resulting in the inhibition of pollen tube growth. The male determinant of SI, proposed to be localized to pollen tube plasma membrane, combines with *S-protein binding protein (SBP)* to regulate calcium channel activity (Hearn et al. 1996). Identification of male determinant of SI is important to understand the role it has in programmed cell death of self-pollen tubes through interaction with the stigmatic *S protein*.

Finally, the third mechanism, also called the degradation of pollen RNA, observed in the upper third region of style, is affected by highly polymorphic pistil-specific glycoprotein *S-RNase (S-locus ribonuclease)* encoded by *S-locus* and secreted into the extracellular matrix lining path of pollen tube growth (Lee et al. 1994; Murfett et al. 1994). Although *S-RNase* RNA degradation activity is non-specific, it directs its activity toward self-pollen tubes when they are grown side by side with non-selfed pollen tubes in a pistil, ultimately, checking the growth of only self-pollen tubes. The role of *S-RNase* affecting *S allele-specific* inhibition of self-pollen tubes inside the pistil region is unknown.

The self-incompatibility in plants, in general cases, depends upon a highly polymorphic single locus called *S-locus* having multiple alleles (Xiaoying et al. 2011). The genetic basis of self-incompatibility on different families: *Brassicaceae, Solanaceae, Rosaceae, Scrophulariaceae,* and *Papaveraceae* relates self-incompatibility with *S-locus* gene that recognizes related (selfed) and unrelated (non-selfed) pollen, resulting in incompatibility reaction. However, the genetic basis of self-incompatibility cannot be explained with a highly polymorphic single locus called *S-locus* in Alfalfa, as Sahni (1957) was unable to explain the self-incompatibility system in two diploid clones of *Medicago* species based on *S-locus*. The absence of well-defined incompatibility relationships coupled with the polyploid nature of the crop makes it difficult to determine the genetic basis of self-incompatibility in alfalfa (Tysdal and Kiesselbach 1944). A simple inheritance pattern has not been established for self-incompatibility as well as self-sterility; however, loss of higher order allelic interaction through inbreeding appears to be of major importance in alfalfa (Viands et al. 1988). On the basis of preceding literatures, the genetics behind self-incompatibility in alfalfa is still unclear and also it cannot be explained by the *S-locus* gene as in *Brassicaceae, Solanaceae, Rosaceae, Scrophulariaceae,* and *Papaveraceae* families.

15.4 Efforts to Overcome Inbreeding Depression

The outcrossing nature of the alfalfa plant coupled with its autotetraploid structure creates complexity in genetic improvement for higher forage and seed production. The prevalence of severe inbreeding depression prevents researchers from capturing heterosis in alfalfa cultivars through hybrid development. Consequently, it has led to modification in breeding strategy for higher yield, with intercrossing of selected parents to produce a synthetic variety (Hill 1987). This strategy is currently more feasible than the development of hybrid cultivar. The intercrossing strategy between plants with a broad genetic base increases heterozygosity thereby increasing the intra-locus interaction, and ultimately the yield in the cultivars. Inbreeding depression reduces the vigor of natural autotetraploid while their performance is improved by maximizing heterozygosity (Dunbier and Bingham 1975). Maximum heterozygosity in alfalfa exploits intra-locus interaction as well as additive variation, eventually improving the performance of the crop.

Study on different years of alfalfa cultivars (1898–1985) indicated genetic improvement of alfalfa in terms of forage yield due to increased frequency of favorable alleles and utilization of non-additive genetic effects; however, they also indicated the increased amount of genetic load possessed by the modern alfalfa cultivars (Holland and Bingham 1994). And outlined that, by reduction of the genetic load and combination with the diverse germplasm, future yield enhancement could be achieved. The broader genetic base of modern alfalfa cultivars masks the harmful deleterious alleles at their heterozygous loci. Genetic purging of these deleterious alleles provides a germplasm source for future improvement of alfalfa cultivars. Kimbeng and Bingham (1998) indicated that inbreeding reduces heterozygosity with no new allelic interaction, and crop improvement through inbreeding and selection is the result of purging of deleterious alleles and selection of the favorable ones.

15.5 Development of Inbred Lines for Hybrids in Alfalfa

Although inbreeding depression is extremely severe in alfalfa, the efforts for developing hybrids are still ongoing. In Alfalfa, the focus on improving traits controlled by major genes such as disease resistance and winter hardiness has clearly limited performance improvement (Volenec et al. 2002). An improved breeding strategy would exploit non-additive gene interaction through heterosis. The deleterious recessive alleles are passed from generation to generation, masked by their dominant counterpart, which makes it difficult to remove inbreeding depression in the succeeding generation of alfalfa (Busbice et al. 1972). As a result, it is difficult to develop inbred lines for hybrid production. However, studies have suggested (Tysdal et al. 1942; Busbice et al. 1972) the potential of hybrids for overall production improvement in alfalfa.

Studies on hybrid alfalfa have shown the potential of hybrids. A hybrid variety developed by Dairyland Seed Company called HybriForce-400 was found performing consistently in the top 10% when tested over 25 different environments (Wiersma 2001) which shows the stability of hybrid variety. Riday and Brummer (2005), with their study from 1998 through 2002, observed higher persistence from the cross between inter-subspecific hybrids (between *Medicago sativa* subspecies *sativa* and *Medicago sativa* subspecies *falcata*) which was equivalent to more persistent parental subspecies over time. They also observed mid-subspecies heterosis between 10–20% for biomass yield in their study. Wagner et al. (2003), in their study of 326 hybrids from 2000 through 2002, observed average mid-parent and high parent heterosis to be 3.4 and 1.6%, respectively. However, values for both mid-parent heterosis and high parent heterosis were observed as high as 13.1 and 9.3%, respectively.

Heterosis is the superior performance of offspring relative to their parents. The genetic basis of heterosis in alfalfa is partial to complete dominance (Gallais 1984; Bingham et al. 1994; Woodfield and Bingham 1995). According to Busbice et al. (1972), hybrids would have full utilization of non-additive gene action in comparison to synthetic counterparts with the potential for higher yield. The first alfalfa hybrid developed in 1968 was based on the concept of cytoplasmic male sterility. However, utilizing male sterility for hybrid seed production is difficult as the tetraploid nature of alfalfa complicates the inheritance of any gene associated with male sterility. Also, the self-incompatibility mechanism in alfalfa for pollen control is not definitive (Barnes et al. 1977) for hybrid seed production. As a result, commercial hybrid seed production relies on utilizing inbred lines to avoid complication in large-scale commercial seed production.

For commercial hybrid seed production in the field, controlled pollination is required as uncontrolled pollination results in varying percentages of hybrid seed (Barnes et al. 1977).

Controlled pollination could be achieved through self-incompatibility and male sterility (Tysdal et al. 1942; Barnes et al. 1977; Busbice et al. 1975). Intensive selection for self-sterility, without taking into account its nature, leads cultivars to low seed production potential; however, selection for self-fertility may increase seed production potential while reducing the vigor of the cultivar (Busbice et al. 1975). As a result, male sterility better suits as an alternative for the production of hybrid cultivars. Expression of genetic male sterility requires homozygous recessive genotypes, so this technique has not been used to produce hybrid alfalfa cultivars. For that reason, cytoplasmic male sterility is an efficient method for hybrid production in alfalfa (Barnes et al. 1977). The first commercial alfalfa hybrid utilized cytoplasmic male sterility in 1968. However, lower seed yield in the male-sterile plant posed serious economic problems in the production of alfalfa hybrids (Viands et al. 1988). An alternative to this method is to utilize female-sterile plants (Bingham et al. 1994) as a source of pollinator for male-sterile plants. However, female sterility is controlled by single recessive genes and its maintenance requires outcrossing with female-fertile plants or clones, so employment of this system is also limited for a larger application.

15.6 Hybrid Alfalfa

Genetic gain in crop breeding is achieved through selection, recombination, and hybridization. Selection and recombination result 1–5% of genetic gain while hybridization results in more than 10% genetic gain. In maize, significant yield improvement was achieved by shifting breeding strategy from open-pollinated varieties to hybrid varieties (Hallauer et al. 1988). Hybrid varieties efficiently utilize inputs like water, sunlight, and nutrients for higher yield. Current Alfalfa varieties are synthetic varieties developed by intercrossing selected parents. Genetic gain from these synthetic varieties of alfalfa is minimum (Fig. 15.1) for the past 30 years. Following the footsteps of maize breeding, Alfalfa breeding should focus on hybrid development for higher forage yield.

Dairyland Seed Corporation introduced the first Alfalfa hybrid, called HybriForce-400, in 2001 that performed in the top 10% over 25 different environments (Wiersma 2001) in University trials from 1998–2000. Similarly, Hybrid Alfalfa showed 8–15% yield advantages over the best synthetic varieties in the University trials. However, there are challenges to the development of hybrid Alfalfa. Unlike Maize, Alfalfa flowers contain both male and female parts in the same flower in close proximity, which makes it difficult to remove male parts for hybrid seed production. Similarly, pollination in maize is carried out by winds from tassel to silk, while insect pollinators like bees are required in Alfalfa. So, controlled pollination is required in Alfalfa for the production of hybrid seed at the commercial level. Based on this concept, the first commercial hybrid Alfalfa was developed utilizing a system of male sterility. HybriForce-400, the first commercial Alfalfa hybrid, was developed by utilizing a male sterility system of Hybridization called msSUNSTRA (Wiersma 2001). The first generation of Alfalfa hybrids was vigorous and had faster regrowth and stronger plants that provided increased forage yield for 3–4 years (Velde 2009).

With the advancement in breeding technology, second-generation Alfalfa hybrids were introduced in June 2009 (Velde 2009). The second-generation hybrids yielded 5% more forage than the first-generation ones. Further, they also provide uniformity and consistency with their performance. The consistency and uniformity were found after testing these hybrids across various soil types and weather as well as different cutting regimes. Additionally, hybrid Alfalfa also grew a finer stem which increased palatability through shortening dry-down time. With continued genetic progress, higher genetic gain per year can be achieved for Alfalfa in the future (Fig. 15.2).

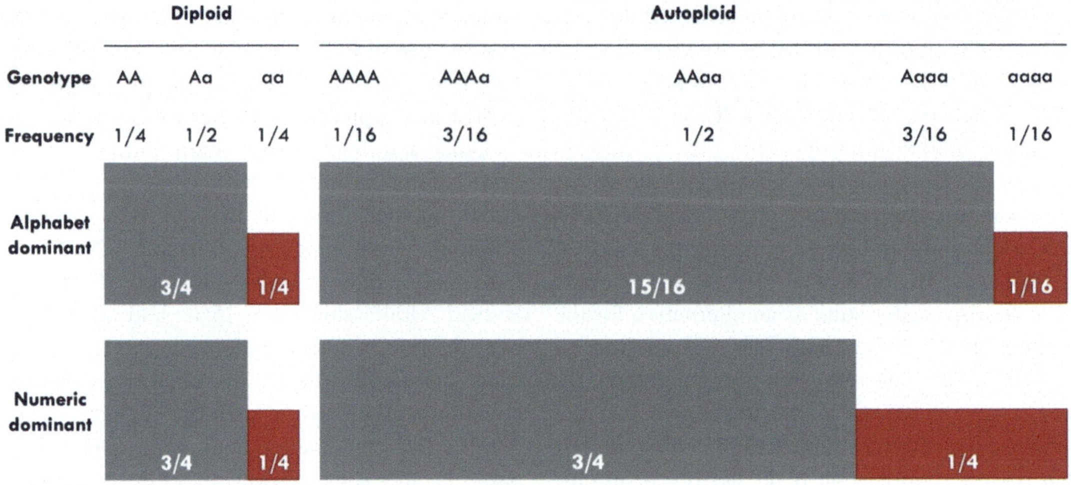

Fig. 15.2 The alphabet and numerical dominant models of inheritance. A genotype exhibits a dominant phenotype as long as the genotype contains one dominant allele under the alphabet dominant model, while a genotype exhibits a dominant phenotype if the genotype with the number of dominant alleles that is equal to or more than the number of recessive allele under the numeric dominant model. The alphabet dominant model and the numeric dominant model do not perform differently among the progeny out of selfing a heterozygous with the genotype of Aa, where A is the dominant allele and a is the recessive allele. The expected frequencies of the genotypes of AA, Aa, and aa are ¼, ½, and ¼, respectively, among the progeny. The dominant phenotype (dark gray or tall bar) and recessive phenotype (light gray or short bar) have a conventional Mendelian ratio of 3:1 under either the alphabet or numeric dominant models. However, the two models perform differently for polyploids such as autoploid. Among the progeny out of selfing the heterozygous of AAaa, the dominant phenotype (gray or tall bar) and recessive phenotype (crimson, or short bar) have a ratio of 15:1 under the alphabet model and a conventional Mendelian ratio of 3:1 under the numeric model

15.7 Prospective

Conventional breeding methods are lengthy which can be shortened by marker-assisted selection (Yu 2017) and genomic selection through rapid breeding cycles and fewer necessary phenotypic evaluation (Hawkins and Yu 2018). Additionally, advanced sequencing technology and bioinformatics will provide in-depth knowledge of the molecular basis of inbreeding depression. Pryce et al. (2014) identified genomic regions associated with inbreeding depression in cattle. Further advancement of genotyping technology coupled with genome-wide association mapping of deleterious alleles will enhance understanding of inbreeding depression.

Natural selection by inbreeding depression (NSID) restricts seed production past the fourth generation of inbreeding in alfalfa. Past studies used NSID alone to remove harmful deleterious alleles, which seriously undermined fertility and survival ability in the progeny. Through MAS, deleterious alleles responsible for reduced fitness can be identified and purged from the population before being fixed to their homozygous recessive form (Fig. 15.3). MAS helps in the identification of change in allele frequencies within the genome to monitor specific alleles or haplotypes (Steele et al. 2004), which can be used to develop lines with specific allele combinations through

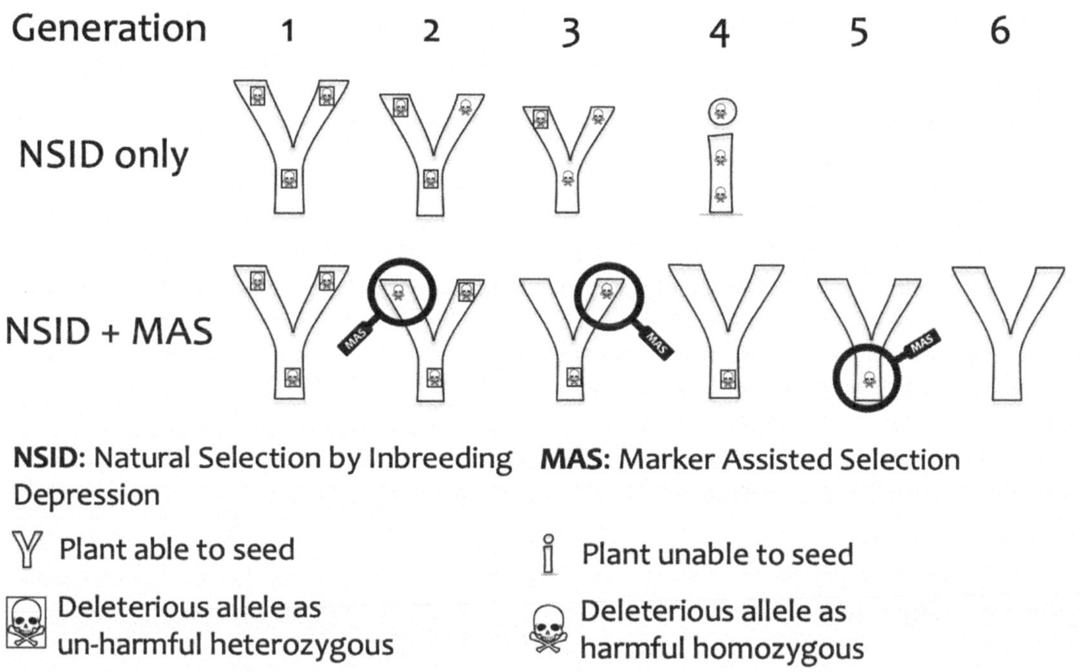

Fig. 15.3 Marker-assisted natural selection by inbreeding depression to purge deleterious alleles. A fertile plant (Y) becomes less fertile during inbreeding when deleterious alleles appear increasingly homozygous, reducing fertility and eventually turning into an infertile plant (i). The addition of MAS reduces inbreeding depression by efficiently purging deleterious alleles

the early generation of selection (Ribaut et al. 2001; Steele et al. 2004). So, the combination of MAS and NSID is important to purge deleterious recessive alleles in the early generation of selfing and retaining seeded individuals. This will ultimately increase the seed production equilibrium. Consequently, selection for seeded individuals will generate deleterious allele free inbred lines.

In Alfalfa, the proportion of heterozygous genes are higher than homozygous genes due to their high outcrossing nature. As a result, deleterious alleles can survive and pass through generation to generation hidden within their dominant counterpart. Inbreeding alfalfa increases the expression of deleterious recessive alleles in their homozygous form (Fig. 15.3) and applying natural selection by inbreeding depression alone does not produce seed after the fourth generation in the inbreeding progeny. So, it is necessary to map genetic loci associated with inbreeding depression in Alfalfa and purge these loci responsible for inbreeding depression. For this, the phenotypic variation of fertility in Alfalfa, producing high and low seeded individuals due to inbreeding, allows to identify genetic loci associated with fertility. Genotyping inbreeding individuals with differential rates of fertility allows them to identify both genetic loci and genetic variants associated with the loci. These identified genetic variants can be used as molecular markers in the next generation of selection to purge deleterious recessive alleles from the population. The process of eliminating individuals with deleterious recessive alleles from the inbreeding population is presented in Fig. 15.3.

Figure 15.3 presents the application of MAS coupled with NSID to purge harmful deleterious recessive alleles from the population of inbreeding individuals. Inbreeding increases homozygosity and reduces heterozygosity by half, respectively, and applying natural selection alone cannot identify the presence of deleterious alleles in their heterozygous form (Fig. 15.3, NSID only 1, 2, and (3). Once these deleterious recessive alleles are expressed in their homozygous form (Fig. 15.3, NSID only, (4) reduced fitness in individual results in no seed production. However, the addition of marker-assisted selection helps to identify these deleterious recessive alleles even when they are masked by their dominant counterpart (Fig. 15.3, NSID + MAS, 1, 2, 3, 4, 5, and (6) in the heterozygous form. As a result, they are subsequently purged from the inbreeding individuals leading to the development of pure inbred lines through inbreeding up to the sixth generation. These inbred lines could be used directly in the breeding programs for developing hybrids or could be used further to develop novel recombination lines.

15.8 Conclusion

Alfalfa is the major forage crop around the world. Genetic improvement in alfalfa is at a slower pace as conventional breeding approaches are simply insufficient to exploit full yield potential. Besides, genomic complexity, severe inbreeding depression, and high outcrossing nature make it difficult to effectively improve complex agronomic traits associated with production. The use of synthetic varieties from multiple crosses for higher production simply is a short-term solution toward genetic improvement. Similar to maize in which significant genetic gain was achieved through exploitation of heterosis, alfalfa breeding programs should shift from synthetic varieties to hybrid varieties. The major setback for hybrid production is the inability to produce pure inbred lines due to severe inbreeding depression and partial self-incompatibility. Recent advancements in bioinformatics and sequencing technology have started to inquire about the molecular basis of inbreeding depression and self-incompatibility. Mapping deleterious alleles through advanced genotyping and genome-wide association study will provide a better understanding of inbreeding depression. The identified markers will help to purge deleterious alleles in the early generation of inbreeding, consequently, increase the chance to produce fertile inbred lines in the succeeding generations. With the successful development of an inbred line, there will be significant changes in the Alfalfa industry that will provide benefits to the Alfalfa breeders. The development of inbred lines will immediately facilitate Alfalfa breeders to produce superior hybrids. The successful mapping of genetic loci associated with alfalfa opens up avenues for breeding crops with high inbreeding depression and self-incompatibility.

Acknowledgements This project was partially supported by the USDA National Institute of Food and Agriculture (Hatch project 1014919 and Award #s 2018-70005-28792), and the Washington Grain Commission (Endowment and Award #s 126593 and 134574). The authors thank Cari Park for helpful comments and editing the manuscript.

References

Allard RW (1960) Principles of plant breeding. Wiley, New York

Barnes DK, Bingham ET, Murphy RP, Hunt OJ, Beard DF, Skrdla WH, Teuber LR (1977) Alfalfa Germplasm in the United States: Genetic Vulnerability, Use, improvement and Maintenance. USDA Technical Bulletin. 1571. Government Printing Office, Washington DC

Bauchan GR, Campbell TA, O'Neill NR Jr, Elgin JH (1990) Self-incompatibility in two alfalfa populations. Crop Sci 30:1205–1210

Bhandari HS, Pierce CA, Murray LW, Ray IM (2007) Combining ability and heterosis for forage yield among high-yielding accessions of the alfalfa core collection. Crop Sci 47:665–673

Bingham ET, Goose RW, Woodfield DR, Kidwell KK (1994) Complementary gene interactions in alfalfa are greater in autotetraploids than diploids. Crop Sci 34:823–829

Bolton JL (1948) A study of combining ability of alfalfa in relation to certain methods of selection. Sci Agri 28:97–126

Brink RA, Cooper DC (1936) The mechanism of pollination in alfalfa (*Medicago sativa*). Amer J Bot 23:678–683

Brink RA, Cooper DC (1938) Partial self-incompatibility in *Medicago sativa*. Proc Natl Acad Sci USA 24:497–499

Burton JW, Stuber CW, Moll Rh (1978) Variability of response to low levels of inbreeding in a population of Maize. Crop Sci 18:65–68

Busbice TH, Wilsie CP (1966) Inbreeding depression and heterosis in autotetraploid with application in *Medicago sativa* L. Euphytica 15:52–67

Busbice TH, Hill RR, Carnahan HL (1972) Genetics and breeding procedure. In: Hanson CH (ed) Alfalfa Science and Technology. Agron 15:283:318
Busbice TH, Gurgis RY, Collins HB (1975) Effect of selection for self-fertility and self-sterility in Alfalfa and related characters. Crop Sci 15:471–475
Casler MD, Brummer EC (2008) Theoretical expected genetic gain for among-and-within-family selection methods in perennial forage crops. Crop Sci 48:890–902
Charlesworth B, Charlesworth D (1999) The genetic basis of inbreeding depression. Genet Res 71:329–334
Charlesworth D, Wills JH (2009) The genetics of inbreeding depression. Nat Rev Genet 10:783–796
Chen H, Zeng Y, Yang Y, Huang L, Tang B, Zhang H, Hao F, Liu W, Li Y, Liu Y, Zhang X, Zhang R, Zhang Y, Li Y, Wang K, He H, Wang Z, Fan G, Yang H, Bao A, Shang Z, Chen J, Wang W, Qiu Q (2020) Allele-aware chromosome-level genome assembly and efficient transgene-free genome editing for the autotetraploid cultivated Alfalfa. Nat Commun 11:2494
Cooper DC (1935) Microsporogenesis and Embryology of *Medicago*. J Agric Res 51:471–477
Cooper DC, Brink RA, Albrecht HR (1937) Embryo mortality in relation to seed formation in alfalfa (*Medicago sativa*). Amer J Bot 24:203–213
Cooper DC, Brink RA (1940) Partial self-incompatibility and the collapse of fertile ovules as factors affecting seed formation in alfalfa. J Agric Res 60:453–472
Darwin C (1867) The effects of cross and self-fertilization in the vegetable Kingdom.
Davenport CB (1908) Degeneration, albinism and inbreeding. Science 28:454–455
Dewey DR (1966) Inbreeding Depression in diploid, tetraploid and hexaploid crested wheatgrass. Crop Sci 6:141–147
Dewey DR (1969) Inbreeding depression in diploid and tetraploid crested wheatgrass. Crop Sci 9:592–595
Dudley JW, Busbice TH (1969) Estimates of genetic variance in 'Cherokee' Alfalfa (*Medicago sativa* L). Crop Sci 9:228–231
Dunbier MW, Bingham ET (1975) Maximum Heterozygosity in Alfalfa: results using haploid-derived autotetraploids. Crop Sci 15:527–531
Duvick DN (1992) Genetic contribution to advances in yield of US maize. Maydica 37:69–79
Falconer DS, Mackay TFC (1996) Introduction to quantitative genetics. Longman, Essex, England
Frankham R, Ballou JD, Briscoe DA (2002) Introduction to conservation genetics. Cambridge University Press, Cambridge
Franklin-Tong VE, Holdaway-Clarke TE, Straatman KR, Kunkel JG (2002) Involvement of extracellular calcium influx in the self-incompatibility response of *Papaver Rhoeas*. Plant J 29:333–345
Galiolla M, Serik K, Sakysh Y, Serik A, Saltanat T (2017) Results of selection studies of Alfalfa based on inbred lines. J Agric Sci Technol A 7:309–316
Gallais A (1984) An analysis of heterosis versus inbreeding effects with an autotetraploid cross-fertilized plants: *Medicago sativa* L. Genetics 106:123–137
Geitmann A (1999) Cell death of self-incompatible pollen tubes: necrosis or apoptosis? In: Cresti M, Cai G, Moscatelli A (eds) Fertilization in higher plants: molecular and cytological aspects. Springer Berlin, Heidelberg, pp 113–137
Genter CF (1973) Comparison of S1 and testcross evaluation after two cycles of recurrent selection in maize. Crop Sci 13:524–527
Hawkins C, Yu L-X (2018) Recent progress in Alfalfa (*Medicago sativa* L.) genomics and genomics selection. Crop J 6:565–575
Hill RR Jr (1987) Alfalfa. In: Fehr WR (ed) Principles of cultivar development. vol1 theory and technique. Macmillan, New York, pp 11–39
Hill RR Jr, Shenk JS, Barnes RF (1988) Breeding for yield and quality. In: Hanson AA, Barnes DK, Jr Hill RR (eds) Alfalfa and Alfalfa improvement. Agronomy Monographs. 29. ASA, CSSA, and SSSA, Madison, WI, pp 809–825.
Holland JB, Bingham ET (1994) Genetic improvement for yield and fertility of Alfalfa cultivars representing different eras of breeding. Crop Sci 34:953–857
Jones JS, Bingham ET (1995) Inbreeding depression in Alfalfa and cross-pollinated crops. Wiley, New York
Julier B, Flajoulot S, Barre P, Cardinet G, Santonni S, Huguet T, Huyghe C (2003) Construction of two genetic linkage maps in cultivated tetraploid Alfalfa (Medicago sativa) using microsatellite and AFLP markers. BMC Plant Biol 3:1–19
Keller L, Waller DM (2002) Inbreeding effects in wild populations. Trends Ecol Evol 68:252–258
Kimbeng CA, Bingham ET (1998) Population improvement in Alfalfa: fertility and S_1 forage yield performance in original and improved populations. Crop Sci 38:1509–1513
Kirk LE (1927) Self-fertilization in relation to forage crop improvement. Sci Agri 8:1–35
Knowles RP (1943) The role of insects, weather condition and plant character in seed setting of Alfalfa. Sci Agric 24:29–50
Kumar S (2011) Biotechnological advancements in Alfalfa improvement. J Appl Genet 52:111–124
Lamb JFS, Sheaffer CC, Rhodes LH, Sulc RM, Undersander DJ, Brummer EC (2006) Five decades of Alfalfa cultivar improvement: impact on forage yield, persistence, and nutritive value. Crop Sci 46:902–909
Lawati AHA, Pierce CA, Murray LW, Ray IM (2011) Combining ability and heterosis for forage yield among elite alfalfa core collection accessions with different fall dormancy responses. Crop Sci 50:150–158
Lee HS, Huang S, Kao Th (1994) S Proteins controls rejection of incompatible pollen in *Petunia inflata*. Nat 367:560–563
Li X, Brummer EC (2012) Applied genetics and genomics in Alfalfa breeding. Agron 2:40–61

Mather K (1943) Specific differences in petunia: I Incompatibility. J Genet 45:215–235
McKay TFC (2001) The genetic architecture of quantitative traits. Annu Rev Genet 35(303):339
Murfett J, Atherton TL, Mou B, Gassert CS, McClure BA (1994) S-RNase expressed in transgenic *Nicotiana* causes S-allele-specific pollen rejection. Nat 367:563–566
Nasrallah JB (2017) Plant mating systems: self-incompatibility and evolutionary transition to self-fertility in the mustard family. Curr Opin Genet Dev 47:54–60
Ness BD, Knight JA (eds) (2004) Encyclopedia of genetics. Salem Press, New York
Nettancourt DD (eds) (2001) Incompatibility and incongruity in wild and cultivated plants. Springer, Berlin, Heidelberg
Paige KN (2010) The functional genomics of inbreeding depression: A new approach to an old problem. Bio Sci 60:267–277
Pederson MW (1953) Preliminary studies on breeding Alfalfa for seed production in Utah. Agron J 45:179–182
Posler GL, Wilsie CP, Atkins RE (1972) Inbreeding Medicago sativa L by selfing, sib-mating and inter-generational crossing. Crop Sci 12:49–52
Pryce JE, Haile-Mariam M, Goddard ME, Hayes BJ (2014) Identification of genomic regions associated with inbreeding depression in holstein and Jersey Dairy Cattle. Genet Select Evol 46:71
Ray IM, Bingham ET (1992) Inbreeding cultivated Alfalfa at the Diploid level by selfing and sib-mating. Crop Sci 32:336–339
Rea AC, Nasrallah JB (2008) Self-incompatibility systems: barriers to self-fertilization in flowering plants. Int J Dev Biol 52:627–636
Ribaut JM, William HM, Khairallah M, Worland AJ, Hoisington D (2001) Genetic basis of physiological traits. In: Reynolds MP, Ortiz-Monasterio JI, McNab A (eds). Application of physiology in wheat breeding. CIMMYT, Mexico
Riday H, Brummer EC (2002) Forage yield heterosis in Alfalfa. Crop Sci 42:716–723
Riday H, Brummer EC (2005) Heterosis in a broad range of Alfalfa germplasm. Crop Sci 45:8–11
Rowe DE, Hill RR (1981) Inter-population improvement procedures for Alfalfa. Crop Sci 21:392–397
Schopfer CR, Nasrallah ME, Nasrallah JB (1999) The male determinant of self-incompatibility in *Brassica*. Sci 286:1697–1700
Shen C, Du H, Chen Z, Lu H, Zhu F, Chen H, Meng Z, Liu Q, Liu P, Zheng L, Li X, Dong J, Liang C, Wang T (2020) The Chromosome-Level Genome Sequence of the Autotetraploid Alfalfa and Resequencing of Core Germplasms Provide Genomic Resources for Alfalfa Research. Mol Plant 13:1250–1261
Shimosato H, Yokota N, Shiba H, Iwano M, Entani T, Che F (2007) Characterization of the SP11/SCR high-affinity binding site involved in self/nonself recognition in *Brassica* self-incompatibility. Plant Cell 19:107–117
Snowman BN, Kovar DR, Shevchenko G, Franklin-Tong VE, Staiger CJ (2002) Signal-mediated depolymerization of actin in pollen during the self-incompatibility response. Plant Cell 14:2613–2626
Steele KA, Edwards G, Zhu J, Witcombe J (2004) RMarker-evaluated selection in rice: shifts in allele frequency among bulks selected in contrasting agricultural environments identify genomic regions of importance to rice adaptation and breeding. Theor Appl Genet 109:1247–1260
Takasaki T, Hatakeyama K, Suzuki G (2000) The S receptor kinase determines self-incompatibility in *Brassica* stigma. Nat 403:913–916
Thomas SG, Huang S, Li S, Staiger CJ, Franklin-Tong VE (2006) Actin depolymerization is sufficient to induce programmed cell death in self-incompatible pollen. J Cell Biol 174:221–229
Tucak M, Popovic S, Cupic T, Simic G, Gantner R, Meglic V (2009) Evaluation of Alfalfa germplasm collection by multivariate analysis based on phenotypic traits. Romanian Agric Res 26:47–52
Tucak M, Cupic T, Spanic V, Meglic V (2012) Combining abilities and heterosis for dry matter yield in Alfalfa diallel crosses. Romanian Agric Res 29:71–77
Tysdal HM, Kiesselbach TA, Westover HL (1942) Alfalfa breeding. Agricultural Experiment Station Research Bulletin: 124
Tysdal HM, Kiesselbach TA (1944) Hybrid Alfalfa. Agron 36:649–667
United States Department of Agriculture (2018) Crop production historical track records. National Agricultural Statistics Service, 91–96
Velde M (2009) The next generation of hybrid Alfalfa. Progressive forage. https://www.progressiveforage.com/forage-types/alfalfa/the-next-generation-of-hybrid-alfalfa
Viands DR, Sun P, Barnes DK (1988) Pollination control: mechanical and sterility. In Hanson AA et al (ed) Alfalfa and Alfalfa improvement. Agronomy monograph. 29. ASA, CSSA and SSSA, Madison WI, pp 931–960
Volenec JJ, Cunningham SM, Haagenson DM, Berg WK, Joern BC, Wiersma DW (2002) Physiological genetics of Alfalfa improvement: past failures, future prospects. Field Crop Res 75:97–110
Wagner S, Sun P, Velde M, Gardner D (2003) Heterosis in hybrid alfalfa. Centre Alfalfa improvement conference (CAIC) abstracts. North American Alfalfa Improvement Conference
Wiersma DW, Undersander DJ, Lauer JG, Grau CR (1997) Lack of Alfalfa yield progress in midwest. In: Proceedings of the 25th central Alfalfa improvement conference lacrosse, 16–18 July 1997. University of Wisconsin Madison
Wiersma DW (2001) Are hybrids the new yield force in Alfalfa? Focus on Forage. 2:12

Williams RD (1931) Self and cross fertility and flowering habits of certain herbage grasses and legumes. Welsh Plant Breed Stat Bull: 5243

Wilsie CP (1958) Effect of inbreeding on fertility and Vigor of Alfalfa. Agron J 50(4):182–185

Woodfield DR, Bingham ET (1995) Improvement in two allele autotetraploid populations of Alfalfa Explained by accumulation of favorable alleles. Crop Sci 35:988–994

Xiaoying M, Sun P, Kao T (2011) S-RNase-based self-incompatibility in *Petunia inflata*. Ann Bot 108:637–646

Yu L-X (2017) Identification of single-nucleotide polymorphic loci associated with biomass yield under water deficit on Alfalfa (*Medicago sativa* L) using genome-wide sequencing and association mapping. Front Plant Sci 8:1–11

Targeted Mutagenesis of Alfalfa

16

Shaun J. Curtin, Susan S. Miller, Melinda R. Dornbusch, Andrew D. Farmer, and Juan Gutierrez-Gonzalez

Abstract

Medicago sativa (alfalfa) is a perennial outcrossing tetraploid flowering legume plant and the third most widely grown crop in the US, approximately 80 million acres. Alfalfa is an important cover and forage crop used for grazing, hay, and silage in many parts of the world and is the main forage crop for meat and milk production. Not surprising, there is interest in the use of new breeding technologies (NBT) including gene editing to improve various traits such as yield and stress responses. The CRISPR/Cas9 reagent is a highly effective gene editing tool that can generate site-directed double-stranded breaks (DSB) that results in frame-shift mutations. In this chapter, we describe the construction of a CRISPR/Cas9 reagent that targets the alfalfa *Pho2* genes. Plants recovered from tissue culture were screened for targeted mutations using either cleaved amplified polymorphic sequences (CAPS) or the target amplicon clone and sequencing assays (TACAS). We found these techniques to be sufficient for screening targets in diploid plants but less effective at screening multiple tetraploid targets. Therefore, PacBio® long amplicon sequencing was evaluated for mutant characterization in alfalfa. The sequence data was processed using the SMRT link platform and analyzed using a workflow on the Geneious® software. Using these tools, hundreds of plants could be rapidly genotyped with approximately 57% of plants exhibiting targeted mutations.

S. J. Curtin (✉) · S. S. Miller · M. R. Dornbusch
United States Department of Agriculture, Plant Science Research Unit, St. Paul, MN 55108, USA
e-mail: shaun.curtin@usda.gov; curti242@umn.edu

S. J. Curtin
Department of Plant Pathology, University of Minnesota, St. Paul, MN 55108, USA

A. D. Farmer
National Center for Genome Resources, Santa Fe, NM, USA

J. Gutierrez-Gonzalez
Departamento de Biología Molecular, Universidad de León, León, Spain

16.1 Introduction

Site-directed mutagenesis technologies include reagents that induce *in planta* double-stranded breaks (DSBs) at specific genetic loci, resulting in random sequence modifications. The DSBs are often seamlessly repaired by the plant host's non-homologous end-joining (NHEJ)repair pathway; however, on occasion, the DSB repair results in the introduction of nucleotide insertion/deletions (indels) that can disrupt gene function.

The CRISPR/Cas9 is an easy-to-use genome engineering reagent that can generate targeted DSBs in plants. It is a two-component system

L.-X. Yu and C. Kole (eds.), *The Alfalfa Genome*, Compendium of Plant Genomes,
https://doi.org/10.1007/978-3-030-74466-3_16

consisting of the Cas9 protein with DNA cleavage capability and a sequence determinant single-stranded guide RNA (gRNA). It differs from previous platforms such as the zinc finger nuclease (ZFN) and TAL effector nuclease (TALEN) (Christian et al. 2010; Curtin et al. 2011) in that it is easily modified to target specific DNA by altering a 20 nucleotide sequence located at 5' end of the gRNA molecule. Moreover, gRNAs can be multiplexed and delivered simultaneously to the plant to generate either multi-target or larger deletions (Čermák et al. 2017).

Targeted mutagenesis of alfalfa has been successfully demonstrated in a number of alfalfa genotypes including XinJiangDaYe, M557, and R2336 with reported transformation efficiencies (TF%) ranging from 0.57–2.5% (Gao et al. 2018; Chen et al. 2020; Wolabu et al. 2020). In these reports, the Cas9 enzyme was expressed by either a 35S or the *Arabidopsis* UBQ10 promoter, with the latter shown to be more active in alfalfa (Wolabu et al. 2020). The guide RNAs in these examples were expressed by either the *Medicago truncatula* or the *Arabidopsis* U6 polIII promoter expressing a single or tRNA-spliced multiplexed gRNA array.

In this protocol, we describe the construction of a CRISPR/Cas9 reagent using components from the Voytas Lab Plant Genome Engineering Toolkit (Čermák et al. 2017; Miller et al. 2021). In this example, a UBC24 encoding ubiquitin-conjugating E2 enzyme (*Pho2*) previously characterized in *Arabidopsis* and *Medicago truncatula* was targeted. Loss of function mutation in the *Pho2* gene causes the plant to hyper-accumulate phosphate (Pi) (Curtin et al. 2017; Delhaize and Randall 1995; Park et al. 2014). The first step in reagent construction is the selection of an appropriate binary vector (pTRANS) suitable for the proposed transformation assay (Samac and Austin-Phillips 2006). Here, the pTRANS_220 binary vector with a 35S:*npt*II selectable marker that can incorporate three module vectors (A, B, and C) was used. The Module 'A' component houses the Cas9 expression cassette and can be easily interchanged with different promoter combinations such as a 35S-expressed *Arabidopsis* codon-optimized Cas9 (pMOD_A0101) or the *Arabidopsis* UBQ10:Cas9 (pMOD_A0102). The Module 'B' component incorporates the target sgRNAs and can be expressed by any pol II or pol III promoter of choice. In this example, the Cestrum yellow leaf curling virus (CmYLCV) pol II promoter (Stavolone et al. 2003) was used to drive a tRNA-processing system for releasing multiple gRNAs from a single transcript (Xie et al. 2015). The final module 'C' component can be empty (pMOD_C0000) or incorporate either a visual reporter or an exonuclease gene such as TREX2 (pMOD_C2911) to increase mutagenesis efficiency (Čermák et al. 2017) (Fig. 16.1). We have observed significantly higher mutagenesis rates using the TREX2 module driven by the *Agrobacterium* rol*D* promoter in alflafa (Wally et al. 2008).

It is not uncommon to recover hundreds of transgene positive plants from tissue culture when using the highly regenerable RegenSY genotype (Miller et al. 2021). This has led to increasing challenges in the characterization of alfalfa mutant plants and is further compounded by the now common practice of delivering multiple gRNAs to improve mutagenesis frequencies (Miller et al. 2021). Screening assays used on diploid targets such as the CAPS or TACAS assays will quickly over-burden the operator and the data produced will be time-consuming to analyze and interpret. Therefore, this chapter describes a new screening option, PacBio® amplicon sequencing, in combination with an intuitive and user-friendly workflow processed on the Geneious® bioinformatic software platform for rapid screening and analysis of mutant plants.

16.2 Materials for Genome Engineering Reagent

16.2.1 Plasmids Reagents

1. Reagents from the Voytas Laboratory Multi-Purpose Plant Genome Engineering Kit were used. The kit and/or select individual vectors are available from Addgene, Cambridge, MA, (https://www.addgene.org/). The vectors used in this chapter include the backbone binary vector with 35S:*npt*II selection,

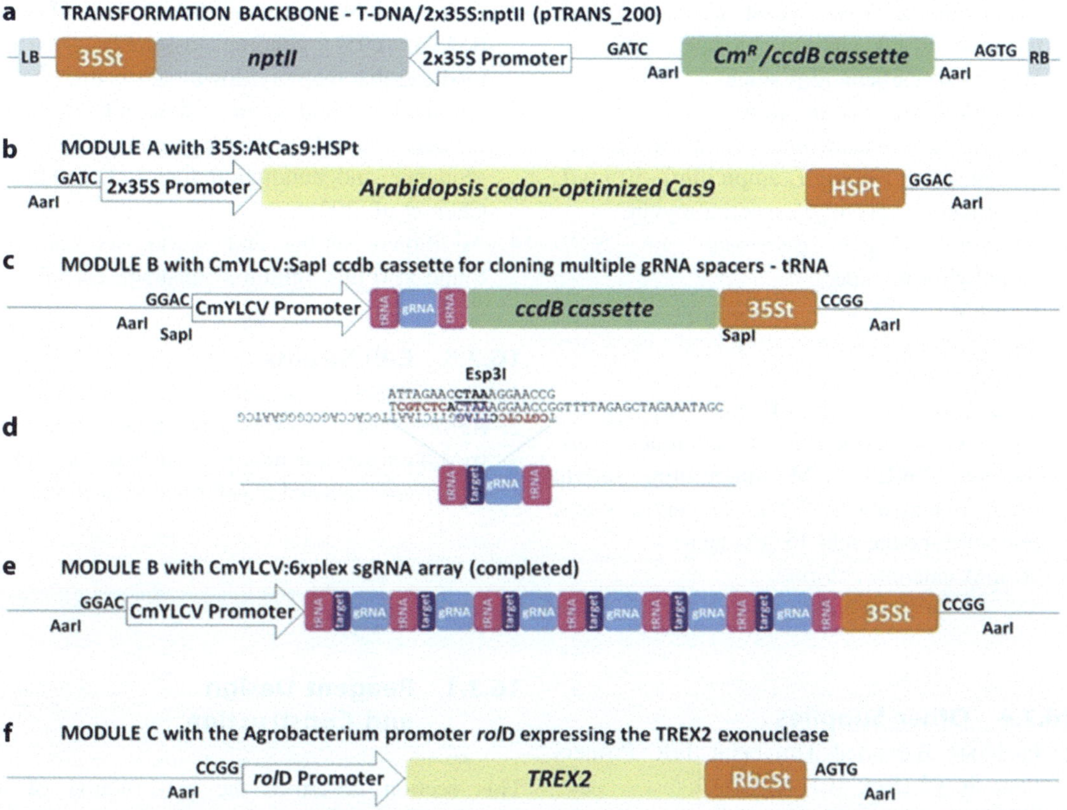

Fig. 16.1. A schematic representation of the modular component system used in this protocol. **a** The binary vector used for transformation is constructed by combining the modules into the AarI sites of the empty backbone T-DNA vector by golden gate cloning. **b** The 35S:Cas9 Module A expression cassette. **c** A representation of an empty Module B vector with its ccdB selection cassette. (**d**) The assembly of the first target and gRNA with the reverse primer from PCR reaction #1 and the forward primer from PCR reaction #2. These two amplicons are cloned together by a Esp3I golden gate reaction. **e** A schematic representation of the completed 6xplex tRNA array in the Module B vector driven by the CmYLCV promoter. **f** The Module C *rol*D:TREX2 expression cassette used to improve mutagenesis efficiency

(pTRANS_220(Plasmid #91,113)), the 35S: Cas9:HSPtcassette (pMOD_A0101 (Plasmid #90,998)), the tRNA:gRNA cassette (pMOD_B2303 (Plasmid #91,068)), and the exonuclease *rol*D:TREX2 (pMOD_C2911 (Plasmid #161,764)).

3. Esp3I restriction enzyme(NEB #R0734S).
4. SapIrestriction enzyme (NEB #R0569S).
5. BanI restriction enzyme (NEB #R0118S).
6. T7 DNA ligase and 2X T7 ligase buffer.
7. T4 DNA ligase (NEB #M0202S) and 10X T4 DNA ligase buffer(NEB #B0202S).

16.2.2 Enzymes and Buffers

1. Proofreading DNA polymerase and buffer such as NEB Q5®(NEB #M0491).
2. AarI restriction enzyme and AarI oligonucleotide(Thermo #ER1581).

16.2.3 Stock Solutions

1. SOC medium: 5 g/l yeast extract, 20 g/l tryptone, 20 mM dextrose, 10 mM sodium chloride, 2.5 mM potassium chloride, 10 mM magnesium chloride.

2. LB medium: 5 g/L yeast extract, 20 g/L tryptone, 10 g/L sodium chloride.
3. 50 mg/mL carbenicillin stock.
4. 50 mg/mL kanamycin stock.
5. LB plates/liquid media with 50 mg/L carbenicillin (or 100 mg/L ampicillin), 50 mg/L kanamycin or 50 mg/L spectinomycin.
6. 40 mg/ml X-gal dissolved in *N′*,*N*-dimethylformamide.
7. 100 mM IPTG (isopropyl β-d-1-thiogalactopyranoside) dissolved in water and filter-sterilized.
8. YEB medium: 5 g/L Bacto beef extract, 1 g/L yeast extract, 5 g/L peptone, 5 g/L sucrose, 2 mL/L 1 M magnesium sulfate stock, pH media to 7.2 before autoclaving. For solid media, add 15 g/L agar.
9. 20 mM calcium chloride.

16.2.4 Other Supplies

1. Pacbio® Barcoded Universal F/R Primers Plate 96 v2 (Ref: 101–629-100).
2. dNTPs.
3. GoTaq® DNA polymerase (M791A).
4. Primers to screen pTRANS_220 and Module Bplasmids.
   TC320: 5′-CTAGAAGTAGTCAAGGCGGC-3′
   TC089R: 5′- GGAACCCTAATTCCCTTATC TGG-3′
   M13R: 5′-CGGATAACAATTTCACACAG-3′
   TC430: 5′-GTTGGATCTCTTCTGCAGCA-3′
5. Gel electrophoresis equipment.
6. QIAprep Spin Miniprep Kit (Qiagen).
7. QIAquick Gel Extraction Kit (Qiagen).
8. Geneious™Bioinformatics Software for Sequence Data Analysis.
9. A Unix-based operating system for running blast and PacBio's long amplicon analysis (LAA) platform.
10. PacBio's open-source SMRT Analysis v9.0 software suite (www.pacb.com/support/software-downloads/).
11. Alfalfa cultivar RegenSY (PI 537,440), seeds available from (www.ars-grin.gov/)
12. Cultivated Alfalfa and the Diploid Level (CADLv0.95) genome assembly (www.medicagohapmap.org/downloads/cadl).
13. Alfalfa tetraploid genome assembly (https://figshare.com/articles/dataset/genome_fasta_sequence_and_annotation_files/12327602)(Chen et al. 2020).
14. Additional scripts and workflows can be found at (https://github.com/shaun-curtin).

16.2.5 Cell Strains

1. *E. coli* DH5α chemically competent cells.
2. *Agrobacterium tumefaciens* strain LBA4404 (pAL4404)(commercially available) for *Agro bacterium*-mediated transformation of alfalfa.

16.3 Methods

16.3.1 Reagent Design and Construction

This section describes the construction of a CRISPR/Cas9 reagent for targeted mutagenesis of tetraploid alfalfa. There are several reagent vector options to choose from including a single or multiplex gRNA 'pDIRECT' system that enables rapid one-step cloning as well as a modular system that allows for single or multiplex gRNA cloning and additional components such as a visual reporter or an exonuclease module (Miller et al. 2021). This protocol describes the assembly of the module-based reagent for the delivery of multiple tRNA-spliced guide RNAs and an exonuclease module.

16.3.2 Target Site Selection

1. Alfalfa *Pho2* genes were identified by querying the CADL(v0.95) assembly (www.medicagohapmap.org/downloads/cadl) using ortholog amino acid sequences from *Medicago truncatula* (Medtr2g013650, Medtr4g088835, and Medtr4g020620 (*see* **Note 1**). Sequence hits from these queries were used to design primers to amplify and validate the

target regions in the transformation cultivar (RegenSY). Targets can be selected manually using the 5′-N_{20}-NGG-3′ template (N_{20} and the NGG indicate the protospacer and PAM sequence, respectively), without any requirements for the first nucleotide, although gRNA spacers that contain AarI sites should be avoided to ensure correct assembly when using Module B plasmids. In addition, online tools such as sgRNA Scorer2.0 https://portals.broadinstitute.org/gpp/public/analysis-tools/sgrna-design or the standalone version of sgRNA Scorer2.0 can be used on the command line with a script called 'identifyAndScore.py' to identify highly active sgRNAs(Doench et al. 2016) (*see* **Note 2**).

2. Using the sgRNA Scorer2.0 tool, three targets were identified for each gene, two of the targets were located in the first exon, and one of the targets located toward the 3′ end of the gene in the sixth exon for all four haplo-alleles of the *Pho2-1* and *Pho2-2* genes.

```
>target1-Pho2-1-1
ATTAGAACCTAAAGGAACCG
>target2-Pho2-1-2
AATTGGGAAGATAAAGACCA
>target3-Pho2-1-3
GTTTGAGTCTCCTAAATACG
>target4-Pho2-2-1
ATTATGTAACTAAGGGGCCA
>target5-Pho2-2-2
TCACACGCAAACTGGCAGCT
>target6-Pho2-2-3
GTTCAAAGGCCGCTCGAGGA
```

3. The target sequences are copied to a text file in 'FASTA' format (see above). Only the first 20 nucleotides of each target are used without the NGG PAM sequence.
4. Open the 'Webtools for the Voytas Lab Plant Genome Engineering Toolkit' (https://crispr-multiplex.cbs.umn.edu//) and click on the 'Primer Design and Map Construction' tab.
5. Use the 'Browse' link to open the target file, then select the 'Target Vector' in the pull-down menu, in this example use pMOD_B2303 with the CmYLCV 'Promoter System', the Esp3I 'Restriction Enzyme' and the tRNA 'Splicing System'are selected, and 'Submit' tab is hit to generate the following primer sequence output.

```
PCR Reaction 1
>oCmYLCV
TGCTCTTCGCGCTGGCAGACATACTGTCCCAC
>TRNA_target1-Pho2-1-1
TCGTCTCCTTAGGTTCTAATTGCACCAGCCG
GGAATCG
PCR Reaction 2
>REP_target1-Pho2-1-1
TCGTCTCACTAAAGGAACCGGTTTTAGAGC
TAGAAATAGC
>TRNA_target2-Pho2-1-2
TCGTCTCCATCTTCCCAATTTGCACCAGCC
GGGAATCG
PCR Reaction 3
>REP_target2-Pho2-1-2
TCGTCTCAAGATAAAGACCAGTTTTAGAGC
TAGAAATAGC
>TRNA_target3-Pho2-1-3
TCGTCTCCGGAGACTCAAACTGCACCAGCC
GGGAATCG
PCR Reaction 4
>REP_target3-Pho2-1-3
TCGTCTCACTCCTAAATACGGTTTTAGAGC
TAGAAATAGC
>TRNA_target4-Pho2-2-1
PCR Reaction 5
>REP_target4-Pho2-2-1
TCGTCTCAACTAAGGGGCCAGTTTTAGAGC
TAGAAATAGC
>TRNA_target5-Pho2-2-2
TCGTCTCCGTTTGCGTGTGATGCACCAGCC
GGGAATCG
PCR Reaction 6
>REP_target5-Pho2-2-2
TCGTCTCAAAACTGGCAGCTGTTTTAGAGC
TAGAAATAGC
>TRNA_target6-Pho2-2-3
TCGTCTCCCGGCCTTTGAACTGCACCAGCC
GGGAATCG
PCR Reaction 7
>REP_target6-Pho2-2-3
```

TCGTCTCAGCCGCTCGAGGAGTTTTAGAG
CTAGAAATAGC
>TRNA_term
TGCTCTTCTGACTGCACCAGCCGGGAATCG

16.3.3 Reagent Construction

1. Synthesize the primers from the list above and click on the link to download the constructed vector map for future reference. Order pTRANS_220, pMOD_A0101, pMOD_2303, pMOD_C2911 plasmids from Addgene and purify plasmid DNA from 5 mL LB cultures using the QIAprep Spin Miniprep Kit.
2. Prepare the PCR reaction #1 template digesting the pMOD_B2303 with BanI (*see* **Note 3**).

   2 μg pMOD_B2303 plasmid DNA.
   2 μL NEB CutSmart® enzyme buffer.
   1 μL BanI restriction enzyme.
   H2O up to 20 μL.
   Incubate at 37 °C for 1 h and gel elute the 1.6kbp fragment.
3. PCR amplifies the gRNA units using a proofreading DNA polymerase such as Q5®. Set up the reaction for each primer pair from Sect. 16.3.2 Step 5. Use the BanI-digested fragment isolated in Step 2 as the template for the first PCR reaction (optional). Use the undigested pMOD_B2303 vector for all remaining reactions.

   10 μL 5 × Q5® Reaction buffer.
   1 μL 10 mM dNTPs.
   2.5 μL 10 mM Forward primer.
   2.5 μL 10 mM Reverse primer.
   5–20 ng Template DNA.
   0.5 μL High-fidelity DNA polymerase.
   dH_2O up to 50 μL.
   Run the following PCR program: 98 °C/1 min + 30 x (98 °C/10 s + 60 °C/20 s + 72 °C/15 s) + 72 °C/2 min + soak @ 15 °C.
4. Confirm successful PCR of gRNA targets by running 5 μL of each PCR product on a 1.5% agarose. The amplicon from the first reaction (lanes 1 and 2) contains the 465-bp CmYLCV promoter and should be approximately 615-bp in length. The remaining reactions should all be 193-bp long (lanes 3 and 4) (Fig. 16.2).
5. Dilute each PCR product ten times with H_2O, for example, 1 μL of the reaction + 9 μL H_2O) (*see* **Note 4**).
6. Assemble the diluted PCR gRNA amplicons into the pMOD_B2303 vector using the following golden gate reaction.

   10 μL 2 × T7 DNA ligase buffer.
   50 ng of pMOD_B2303 vector.
   0.5 μL of each 10 × diluted PCR product.
   0.5 μL SapI (*see* **Note 5**).
   0.5 μL of Esp3I.
   1 μL T7 DNA ligase.
   H2O up to 20 μL.
   Gently mix reaction by pipetting up and down several times.
   Place the Golden Gate reaction in a thermocycler and incubate using the following reaction conditions:
   (37 °C/5 min + 25 °C/10 min + soak @ 15 °C/∞ (*see* **Note 6**).

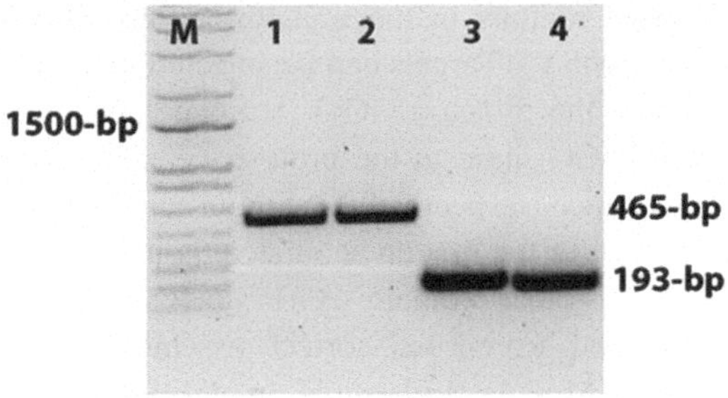

Fig. 16.2. Lanes 1 and 2 are amplicons generated from PCR reaction #1 using the undigested and BanI-digested pMOD_B2303 template, respectively. It has been previously recommended to use the BanI-digested template for PCR reaction# 1 in order to prevent the amplification of larger PCR amplicon due to the presence of the second tRNA repeat. However, according to the above gel, both templates have generated a identical sized amplicons, suggesting the use of the Ban-digested template is not necessary. Lanes 3 and 4 are 193-bp amplicons from PCR reactions #2 and #3. The remaining PCR reactions were not included

7. Transform 5 μL of the Golden Gate reaction into *E. coli* (DH5α or similar cells sensitive to the presence of *ccdB* gene) and plate on LB 100 mg/L ampicillin. Incubate at 37 °C overnight.
8. The following day, PCR screen approximately 2–6 colonies to confirm correctly assembled gRNA array with TC320 and TC089R primers. For 6 × sgRNAs in pMOD_B2303, the reaction will generate ~1.3 kb amplicon.
9. The following day purify plasmids from the LB cultures and sequence confirm correct assembly using the TC320 and TC089R primers.
10. Assemble the final T-DNA vector using a Golden Gate reaction:

    2 μL 10 × T4 DNA ligase buffer.
    75 ng pTRANS_220 (transformation backbone with 35S:*nptII*).
    150 ng pMOD_A0101 (35S:Cas9).
    150 ng of sequence confirmed pMOD_B2303 (with the 6 × sgRNAs targeting Pho2).
    150 ng *rol*D:Trex2 exonuclease (pMOD_C2911).
    0.5 μL AarI.
    0.4 μL AarI oligonucleotide (included with the AarI enzyme).
    1 μL T4 DNA ligase.
    H_2O up to 20 μL.
    Combine in a PCR tube and place the reaction in a thermocycler for the following conditions: 10 x (37 °C/5 min + 16 °C/10 min) + 37 °C/15 min + 80 °C/5 min + soak 15 °C/∞.

11. Transform 5 μL of the Golden Gate reaction into competent *E. coli* and select on a LB plate with 50 mg/L kanamycin. Incubate overnight at 37 °C.
12. The following day inoculate 1–2 colonies into 5 mL LB + 50 mg/L kanamycin and incubate/shake overnight at 37 °C.
13. The following day purify T-DNA plasmid and sequence confirm with M13R, TC430, TC320, and TC089R primers.

16.3.4 Transformation of Reagent into Agrobacterium

1. Prepare competent cells by growing *Agrobacterium tumefaciens* strain LBA4404 plus rifamycin 25 mg/L 5 mL of YEB liquid media overnight at 28–30°C. Add 2 mL of overnight culture to 50 mL of the same medium in a 250 mL flask and shake at the same temperature at 250 rpm until the culture reaches OD600 = 0.5–1.0. Chill the culture on ice and centrifuge at 3500 g for 10 min at 10 °C.
2. Discard supernatant without disrupting pellet and resuspend gently with 1 mL of 20 mM Calcium Chloride (ice-cold). Dispense 100μL aliquots into prechilled tubes. The cells can be frozen in liquid nitrogen and stored at −80 C for future use.
3. Transform the *Agrobacterium* competent cells by adding 1 μg of plasmid DNA and mix by gentle finger-tapping. Completely freeze cells in liquid Nitrogen until liquid Nitrogen stops bubbling. Transfer the cells to a 37 °C incubator to thaw without shaking for 4–5 min. Transfer to ice and add 900 μL SOC media to rescue; incubate and shake at 30° for 2–4 h. Plate on YEB plates with 50 mg/L kanamycin and 25 mg/L rifamycin.
4. Transform the T-DNA vector into LBA4404 competent cells using a freeze–thaw *Agrobacterium* transformation protocol (Winicov I and Bastola 1999).
5. Confirm the presence of the reagent T-DNA in the transformed *Agrobacterium* cells by colony PCR using T-DNA specific primers and GoTaq® DNA polymerase.
6. Transform the alfalfa genotype RegenSY according to the following protocol (Samac and Austin-Phillips 2006).

16.3.5 *Screening of T_0 Plants to Confirm Mutant Status*

The screening of gene-edited plants in tetraploid alfalfa can be challenging due to homology between haplo-alleles. To assist with mutant characterization, sequence data spanning the Pho2 loci was obtained from the Noble foundation's unpublished NECS-141 tetraploid assembly. Conserved primer target sequences were identified and 7-kbp amplicons were amplified for each transgene positive sample. SMRTbell® libraries were prepared using PacBio® barcoded primers for multiplexing amplicons for sequencing on the PacBio® SequelI System. Long amplicon analysis (LAA) was used to generate highly accurate, phased, and full-length consensus sequences of mutant haplo-alleles in a single sequencing run and the processed reads were analyzed using a Geneious®workflow. We prepared the SMRTbell® libraries according to the guidelines of the manufacturer (see link below) and in consultation with the sequencing provider. This section briefly describes this protocol. (https://www.pacb.com/wp-content/uploads/Procedure-Checklist-Preparing-SMRTbell-Libraries-using-PacBio-Barcoded-Universal-Primers-for-Multiplexing-Amplicons.pdf).

16.3.6 Preparation of SMRT Bell® Libraries Using PacBio® Barcoded Universal Primers for Multiplexing Amplicons

1. Design conserved primers sites spanning your GOI and confirm amplification of expected amplicon size (7–10 kbp). Ideally, the PCR reaction should capture all four haplo-alleles (*see* **Note #7**).

   CLPho2-1-F4 5'-GTATCTCTTGTCTCATACTGTTC-3'
   CLPho2-1-R3 5'-GATTGAGGCTCGAACTCTTTCTCTTCC-3'
   CLPho2-2-F4 5'-GTTCTTCACAATGATGTCCCAAC-3'
   CLPho2-2-R3 5'-GAAAACTAGTCATGGAACGGCGC-3'

   10 μL 5 × Q5® Reaction buffer.
   1 μL 10 mM dNTPs.
   2.5 μL 10 mM Forward primer.
   2.5 μL 10 mM Reverse primer.
   5–20 ng Template DNA.
   0.5 μL High-fidelity DNA polymerase.
   dH_2O up to 50 μL.
   Run the following PCR program: 98 °C/1 min + 30 x (98 °C/10 s + 60 °C/20 s + 72 °C/4 min) + 72 °C/2 min + soak @ 15 °C.

2. Design a new set of primers based on the confirmed sequences from Step 2 but with additional universal adaptors sequences for the first-round PCR. The 5'-end of the primer should be blocked (e.g., 5AmMC6) to prevent amplicons from being carried over from this first round of PCR and forming SMRT bell templates during library construction.

   CLPho2-1F4_Rnd1_Internal /5AmMC6/GCAGTCGAACATGTAGCTGACTCAGGTCACGTATCTCTTGTCTCATACTGTTC
   CLPho2-1R3_Rnd1_Internal /5AmMC6/TGGATCACTTGTGCAAGCATCACATCGTAGGATTGAGGCTCGAACTCTTTCTCTTCC
   CLPho2-2F4_Rnd1_Internal /5AmMC6/GCAGTCGAACATGTAGCTGACTCAGGTCACGTTCTTCACAATGATGTCCCAAC
   CLPho2-2R3_Rnd1_Internal /5AmMC6/TGGATCACTTGTGCAAGCATCACATCGTAGGAAAACTAGTCATGGAACGGCGC

   10 μL 5 × Q5® Reaction buffer.
   1 μL 10 mM dNTPs.
   2.5 μL 10 mM Forward primer.
   2.5 μL 10 mM Reverse primer.
   5–20 ng Template DNA.
   0.5 μL High-fidelity DNA polymerase.
   dH_2O up to 50 μL.
   Run the following PCR program: 98 °C/1 min + 20 x (98 °C/10 s + 60 °C/20 s + 72 °C/4 min) + 72 °C/2 min + soak @ 15 °C.

3. Next, a ready-to-use reagent kit containing 96 barcoded universal primers (BUP) in a plate (Pacific Biosciences) is used to perform the second-round barcoding PCR for each sample. Avoid over-amplification by using the

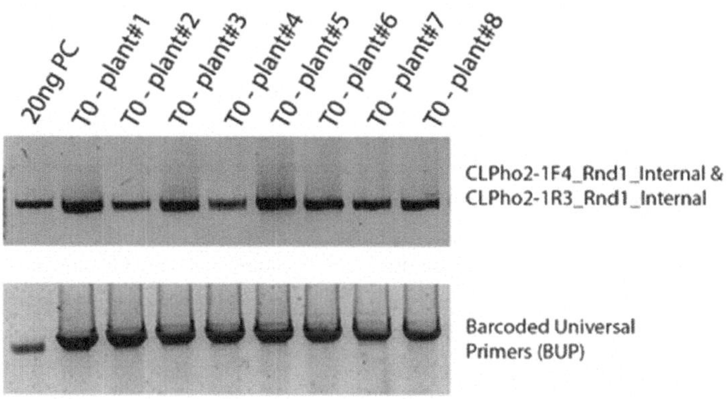

Fig. 16.3. Amplicons from the first-round PCR to incorporate the universal primer sequence followed by the second-round barcoding PCR. Amplicons were visualized on an agarose gel to crudely adjust their concentration

lowest number of cycles required for obtaining adequate yields (ng). Verify the presence of the amplicon and its approximate concentration by running on a gel. Adjust the concentration accordingly and submit for sequencing (Fig. 16.3) (*see* **Note 8**). All QC and clean-up steps are performed by the sequencing provider.

16.3.7 Data Processing and Analysis

1. When the sequencing is complete (typically 4–8 weeks), the demultiplexed data files can be accessed by the user. In this example, a link to a secure download website is sent to the user, and the following commands are used to access the data. Using the terminal, open a new folder and download the data directly to a folder. The command example below was sent to us by our sequencing provider.

```
wget -nH -np -N -r --cut-dirs 2
--no-check-certificate \
--user #### --password ####
https://sequencing-
provider/r##-A01/
```

2. Download the sequencing data from the sequence provider to a project folder. View the demultiplexed data files, and there should be eight files for each of the 96 samples.

```
/Project/r##-
A01/demultiplex_example
sequel-demultiplex.plant#1.
bam
sequel-demultiplex.plant#1.
bam.pbi
sequel-demultiplex.plant#1.
ccs.bam
sequel-demultiplex.plant#1.
ccs.bam.pbi
sequel-demultiplex.plant#1.
ccs.fasta.gz
sequel-demultiplex.plant#1.
ccs.fastq.gz
sequel-
demultiplex.plant#1.fasta.gz
sequel-
demultiplex.plant#1.fastq.gz
```

3. To generate the high-quality consensus sequences, the samples are processed using the long amplicon analysis (LAA) program from the SMRT Link 9.0.0 program. In the terminal load SMRT Link module using the `module load smrtlink` command if available, otherwise, the software can be downloaded from (www.pacb.com/support/software-downloads/).
4. Laa accepts PacBio-compatible BAM files or data set.xml files as input. Prior to the laa analysis run the `data create` command to generate an.xml file for each sample.

dataset create /Project/r##-A01/demultiplex_example/sequel-demultiplex.plant#1.subreadset.xml \
/Project/r##-A01/demultiplex_example/sequel-demultiplex.plant#1.bam

5. Run the laa command to generate the consensus reads. The laa program is computationally intensive and will take around 20–30 min to process each sample. Both Steps 4 and 5 can be automated using a shell script (recommended). The following alfalfa-laa.job example script can be download from (https://github.com/shaun-curtin). If manually processing the sequence data, first generate a config file using the command below and then use the laa command to start processing the raw sequence files.
pbcromwell configure --default-backend slurm \
--output-file ~ /pbcromwell.conf(*see* **Note 9**).
pbcromwell run pb_laa -e \
/Project/r##-A01/demultiplex_example/sequel-demultiplex.plant#1.subreadset.xml \
--config ~/cromwell.conf \
--output-dir ~/outputdir-plant#1 --nproc 8
6. When the laa program is complete, the consensus sequences are retrieved by searching the outputdir-plant#1 folder for the amplicon_analysis.fastq file. This file will contain 3–4 high-quality consensus sequences of GOI haplo-allele sequences. Alternatively, if the laa analysis was carried out using the shell script, a folder labeled 'PROCESSED' will be populated with all of the processed amplicon_analysis.fastq files, but will have sample name information incorporated into the data file, such as amplicon_analysis.plant#1.fastq.
7. To visualize the consensus sequences for mutant characterization, a workflow titled 'Targeted mutagenesis of alfalfa' was prepared for the Geneious® platform and can be download from (https://github.com/shaun-curtin). This workflow can also be programmed manually using the guide in Fig. 16.4.
8. Create a project destination folder in Geneious® and copy to this folder the four annotated haplo-allele sequence files of the GOI. Highlight each sequence file and right-click to bring up a menu. Scroll down to the 'Group Sequences into a List', click it and name your list appropriately, such as 'GOI-ReferenceGenes'.
9. Copy and paste or drag the 96 × amplicon_analysis.plant#1-96.fastq files from the 'PROCESSED' folder to the Geneious® destination folder from Step 8 and click ctrl 'A' to highlight all the files in the folder.
10. Click on the 'Tools' tab and scroll down to 'Workflows'. Click 'Run Workflow', select from the pull-down menu to find 'Targeted Mutagenesis of Alfalfa' workflow. Click 'OK'.
11. Next, select the 'References' from the pull-down menu. Select 'GOI-ReferenceGenes' and click 'OK'.
12. The workflow will extract the name of each sample from the list of processed fastq files in Step 6 such as 'plant#1' and create an individual folder for that sample name. It then copies the four annotated reference files along with the consensus fastq sequences; aligns and sorts each sequence to one of the four annotated references

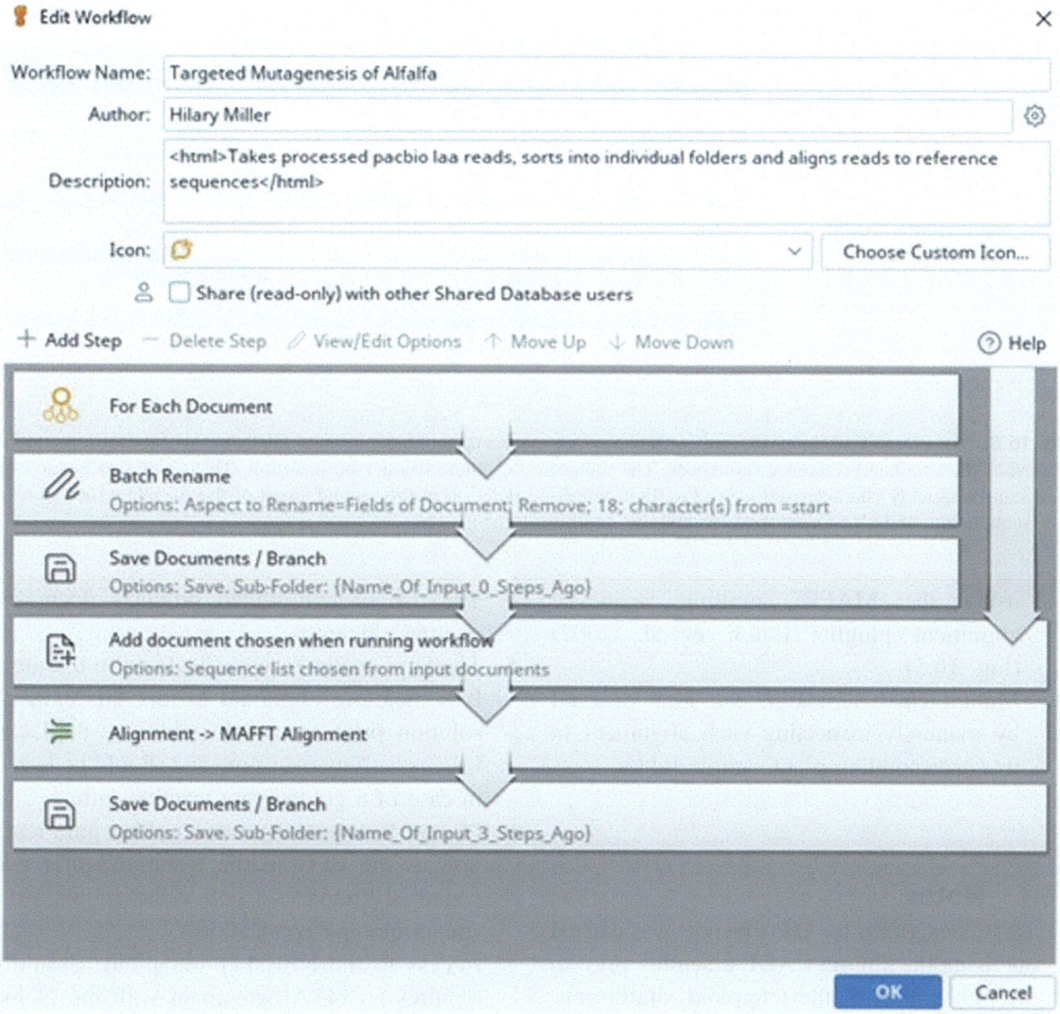

Fig. 16.4. Programming the Geneious® workflow is user-friendly and can be done with the assistance of the settings pictured in the image above (Fig. 16.4) and the following description. First use the Tools tab > Workflow > Manage Workflows > New Work Workflow. Click > Add Step > For each document (sequence files) > Click Add Step > Add Operation > Batch Rename > (double-click) Check 'Expose No options' > Check 'Remove' > add 18 characters from start > OK. Click > Add Step > Save Documents/Branch > Check 'Save these documents as output from workflow' > Check 'Save in sub-folder called "{Name_Of_Input_0_Steps_Ago}"' > 'And then' Check 'Continue' > OK. Next Click> Add Step> Add document chosen when running workflow > In Option Name add 'References' and Document Type 'Sequence List' > OK. Click> Add Step > Add Operation > Alignment -> MAFFT Alignment > OK (double-click) > Check 'Expose No Options' > Check Automatically determine sequence's direction > OK. Click > Add Step > 'Save Documents / Branch' > Check 'Save these documents as output from workflow' > Check 'Save in sub-folder called "{Name_Of_Input_3_Steps_Ago}"' > And then Check 'Continue'. Alternatively, the workflow can be downloaded from (https://github.com/shaun-curtin)

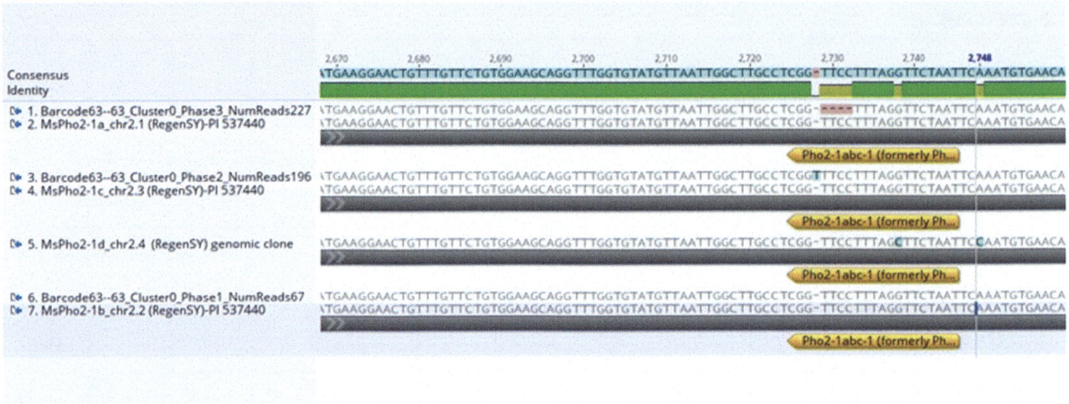

Fig. 16.5. An example of the processed consensus reads aligned to the annotated reference sequences. The mutant plant can be quickly characterized using PacBio® amplicon sequencing and the Geneious® workflow. In this example, the putative 'a' allele has a 4-bp deletion and the 'c' allele has a 1-bp insertion. This plant also has a 1-bp deletion in the second target of the 'b' allele (not shown)

using the MAFFT multiple sequence alignment plugin (Katoh et al. 2002) (Fig. 16.5).

13. Mutant characterization can now proceed by manually inspecting each alignment in its corresponding plant sample folder.

16.4 Notes

1. Target design for the Pho2 project was carried out using the diploid CADL assembly prior to the publication of the tetraploid alfalfa genome assembly (Chen et al. 2020). For new projects, the tetraploid genome assembly is a preferred genomics resource.
2. Targets were selected based on a high activity score and the presence of a restriction enzyme site for future screening efforts.
3. Due to the presence of two tRNA repeats in each template vector, two products can be amplified when amplifying the first gRNA unit since the reverse primer binds to the repeated sequence. This is prevented by using BanI-digested plasmid as the template. BanI cleaves the gRNA repeat sequence and separates the two tRNA repeats. However, as Fig. 16.2 indicates, this amplification of a second product appears not to be a problem.
4. The use of non-diluted products decreases cloning efficiency.
5. The SapI enzyme can settle down in the tube. It is therefore important to mix this enzyme solution prior to use. In addition, SapI can lose its activity over time and should be tested in case of a golden gate reaction failure.
6. Do not heat-inactivate this golden gate reaction as the PEG in the reaction buffer can negatively impact *E. coli* viability in downstream cloning applications.
7. Access to the NEC-141 tetraploid assembly requires an MTA agreement with the Noble Foundation and was initially used in this project. However, the recent publication of the tetraploid assembly of cultivar XinJiang-DaYe is an excellent genome resource that can be accessed without MTA (Chen et al. 2020).
8. Prior to sequencing, several purification and QC steps are carried out by the sequencing provider. Refer to this linked document for more information (www.pacb.com/wp-content/uploads/Procedure-Checklist-Preparing-SMRTbell-Libraries-using-PacBio-Barcoded-Universal-Primers-for-Multiplexing-Amplicons.pdf).
9. The pbcromwell.conf file generated by the pbcromwell configure command was initially

used in this analysis. However, for whatever reason, it stopped working and would return empty processed files. We then obtained an alternative file called 'cromwell.conf' from John Garbe at the University of Minnesota and the laa processing worked without a problem. This alternative cromwell.conf can be downloaded from (https://github.com/shaun-curtin).

Acknowledgements This research was supported by the U.S. Department of Agriculture. The authors acknowledge the Minnesota Supercomputing Institute (MSI) at the University of Minnesota for providing resources that contributed to the research results reported within this chapter. The authors also acknowledge support from the University of Minnesota Genomics Center (UMGC) in particular, Matt Arriola, John Garbe, and Juan Abrahante. Maria Monteros and Michael Udvardi for access to the alfalfa assembly. Special thanks for the technical assistance of Hilary Miller from Biomatters for writing the Geneious® workflow and Curtis Shimamoto from Pacific Biosciences for excellent technical and trouble-shooting assistance with the SMRT Link and laa program.

References

Čermák T, Curtin SJ, Gil-Humanes J, Čegan R, Kono TJY, Konečná E et al (2017) A multipurpose toolkit to enable advanced genome engineering in plants. Plant Cell 29(6):1196–1217

Chen H, Zeng Y, Yang Y, Huang L, Tang B, Zhang H et al (2020) Allele-aware chromosome-level genome assembly and efficient transgene-free genome editing for the autotetraploid cultivated alfalfa. Nat Commun 11(1)

Christian M, Cermak T, Doyle EL, Schmidt C, Zhang F, Hummel A et al (2010) TAL effector nucleases create targeted DNA double-strand breaks. Genetics 186 (2):757–761

Curtin SJ, Tiffin P, Guhlin J, Trujillo DI, Burghardt LT, Atkins P et al (2017) Validating genome-wide association candidates controlling quantitative variation in nodulation. Plant Physiol 173(2):921–931

Curtin SJ, Zhang F, Sander JD, Haun WJ, Starker C, Baltes NJ et al (2011) Targeted mutagenesis of duplicated genes in soybean with zinc-finger nucleases. Plant Physiol 156(2):466–473

Delhaize E, Randall P (1995) Characterization of a phosphate-accumulator mutant of a rabidopsis thaliana. Plant Physiol 107:207–213

Doench JG, Fusi N, Sullender M, Hegde M, Vaimberg EW, Donovan KF et al (2016) Optimized sgRNA design to maximize activity and minimize off-target effects of CRISPR-Cas9. Nat Biotechnol 34 (2):184–191

Gao R, Feyissa BA, Croft M, Hannoufa A (2018) Gene editing by CRISPR/Cas9 in the obligatory outcrossing *Medicago sativa*. Planta 247(4):1043–1050

Katoh K, Misawa K, Kuma K, Miyata T (2002) MAFFT: a novel method for rapid multiple sequence alignment based on fast Fourier transform. Nucleic Acids Res 30 (14):3059–3066

Miller SS, Dornbusch MR, Farmer AD, Gutierrez-Gonzalez JJ, Samac DA, Curtin SJ (2021) Targeted mutagenesis of Pho2 ortholog genes in *Medicago sativa* L. (alfalfa). In preparation

Park BS, Seo JS, Chua NH (2014) NITROGEN LIMITATION ADAPTATION recruits PHOSPHATE2 to target the phosphate transporter PT2 for degradation during the regulation of Arabidopsis phosphate homeostasis. Plant Cell 26(1):454–464

Samac DA, Austin-Phillips S (2006) Alfalfa (*Medicago sativa* L.). In: Wang K (ed) *Agrobacterium* protocols. Humana Press, Totowa, NJ, pp 301–312

Stavolone L, Kononova M, Pauli S, Ragozzino A, de Haan P, Milligan S et al (2003) Cestrum yellow leaf curling virus (CmYLCV) promoter: a new strong constitutive promoter for heterologous gene expression in a wide variety of crops. Plant Mol Biol 53 (5):703–713

Wally O, Jayaraj J, Punja ZK (2008) Comparative expression of β-glucuronidase with five different promoters in transgenic carrot (*Daucus carota* L) root and leaf tissues. Plant Cell Rep 27(2):279–287

Winicov II, Bastola DR (1999) Transgenic overexpression of the transcription factor alfin1 enhances expression of the endogenous MsPRP2 gene in alfalfa and improves salinity tolerance of the plants. Plant Physiol 120(2):473–480

Wolabu TW, Cong L, Park JJ, Bao Q, Chen M, Sun J et al (2020) Development of a highly efficient multiplex genome editing system in outcrossing tetraploid alfalfa (*Medicago sativa*). Front Plant Sci 11:1–9

Xie K, Minkenberg B, Yang Y (2015) Boosting CRISPR/Cas9 multiplex editing capability with the endogenous tRNA-processing system. Proc Natl Acad Sci USA 112(11):3570–3575

The manufacturer's authorised representative in the EU is Springer Nature Customer Service Centre GmbH, Europaplatz 3, 69115 Heidelberg, Germany. If you have any concerns regarding our products, please contact ProductSafety@springernature.com

Printed and bound by CPI Group (UK) Ltd, Croydon, CR0 4YY
30/09/2026
02225687-0019